Lactic Acid Bacteria and Bifidobacteria

Current Progress in Advanced Research

Edited by

Kenji Sonomoto

Laboratory of Microbial Technology
Faculty of Agriculture
Kyushu University
Japan

and

Atsushi Yokota

Laboratory of Microbial Physiology
Research Faculty of Agriculture
Hokkaido University
Japan

Caister Academic Press

Caister Academic Press
Norfolk, UK

www.caister.com

British Library Cataloguing-in-Publication Data
A catalogue record for this book is available from the British Library

ISBN: 978-1-904455-82-0

Cover image adapted from Figure 10.1

Printed and bound in Great Britain

Contents

Contributors

Keietsu Abe
New Industry Creation Hatchery Center
Tohoku University
Sendai, Miyagi
Japan
Laboratory of Applied Microbiology
Department of Microbial Biotechnology
Graduate School of Agricultural Science
Tohoku University
Sendai, Miyagi
Japan

kabe@biochem.tohoku.ac.jp

Aleksandr Barinov
INRA, UMR1319 Institut Micalis
Jouy-en-Josas
France

abarinov@jouy.inra.fr

Alexander Bolotin
INRA, UMR1319 Institut Micalis
Jouy-en-Josas
France

alexandre.bolotine@jouy.inra.fr

Satoru Fukiya
Laboratory of Microbial Physiology
Research Faculty of Agriculture
Hokkaido University
Sapporo, Hokkaido
Japan

s-fukiya@chem.agr.hokudai.ac.jp

Yoichi Fukushima
Nestlé Research Center, Nestec Ltd.
Vers-chez-les-Blanc
Lausanne
Switzerland

yoichi.fukushima@rdls.nestle.com, yoichi.fukushima@jp.nestle.com

Philippe Gaudu
INRA, Micalis UMR1319
Jouy en Josas
France

phillipe.gaudu@jouy.inra.fr

Alexandra Gruss
INRA, Micalis UMR1319
Jouy en Josas
France

alexandra.gruss@jouy.inra.fr

Amira M. Hamdan
Laboratory of Microbial Technology
Division of Applied Molecular Microbiology and Biomass Chemistry
Department of Bioscience and Biotechnology
Faculty of Agriculture
Graduate School
Kyushu University
Higashi-ku, Fukuoka
Japan
Oceanography Department
Faculty of Science
Alexandria University
Alexandria
Egypt

amira_hamdan1978@yahoo.com

Eva Hurt
Regulatory and Scientific Affairs, Nestec Ltd.
Vevey
Switzerland
eva.hurt@nestle.com

Noriyuki Iwabuchi
Food Science and Technology Institute
Morinaga Milk Industry Co., Ltd
Higashihara, Zama
Japan
n-iwabuchi@morinagamilk.co.jp

Yasunobu Kano
Department of Molecular Genetics
Kyoto Pharmaceutical University
Yamashina-ku, Kyoto
Japan
ykano@mta.biglobe.ne.jp

Takane Katayama
Research Institute for Bioresources and Biotechnology
Ishikawa Prefectural University
Nonoichi, Ishikawa
Japan
takarve@ishikawa-pu.ac.jp

Yasushi Kawai
Laboratory of Animal Products Chemistry
Graduate School of Agricultural Science
Tohoku University
Sendai, Miyagi
Japan
ykawai@bios.tohoku.ac.jp

Shinji Kawasaki
Department of Bioscience
Tokyo University of Agriculture
Setagaya-ku, Tokyo
Japan
kawashin@nodai.ac.jp

Katsunori Kimura
Food Science Institute
Division of Research and Development
Meiji Dairies Corporation
Odawara, Kanagawa
Japan
Katsunori_kimura@meiji-milk.com

Motomitsu Kitaoka
National Food Research Institute
National Agriculture and Food Research Organizaition
Tsukuba, Ibaraki
Japan
mkitaoka@affrc.go.jp

Philippe Langella
INRA, UMR1319 Institut Micalis
Jouy-en-Josas
France
phillipe.langella@jouy.inra.fr

Emmanuelle Maguin
INRA, UMR1319 Institut Micalis
Jouy-en-Josas
France
emmanuelle.maguin@jouy.inra.fr

Abelardo Margolles
Instituto de Productos Lácteos de Asturias
Consejo Superior de Investigaciones Científicas
Villaviciosa, Asturias
Spain
amargolles@ipla.csic.es

Kei Nanatani
Department of Biomolecular Engineering
Graduate School of Engineering
Tohoku University
Sendai, Miyagi
Japan
nanatani@biophy.che.tohoku.ac.jp

Toshitaka Odamaki
Food Science and Technology Institute
Morinaga Milk Industry Co., Ltd
Higashihara, Zama
Japan
t-odamak@morinagamilk.co.jp

Tadao Saito
Laboratory of Animal Products Chemistry
Graduate School of Agricultural Science
Tohoku University
Sendai, Miyagi
Japan
saito@bios.tohoku.ac.jp

Kenji Sonomoto
Laboratory of Microbial Technology
Division of Applied Molecular Microbiology and Biomass Chemistry
Department of Bioscience and Biotechnology
Faculty of Agriculture
Graduate School
Kyushu University
Hakozaki
Higashi-ku, Fukuoka
Japan
Laboratory of Functional Food Design
Department of Functional Metabolic Design
Bio-Architecture Center
Kyushu University
Hakozaki
Higashi-ku, Fukuoka
Japan

sonomoto@agr.kyushu-u.ac.jp

Shinya Sugimoto
Department of Bacteriology
Jikei University School of Medicine
Nishi-shimbashi
Minato-ku, Tokyo
Japan

ssugimoto@jikei.ac.jp

Tohru Suzuki
The United Graduate School of Agricultural Science
Gifu University
Anagido, Gifu
Japan

suzuki@gifu-u.ac.jp

Maarten Van de Guchte
INRA, UMR1319 Institut Micalis
Jouy-en-Josas
France

maarten.vandeguchte@jouy.inra.fr

Jin-zhong Xiao
Food Science and Technology Institute
Morinaga Milk Industry Co., Ltd
Higashihara, Zama
Japan

j.xiao@morinagamilk.co.jp

Kenji Yamamoto
Research Institute for Bioresources and Biotechnology
Ishikawa Prefectural University
Nonoichi, Ishikawa

yamamotk@ishikawa-pu.ac.jp

Naoyuki Yamamoto
Food Research Laboratory
Calpis Co., Ltd
Fuchinobe
Sagamihara, Kanagawa
Japan

naoyuki.yamamoto@calpis.co.jp

Yuji Yamamoto
Laboratory of Cellular Microbiology
School of Veterinary Medicine
Kitasato University
Towada, Aomori
Japan

yyamamot@vmas.kitasato-u.ac.jp

Atsushi Yokota
Laboratory of Microbial Physiology
Research Faculty of Agriculture
Hokkaido University
Sapporo, Hokkaido
Japan

yokota@chem.agr.hokudai.ac.jp

Fuminori Yoneyama
Laboratory of Microbial Technology
Division of Applied Molecular Microbiology and Biomass Chemistry
Department of Bioscience and Biotechnology
Faculty of Agriculture
Graduate School
Kyushu University
Hakozaki
Higashi-ku, Fukuoka
Japan

ppp-fy@tk9.50-net.ne.jp

Takeshi Zendo
Laboratory of Microbial Technology
Division of Applied Molecular Microbiology and Biomass Chemistry
Department of Bioscience and Biotechnology
Faculty of Agriculture
Graduate School
Kyushu University
Hakozaki
Higashi-ku, Fukuoka
Japan

zendo@agr.kyushu-u.ac.jp

Preface

Since the discovery of lactic acid bacteria and bifidobacteria early last century, these bacteria have been investigated as intestinal bacteria, starter strains for fermented food and the producers of useful compounds such as lactic acid in fermentation industries. Recently, the use of lactic acid bacteria as probiotics for human health has attracted a lot of attention.

In this book, current progress in the basic science of these bacteria, which includes several studies on genetics, energy generation, stress response and bacteriocins, has been reviewed. As the applied studies, some important examples of the development of these bacteria as probiotics were contributed from food industries. Also, the systems for probiotic health claim in Japan and Europe were compared to understand the criteria needed for the development of the probiotics.

The information provided in this book is expected to show how basic sciences have been contributing to the development of probiotics and related products, and suggests future directions in the studies in both the basic and applied sciences of these bacteria.

In the end, we would like to thank all the authors for their excellent contributions. Thanks are also owed to Annette Griffin, Horizon Scientific Press, and Emma Needs, Prepress Projects Ltd, for their patience and encouragement during the editorial work.

Editors
Kenji Sonomoto and Atsushi Yokota

Part I

Genetics and Energy Metabolism

Genomics of the Genus *Lactobacillus*

1

Aleksandr Barinov, Alexander Bolotin, Philippe Langella, Emmanuelle Maguin* and Maarten Van De Guchte

Abstract

Lactobacilli can be found in very diverse environments, ranging from plants and fermented food products to the mucosal surfaces of the human body, including the gastrointestinal (GI) tract. Like many other lactic acid bacteria, they are traditionally used in food fermentation, serving food preservation as well as flavour and texture development.

In this chapter we present an overview of the phylogenetic diversity, occurrence and industrial applications of lactobacilli. We then focus on genome sequence data and how these recent sources of information changed our view of the genus *Lactobacillus*, largely improving our understanding of these bacteria and their particular properties. To end with, we compare lactobacilli from the viewpoint of surface-exposed and secreted proteins, which may prove particularly important in the emerging field of bacteria–host interactions in the GI tract.

We present a novel vision of what the proteinaceous component of the *Lactobacillus* cell surface may look like, and reveal differences between the different species in an attempt to establish a potentially important link in the interaction of the bacteria with their biotic and abiotic environment.

Introduction

The genus *Lactobacillus* comprises a large number of species from very diverse environments, ranging from plants and fermented food products to the mucosal surfaces of the human oral cavity, the gastrointestinal tract (GI tract), and the vagina. They are Gram-positive, non-spore-forming, rod-shaped bacilli that belong to the heterogeneous group of lactic acid bacteria (LAB) and are characterized by the formation of lactic acid as the sole or main end product of carbohydrate metabolism. Based on their fermentation characteristics, lactobacilli can be divided in homofermentative and heterofermentative species, the latter of which can, in addition to lactic acid, also produce carbon dioxide, ethanol or acetic acid.

Like other LAB, lactobacilli have been used through the centuries for the fermentation of food and feed, with the first records of yogurt (kisim), the product of the fermentation of milk by *Lactobacillus delbrueckii* subsp. *bulgaricus* (*L. bulgaricus*) and *Streptococcus thermophilus*, dating to 3200 BC (Stiles and Holzapfel, 1997). Fermentation has thus since long played an important role in food preservation, through the acidification of the product and the resulting inhibition of spoilage organisms, and flavour development.

More recently, health beneficial effects have been ascribed to fermented milk products. In 1907, Metchnikoff suggested that the living bacteria in yogurt were responsible for the longevity of its consumers (Metchnikoff, 1907), a concept that found its way into the current-day definition of probiotics: 'live microorganisms which when administered in adequate amounts confer a health benefit on the host' (as defined by the Food and Agriculture Organization of the United Nations). A well-documented health benefit of the consumption of yogurt containing live *L. bulgaricus* and *Streptococcus thermophilus* is an attenuation of lactose intolerance (Mercenier *et al.*, 2003).

*corresponding author

In addition, immune modulation and diarrhoea-alleviating effects have been reported (Adolfsson *et al.*, 2004), and both *L. bulgaricus* and *S. thermophilus* have been implicated in these effects (Perdigon *et al.*, 2002; Mercenier *et al.*, 2003). Nowadays, several lactobacilli, notably of the acidophilus group, to which *L. bulgaricus* belongs, are commercialized as probiotics and research into the way in which these and other GI tract bacteria interact with the human host is booming.

Much of recent progress in *Lactobacillus* research has become possible through genome sequencing: 15 complete *Lactobacillus* genome sequences are publicly available to date (NCBI, 2008), which revealed a large diversity in metabolic capacities, yielded invaluable information on the adaptation of some species to specific habitats, and allowed to appreciate the extent of horizontal gene transfer in the GI tract (Nicolas *et al.*, 2007). Comparative genomics of lactobacilli made it possible to evaluate phylogenetic relationships and reconstruct ancestral gene sets, revealing extensive gene loss and acquisition during the evolution of the genus (Makarova *et al.*, 2006; Makarova and Koonin, 2007). In the following paragraphs we will present a short overview of the phylogeny, occurrence and industrial applications of lactobacilli. Special attention will then be given to genome sequence data and how this information improved our understanding of lactobacilli and their particular properties. Finally, we will focus on the prediction of surface exposed and secreted proteins from these bacteria that may prove particularly important in the emerging field of bacteria–host interactions in the GI tract, and a comparison of lactobacilli from this point of view.

Phylogeny of the genus *Lactobacillus*

The genus *Lactobacillus* belongs to the phylum Firmicutes (Gram-positive bacteria with low GC content), class Bacilli, order Lactobacillales, and family Lactobacillaceae. Within the same family, the closest related genera are *Paralactobacillus* and *Pediococcus*.

The first description of the genus *Lactobacillus* dates to 1901, when Martinus Willem Beijerinck introduced this name for bacteria isolated from various fermenting material that shared similar morphologic and phenotypic characteristics (Dellaglio *et al.*, 2004). In 1919 the genus was reorganized by Orla-Jensen and have been divided into several groups based on differences in the optimal growth temperature, catalase and nitrite reduction activities and fermentation of hexoses (Stiles and Holzapfel, 1997). Since that time, the metabolic characteristics of the lactobacilli were used as distinctive features, and the terms obligate and facultative homo- or heterofermentative were introduced to distinguish the members of the group (Stiles and Holzapfel, 1997).

With the development of molecular tools the first exploratory phylogenetic studies of lactobacilli started by analysing the genes encoding ribosomal RNA of the translation machinery (Felis and Dellaglio, 2007). These analyses are based on the observation that rRNA gene sequences are universally present, subject to limited sequence alterations over time, and hardly or not prone to horizontal transfer, and thus permit reconstruction of an image of the evolutional relationships of species (Woese, 1987). In this way, in 1991 Collins and co-workers subdivided the genus into three groups – the *Lactobacillus delbrueckii* group, the *Lactobacillus casei–Pediococcus* group and the *Leuconostoc* group (Felis and Dellaglio, 2007). These phylogenetic studies were confirmed by Schleifer and Ludwig and demonstrated high degree of metabolic diversity and complex phylogeny of lactobacilli (Felis and Dellaglio, 2007). Nowadays, with more than 100 *Lactobacillus* species described, a number of propositions for the phylogenetic re-examination and further subdivision of the genus have been made (Stiles and Holzapfel, 1997; Felis and Dellaglio, 2007; Claesson *et al.*, 2008).

The expansion of full genome sequencing shed new light on the phylogenetic relationships in the *Lactobacillus* genus. This allowed to to perform phylogenetic analyses based on the conserved protein sequences of the DNA dependent RNA polymerase subunits and ribosomal proteins, in addition to 16S rRNA gene sequences (Makarova *et al.*, 2006; Makarova and Koonin, 2007). In this way, for example, it was confirmed that *L. gasseri*, *L. johnsonii*, *L. acidophilus*, and *L. delbrueckii* belong to the same phylogenetic group. The phylogenetic tree of the lactobacilli for which the genome sequences have been determined is presented in Fig. 1.1.

Phylogenetic results revealed that *Lactobacillus* genus is highly heterogeneous and the phylogeny of the genus is far from being finally established as more changes will most probably be introduced with the descriptions of new *Lactobacillus* species.

Occurrence and industrial application

Lactobacilli are widespread in carbohydrate rich environments and are often found in meat and fermented foods, in sewage, and in the gastrointestinal (GI) tract of humans and animals. They are among the first colonizers of the human GI tract in newborns where they tolerate the initially aerobic conditions until the gradual disappearance of oxygen renders the environment acceptable for the growth of other, obligate anaerobic, species (Parracho *et al.,* 2007). In healthy humans, lactobacilli are also present in the oral cavity and breast milk (Olivares *et al.,* 2006), and are dominant bacteria in the vagina. The natural reservoir for many lactobacilli remains unclear, although some of them have been associated with plants (Mundt and Hammer, 1968; Corsetti *et al.,* 2007). Lactobacilli are regarded as safe, beneficial and ancestral companions of humans and animals. They have since long been used in the production of fermented foods, where they contribute to shelf life and to the large repertoire of flavours, aromas and textures of fermented products.

Occurrence in foods

A wide variety of lactobacilli can be found in diverse fermented foods of plant origin (olives, pickles, sauerkraut, sourdough bread, wine), fresh and fermented meat (dry sausages, salami), and fermented milk products (yogurt, fermented milk, cheese) (Bernardeau *et al.,* 2006).

Plant fermentations

Lactobacilli are commonly found in naturally fermented foods of plant origin. Natural fermentation of green olives, for example, involves *L. plantarum, L. casei* and *L. brevis* (Duran Quintana *et al.,* 1999; Randazzo *et al.,* 2004). A large diversity of lactobacilli such as *L. plantarum, L. paraplantarum, L. coryniformis, L. brevis* and *L. curvatus* are found together with other LAB in cabbage fermentations (Plengvidhya *et al.,* 2007). As spontaneous fermentation by indigenous vegetable-associated flora may lead to variations in the properties of the fermented products (Duran Quintana *et al.,* 1999; Gardner *et al.,* 2001), starter cultures have been selected that include species

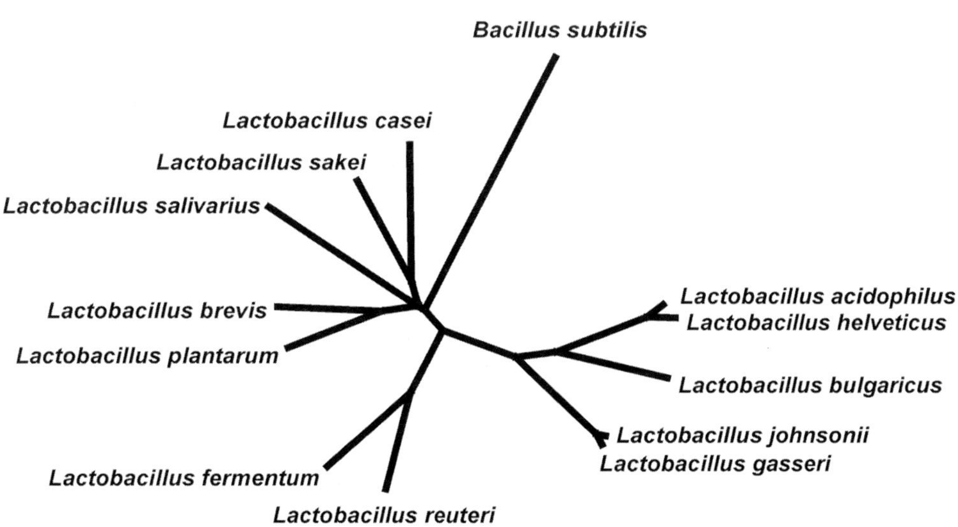

Figure 1.1 Unrooted 16S rRNA based phylogenetic tree. Sequences were extracted from the ribosomal database project (Cole *et al.*, 2009) and the tree drawn using SplitsTree4 software (Huson and Bryant, 2006). *Bacillus subtilis* is used as an outgroup.

commonly found in natural fermentations in order to standardize the fermentation processes. Nowadays *L. plantarum*, *L. brevis* and *L. casei* are widely used for vegetable fermentations, such as cabbage, carrots, beets and cucumber, allowing manufacturers to manufacture food products with high quality standards (Gardner *et al.*, 2001; Tamminen *et al.*, 2004). Lactobacilli also play an important role in winemaking where they can take part in malolactic fermentation, which follows the alcoholic fermentation by yeasts (Lonvaud-Funel, 1999). Among the facultative heterofermentative lactobacilli important in winemaking, *L. plantarum* and *L. casei* are the most important, while *L. hilgardii*, *L. brevis*, and *L. fructivorans* are among the frequently found species from the obligatory heterofermentative group.

Cheese

Cheese-making is a traditional process already mentioned by Homer and Aristotle in the ancient Greece (Bernardeau *et al.*, 2006). Recent studies documented the presence of a wide range of lactobacilli in cheese fermentations where they constitute a significant proportion of the microflora and influence the properties of the product (Beresford *et al.*, 2001). Lactobacilli found in cheese can be divided in starter and non-starter cultures. The key function of starter cultures is to produce lactic acid during the fermentation process. In addition, they contribute to cheese ripening where their enzymes are involved in proteolysis and conversion of amino acids into flavour compounds. They are either intentionally added at the beginning of the cheese manufacturing process, or may be natural contaminants of the raw milk. Among them are the thermophilic lactobacilli such as *L. delbrueckii* (subsp. *delbrueckii*, *bulgaricus*, and *lactis*), and *L. helveticus* (Beresford *et al.*, 2001). The non-starter, mesophilic, lactobacilli do usually not grow well in milk and do not contribute to acid production in the first steps of cheese production. Initially present in small numbers, they constitute a significant portion of the microbial flora of most cheese varieties during the ripening process, and are often major contributors to flavour and texture of the cheese (Beresford *et al.*, 2001; Bernardeau *et al.*, 2006). Among the most frequently encountered non-starter lactobacilli are *L. casei*, *L. paracasei*, *L. plantarum*, *L. rhamnosus* and *L. curvatus*, some of which can inhibit the growth of non-desired or food-spoiling bacteria.

Fermented milk products

Apart from cheese, lactobacilli are widely found in national fermented milk products such as kefir, kule naoto, kwerionik, laban zeer, koumiss, zincica and many others (Bernardeau *et al.*, 2006). Kefir, for example, is a traditional fermented milk produced in the Caucasus region and the result of lactic acid and alcoholic fermentations. A wide diversity of *Lactobacillus* species can be found in fermented milk products, among which *L. acidophilus*, *L. rhamnosus*, *L. reuteri*, *L. casei*, *L. plantarum*, *L. johnsonii*, *L. crispatus*, *L. paracasei* and *L. gasseri* are the most abundant and, for a number of them, often regarded as health beneficial (Mercenier *et al.*, 2003; Bernardeau *et al.*, 2008). Other species found are *L. kefir*, *L. parakefir*, *L. brevis*, *L. kefirgranum*, *L. kefiranofaciens*, and *L. bulgaricus*. Despite a long history of use, the beneficial effects of fermented milk products have been described only recently. Ilya Metchnikoff suggested that the living bacteria present in fermented milk (kefir and yogurt) were responsible for the longevity of its consumers, Bulgarians and villagers from the Caucasus mountains, hypothesizing that their colonization of the intestine normalizes bowel movements and combats intestinal disease (Metchnikoff, 1907). Nowadays yogurt containing live *L. bulgaricus* and *S. thermophilus* has become one of the economically most important dairy products worldwide. A now well-documented health benefit of the consumption of yogurt is an attenuation of lactose intolerance (Mercenier *et al.*, 2003). In addition, immune modulation and diarrhoea-alleviating effects have been reported (Adolfsson *et al.*, 2004), and both *L. bulgaricus* and *S. thermophilus* have been implicated in these effects (Perdigon *et al.*, 2002; Mercenier *et al.*, 2003).

Meat fermentations

Lactobacilli are found in meat and meat products, such as dry fermented sausages and salami. The most commonly found species in meat fermentations are *L. sakei*, *L. curvatus* and *L. plantarum*

(Rantsiou and Cocolin, 2006; Urso *et al.*, 2006). *L. sakei* is one of the best studied species used in fermented meat products, and has become an industrially important bacterium used as a starter culture, not only in fermented sausage production but also in seafood fermentations (Chaillou *et al.*, 2005; Najjari *et al.*, 2008). *L. sakei* is a psychotropic *Lactobacillus*, able to grow at low temperatures. It serves food preservation, inhibiting the growth of food spoiling microorganisms through the production of lactic acid and hydrogen peroxide, while some strains are also able to produce bacteriocins (Chaillou *et al.*, 2005).

Animal-associated lactobacilli

Lactobacilli are closely associated with humans and animals. They inhabit the GI tract (Walter, 2008) and are ubiquitous in mucous membranes of the oral cavity (Dal Bello and Hertel, 2006) and the vagina (Falagas *et al.*, 2007). They are among the frequently isolated species of the normal vaginal microbiota where *L. crispatus*, *L. jensenii*, *L. iners* and *L. gasseri* are the most abundant. Some of these *Lactobacillus* strains have an inhibitory effect on the growth of bacteria that cause bacterial vaginosis, which may in part be explained by the production of hydrogen peroxide, lactic acid and bacteriocins (Falagas *et al.*, 2007). Lactobacilli are among the first bacterial colonizers of the GI tract, where they are found soon after birth and make up about 1% of the adult microbiota (Heilig *et al.*, 2002). Seventeen *Lactobacillus* species are associated with the human GI tract, comprising *L. acidophilus*, *L. crispatus*, *L. gasseri*, *L. johnsonii*, *L. salivarius*, *L. ruminis*, *L. reuteri*, and other species (Walter, 2008). In healthy humans, lactobacilli are present in human saliva in populations exceeding 10^5 CFU per ml, and the predominant species from the oral cavity, such as *L. acidophilus*, *L. gasseri*, *L. crispatus*, *L. plantarum*, *L. salivarius*, *L. brevis*, *L. rhamnosus*, *L. paracasei*, and *L. vaginalis*, are also frequently isolated from human faeces (Walter, 2008). The persistence of lactobacilli in the human GI tract tract is often thought to depend on the adherence of these bacteria to epithelial cells or mucus, and several putative protein adherence factors have been identified in lactobacilli (Velez *et al.*, 2007). It was shown that several lactobacilli of human origin such as *L. casei*, *L. johnsonii*, *L. rhamnosus* and *L. plantarum* are beneficial for their hosts in decreasing the duration of diarrhoea in children with acute gastroenteritis, strengthening the mucosal barrier and in competitive pathogen exclusion (Guarner and Malagelada, 2003; Saxelin *et al.*, 2005). Specific interest was given to probiotic GI tract bacteria which can stimulate or modulate the immune system. These bacteria–host interactions comprise modulation of signal transduction pathways and gene expression in epithelial and immune cells (Tappenden and Deutsch, 2007; Walter, 2008). Health beneficial effects of individual *Lactobacillus* strains have been documented in a number of studies, indicating the possible applications in health-related areas such as intestinal inflammation, maintenance of remission in Crohn's disease, their value in treating infections during pregnancy, their therapeutic role in gastroenterology, management of allergic diseases, control of antibiotic-related diarrhoea, prevention of urinary tract infections, and a number of other medical applications (Mercenier *et al.*, 2003). Although the mechanisms of these host beneficial effects are not clearly elucidated yet, they raise a constantly growing interest for the *Lactobacillus* genus.

Industrial applications

The selection of lactobacilli with specific properties contributing to flavour and texture development and nutritional value is extremely important in industry, where these bacteria are used in numerous fermentation processes. A large number of compounds produced by lactobacilli provide specific characteristics to the fermented products (Teusink and Smid, 2006). Flavour compounds produced by lactobacilli are usually divided in two groups depending on their metabolic origin. Compounds derived from glycolysis include diacetyl, acetoin, acetic acid and acetaldehyde, some of which provide the typical yogurt flavour (Teusink and Smid, 2006). Compounds derived from proteolysis constitute the second group. These include peptides and amino acids responsible for the flavours associated with fermented dairy products. Further degradation of these compounds during fermentation, and specifically cheese ripening, results in formation of aldehydes, alcohols, esters and sulfur compounds which in turn give characteristic flavours to the end products.

Many lactobacilli produce exopolysaccharides (EPS), long-chain polysaccharides consisting of branched, repeating units of sugars (mainly glucose, galactose and rhamnose) or their derivatives (Welman and Maddox, 2003). EPS are not attached to the surface of the microbial cell and are secreted into bacterial surroundings, a feature that distinguishes them from the structurally similar capsular polysaccharides (CPS) which remain attached to the bacterial cell surface (Welman and Maddox, 2003). EPS produced by lactobacilli are widely used in dairy product manufacturing and contribute to the texture, viscosity and 'mouthfeel', playing an emulsifying or gelling role in these products. Interestingly, polysaccharides derived from different lactobacilli show a large variation in composition, and no defining correlation between EPS concentration and viscosity has been established. The industrial importance of lactobacilli producing EPS is based on the consumer demand for products with low fat or sugar content. This demand can in part be achieved by the utilization of high yield EPS producing *Lactobacillus* strains as EPS may confer good 'mouthfeel' and thus substitute fat, sugars, proteins or stabilizers such as pectin, starch, alginate or gelatin which are deliberately added in yoghurt for controlling texture and viscosity (Welman and Maddox, 2003). Hence, the selection of these strains is important in industrial production of fermented milk products with desired specific characteristics. Interestingly, EPS of some of the lactobacilli have been claimed to affect host-mediated antitumour activity and thus may have host beneficial effects (Kitazawa *et al.*, 1998).

Another significant industrial application aspect of the lactobacilli is the production of lactic acid, an important building block for the production of biodegradable polymers (Teusink and Smid, 2006). Lactic acid is widely used for the production of different organic compounds like propionic acid, acrylic acid, ethanol, propylene glycol and acetaldehyde, which are widely used in chemical industries. Economical production of lactic acid by lactobacilli has its weak points, as these bacteria require nutrient rich and usually expensive media like yeast extracts. An economical solution of this problem was achieved by Kwon *et al.* (2000), who used *L. rhamnosus* grown on an enzyme hydrolysate of soybean meal and vitamins and achieved good yields of lactic acid. Similarly, *Lactobacillus amylovorus* was adapted to produce lactic acid by fermenting raw maize, rice and starch media, which makes the production of lactic acid economically and ecologically attractive (Xiaodong, 1997).

The diversity and variety of the niches inhabited by the lactobacilli provides a tremendous potential for their application in industrial areas. Generally regarded as safe, lactobacilli became extremely attractive objects for their feature-specific selection in the food industry. A huge variety of lactobacilli-fermented products provides consumers with a wide repertoire of products such as cheese, yogurt, fermented milk products and many more. All of them differ in taste, flavour and 'mouthfeel', creating an extreme variety for gourmands with specific demands. In addition, dairy products are now becoming well established sources of probiotic strains of lactobacilli. A great variety of products containing probiotic *Lactobacillus* strains isolated from the human intestinal flora have been brought to the market. A good example is *L. acidophilus,* isolated from infant faeces and now commercially used as a probiotic food supplement. Lactobacilli are now becoming more attractive for the utilization of starch medium to produce lactic acid, a very important chemical used by the food, beverages, plastics, textile and leather industries.

Improved knowledge through genome sequencing

The traditional importance of lactobacilli in the food industry, and the more recent allegations of health beneficial effects for a number of related species, stimulated the in-depth study of several of these organisms. The sequencing of bacterial genomes now makes part of these studies, and provides a comprehensive view of principal characteristics that is of great value in comparative and functional studies. Today, 15 genome sequences (covering 12 species) are publicly available (NCBI, 2008). Another 12 are in progress and expected to be available shortly, and four are announced

Table 1.1 General features of the sequenced *Lactobacillus* genomes

Lactobacillus species/ strain	Genome sequencing status	Genome size (Mb)	GC (%)	Prot (nr)	Reference/sequencing centre
acidophilus NCFM	a	2.0	34.7	1862	Altermann *et al.* (2005)
brevis ATCC 367	a	2.3	46.2	2185	Makarova *et al.* (2006)
casei ATCC 334	a	2.9	46.6	2751	Makarova *et al.* (2006)
casei BL23	a	3.1	46.3	3044	AgroParisTech
bulgaricus ATCC 11842	a	1.9	49.7	1562	van de Guchte *et al.* (2006)
bulgaricus ATCC BAA-365	a	1.9	49.7	1721	Makarova *et al.* (2006)
fermentum IFO 3956	a	2.1	51.5	1844	Morita *et al.* (2008)
gasseri ATCC 33323	a	1.9	35.3	1755	Azcarate-Peril *et al.* (2008)
helveticus DPC 4571	a	2.1	37.1	1610	Callanan *et al.* (2008)
johnsonii NCC 533	a	2.0	34.6	1821	Pridmore *et al.* (2004)
plantarum WCFS1	a	3.3	44.5	3052	Kleerebezem *et al.* (2003)
reuteri DSM 20016	a	2.0	38.9	1900	DOE Joint Genome Institute
reuteri JCM 1112	a	2.0	38.9	1820	Morita *et al.* (2008)
sakei 23K	a	1.9	41.3	1879	Chaillou *et al.* (2005)
salivarius UCC118	a	1.8	32.9	1717	Claesson *et al.* (2006)
gasseri MV-22	b	1.9	35	–	The Broad Institute Genome Sequencing Platform
jensenii 1153	b	1.7	34.5	–	The Broad Institute Genome Sequencing Platform
paracasei 8700:2	b	3.0	46	–	The Broad Institute Genome Sequencing Platform
reuteri 100–23	b	2.2	38	~2181	DOE Joint Genome Institute
rhamnosus HN001	b	2.8	46	~2758	Fonterra Research Centre
buchneri NRRL B-30929	c	–	–	–	DOE Joint Genome Institute
casei str. Zhang	c	–	–	–	Beijing Institute of Genomics
fermentum ATCC 14931	c	–	–	–	Baylor College of Medicine
crispatus JV V101	c	–	–	–	Baylor College of Medicine
helveticus CNRZ32	c	~2.3	–	–	University of Wisconsin
reuteri SD2112	c	–	–	–	Baylor College of Medicine
rhamnosus GG	c	~2.3	-	–	University of Helsinki
casei DN 114001	d	3.1	–	–	
casei Shirota	d	3.0	–	–	
bulgaricus DN 100107	d	2.1	–	–	
helveticus CM4	d	2.0	–	–	

a, genome sequence is publically available; b, genome sequence is in draft assembly; c, genome sequencing project announced; d, genome sequence is unpublished (private); Prot (nr), total number of proteins predicted for the genome.

Table represents the general features of the (being) sequenced genomes of lactobacilli.

as private projects (Table 1.1). The genome sequences of these bacteria enable researchers to take a closer view at, for example, the adaptation of lactobacilli to the environmental niches they occupy, at their metabolic capacities, and at the evolutionary relationships between species.

Lactobacilli have relatively small genomes, with a size of about 2 Mb for most of the sequenced species, and maxima of 3.1 and 3.3 Mb for *L. casei* and *L. plantarum*, respectively (Makarova *et al.*, 2006; Makarova and Koonin, 2007; NCBI, 2008). The results of comparative genomic analyses indicate that adaptation to growth in nutrient-rich environments has been a major driving force in the evolution of the *Lactobacillales*, which was accompanied by a massive gene loss compared with the ancestor of all bacilli (Makarova *et al.*, 2006). In present-day *Lactobacillus* genomes, a clear correlation between genome features and niche adaptation can be observed, as will become clear in the following paragraphs.

Versatile lactobacilli

The first *Lactobacillus* genome to be completely sequenced was that of *L. plantarum*, a versatile bacterium able to adapt to a variety of environments including plant, dairy and meat fermentations, and frequently encountered as a natural inhabitant of the human gastrointestinal tract (Kleerebezem *et al.*, 2003). With 3.3 Mb, this is the biggest *Lactobacillus* genome known to date, thought to reflect the ecological flexibility of this bacterium. Among the notable features leading to this conclusion are the large number of regulatory and transport functions, including 25 complete phosphoenolpyruvate-dependent phosphotransferase systems (PTS) for sugar transport and 30 other transport systems predicted to be involved in the transport of carbon sources (Kleerebezem *et al.*, 2003). *L. plantarum* possesses a number of peptide uptake and degradation systems, while it is also capable of synthesizing most amino acids *de novo*, at a difference with most other lactobacilli (Boekhorst *et al.*, 2004; Canchaya *et al.*, 2006; Claesson *et al.*, 2007). Interestingly, many of the genes for sugar transport and metabolism are clustered near the origin of replication, occupying a 213-kb region with a lower GC content (41.5%) than the rest of the genome (44.5%), suggesting that many genes may have been acquired by horizontal gene transfer.

The ecological adaptability of *L. casei* to diverse habitats ranging from raw and fermented dairy products, fresh and fermented plant products, sourdough, and silage to the human oral cavity, intestinal contents, stools, and vagina, raised a special interest in this bacterium for studies aiming to understand its ecological flexibility (Cai *et al.*, 2007). To date, the genome sequences of four *L. casei* strains have been established with sizes ranging from 2.92 to 3.14 Mb (Table 1.1). Most analyses have been performed on the genome of *L. casei* strain ATCC 334 (Makarova *et al.*, 2006), which was originally isolated from Emmental cheese. Niche versatility of this bacterium may be explained by the presence of many insertion sequences and several bacteriophage-associated genomic regions (Makarova *et al.*, 2006; Ventura *et al.*, 2006). These are suggested to participate in intra-species recombination events leading to a recombinatorial population structure (Cai *et al.*, 2007) which would facilitate niche adaptation. Similar indications for intraspecies recombination were observed in *L. plantarum* (de Las Rivas *et al.*, 2006). Genome comparison of the two strains *L. casei* ATCC 334 and *L. casei* BL23 revealed that the latter carries numerous DNA insertions, which accounts for the difference in genome size (Yebra *et al.*, 2007). Among these is a 12.8 kbp insertion, which at both ends is flanked by a 19-bp sequence, and only a single copy of the 19-bp sequence was found in the ATCC 334 strain, suggesting that insertion in BL23 took place by recombination or that excision in ATCC 334 was caused by homologous recombination between the two direct repeats. The 12.8-kb fragment contains genes involved in the catabolism of the cyclic polyol myo-inositol, a constituent of phytic acid, which in the form of various salts (phytates) constitutes a major phosphate storage molecule in plant seeds. The presence of these genes enables this strain to utilize myo-inositol as a carbon source (Yebra *et al.*, 2007), thus rendering it metabolically different from strain ATCC 334.

Two versatile species with smaller genomes are *L. brevis* and *L. fermentum*. *L. brevis* is an obligately heterofermentative bacterium with a genome of 2.29 Mb. It can be isolated from many different environments like cheese (Abriouel *et al.*, 2008; Van Hoorde *et al.*, 2008), kefir (Bosch *et al.*, 2006) and plants (Pulido *et al.*, 2007). *L. brevis* is the dominant obligately heterofermentative species in the vagina (Ocana *et al.*, 1999), and is commonly detected in human saliva and faeces (Walter, 2008). Although the genetic basis for the ecological flexibility of *L. brevis* is not fully understood, it has been suggested that a gene cluster encoding cell-surface proteins potentially dedicated to carbohydrate utilization plays a role (Siezen *et al.*, 2006). *L. brevis* ATCC 367 contains two plasmids, one of 13.4 kb and one of 35.6 kb (Makarova *et al.*, 2006). The plasmid pRH45 contributes to the versality of *L. brevis* ABBC45 through the presence of the horA gene, which codes for an ATP-dependent multidrug transporter capable of extruding toxic compounds from the cell (Sakamoto *et al.*, 2001). This transporter allows *L. brevis* to resist to hop compounds and grow in beer, where it is one of the most important spoiling species.

L. fermentum belongs to the *Lactobacillus reuteri* phylogenetic group (Morita *et al.*, 2008), together with *L. reuteri*, an resident GI tract species. The genome size of *L. fermentum* is 2.09 Mb. Compared with other lactobacilli, *L. fermentum* contains fewer glycosidases and PTS systems, suggesting a reduced capacity to utilize carbohydrates. Nevertheless, *L. fermentum* is a species widely distributed in nature, often isolated from fermenting plant material (Nielsen *et al.*, 2007), dairy products (Coton *et al.*, 2008; Gala *et al.*, 2008), naturally fermented sausages (Kaban and Kaya, 2008), breast milk (Olivares *et al.*, 2006) and saliva (Dal Bello and Hertel, 2006).

Versatile lactobacilli are found in various environmental niches. The genome sequences reveal certain characteristics that may play a role in their ecological flexibility, although our understanding is still far from complete. For some of them, relatively large genomes and abundant carbohydrate utilization systems seem to be indicative, while horizontal gene transfer and recombinatorial population structures may also be important in facilitating the adaptation to a variety of niches.

Lactobacilli adapted to specific niches

A number of lactobacilli are (so far) mainly or exclusively found in one particular environment to which they seem to be specifically adapted, and signs of this adaptation can sometimes be found in the genome sequence. The 1.88 Mb genome sequence of *L. sakei*, for example, revealed a number of specific adaptations to the (fermented) meat environment where this bacterium is typically found (Chaillou *et al.*, 2005). Different from other lactobacilli, *L. sakei* is able to catabolize arginine, an amino acid abundant in meat. It contains a multiplicity of catabolic genes involved in exogenous nucleoside salvage pathways which can be used for energy production, and thus may improve survival on meat when the scarcely available glucose has been exhausted. It contains a haem dependent catalase, but cannot synthesize haem, which it would take up from meat where this compound is abundant. *L. sakei* is capable of growing on meat during refrigeration and in the presence of curing salts (3–9% NaCl). This may be linked to its ability to efficiently accumulate osmo- and cryoprotective solutes such as betaine and carnitine, probably driven by three ABC uptake systems and a Na^{+}-dependent symporter, and the presence of four cold shock proteins.

Other interesting examples of adaptation to a specific niche are found in *L. bulgaricus* and *L. helveticus*. These milk fermenting bacteria are closely related to a number of GI tract bacteria (see below) with which they constitute the acidophilus group of lactobacilli. Consequently, a clear synteny can be observed between the genomes of the bacteria in this group (van de Guchte *et al.*, 2006; Berger *et al.*, 2007). The *L. bulgaricus* 1.86 Mb size genome appears to contain remarkably high numbers of ribosomal RNA operons and tRNA genes with regard to its size, as well as 270 pseudogenes, both of which can be interpreted as indications for a recent reduction of genome size (van de Guchte *et al.*, 2006). Another feature indicating that this genome is going through an active phase of rapid evolution is its aberrant GC content at the third codon position (Nicolas *et al.*, 2007). The functions that are inactivated (present as pseudogenes) or completely missing, of which many are related to the biosynthesis of amino acids, suggest that the evolution of this genome is associated with an adaptation to the milk environment where amino acids can be readily assimilated after the proteolytic degradation of milk proteins. A similar observation of high numbers of pseudogenes and missing genes inactivating, among other functions, amino acid metabolism was made in *Lactobacillus helveticus* DPC 4571, a Swiss cheese isolate (Callanan *et al.*, 2008). *L. helveticus* is commonly found in fermented milk products and characterized by a limited sugar fermentation capacity, a feature that is reflected in a low number (nine) of PTS systems involved in sugar transport compared with its closest relative, the GI tract bacterium *L. acidophilus* which contains 20 PTS systems (Callanan *et al.,* 2008). A comparison between these two species also revealed a reduced number of cell wall-anchored proteins and the absence or fragmentation of predicted mucus binding proteins, potentially involved in adhesion to the mucus of the GI tract, in *L. helveticus.*

Gastrointestinal tract-associated lactobacilli from the acidophilus group

In addition to the above mentioned milk fermenting *L. delbrueckii* and *L. helveticus,* the acidophilus group contains a number of GI tract-associated lactobacilli. The genome sequences of three of these, *L. acidophilus, L. johnsonii* and *L. gasseri* have been published recently (Altermann *et al.,* 2005; Azcarate-Peril *et al.,* 2008; Pridmore *et al.,* 2004). Originating from different ecological niches, the closely related acidophilus group lactobacilli are of particular interest to study environmental adaptation and potential differences in bacteria–host interactions, which may in turn allow the elucidation of the mechanisms involved in the latter.

It appears that acidophilus group lactobacilli are limited in synthesizing most of the amino acids, a feature not restricted to milk fermenting species. This group possess high number of genes encoding peptidases which are supposed to compensate the lack of the genes for the *de novo* synthesis of amino acids, but at a difference to milk fermenting bacteria a high number of transporters for the carbohydrate utilization are observed in *L. acidophilus, L. gasseri* and *L. johnsonii* (from 16 to 21 and 9 PTSs in *L. helveticus* and *L. bulgaricus* BAA-365 strain). The three GI tract-associated strains also possess high number of proteins with predicted LPXTG motif, among which are several protein adhesion factors with mucus binding domains (important for binding to intestinal mucus), and *L. gasseri* encodes the highest number of putative mucus-binding proteins among all the lactobacilli sequenced to date. Milk-fermenting bacteria of the acidophilus group contain only few predicted cell wall-anchored proteins and fragmented or none of the mucus-binding proteins. Extensive studies of several adhesion proteins of *L. acidophilus* revealed multiple cell surface proteins that contribute to the ability of the bacterium to adhere to the intestinal cells (Buck *et al.,* 2005). Both *L. acidophilus* and *L. johnsonii* encode genes involved in production of bacteriocins, whereas *L. gasseri* ATCC 33323 does not (Azcarate-Peril *et al.,* 2008). The genome sizes and the GC content of the three GI tract species are similar (*L. gasseri* ATCC 33323 1.89 Mb, 35.3%, *L. johnsonii* NCC 533 1.99 Mb, 34% and *L. acidophilus* NCFM 1.99, 34%), and both phylogenetic and whole genome comparisons of these bacteria confirmed their close relationship to *L. bulgaricus* and *L. helveticus* (van de Guchte *et al.,* 2006; Callanan *et al.,* 2008).

Other gastrointestinal tract-associated lactobacilli

L. reuteri is naturally found in the GI tract of humans and animals (Morita *et al.,* 2008), where several strains have been reported to exhibit probiotic effects. These include diarrhoea therapeutic effects, inhibition of bacterial pathogens and immune modulation of the gastrointestinal mucosa (Morita *et al.,* 2008). In *L. reuteri,* an interesting observation of host specificity has been made: while strain 100–23, a rodent-specific strain, is often used as a model organism to study bacteria–host interactions in mice, it appears that strain DSM20016 of human origin does not persist in the mouse GI tract. The reason for this difference is unknown. Sequencing and comparison of the two genomes is ongoing, and expected to yield more information on niche adaptation in *L. reuteri* (JGI, 2009). At the time of writing, the genome sequences of two *L. reuteri* strains isolated from human faeces are publically available (Morita *et al.,* 2008; NCBI, 2008). Their sizes, 2.03 Mb (strain JCM1112) and 1.99 Mb (strain DSM20016), are similar to the phylogenetically closely related versatile *L. fermentum* (2.09 Mb). Similar to *L. fermentum, L. reuteri* strains contain relatively few glycosidases and PTS systems, indicating a potentially reduced capacity of carbohydrate utilization. Genome comparison of *L. fermentum* and *L. reuteri* JCM 1112, revealed a unique cluster of 58 genes in the latter species, among which several encode proteins for the biosynthesis of the bacteriocin reuterin and vitamin B12, that may contribute to the probiotic properties of the bacterium. No obvious reasons for the difference in niche specificity have been reported.

Lactobacillus salivarius subsp. *salivarius* UCC118 is generally found in the intestinal mucosa, faeces and saliva of humans and animals. Genome sequencing (Claesson *et al.*, 2006) revealed a circular chromosome of 1.82 Mb and a 242 kb megaplasmid (pMP118), in addition to the two previously described smaller plasmids, pSF118–20 (20.4 kb) and pSF118–44 (44 kb). The megaplasmid, which appears to vary in size between different strains, was suggested to play a role in the adaptation to different niches in the GI tract or to different hosts, conferring metabolic flexibility and the capacity to produce a bacteriocin and a bile salt hydrolase. Genome analysis indicated that *L. salivarius* has the capacity to synthesize nine amino acids, rendering this bacterium less auxotrophic than *L. acidophilus* and *L. johnsonii*, but more dependent on external amino acids than *L. plantarum*. An in-depth *in silico* study of potential surface proteins identified several sortase-dependent LPXTG proteins. A sortase mutant was found to be impaired in adhesion to human intestinal epithelial cells (van Pijkeren *et al.*, 2006), showing the importance of genome-wide studies of surface proteins as these can broaden our understanding of bacteria–host interaction.

From genome sequence to deciphering bacteria–host interactions

While bacteria–host interactions in the GI tract receive growing attention, still little is known about the underlying mechanisms and the effectors involved, especially when regarding non-pathogenic bacteria. When restricting ourselves to interactions at the eukaryotic cell surface (i.e. omitting invasive bacterial effector proteins that have been described for several pathogens but are beyond the scope of this chapter), bacterial effectors of different nature have been described. These include lipoproteins and lipoteichoic acids, lipopolysaccharides (from Gram-negative bacteria), flagelin, double- and single-stranded RNA, and CpG motifs in DNA that bind to eukaryotic Toll-like surface receptors (TLRs) (TLR2, 4, 5, 3 and 7, 9, respectively) (Iwasaki and Medzhitov, 2004), inducing distinct patterns of gene expression in the host cell that guide the activation of innate immunity and initiate the development of antigen-specific acquired immunity (Akira and Takeda, 2004). Bacterial surface proteins have also been implicated in the binding of bacteria to epithelial host cells, mucus or fibronectin, and several motifs have been described that are more or less conserved in these proteins (Sánchez *et al.*, 2008). The picture of these interactions is far from complete, however.

Bacterial effectors

Lipoteichoic acid (LTA) is an amphiphilic (possess both hydrophilic and lipophilic properties) negatively charged glycolipid containing a diacylated moiety with the lipid portion is attached to the membrane. LTA is a surface component of most Gram-positive bacterial cell walls, and possess variable structures in different species (Draing *et al.*, 2006; Ryu *et al.*, 2009). LTA from *L. plantarum* strain KCTC 10887 BP showed a neglectable induction of tumour necrosis factor-alpha (TNF-α, a cytokine involved in systemic inflammation) compared with LTA from *S. aureus* and *B. subtilis*, indicating that the potency of LTA to stimulate TLR2 may be species or even strain dependent (Ryu *et al.*, 2009). D-alanylation of LTA has been reported to affect the anti-inflammatory properties of the probiotic *L. plantarum* strain NCIMB8826 (Grangette *et al.*, 2005). *L. plantarum* mutant which incorporated much less D-Ala in its LTA than the *wt* strain caused a significantly higher production of anti-inflammatory cytokine IL-10 in *in vivo* and *in vitro* models (Grangette *et al.*, 2005). Similar mutants of *L. rhamnosus* GG and *Lactococcus lactis* MG1363 did not increase the anti-inflammatory potential (Perea Velez *et al.*, 2007), indicating that the role of LTA modification in bacteria–host interactions may be more complex than initially thought.

Peptidoglycan (PG) which makes up the body of the bacterial cell wall is among the main surface components of Gram-positive bacteria recognized by the innate immune system (Guan and Mariuzza, 2007). Peptidoglycan recognition is mediated by a number of host receptor molecules which include a family of peptidoglycan recognition proteins (PGRPs): the cytoplasmic proteins NOD1 and NOD2, and the surface receptors CD14 and TLR2 (Guan and Mariuzza,

2007). Recognition of the PG is able to induce various host responses or have direct antimicrobial effects (Dziarski and Gupta, 2005; Guan and Mariuzza, 2007), thus showing the importance of this surface component in host recognition mechanisms.

Bacterial DNA and nucleotides containing unmethylated cytosine phosphoryl guanine DNA (CpG) have been reported to have an immune-stimulatory effect on the cells of the innate immune system such as dendritic cells and macrophages (West *et al.*, 2006). These effects have been shown to be mediated via endoplasmic reticulum localized TLR9, which upon activation stimulates the production of an array of cytokines similar to that induced by LPS. The recognition of unmethylated DNA is important for eliminating intracellular pathogens, such as *Listeria monocytogenes* and *Shigella flexneri*, that replicate in the cytoplasm (West *et al.*, 2006). The lack of stimulatory activity of mammalian DNA on TLR9 is explained by a low occurrence of unmethylated CG dinucleotide motifs in eukaryotic DNA (Rachmilewitz *et al.*, 2002; Krieg, 2003), which allows the discrimination between host and invading DNA. In addition the localization of the receptor plays a role, as host DNA and TLR9 are not localized in the same cell compartment.

Proteins can also interact with TLRs, as has been demonstrated mainly for pathogenic bacteria and protozoan parasites. Bacterial flagellins and lipoproteins, *Mycobacterium tuberculosis* heat shock proteins and protozoan profilin-like proteins are among the best defined examples of proteinaceous effectors interacting with TLRs (Hayashi *et al.*, 2001; Bulut *et al.*, 2005; Yarovinsky *et al.*, 2005; Schroder *et al.*, 2008).

Bacterial flagellin, a structural protein of the flagellum required for bacterial motility, induces a strong immune response mediated uniquely by TLR5 (Hayashi *et al.*, 2001; Nempont *et al.*, 2008). The well studied FliC flagellin from *Salmonella enterica* is a protein of 494 amino acids in which several domains can be identified. The conserved C- and N-terminal domains (comprising about 170 and 90 aa, respectively) are essential for TLR5 activation, whereas the middle region of the protein shows a high variability in length and composition between species and is not essential for TLR5 signalling. The activation of TLR5 by flagellins from both Gram-negative and Gram-positive bacteria activates the NF-κB pathway which regulates the transcription of genes encoding pleiotropic immune mediators (such as TNF-α) involved in innate microbial clearance (Hayashi *et al.*, 2001). Lactobacilli are not flagellated and do not possess structural flagellin genes in their genomes.

Bacterial lipoproteins are characterized by the N-terminal lipidation of the N-terminal cysteine residue which assures their attachment to the cytoplasmic membrane (Sankaran and Wu, 1994). Immune responses to lipoproteins from a wide variety of microorganisms have been reported to be mediated via TLR 1, 2 and 6 (Takeuchi *et al.*, 2001; Alexopoulou *et al.*, 2002). These receptors act together and are able to discriminate a diverse range of bacterial lipoproteins; TLR2/TLR1 heterodimers recognize triacylated lipopeptides (Alexopoulou *et al.*, 2002) while TLR2/TLR6 mediates a response to diacylated lipopeptides (Takeuchi *et al.*, 2001). The best-characterized surface-exposed lipoproteins that induce an immune response originate from pathogenic bacteria, such as *Borrelia burgdorferi*, *Treponema pallidum* and *Francisella tularensis* (Schroder *et al.*, 2008; Thakran *et al.*, 2008; Yang *et al.*, 2008). For BmpA from *Borrelia burgdorferi*, it was shown that the protein moiety of the lipoprotein could induce a full response in the absence of the lipid moiety (Yang *et al.*, 2008). Commensal bacteria and several probiotic lactobacilli in particular can exert immune modulating effects, strengthen the mucosal barrier in the GI tract, suppress intestinal inflammation and competitively exclude pathogenic bacteria (Saxelin *et al.*, 2005). Little attention has been given to the potential role of *Lactobacillus* surface exposed and secreted proteins in the communication with the host cells and their interaction with TLRs or other receptors. Only a limited number of interactions, such as adhesion to epithelial cells and extracellular matrix components, have been reported to involve proteinaceous components of lactobacilli (Sánchez *et al.*, 2008). The sequencing of bacterial genomes allows the identification of surface exposed and secreted proteins that may play a role in bacteria–host interactions, thereby paving a way to a better understanding of the molecular mechanisms of these interactions.

Prediction of surface exposed and secreted proteins in Gram-positive bacteria

The *in silico* prediction of bacterial surface exposed and secreted proteins is of growing interest in the study of bacteria–host relationships, whether pathogenic or host beneficial, as they may mediate physical interactions with the host and/or activate host responses. This interest is driven by the increasing use of DNA sequencing as a major tool in the characterization of bacterial species and strains and, more recently, even of complex ecosystems at the host–environment interface in metagenomics approaches, and a need to rationalize research efforts, focusing on most promising candidate proteins for experimental studies.

Most of the current *in silico* protein localization protocols for Gram-positive bacteria aim to classify the proteins in one of four general classes: cytoplasmic, membrane, cell wall, and secreted. They rely on well-established algorithms to detect transmembrane helices (TMH) characteristic of membrane proteins, signal sequences characteristic of secreted proteins or lipoproteins, and retention signals characteristic of proteins that are covalently or transiently linked to the cell wall. The results of *in silico* protein sequence analysis using these algorithms may be completed by data on protein composition or homology to proteins of known localization (like in PsortB; Gardy *et al.*, 2005), and integrated to predict protein localization. For this purpose a weight factor may be applied to each of the composing data (Gardy *et al.*, 2005; Tjalsma and van Dijl, 2005).

Only some of the more recent methods put some emphasis on cell wall binding domains, lipoproteins and proteins containing an N- or C-terminal transmembrane helix (TMH) to more specifically identify potentially surface exposed proteins (Gardy *et al.*, 2005; Zhou *et al.*, 2008), while the other methods are generally not designed for this purpose. All these methods ignore the potential surface exposition of membrane-associated proteins with two or more TMH, however, which are classified as integral membrane proteins without indication as to whether parts of these proteins may protrude from the cell wall and be exposed at the cell surface. That this possibility is far from hypothetical was recently shown in an elegant series of experiments to reveal the surface proteome of the pathogen *Streptococcus pyogenes*, using a proteomic approach combined with antibody-binding assays (Rodriguez-Ortega *et al.*, 2006; Severin *et al.*, 2007). In addition, proteins in which no signal sequence for protein secretion is detected are often classified as cytoplasmic proteins in the first step of a series of analyses that are organized in a linear pipeline, in which after each analysis only a part of the proteins proceeds to the next step while another part is definitively classified without further analysis.

In our laboratory we therefore developed a new method for the processing of protein sequence data with the particular aim of identification of potentially surface-exposed (PSE) proteins from Gram-positive bacteria: SurfG$^+$ (Barinov *et al.*, 2009).

Surface exposition of membrane-bound proteins

Proteins may be attached to the cell membrane through one or more TMH that are locked in the lipid bi-layer as a result of their hydrophobic properties. Alternatively, proteins that contain a signal peptidase II specific signal peptide can be exported and processed to produce lipoproteins containing N-acyldiacylglyceryl-cysteine as the N-terminal amino acid to assure the association of this N-terminal end with the cell membrane (Sankaran and Wu, 1994). Parts of both types of proteins may protrude from the cell wall and thus be exposed at the cell surface (Fig. 1.2C and D). As becomes clear from the results presented in (Rodriguez-Ortega *et al.*, 2006), also parts of proteins that contain multiple TMH and that are currently classified as membrane proteins may be surface exposed (Fig. 1.2E).

SurfG$^+$ was designed to take all these different possibilities into account. In order to do so, all proteins encoded in the genome of a Gram-positive bacterium are analysed using LipoP (Juncker *et al.*, 2003) and the predicted mature lipoproteins considered as bound to the membrane at their N-terminus. Signal peptides for protein secretion

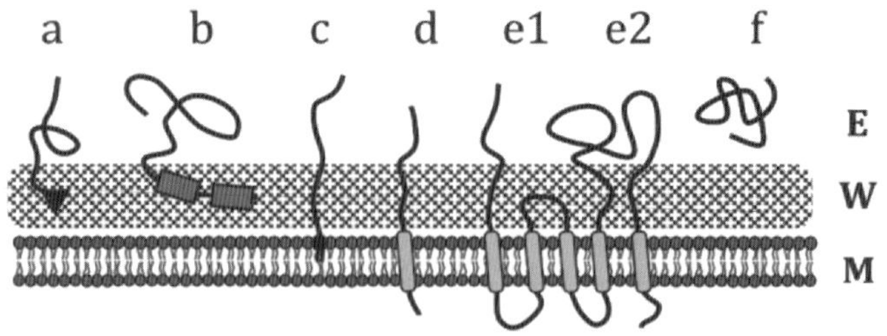

Figure 1.2 Surface exposed proteins. Schematic representation of the cell envelope with different types of PSE proteins. a, covalently bound to the cell wall; b, non-covalently bound to the cell wall; c, membrane-bound lipoprotein; d, membrane anchored through N or C-terminal TMH; e, membrane anchored through several TMH; e1, surface exposed N or C-terminal end; e2, surface exposed loop; f, secreted; M, cell membrane; W, cell wall; E, surface exposed or secreted. (Figure from Barinov *et al.*, 2009.)

are detected using SignalP (Bendtsen *et al.*, 2004) and removed, and the N-termini of the mature proteins considered as free ends. TMH and membrane topology are predicted using TMMOD (Kahsay *et al.*, 2005), a method derived from TMHMM (Krogh *et al.*, 2001), and the length and membrane topology of the N and C-terminal free 'ends' of the (mature) membrane-bound proteins, i.e. upstream of the first detected TMH or downstream of the last TMH or the lipoprotein signal peptide cleavage site, respectively, and of protein 'loops' between two TMH or between the lipoprotein signal peptide cleavage site and a TMH (Fig. 1.2) is determined.

We reasoned that for N- or C-terminal free ends to protrude from the cell wall they should be long enough to traverse the cell wall (peptidoglycan), while protein loops should be long enough to traverse the cell wall twice, putting lower limits to the length of the ends and loops to be taken into consideration for potential surface exposition. Data on the width of the bacterial cell wall (which may vary from 20 to 80 nm; Cabeen and Jacobs-Wagner, 2005) and the length of a single amino acid residue in a linear peptide chain (0.365 nm; Dietz and Rief, 2006) can be used to adjust the cut-off values, which by default are set to 50 and 100 amino acids for PSE ends and loops, respectively, on the basis of our analysis of *S. pyogenes* and several lactobacilli. Only long ends and loops that are predicted to be at the outside of the cytoplasmic membrane are classified as PSE. Proteins for which such ends or loops are predicted to be at the cytoplasmic side of the membrane or shorter than the cut-off value are classified as 'membrane' proteins.

Integrated protein sequence analysis to predict potential surface exposition

SurfG+ analyses protein sequences using various domain search modules that are available as standalone tools or as web based applications. The search modules used include (1) HMM search (Eddy, 1998), to search for a number of motifs that have been described as cell wall anchoring or binding domains (e.g. LPXTG motifs and LysM domains; see legend to Fig. 1.3); (2) LipoP (Juncker *et al.*, 2003), to identify lipoproteins; (3) SignalP (Bendtsen *et al.*, 2004), to identify proteins that are secreted by the Sec pathway; and (4) TMMOD (Kahsay *et al.*, 2005), to identify membrane spanning protein segments (TMH) and predict membrane topology. The results from these analyses and from the analysis of membrane-bound proteins described above are then integrated according to the data flow scheme presented in Fig. 1.3 to predict potential surface exposition. Dataflow through the scheme is directed by logical decisions, and the modular structure of the scheme allows for easy updating through the integration of additional data analysis modules or customized parameter settings.

In order to evaluate the performance of the prediction method, we compared SurfG+ generated predictions for *S. pyogenes* with the experimental data reported in Rodriguez-Ortega *et al.* (2006) and Severin *et al.* (2007). Both studies used a proteomic approach to identify protease accessible surface-associated proteins of *S. pyogenes* (114 proteins in total), and confirmed the surface exposition of a number of these proteins (49 out

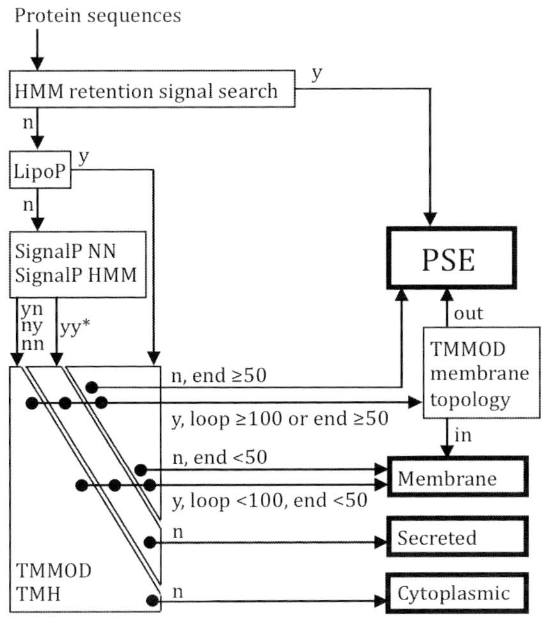

Figure 1.3 SurfG+ flow scheme for protein analysis. Protein sequences are analysed using the methods in the thin lined boxes (see text for details) and, through a series of logical decisions represented by the arrows and accompanying text, classified into four categories (fat lined boxes). y, path if the feature searched for is present; n, path if the feature searched for is not present; for SignalP, the first character represents the result of the NN method and the second character the result of the HMM method; *, mature proteins are analysed using TMMOD; commas represent the Boolean expression 'AND'; end and loop cutoff values are expressed as numbers of amino acids; in, end or loop predicted at the cytoplasmic side of the cell membrane; out, end or loop predicted outside the cytoplasmic membrane; PSE, potentially surface exposed. The retention signals (domains) (Finn *et al.*, 2006; Gough *et al.*, 2001) searched using HMM search are: LPXTG (Boekhorst *et al.*, 2005), GW (Jonquieres *et al.*, 1999), peptidoglycan (PG) binding domain of Type 1 (Foster, 1991), choline binding (Janecek *et al.*, 2000), LysM (Bateman and Bycroft, 2000), cell wall-binding domain of Type 2 (Waligora *et al.*, 2001), S-layer homology domain (Mesnage *et al.*, 2000). (Figure modified from Barinov *et al.*, 2009.)

of 55 tested) by their accessibility to antibodies, using fluorescence activated cell sorting (FACS) or whole cell enzyme-linked immunosorbent assay (ELISA). SurfG+ predicted 71 of the 114 experimentally identified proteins (62%) to be PSE (49 proteins), or secreted and consequently also potentially present at the cell surface (22 proteins). Thirty-seven proteins were found to be accessible to protease treatment although predicted to be cytoplasmic by SurfG+. Nearly all these proteins are annotated with functions that support their cytoplasmic localization. Although the positive results of FACS analyses or ELISAs for six of these proteins (out of six tested) (Rodriguez-Ortega *et al.*, 2006; Severin *et al.*, 2007) and the reports on homologues of a large number of these proteins in literature seem to confirm that these proteins, in addition to their cytoplasmic localization, can also be found at the cell surface, they lack known motifs that would permit the prediction of an extracellular location. For only six proteins the SurfG+ prediction of 'membrane protein' was in disagreement with the experimental data.

In summary, SurfG+ predicted a PSE or secreted localization for nearly all experimentally identified proteins for which a surface exposition can reasonably be predicted on the basis of specific sequence features. In the dataset of 114 experimentally detected proteins used for evaluation, this included 20 membrane-bound proteins that were classified as PSE on the mere basis of the possession of a long end or loop, 10 of which contain two or more TMH and would generally be regarded as membrane proteins without further analysis of potential surface exposition. Remarkably, this number of 20 proteins is 50% higher than the numbers of detected lipoproteins or LPXTG proteins, often considered as the paradigm of surface exposed proteins, in the same dataset.

A second way to compare predicted and experimental results is to determine what part of the predicted PSE proteins has been identified experimentally. In the case of *S. pyogenes*, this appears to be true for 49 out of 140 proteins. Although several factors could explain the discrepancy between the two figures (e.g. low gene expression levels under the experimental conditions used or compact tertiary protein structures), the fact that the results of the two experimental studies that were used are only partially overlapping strongly suggests that the experimental detection of surface exposed proteins has not been exhaustive. Today, the difficulty of obtaining reliable and exhaustive experimental data hampers the direct comparison of bacterial surface proteomes, and emphasizes the value of *in silico* prediction methods to obtain a first impression of potential differences between species.

Comparing lactobacilli from the viewpoint of surface-exposed and secreted proteins

As described in the preceding paragraphs, today 15 *Lactobacillus* genome sequences covering 12 species are publicly available, and several studies reported on the comparison of varying numbers of these genomes with regard to genome synteny and encoded metabolic and other functions. Here we will compare the sequenced lactobacilli from a different viewpoint, through a comparison of predicted surface exposed and secreted proteins. This comparison will give an impression of what the proteinaceous component of the bacterial cell surface may look like, and reveal potential differences between species that may be important in the interaction of the bacteria with their biotic and abiotic environment. The comparison of the predicted secretomes may be equally instructive for the understanding of the way in which bacteria exploit and interact with their environment, as demonstrated in a study of the predicted secretome of *L. plantarum* (Boekhorst *et al.*, 2006b).

We analysed the proteins derived from the 15 *Lactobacillus* genome sequences, using SurfG$^+$ with the default parameters for N- and C-terminal long ends (≥50 amino acids) and long loops (≥ 100 amino acids). The results of this comparative analysis are summarized in Table 1.2.

The SurfG$^+$ analysis revealed a 2-fold variation in the fractions of total proteins in the surface exposed (5.7–9.7%) and secreted (2.2–4.9%) categories among the 15 *Lactobacillus* proteomes. The smallest fraction of surface exposed proteins was predicted for *L. fermentum* (5.7 %), followed by the two *L. reuteri* strains (6.2 and 6.5%, respectively), species that make part of the same phylogenetic group (the reuteri group; Fig. 1.1). These two species also encode a relatively small fraction of secreted proteins (2.8–3.0%), and possess the lowest combined fractions of PSE and secreted proteins among the genomes analysed (8.5–9.5%). The highest fraction of coding capacity dedicated to PSE proteins was observed in some of the acidophilus group lactobacilli (9.7%, 9.1% and 9.0% for *L. johnsonii*, *L. acidophilus* and *L. gasseri*, respectively) that will be discussed in detail below.

The sequenced genomes of lactobacilli different from acidophilus group showed smaller amplitude of the variation regarding PSE proteins. It is interesting to note that in the meat fermenting *L. sakei* the fractions of PSE and secreted proteins (7.4 and 3.9%, respectively) are similar to those found in the milk fermenting lactobacilli, *L. bulgaricus* and *L. helveticus* of the acidophilus group. One may speculate that this similarity is related to the fact that these three species have adapted to specific confined environments.

Comparative analysis of lactobacilli from the acidophilus group

An in depth analysis was made of the results obtained for four closely related lactobacilli of the acidophilus group as these may more readily reveal differences that are directly related to the different environments where these bacteria are found: yogurt for *L. bulgaricus*, and the human GI tract for the three commensal species *L. acidophilus*, *L. johnsonii* and *L. gasseri* (Barinov *et al.*, 2009). Although these lactobacilli have similarly sized genomes, the number of predicted proteins in *L.*

Table 1.2 SurfG⁺ prediction of protein localization in the 15 sequenced lactobacilli

Strain	CYT (%)	MEM (%)	PSE (%)	SEC (%)	PSE + SEC (%)	Nr PROT
L. acidophilus NCFM	69.2	16.9	9.1	4.8	13.9	1862
L. johnsonii NCC 533	68.5	19.6	9.7	2.2	11.9	1821
L. gasseri ATCC 33323	71.9	16.9	9.0	2.2	11.2	1755
L. bulgaricus ATCC 11842	72.5	15.6	7.5	4.4	11.9	1562
L. bulgaricus BAA-365	72.9	16.1	6.6	4.4	11.0	1721
L. helveticus DPC 4571	74.8	14.4	6.8	4.0	10.8	1610
L. sakei 23K	70.1	18.6	7.4	3.9	11.3	1885
L. fermentum IFO 3956	77.4	14.2	5.7	2.8	8.5	1843
L. casei ATCC 334	69.3	18.7	8.3	3.7	12.0	2751
L. casei BL23	69.7	18.0	8.6	3.7	12.3	3044
L. plantarum WCFS1	70.3	18.3	8.4	3.0	11.5	3009
L. brevis ATCC 367	69.2	19.1	6.8	4.9	11.8	2185
L. salivarius UCC118	73.5	16.5	7.2	2.8	10.0	1717
L. reuteri DSM 20016	75.2	15.8	6.2	2.8	9.0	1900
L. reuteri JCM 1112	75.4	15.1	6.5	3.0	9.5	1820

CYT, cytoplasmic; MEM, membrane; PSE, potentially surface exposed; SEC, secreted; Nr PROT, total number of proteins predicted for the genome.

Figures indicate the number of proteins as a percentage of the total number of proteins in the strain analysed.

bulgaricus is much lower than in the other three species owing to the presence of a high number of pseudogenes (van de Guchte *et al.*, 2006). Consequently, the numbers of proteins for *L. bulgaricus* in all localization categories but the secreted proteins are lower than for the other species (Table 1.2). The fraction of proteins predicted to be secreted by *L. bulgaricus* is comparable to that of *L. acidophilus*, the closest related species (Nicolas *et al.*, 2007), and twice as high as for *L. johnsonii* and *L. gasseri* (Table 1.3A). In contrast, the three commensal strains, and especially *L. johnsonii*, appear to have dedicated a higher fraction of their coding capacity to PSE proteins than *L. bulgaricus* (9.0–9.7% against 7.5%, respectively), resulting in 41–60 additional proteins in the PSE category.

A detailed comparison of PSE proteins reveals that the commensal species contain much higher numbers of predicted LPXTG proteins (12–16) and (Lipo)proteins with long C-terminal ends (102–118) than *L. bulgaricus* (2 and 71 proteins, respectively) (Table 1.3B). Together, the differences in the number of LPXTG proteins and lipoproteins account for about 50% of the difference in the number of PSE proteins between the commensal species and *L. bulgaricus*, and it is interesting to note that about 50% of the LPXTG proteins contain mucus binding domains that have been described in proteins involved in adherence to the mucus layer lining the intestine (Boekhorst *et al.*, 2006a). Another important observation is that more than 50% of the PSE ends and loops are found in proteins with three or more TMH, thus supporting the importance of the analysis of what are classically called membrane proteins for potential surface exposition.

Together, these data suggest that PSE proteins are more important in the GI tract environment than in milk fermentation, a conclusion that seems to be confirmed by the important difference in the number of LPXTG proteins, of which about 50% contain predicted mucus binding domains. This raises the question whether the closely related *Lactobacillus* species share a common core of PSE proteins, in addition to which the commensal species contain a number of proteins that are not found in *L. bulgaricus*, and whether the commensal species share the additional proteins among them. Similar questions may be asked regarding the proteins in the other localization categories. In order to provide an answer to these questions, we analysed the distribution of protein families in

Table 1.3A SurfG+ prediction of protein localization in lactobacilli from the acidophilus group

	L. acidophilus		*L. johnsonii*		*L. gasseri*		*L. bulgaricus*	
Cytoplasmic	1289	69.2%	1248	68.5%	1262	71.9%	1132	72.5%
Membrane	315	16.9%	356	19.6%	297	16.9%	244	15.6%
PSE	169	9.1%	177	9.7%	158	9.0%	117	7.5%
Secreted	89	4.8%	40	2.2%	38	2.2%	69	4.4%
Total	**1862**		**1821**		**1755**		**1562**	

%, number of proteins as a percentage of the total number of proteins in the strain analysed; PSE, potentially surface exposed.

Figures indicate the number of proteins in each localization category. In each localization category the numbers of proteins significantly differ between species (chi-squared test with df = 3, $P<0.05$). (Table modified from Barinov *et al.*, 2009.)

Table 1.3B Subdivision of predicted PSE proteins in lactobacilli from the acidophilus group

	L. acidophilus	*L. johnsonii*	*L. gasseri*	*L. bulgaricus*
LPXTG	12	16	12	2
LysM	1	1	1	2
GW	2	–	–	–
Choline binding	1	–	–	–
PG binding	–	1	1	–
N	24	21	22	21
C	112 (42)	118 (39)	102 (34)	71 (23)
L	14	18	16	14
N+C	1	2	2	2
N+L	1	–	1	4
C+L	1	–	1	1
N+C+L	–	–	–	–
Total	**169**	**177**	**158**	**117**

LPXTG, LysM, GW, Choline binding, PG binding, cell wall anchoring motifs; N, C, outward N or C-terminal ends, respectively, ≥50 amino acids; L, outward loops ≥100 amino acids; N+C, N+L, C+L, N+C+L, proteins that contain a combination of the aforementioned features.

Proteins with long C-terminal ends include the numbers of lipoproteins specified in parentheses. Figures indicate the number of proteins in each category of PSE proteins. For proteins with an LPXTG motif or a long C-terminal end the numbers of proteins significantly differ between species (chi-squared test with df = 3, $P<0.05$). (Table modified from Barinov *et al.*, 2009.)

L. bulgaricus, L. acidophilus and *L. johnsonii.* The results presented in Fig. 1.4 show that:

1 While protein families common to the three species account for 52–60% of the cytoplasmic protein families in each species, the part of commonly shared protein families is much less important in the membrane, the PSE, and in particular the secreted categories.
2 *L. acidophilus* and *L. johnsonii* share a significant number of protein families (15–24% of the species' proteins in the category concerned) that are not represented in *L. bulgaricus,* in all categories but the secreted proteins.
3 The part of the protein families shared by *L. bulgaricus* and either *L. johnsonii* or *L. acidophilus* is negligible except for the secreted

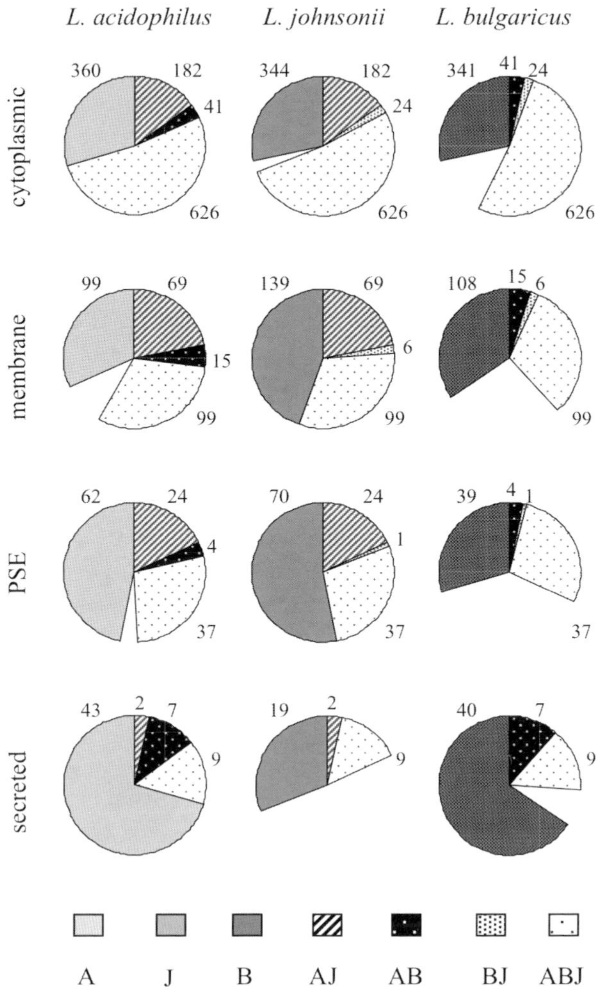

Figure 1.4 Protein conservation in three lactobacilli from the acidophilus group. Each diagram presents the numbers of conserved and strain specific proteins in the species indicated at the top of the figure, for the SurfG⁺ localization category indicated at the left of the figure. Within each localization category, the strain with the highest number of proteins in that category is represented by a full circle and the same scale is applied to the other strains. Consequently, the total areas covered by the diagrams for a given localization category reflect the relative numbers of proteins in the different strains for that category. A, J, B, proteins only present in *L. acidophilus*, *L. johnsonii* or *L. bulgaricus*, respectively; AJ, AB, BJ, proteins shared by two strains but absent in the third strain; ABJ, proteins shared by the three strains. (Figure modified from Barinov *et al.*, 2009.)

proteins shared by *L. bulgaricus* and *L. acidophilus*.

4 The higher numbers of PSE protein families in *L. acidophilus* and *L. johnsonii* as compared with *L. bulgaricus* are explained in part by protein families shared by the first two species, and in part by species specific proteins. Notably, all the proteins that are predicted to be LPXTG proteins are species specific.

5 Roughly 50% of the protein families in the PSE category of each species are species specific.

6 Although *L. acidophilus* and *L. bulgaricus* share the property of a relatively high number of secreted proteins, this property can be ascribed only for a very small part to proteins shared between the two species, while the large majority of proteins in this category are species specific.

Thus, although PSE proteins seem to be relatively more important in the GI tract environment, an important part of these proteins appears to be species specific. In particular, the LPXTG proteins that are often considered as important for adherence to host tissues (Boekhorst *et al.*, 2006a; van Pijkeren *et al.*, 2006), differ between *L. acidophilus* and *L. johnsonii* in spite of the close relationship of, and the extensive horizontal gene transfer between, these species (Nicolas *et al.*, 2007). This may indicate that *L. acidophilus* and *L. johnsonii* occupy different niches in the GI tract. Interestingly, among the PSE and secreted proteins that are shared by *L. acidophilus* and *L. johnsonii*, but absent from *L. bulgaricus*, are a number of ABC and other transporters that may play a particular role in the utilization of GI tract resources that are not present in milk, as exemplified by a maltose ABC transporter. Among the proteins specific for *L. acidophilus* are the PSE LPXTG mucus-binding protein LBA1392 and the secreted S-layer protein LBA0169, which both have been implicated in bacterial adherence to eukaryotic cells (Buck *et al.*, 2005).

Prevalence of PSE protein categories in lactobacilli

The analysis of the data for the acidophilus group lactobacilli revealed the existence of numerical differences between species when regarding PSE protein subcategories. In the next paragraphs we will present the results of a similar analysis for all 15 available *Lactobacillus* genomes.

The distribution of PSE proteins with predicted retention signals varies considerably between different species (Table 1.4). As discussed above, important differences in the number of predicted LPXTG proteins, lipoproteins, and membrane bound proteins with surface exposed C-terminal ends are observed among the members of the acidophilus group, clearly separating the GI tract isolates and the milk-fermenting bacteria. A common feature that distinguishes the acidophilus group from most other lactobacilli is the relatively low number of proteins that are predicted to be bound to the cell wall in a non-covalent manner. A distribution of retention signals, loops and terminal ends similar to that of *L. bulgaricus* and *L. helveticus* is found in *L. sakei* and *L. fermentum*. *L. casei* and *L. plantarum* constitute a group characterized by relatively high numbers of proteins in each of the categories considered, which in part may be a consequence of their big genome size and potentially important for their versatility.

Conserved domain analysis in the PSE and secreted proteins

In order to obtain an impression of potentially different functionalities of the bacterial surface and its close environment, we subjected the predicted PSE and secreted proteins to a domain analysis using InterPro (Hunter *et al.*, 2009). It should be mentioned that the number of repetitive domains, like mucus binding domains, can vary dramatically between proteins from the same or different species (Boekhorst *et al.*, 2006a). The analysis of protein families which takes the length of the proteins into consideration may classify such functionally related proteins as species specific, which seems correct, as duplication of mucus binding domains within the same protein has previously been suggested to increase the affinity of adhesins for mucins (Boekhorst *et al.*, 2006a). The conservation of domains can provide a first indication for the conservation of function, however. This analysis allowed us to identify a total of 587 different domains with varying distributions among the analysed species. In the following paragraphs we first focus on domains of which the observed frequency varies most between species. Then, special attention is given to domains that have been described in proteins that have been shown experimentally to be involved in adhesion effects when in contact with host cells. Finally, we look at the domains common in all the lactobacilli to identify conserved domain patterns in the 15 genomes.

Variability of domain representation

A total of 587 different domains were detected for which we determined the frequency of occurrence in the PSE and secreted proteins of each species. The domains for which the frequency varied most between genomes are presented in Table 1.5. The results of this analysis give a first impression of differing abilities of the lactobacilli analysed to interact with their environment.

Table 1.4 Subdivision of predicted PSE proteins in 15 *Lactobacillus* genomes

Strain	LPXTG	RET	N	L	C	Lipo	Total
L. acidophilus NCFM	12	4	25	14	72	42	169
L. johnsonii NCC 533	16	2	21	18	81	39	177
L. gasseri ATCC 33323	12	2	23	16	71	34	158
L. bulgaricus ATCC 11842	2	2	25	14	51	23	117
L. bulgaricus BAA-365	1	2	26	15	47	24	115
L. helveticus DPC 4571	2	4	14	12	55	23	110
L. sakei 23K	3	4	30	16	59	28	140
L. fermentum IFO 3956	5	6	23	13	48	10	105
L. casei ATCC 334	13	8	61	24	80	44	230
L. casei BL23	14	8	74	32	84	52	264
L. plantarum WCFS1	25	14	47	29	89	50	254
L. brevis ATCC 367	9	12	27	20	57	25	150
L. salivarius UCC118	3	14	22	19	53	19	130
L. reuteri DSM 20016	5	12	24	11	53	16	121
L. reuteri JCM 1112	6	8	23	11	54	16	118

Figures indicate the number of proteins in each category of PSE proteins. LPXTG, anchoring motif for covalent attachment of the proteins to the bacterial cell wall; RET, number of proteins with the motifs for non-covalent attachment to the cell wall. N, C, outward N-or C-terminal end respectively, ≥ 50 amino acids, proteins with long C-terminal ends exclude the numbers of lipoproteins; L, outward loops ≥ 100 amino acids; Lipo, the numbers of PSE lipoproteins; black and grey shadings indicate frequencies of occurrence that are higher or lower than the average.

Among the domains for which the frequency varies most, several domains are found that have been associated with binding to mucus (IPR009459), collagen (IPR008454), or to eukaryotic cells (S-layer domain, IPR004903). These and other adhesion domains identified in this study will be presented in detail in the next paragraph.

Domains specific for phosphotransferase (PTS) sugar transport systems (PTS EIIA and EIIB) appear to be particularly abundant in PSE proteins from *L. plantarum,* followed by *L. acidophilus,* and *L. johnsonii*. For *L. plantarum,* this feature has been interpreted as indicative for the versatility of this bacterium, which is found in very diverse environments (Kleerebezem *et al.,* 2003). ABC transporter domains are found to be frequent in many species, which is of particular interest as recent observations show that ABC transporters may, in addition to their direct function in transport processes, participate in bacterial adhesion to collagen and epithelial cells. This is the case for *L. reuteri* CnBP (Roos *et al.,* 1996; Bath *et al.,* 2005) and ABC transporters involved in the transport of metal ions in oral streptococci (Jenkinson, 1994). In addition to domains with assigned functions, a number of domains with unknown functions have been identified (32 entries out of 587, data not shown). Interestingly, among the most variable, DUF1542 has been found to be abundant in and specific for *L. acidophilus* and *L. johnsonii*. There are limited and controversial data describing the potential function of this domain. It is found in several cell surface proteins from Staphylococci that are involved in antibiotic resistance (Wu and De Lencastre, 1999) and/or cellular adhesion (Clarke *et al.,* 2002). Group specific distribution has also been observed for the Rib/alpha/Esp surface antigen domain (IPR012706) which has been described in proteins associated with

Table 1.5 InterPro domains identified in the PSE and secreted proteins from 15 lactobacilli

No.	IPR	*L. acidophilus* NCFM	*L. johnsonii* NCC 533	*L. gasseri* ATCC 33323	*L. bulgaricus* ATCC 11842	*L. bulgaricus* BAA-365	*L. helveticus* DPC 4571	*L. sakei* 23K	*L. fermentum* IFO 3956
1	009459	42	36	52	1	1	0	0	4
2	011439	21	38	0	0	0	0	0	0
3	001996	24	24	10	0	0	0	5	0
4	001127	14	14	7	0	0	3	4	0
5	004903	23	0	0	1	0	17	0	0
6	012706	22	20	16	4	4	0	0	0
7	008454	0	0	0	0	0	0	0	0
8	003439	16	24	22	16	18	10	8	4
9	011439	21	38	0	0	0	0	0	0
10	006128	3	7	7	3	3	6	9	0
11	006129	3	6	6	0	3	6	9	0
12	008966	1	0	1	0	0	2	2	0
13	006179	0	0	6	5	5	5	0	0
14	000421	0	0	0	0	0	0	0	0
15	002035	0	0	0	0	0	0	0	0
16	003343	0	0	0	2	2	0	0	0
17	003961	3	0	0	0	0	3	0	0
18	008334	0	0	3	3	3	3	0	0
19	008985	0	0	0	0	0	0	1	0
20	008964	0	0	0	1	1	0	0	0
21	011009	1	1	1	1	1	1	1	1
22	015919	2	0	0	0	0	0	0	0
23	001876	0	0	0	0	0	0	0	0
24	011032	0	0	0	0	0	0	0	0
25	011050	0	0	0	0	0	0	0	0
26	011252	0	0	0	0	0	0	0	0
27	013574	1	0	0	0	0	0	0	0
28	001254	1	1	1	1	1	1	1	1

Domains for which the frequency varied most between genomes are listed in the first part of the table (lines 1 to 8). LPXTG, LysM and putative cell wall binding domains involved in binding to the bacterial cell wall were omitted from the table as they are presented in Table 1.4 ('LPXTG' and 'RET', respectively). Numbers indicate the predicted occurrence of the domain in the PSE and secreted categories; IPR, InterPro domain entry number; MAX-MIN, maximal observed difference in frequency of occurrence between two species; black and

L. casei ATCC 334	*L. casei* BL23	*L. plantarum* WCFS1	*L. brevis* ATCC 367	*L. salivarius* UCC118	*L. reuteri* DSM 20016	*L. reuteri* JCM 1112	MAX-MIN	IPR domain description
5	3	38	9	12	1	1	52	Protein of unknown function DUF1085
0	0	0	0	0	0	0	38	Protein of unknown function DUF1542
5	0	30	0	5	0	0	30	Phosphotransferase system, EIIB
4	0	23	0	4	0	0	23	Phosphotransferase system, sugar-specific permease EIIA 1 domain
0	0	0	0	0	0	0	23	Bacterial surface layer protein
0	0	0	0	0	0	0	22	Rib/alpha/Esp surface antigen
21	20	5	3	0	0	0	21	Collagen-binding surface protein Cna-like, B region
17	20	18	8	8	4	4	20	ABC transporter-like
0	0	0	0	0	0	0	38	Protein of unknown function DUF1542
6	6	13	0	3	0	0	13	Adhesion lipoprotein
9	9	9	0	2	0	0	9	Adhesin B
5	5	7	6	1	0	0	7	Adhesion, bacterial
0	0	0	6	0	6	6	6	5'-Nucleotidase and apyrase
2	0	3	0	0	0	0	3	Coagulation factor 5/8 type, C-terminal
3	3	0	0	0	0	0	3	von Willebrand factor, type A
0	0	3	0	0	0	0	3	Bacterial Ig-like, group 2
3	0	0	0	0	0	0	3	Fibronectin, type III
0	0	0	3	0	3	3	3	5'-Nucleotidase, C-terminal
3	3	3	2	0	0	0	3	Concanavalin A-like lectin/glucanase
0	0	2	0	0	0	0	2	Invasin/intimin cell-adhesion
1	1	1	2	0	1	1	2	Protein kinase-like
0	0	0	0	0	0	0	2	Cadherin-like
0	0	1	0	1	0	0	1	Zinc finger, RanBP2-type
0	1	0	0	0	1	1	1	GroES-like
1	1	0	1	0	0	0	1	Pectin lyase fold/virulence factor
0	0	1	0	0	0	0	1	Fibrogen-binding, domain 1
0	0	0	1	0	0	0	1	Glucan-binding protein C/ Surface antigen I/II, V-region
1	1	1	1	1	1	1	0	Peptidase S1 and S6, chymotrypsin/Hap

grey shadings indicate frequencies of occurrence that are at least 30 % higher or lower than the average of the non-zero values for the identified domain, respectively. Second part of the table represents the potential adhesion domains identified in PSE and secreted proteins in 15 lactobacilli (lines 9–28). Black shading indicates the presence of the domain; grey shadings indicate absence of the domain.

biofilm formation (Toledo-Arana *et al.*, 2001). Here, this domain is found only in the acidophilus group, with the exception of *L. helveticus*. High-frequency occurrence of this domain was found in the GI tract-associated *L. acidophilus*, *L. johnsonii* and *L. gasseri* (from 16 to 22 occurrences), while it is less represented in both *L. bulgaricus* strains (four domains) and as mentioned absent from *L. helveticus*.

Adhesins

Adhesion factors play an important role in host–microbe interactions. One of the most abundant domains identified in this study is the mucus-binding domain (DUF1085, IPR009459), which is particularly abundant in the GI tract-associated lactobacilli of the acidophilus group (*L. acidophilus, L. johnsonii, L. gasseri*) and in *L. plantarum*, and these bacteria have indeed been shown to adhere to mucus producing epithelial cells (Buck *et al.*, 2005; Boekhorst *et al.*, 2006a). Interestingly, a neglectable number of mucus binding domains was detected in the PSE and secreted proteins from *L. reuteri*, a natural inhabitant of the GI tract, when compared with other lactobacilli from similar environments (only one predicted in the analysed categories). This suggests that another mechanisms for the persistence of this bacterium in the GI tract may exist. This is in part confirmed by the role of the ABC-transporter/collagen binding protein (CnBP) and the MapA protein, both involved in adhesion in this bacterium (Roos *et al.*, 1996; Miyoshi *et al.*, 2006; Wang *et al.*, 2008). Interestingly, both CnBP and MapA carry a bacterial extracellular solute-binding domain (IPR001638), which has also been identified in the PSE category of other tested lactobacilli, and it will be interesting to see whether the presence of this domain is indicative for adherence. Among the other well characterized adhesion domains, a collagen-binding protein domain (IPR008454) has been identified uniquely in *L. casei, L. plantarum*, and *L. brevis* These observations are in agreement with, and may explain, the experimental observations showing the potential of these bacteria to bind collagen (Boekhorst *et al.*, 2006b; de Leeuw *et al.*, 2006; Munoz-Provencio *et al.*, 2008).

The role of surface layer (S-layer) proteins in adhesion is well documented (Velez *et al.*, 2007). They mediate the adhesion to intestinal epithelial cells, extracellular matrices and/or to LTA of other bacterial species (Velez *et al.*, 2007). Our analysis revealed the presence of bacterial surface layer protein domains (IPR004903) in *L. acidophilus, L. helveticus* and *L. bulgaricus* ATCC 11842, with a high representation in the first two species (23 and 17 copies in *L. acidophilus* and *L. helveticus*, respectively, versus one copy of the domain in *L. bulgaricus*). The previously reported S-layer protein of *L. brevis* (Vidgren *et al.*, 1992) was not detected in our analysis, which may be explained by the fact that S-layer proteins show low overall sequence similarities (Avall-Jaaskelainen and Palva, 2005) and not all of them have been assigned to the domain records (Velez *et al.*, 2007). 20 additional domains potentially involved in adhesion were identified by keyword search in the descriptions of the InterPro domains (Table 1.5). These include fibrogen-binding domain 1 (involved in binding to human fibrinogen), bacterial immunoglobulin-like domains and phospholipid-binding domains (potentially important in host-cell adhesion).

Only few *Lactobacillus* proteins have been experimentally evaluated for interactions with the host, thus leaving a great potential for further studies aiming to provide answers to fundamental questions on how lactobacilli interact with their surroundings and 'communicate' with the host. Results of the SurfG^{+} analysis coupled with the domain predictions allowed to identify both experimentally verified and new putative adhesion proteins present at the surface of the sequenced lactobacilli.

Conserved domains present in all species analysed

60 domains were found to be present in all the analysed species. These include a number of ABC transporter domains, which are found in transporter systems vital for the import of essential nutrients and the export of toxic molecules (Lorca *et al.*, 2007; Davidson *et al.*, 2008). We identified the domains of signal transduction histidine kinases specific for two-component

signal transduction systems which enable bacteria to sense, respond, and adapt to a wide range of environments and growth conditions (Skerker *et al.*, 2008). This analysis allowed us to identify domains present in proteins essential, for example, in cell wall metabolism, as exemplified by the penicillin-binding proteins involved in the final stages of the synthesis of peptidoglycan (Sauvage *et al.*, 2008) and murein polymerases which catalyses the polymerization of murein glycan chains, as exemplified by glycosyl transferase family 51 domains. A number of peptidase domains identified in all species matched to the HtrA (high-temperature requirement A) family protease, which constitutes a group of housekeeping heat shock-induced serine proteases crucial in natural processing of exported proteins and protein quality control mechanisms (Smeds *et al.*, 1998). Three domains corresponding to the N-terminal, C-terminal and central region of the DltD protein identified in the PSE protein category were identified. This membrane-associated protein is important in the esterification of lipoteichoic acid (LTA) by D-alanine (Debabov *et al.*, 2000) and as mentioned above, an *L. plantarum* mutant with reduced D-alanylation of its LTA exerted anti-inflammatory effects (Grangette *et al.*, 2005).

Concluding remarks

The comparative analyses of *Lactobacillus* genomes from diverse environments using SurfG$^+$ coupled with protein domain analysis proved instrumental in conceiving a new view of the high complexity and variety of bacterial surfaces. The results indicate that the complexity of the proteinaceous component of the bacterial cell surface is far from completely understood, and has been highly underestimated so far. Further investigations of surface exposed and secreted proteins may provide important leads to a better understanding of how bacteria adapt to and interact with their biotic and abiotic environment.

References

Abriouel, H., Martin-Platero, A., Maqueda, M., Valdivia, E., and Martinez-Bueno, M. (2008). Biodiversity of the microbial community in a Spanish farmhouse cheese as revealed by culture-dependent and culture-independent methods. Int. J. Food Microbiol. *127*, 200–208.

Adolfsson, O., Meydani, S.N., and Russell, R.M. (2004). Yogurt and gut function. Am. J. Clin. Nutr. *80*, 245–256.

Akira, S., and Takeda, K. (2004). Toll-like receptor signalling. Nat. Rev. Immunol. *4*, 499–511.

Alexopoulou, L., Thomas, V., Schnare, M., Lobet, Y., Anguita, J., Schoen, R.T., Medzhitov, R., Fikrig, E., and Flavell, R.A. (2002). Hyporesponsiveness to vaccination with *Borrelia burgdorferi* OspA in humans and in TLR1- and TLR2-deficient mice. Nat. Med. *8*, 878–884.

Altermann, E., Russell, W.M., Azcarate-Peril, M.A., Barrangou, R., Buck, B.L., McAuliffe, O., Souther, N., Dobson, A., Duong, T., Callanan, M., *et al.* (2005). Complete genome sequence of the probiotic lactic acid bacterium *Lactobacillus* acidophilus NCFM. Proc. Natl. Acad. Sci. U.S.A *102*, 3906–3912.

Avall-Jaaskelainen, S., and Palva, A. (2005). *Lactobacillus* surface layers and their applications. FEMS Microbiol. Rev. *29*, 511–529.

Azcarate-Peril, M.A., Altermann, E., Goh, Y.J., Tallon, R., Sanozky-Dawes, R.B., Pfeiler, E.A., O'Flaherty, S., Buck, B.L., Dobson, A., Duong, T., *et al.* (2008). Analysis of the genome sequence of *Lactobacillus* gasseri ATCC 33323 reveals the molecular basis of an autochthonous intestinal organism. Appl. Environ. Microbiol. *74*, 4610–4625.

Barinov, A., Loux, V., Hammani, A., Nicolas, P., Langella, P., Ehrlich, D., Maguin, E., and van de Guchte, M. (2009). Prediction of surface exposed proteins in *Streptococcus* pyogenes, with a potential application to other Gram-positive bacteria. Proteomics *9*, 61–73.

Bateman, A., and Bycroft, M. (2000). The structure of a LysM domain from *E. coli* membrane-bound lytic murein transglycosylase D (MltD). J. Mol. Biol. *299*, 1113–1119.

Bath, K., Roos, S., Wall, T., and Jonsson, H. (2005). The cell surface of *Lactobacillus* reuteri ATCC 55730 highlighted by identification of 126 extracellular proteins from the genome sequence. FEMS Microbiol. Lett. *253*, 75–82.

Bendtsen, J.D., Nielsen, H., von Heijne, G., and Brunak, S. (2004). Improved prediction of signal peptides: SignalP 3.0. J. Mol. Biol. *340*, 783–795.

Beresford, T.P., Fitzsimons, N.A., Brennan, N.L., and Cogan, T.M. (2001). Recent advances in cheese microbiology. Int. Dairy J. *11*, 259–274.

Berger, B., Pridmore, R.D., Barretto, C., Delmas-Julien, F., Schreiber, K., Arigoni, F., and Brussow, H. (2007). Similarity and differences in the *Lactobacillus* acidophilus group identified by polyphasic analysis and comparative genomics. J. Bacteriol. *189*, 1311–1321.

Bernardeau, M., Guguen, M., and Vernoux, J.P. (2006). Beneficial lactobacilli in food and feed: long-term use, biodiversity and proposals for specific and realistic safety assessments. FEMS Microbiol. Rev. *30*, 487–513.

Bernardeau, M., Vernoux, J.P., Henri-Dubernet, S., and Gueguen, M. (2008). Safety assessment of dairy microorganisms: The *Lactobacillus* genus. Int. J. Food. Microbiol. *126*, 278–285.

Boekhorst, J., de Been, M.W., Kleerebezem, M., and Siezen, R.J. (2005). Genome-wide detection and analysis of cell wall-bound proteins with LPXTG-like sorting motifs. J. Bacteriol. *187*, 4928–4934.

Boekhorst, J., Helmer, Q., Kleerebezem, M., and Siezen, R.J. (2006a). Comparative analysis of proteins with a mucus-binding domain found exclusively in lactic acid bacteria. Microbiology *152*, 273–280.

Boekhorst, J., Siezen, R.J., Zwahlen, M.C., Vilanova, D., Pridmore, R.D., Mercenier, A., Kleerebezem, M., de Vos, W.M., Brussow, H., and Desiere, F. (2004). The complete genomes of *Lactobacillus plantarum* and *Lactobacillus johnsonii* reveal extensive differences in chromosome organization and gene content. Microbiology *150*, 3601–3611.

Boekhorst, J., Wels, M., Kleerebezem, M., and Siezen, R.J. (2006b). The predicted secretome of *Lactobacillus* plantarum WCFS1 sheds light on interactions with its environment. Microbiology *152*, 3175–3183.

Bosch, A., Golowczyc, M.A., Abraham, A.G., Garrote, G.L., De Antoni, G.L., and Yantorno, O. (2006). Rapid discrimination of lactobacilli isolated from kefir grains by FT-IR spectroscopy. Int. J. Food. Microbiol. *111*, 280–287.

Buck, B.L., Altermann, E., Svingerud, T., and Klaenhammer, T.R. (2005). Functional analysis of putative adhesion factors in *Lactobacillus acidophilus* NCFM. Appl. Environ. Microbiol. *71*, 8344–8351.

Bulut, Y., Michelsen, K.S., Hayrapetian, L., Naiki, Y., Spallek, R., Singh, M., and Arditi, M. (2005). Mycobacterium tuberculosis heat shock proteins use diverse Toll-like receptor pathways to activate pro-inflammatory signals. J. Biol. Chem. *280*, 20961–20967.

Cabeen, M.T., and Jacobs-Wagner, C. (2005). Bacterial cell shape. Nat. Rev. Microbiol. *3*, 601–610.

Cai, H., Rodriguez, B.T., Zhang, W., Broadbent, J.R., and Steele, J.L. (2007). Genotypic and phenotypic characterization of *Lactobacillus casei* strains isolated from different ecological niches suggests frequent recombination and niche specificity. Microbiology *153*, 2655–2665.

Callanan, M., Kaleta, P., O'Callaghan, J., O'Sullivan, O., Jordan, K., McAuliffe, O., Sangrador-Vegas, A., Slattery, L., Fitzgerald, G.F., Beresford, T., *et al.* (2008). Genome sequence of *Lactobacillus helveticus*, an organism distinguished by selective gene loss and insertion sequence element expansion. J. Bacteriol. *190*, 727–735.

Canchaya, C., Claesson, M.J., Fitzgerald, G.F., van Sinderen, D., and O'Toole, P.W. (2006). Diversity of the genus *Lactobacillus* revealed by comparative genomics of five species. Microbiology *152*, 3185–3196.

Chaillou, S., Champomier-Verges, M.C., Cornet, M., Crutz-Le Coq, A.M., Dudez, A.M., Martin, V., Beaufils, S., Darbon-Rongere, E., Bossy, R., Loux, V., *et al.* (2005). The complete genome sequence of the meat-borne lactic acid bacterium *Lactobacillus sakei* 23K. Nat. Biotechnol. *23*, 1527–1533.

Claesson, M.J., Li, Y., Leahy, S., Canchaya, C., van Pijkeren, J.P., Cerdeno-Tarraga, A.M., Parkhill, J., Flynn, S., O'Sullivan, G.C., Collins, J.K., *et al.* (2006). Multireplicon genome architecture of *Lactobacillus salivarius*. Proc. Natl. Acad. Sci. U.S.A *103*, 6718–6723.

Claesson, M.J., van Sinderen, D., and O'Toole, P.W. (2007). The genus *Lactobacillus* – a genomic basis for understanding its diversity. FEMS Microbiol. Lett. *269*, 22–28.

Claesson, M.J., van Sinderen, D., and O'Toole, P.W. (2008). *Lactobacillus* phylogenomics – towards a reclassification of the genus. Int. J. Syst. Evol. Microbiol. *58*, 2945–2954.

Clarke, S.R., Harris, L.G., Richards, R.G., and Foster, S.J. (2002). Analysis of Ebh, a 1.1-megadalton cell wall-associated fibronectin-binding protein of *Staphylococcus aureus*. Infect. Immun. *70*, 6680–6687.

Cole, J.R., Wang, Q., Cardenas, E., Fish, J., Chai, B., Farris, R.J., Kulam-Syed-Mohideen, A.S., McGarrell, D.M., Marsh, T., Garrity, G.M., *et al.* (2009). The Ribosomal Database Project: improved alignments and new tools for rRNA analysis. Nucleic Acids Res. *37*, D141–145.

Corsetti, A., Settanni, L., Chaves Lopez, C., Felis, G.E., Mastrangelo, M., and Suzzi, G. (2007). A taxonomic survey of lactic acid bacteria isolated from wheat (Triticum durum) kernels and non-conventional flours. Syst. Appl. Microbiol. *30*, 561–571.

Coton, M., Berthier, F., and Coton, E. (2008). Rapid identification of the three major species of dairy obligate heterofermenters *Lactobacillus brevis*, *Lactobacillus fermentum* and *Lactobacillus parabuchneri* by species-specific duplex PCR. FEMS Microbiol. Lett. *284*, 150–157.

Dal Bello, F., and Hertel, C. (2006). Oral cavity as natural reservoir for intestinal lactobacilli. Syst. Appl. Microbiol. *29*, 69–76.

Davidson, A.L., Dassa, E., Orelle, C., and Chen, J. (2008). Structure, function, and evolution of bacterial ATP-binding cassette systems. Microbiol. Mol. Biol. Rev. *72*, 317–364, table of contents.

Debabov, D.V., Kiriukhin, M.Y., and Neuhaus, F.C. (2000). Biosynthesis of lipoteichoic acid in *Lactobacillus rhamnosus*: role of DltD in D-alanylation. J. Bacteriol. *182*, 2855–2864.

Dellaglio, F., Torriani, S., and Felis, G.E. (2004). Reclassification of *Lactobacillus cellobiosus* Rogosa *et al.* 1953 as a later synonym of *Lactobacillus fermentum* Beijerinck 1901. Int. J. Syst. Evol. Microbiol. *54*, 809–812.

Dietz, H., and Rief, M. (2006). Protein structure by mechanical triangulation. Proc. Natl. Acad. Sci. U.S.A *103*, 1244–1247.

Draing, C., Pfitzenmaier, M., Zummo, S., Mancuso, G., Geyer, A., Hartung, T., and von Aulock, S. (2006). Comparison of lipoteichoic acid from different serotypes of *Streptococcus pneumoniae*. J. Biol. Chem. *281*, 33849–33859.

Duran Quintana, M.C., Garcia Garcia, P., and Garrido Fernandez, A. (1999). Establishment of conditions for green table olive fermentation at low temperature. Int. J. Food. Microbiol. *51*, 133–143.

Dziarski, R., and Gupta, D. (2005). Peptidoglycan recognition in innate immunity. J. Endotoxin Res. *11*, 304–310.

Eddy, S.R. (1998). Profile hidden Markov models. Bioinformatics *14*, 755–763.

Falagas, M.E., Betsi, G.I., and Athanasiou, S. (2007). Probiotics for the treatment of women with bacterial vaginosis. Clin. Microbiol. Infect. *13*, 657–664.

Felis, G.E., and Dellaglio, F. (2007). Taxonomy of Lactobacilli and Bifidobacteria. Curr. Issues Intest. Microbiol. *8*, 44–61.

Finn, R.D., Mistry, J., Schuster-Bockler, B., Griffiths-Jones, S., Hollich, V., Lassmann, T., Moxon, S., Marshall, M., Khanna, A., Durbin, R., *et al.* (2006). Pfam: clans, web tools and services. Nucleic Acids Res. *34*, D247–251.

Foster, S.J. (1991). Cloning, expression, sequence analysis and biochemical characterization of an autolytic amidase of *Bacillus subtilis* 168 trpC2. J. Gen. Microbiol. *137*, 1987–1998.

Gala, E., Landi, S., Solieri, L., Nocetti, M., Pulvirenti, A., and Giudici, P. (2008). Diversity of lactic acid bacteria population in ripened Parmigiano Reggiano cheese. Int. J. Food. Microbiol. *125*, 347–351.

Gardner, N.J., Savard, T., Obermeier, P., Caldwell, G., and Champagne, C.P. (2001). Selection and characterization of mixed starter cultures for lactic acid fermentation of carrot, cabbage, beet and onion vegetable mixtures. Int. J. Food. Microbiol. *64*, 261–275.

Gardy, J.L., Laird, M.R., Chen, F., Rey, S., Walsh, C.J., Ester, M., and Brinkman, F.S. (2005). PSORTb v.2.0: expanded prediction of bacterial protein subcellular localization and insights gained from comparative proteome analysis. Bioinformatics *21*, 617–623.

Gough, J., Karplus, K., Hughey, R., and Chothia, C. (2001). Assignment of homology to genome sequences using a library of hidden Markov models that represent all proteins of known structure. J. Mol. Biol. *313*, 903–919.

Grangette, C., Nutten, S., Palumbo, E., Morath, S., Hermann, C., Dewulf, J., Pot, B., Hartung, T., Hols, P., and Mercenier, A. (2005). Enhanced antiinflammatory capacity of a *Lactobacillus plantarum* mutant synthesizing modified teichoic acids. Proc. Natl. Acad. Sci. U.S.A *102*, 10321–10326.

Guan, R., and Mariuzza, R.A. (2007). Peptidoglycan recognition proteins of the innate immune system. Trends Microbiol. *15*, 127–134.

Guarner, F., and Malagelada, J.R. (2003). Gut flora in health and disease. Lancet *361*, 512–519.

van de Guchte, M., Penaud, S., Grimaldi, C., Barbe, V., Bryson, K., Nicolas, P., Robert, C., Oztas, S., Mangenot, S., Couloux, A., *et al.* (2006). The complete genome sequence of *Lactobacillus bulgaricus* reveals extensive and ongoing reductive evolution. Proc. Natl. Acad. Sci. U.S.A *103*, 9274–9279.

Hayashi, F., Smith, K.D., Ozinsky, A., Hawn, T.R., Yi, E.C., Goodlett, D.R., Eng, J.K., Akira, S., Underhill, D.M., and Aderem, A. (2001). The innate immune response to bacterial flagellin is mediated by Toll-like receptor 5. Nature *410*, 1099–1103.

Heilig, H.G., Zoetendal, E.G. Vaughan, E.E., Marteau, P., Akkermans, A.D., and de Vos, W.M. (2002). Molecular diversity of *Lactobacillus* spp. and other lactic acid bacteria in the human intestine as determined by specific amplification of 16S ribosomal DNA. Appl. Environ. Microbiol. *68*, 114–123.

Hunter, S., Apweiler, R., Attwood, T.K., Bairoch, A., Bateman, A., Binns, D., Bork, P., Das, U., Daugherty, L., Duquenne, L., *et al.* (2009). InterPro: the integrative protein signature database. Nucleic Acids Res. *37*, D211–215.

Huson, D.H., and Bryant, D. (2006). Application of phylogenetic networks in evolutionary studies. Mol. Biol. Evol. *23*, 254–267.

Iwasaki, A., and Medzhitov, R. (2004). Toll-like receptor control of the adaptive immune responses. Nat. Immunol. *5*, 987–995.

Janecek, S., Svensson, B., and Russell, R.R. (2000). Location of repeat elements in glucansucrases of *Leuconostoc* and *Streptococcus* species. FEMS Microbiol. Lett. *192*, 53–57.

Jenkinson, H.F. (1994). Cell surface protein receptors in oral streptococci. FEMS Microbiol. Lett. *121*, 133–140.

JGI (2009). Why sequence *Lactobacillus reuteri*? http://www.jgi.doe.gov/sequencing/why/3496.html

Jonquieres, R., Bierne, H., Fiedler, F., Gounon, P., and Cossart, P. (1999). Interaction between the protein InlB of Listeria monocytogenes and lipoteichoic acid: a novel mechanism of protein association at the surface of Gram-positive bacteria. Mol. Microbiol. *34*, 902–914.

Juncker, A.S., Willenbrock, H., Von Heijne, G., Brunak, S., Nielsen, H., and Krogh, A. (2003). Prediction of lipoprotein signal peptides in Gram-negative bacteria. Protein Sci. *12*, 1652–1662.

Kaban, G., and Kaya, M. (2008). Identification of lactic acid bacteria and Gram-positive catalase-positive cocci isolated from naturally fermented sausage (sucuk). J. Food Sci. *73*, M385–388.

Kahsay, R.Y., Gao, G., and Liao, L. (2005). An improved hidden Markov model for transmembrane protein detection and topology prediction and its applications to complete genomes. Bioinformatics *21*, 1853–1858.

Kitazawa, H., Harata, T., Uemura, J., Saito, T., Kaneko, T., and Itoh, T. (1998). Phosphate group requirement for mitogenic activation of lymphocytes by an extracellular phosphopolysaccharide from *Lactobacillus delbrueckii* subsp. *bulgaricus*. Int. J. Food. Microbiol. *40*, 169–175.

Kleerebezem, M., Boekhorst, J., van Kranenburg, R., Molenaar, D., Kuipers, O.P., Leer, R., Tarchini, R., Peters, S.A., Sandbrink, H.M., Fiers, M.W., *et al.* (2003). Complete genome sequence of *Lactobacillus plantarum* WCFS1. Proc. Natl. Acad. Sci. U.S.A *100*, 1990–1995.

Krieg, A.M. (2003). CpG motifs: the active ingredient in bacterial extracts? Nat. Med. *9*, 831–835.

Krogh, A., Larsson, B., von Heijne, G., and Sonnhammer, E.L. (2001). Predicting transmembrane protein topology with a hidden Markov model: application to complete genomes. J. Mol. Biol. *305*, 567–580.

Kwon, S., Lee, P.C., Lee, E.G. Keun Chang, Y., and Chang, N. (2000). Production of lactic acid by *Lactobacillus rhamnosus* with vitamin-supplemented soybean hydrolysate. Enzyme Microb. Technol. *26*, 209–215.

de Las Rivas, B., Marcobal, A., and Munoz, R. (2006). Development of a multilocus sequence typing method for analysis of *Lactobacillus plantarum* strains. Microbiology *152*, 85–93.

de Leeuw, E., Li, X., and Lu, W. (2006). Binding characteristics of the *Lactobacillus brevis* ATCC 8287 surface layer to extracellular matrix proteins. FEMS Microbiol. Lett. *260*, 210–215.

Lonvaud-Funel, A. (1999). Lactic acid bacteria in the quality improvement and depreciation of wine. Antonie Van Leeuwenhoek *76*, 317–331.

Lorca, G.L., Barabote, R.D., Zlotopolski, V., Tran, C., Winnen, B., Hvorup, R.N., Stonestrom, A.J., Nguyen, E., Huang, L.W., Kim, D.S., *et al.* (2007). Transport capabilities of eleven Gram-positive bacteria: comparative genomic analyses. Biochim. Biophys. Acta *1768*, 1342–1366.

Makarova, K.S., and Koonin, E.V. (2007). Evolutionary genomics of lactic acid bacteria. J. Bacteriol. *189*, 1199–1208.

Makarova, K., Slesarev, A., Wolf, Y., Sorokin, A., Mirkin, B., Koonin, E., Pavlov, A., Pavlova, N., Karamychev, V., Polouchine, N., *et al.* (2006). Comparative genomics of the lactic acid bacteria. Proc. Natl. Acad. Sci. U.S.A. *103*, 15611–15616.

Mercenier, A., Pavan, S., and Pot, B. (2003). Probiotics as biotherapeutic agents: present knowledge and future prospects. Curr. Pharm. Des. *9*, 175–191.

Mesnage, S., Fontaine, T., Mignot, T., Delepierre, M., Mock, M., and Fouet, A. (2000). Bacterial SLH domain proteins are non-covalently anchored to the cell surface via a conserved mechanism involving wall polysaccharide pyruvylation. Embo J. *19*, 4473–4484.

Metchnikoff, I.I. (1907). The Prolongation of Life (Heinemann, London).

Miyoshi, Y., Okada, S., Uchimura, T., and Satoh, E. (2006). A mucus adhesion promoting protein, MapA, mediates the adhesion of *Lactobacillus reuteri* to Caco-2 human intestinal epithelial cells. Biosci. Biotechnol. Biochem. *70*, 1622–1628.

Morita, H., Toh, H., Fukuda, S., Horikawa, H., Oshima, K., Suzuki, T., Murakami, M., Hisamatsu, S., Kato, Y., Takizawa, T., *et al.* (2008). Comparative genome analysis of *Lactobacillus reuteri* and *Lactobacillus fermentum* reveal a genomic island for reuterin and cobalamin production. DNA Res. *15*, 151–161.

Mundt, J.O., and Hammer, J.L. (1968). Lactobacilli on plants. Appl. Microbiol. *16*, 1326–1330.

Munoz-Provencio, D., Llopis, M., Antolin, M., de Torres, I., Guarner, F., Perez-Martinez, G., and Monedero, V. (2008). Adhesion properties of *Lactobacillus casei* strains to resected intestinal fragments and components of the extracellular matrix. Arch. Microbiol. *191*, 153–161.

Najjari, A., Ouzari, H., Boudabous, A., and Zagorec, M. (2008). Method for reliable isolation of *Lactobacillus sakei* strains originating from Tunisian seafood and meat products. Int. J. Food. Microbiol. *121*, 342–351.

NCBI (2008). Complete Microbial Genomes. http://www.nobi.nlm.nih.gov/genomes/lproks.cgi

Nempont, C., Cayet, D., Rumbo, M., Bompard, C., Villeret, V., and Sirard, J.C. (2008). Deletion of flagellin's hypervariable region abrogates antibody-mediated neutralization and systemic activation of TLR5-dependent immunity. J. Immunol. *181*, 2036–2043.

Nicolas, P., Bessieres, P., Ehrlich, S.D., Maguin, E., and van de Guchte, M. (2007). Extensive horizontal transfer of core genome genes between two *Lactobacillus* species found in the gastrointestinal tract. BMC Evol. Biol. 7, 141.

Nielsen, D.S., Teniola, O.D., Ban-Koffi, L., Owusu, M., Andersson, T.S., and Holzapfel, W.H. (2007). The microbiology of Ghanaian cocoa fermentations analysed using culture-dependent and culture-independent methods. Int. J. Food. Microbiol. *114*, 168–186.

Ocana, V.S., Bru, E., De Ruiz Holgado, A.A., and Nader-Macias, M.E. (1999). Surface characteristics of lactobacilli isolated from human vagina. J. Gen. Appl. Microbiol. *45*, 203–212.

Olivares, M., Diaz-Ropero, M.P., Martin, R., Rodriguez, J.M., and Xaus, J. (2006). Antimicrobial potential of four *Lactobacillus* strains isolated from breast milk. J. Appl. Microbiol. *101*, 72–79.

Parracho, H., McCartney, A.L., and Gibson, G.R. (2007). Probiotics and prebiotics in infant nutrition. Proc. Nutr. Soc. *66*, 405–411.

Perdigon, G., Maldonado Galdeano, C., Valdez, J.C., and Medici, M. (2002). Interaction of lactic acid bacteria with the gut immune system. Eur. J. Clin. Nutr. *56* (Suppl .4), S21–26.

Perea Velez, M., Verhoeven, T.L., Draing, C., Von Aulock, S., Pfitzenmaier, M., Geyer, A., Lambrichts, I., Grangette, C., Pot, B., Vanderleyden, J., *et al.* (2007). Functional analysis of D-alanylation of lipoteichoic acid in the probiotic strain *Lactobacillus rhamnosus* GG. Appl. Environ. Microbiol. *73*, 3595–3604.

van Pijkeren, J.P., Canchaya, C., Ryan, K.A., Li, Y., Claesson, M.J., Sheil, B., Steidler, L., O'Mahony, L., Fitzgerald, G.F., van Sinderen, D., *et al.* (2006). Comparative and functional analysis of sortase-dependent proteins in the predicted secretome of *Lactobacillus salivarius* UCC118. Appl. Environ. Microbiol. 72, 4143–4153.

Plengvidhya, V., Breidt, F., Jr., Lu, Z., and Fleming, H.P. (2007). DNA fingerprinting of lactic acid bacteria in sauerkraut fermentations. Appl. Environ. Microbiol. *73*, 7697–7702.

Pridmore, R.D., Berger, B., Desiere, F., Vilanova, D., Barretto, C., Pittet, A.C., Zwahlen, M.C., Rouvet, M., Altermann, E., Barrangou, R., *et al.* (2004). The genome sequence of the probiotic intestinal bacterium *Lactobacillus johnsonii* NCC 533. Proc. Natl. Acad. Sci. U.S.A. *101*, 2512–2517.

Pulido, R.P., Omar, N.B., Abriouel, H., Lopez, R.L., Canamero, M.M., Guyot, J.P., and Galvez, A. (2007). Characterization of lactobacilli isolated from caper berry fermentations. J. Appl. Microbiol. *102*, 583–590.

Rachmilewitz, D., Karmeli, F., Takabayashi, K., Hayashi, T., Leider-Trejo, L., Lee, J., Leoni, L.M., and Raz, E. (2002). Immunostimulatory DNA ameliorates experimental and spontaneous murine colitis. Gastroenterology *122*, 1428–1441.

Randazzo, C.L., Restuccia, C., Romano, A.D., and Caggia, C. (2004). *Lactobacillus casei*, dominant species in naturally fermented Sicilian green olives. Int. J. Food. Microbiol. *90*, 9–14.

Rantsiou, K., and Cocolin, L. (2006). New developments in the study of the microbiota of naturally fermented sausages as determined by molecular methods: a review. Int. J. Food. Microbiol. *108*, 255–267.

Rodriguez-Ortega, M.J., Norais, N., Bensi, G., Liberatori, S., Capo, S., Mora, M., Scarselli, M., Doro, F., Ferrari, G., Garaguso, I., *et al.* (2006). Characterization and identification of vaccine candidate proteins through analysis of the group A *Streptococcus* surface proteome. Nat. Biotechnol. *24*, 191–197.

Roos, S., Aleljung, P., Robert, N., Lee, B., Wadstrom, T., Lindberg, M., and Jonsson, H. (1996). A collagen binding protein from *Lactobacillus reuteri* is part of an ABC transporter system? FEMS Microbiol. Lett. *144*, 33–38.

Ryu, Y.H., Baik, J.E., Yang, J.S., Kang, S.S., Im, J., Yun, C.H., Kim, D.W., Lee, K., Chung, D.K., Ju, H.R., *et al.* (2009). Differential immunostimulatory effects of Gram-positive bacteria due to their lipoteichoic acids. Int. Immunopharmacol. *9*, 127–133.

Sakamoto, K., Margolles, A., van Veen, H.W., and Konings, W.N. (2001). Hop resistance in the beer spoilage bacterium *Lactobacillus brevis* is mediated by the ATP-binding cassette multidrug transporter HorA. J. Bacteriol. *183*, 5371–5375.

Sánchez, B., Bressollier, P., and Urdaci, M.C. (2008). Exported proteins in probiotic bacteria: adhesion to intestinal surfaces, host immunomodulation and molecular cross-talking with the host. FEMS Immunol. Med. Microbiol. *54*, 1–17.

Sankaran, K., and Wu, H.C. (1994). Lipid modification of bacterial prolipoprotein. Transfer of diacylglyceryl moiety from phosphatidylglycerol. J. Biol. Chem. *269*, 19701–19706.

Sauvage, E., Kerff, F., Terrak, M., Ayala, J.A., and Charlier, P. (2008). The penicillin-binding proteins: structure and role in peptidoglycan biosynthesis. FEMS Microbiol. Rev. *32*, 234–258.

Saxelin, M., Tynkkynen, S., Mattila-Sandholm, T., and de Vos, W.M. (2005). Probiotic and other functional microbes: from markets to mechanisms. Curr. Opin. Biotechnol. *16*, 204–211.

Schroder, N.W., Eckert, J., Stubs, G., and Schumann, R.R. (2008). Immune responses induced by spirochetal outer membrane lipoproteins and glycolipids. Immunobiology *213*, 329–340.

Severin, A., Nickbarg, E., Wooters, J., Quazi, S.A., Matsuka, Y.V., Murphy, E., Moutsatsos, I.K., Zagursky, R.J., and Olmsted, S.B. (2007). Proteomic analysis and identification of *Streptococcus* pyogenes surface-associated proteins. J. Bacteriol. *189*, 1514–1522.

Siezen, R., Boekhorst, J., Muscariello, L., Molenaar, D., Renckens, B., and Kleerebezem, M. (2006). *Lactobacillus plantarum* gene clusters encoding putative cell-surface protein complexes for carbohydrate utilization are conserved in specific Gram-positive bacteria. BMC Genomics 7, 126.

Skerker, J.M., Perchuk, B.S., Siryaporn, A., Lubin, E.A., Ashenberg, O., Goulian, M., and Laub, M.T. (2008). Rewiring the specificity of two-component signal transduction systems. Cell *133*, 1043–1054.

Smeds, A., Varmanen, P., and Palva, A. (1998). Molecular characterization of a stress-inducible gene from *Lactobacillus helveticus*. J. Bacteriol. *180*, 6148–6153.

Stiles, M.E., and Holzapfel, W.H. (1997). Lactic acid bacteria of foods and their current taxonomy. Int. J. Food. Microbiol. *36*, 1–29.

Takeuchi, O., Kawai, T., Muhlradt, P.F., Morr, M., Radolf, J.D., Zychlinsky, A., Takeda, K., and Akira, S. (2001). Discrimination of bacterial lipoproteins by Toll-like receptor 6. Int. Immunol. *13*, 933–940.

Tamminen, M., Joutsjoki, T., Sjoblom, M., Joutsen, M., Palva, A., Ryhanen, E.L., and Joutsjoki, V. (2004). Screening of lactic acid bacteria from fermented vegetables by carbohydrate profiling and PCR-ELISA. Lett. Appl. Microbiol. *39*, 439–444.

Tappenden, K.A., and Deutsch, A.S. (2007). The physiological relevance of the intestinal microbiota--contributions to human health. J. Am. Coll. Nutr. *26*, 679S–683S.

Teusink, B., and Smid, E.J. (2006). Modelling strategies for the industrial exploitation of lactic acid bacteria. Nat. Rev. Microbiol. *4*, 46–56.

Thakran, S., Li, H., Lavine, C.L., Miller, M.A., Bina, J.E., Bina, X.R., and Re, F. (2008). Identification of *Francisella tularensis* lipoproteins that stimulate the toll-like receptor (TLR) 2/TLR1 heterodimer. J. Biol. Chem. *283*, 3751–3760.

Tjalsma, H., and van Dijl, J.M. (2005). Proteomics-based consensus prediction of protein retention in a bacterial membrane. Proteomics *5*, 4472–4482.

Toledo-Arana, A., Valle, J., Solano, C., Arrizubieta, M.J., Cucarella, C., Lamata, M., Amorena, B., Leiva, J., Penades, J.R., and Lasa, I. (2001). The enterococcal surface protein, Esp, is involved in *Enterococcus faecalis* biofilm formation. Appl. Environ. Microbiol. *67*, 4538–4545.

Urso, R., Comi, G., and Cocolin, L. (2006). Ecology of lactic acid bacteria in Italian fermented sausages: isolation, identification and molecular characterization. Syst. Appl. Microbiol. *29*, 671–680.

Van Hoorde, K., Verstraete, T., Vandamme, P., and Huys, G. (2008). Diversity of lactic acid bacteria in two Flemish artisan raw milk Gouda-type cheeses. Food Microbiol. *25*, 929–935.

Velez, M.P., De Keersmaecker, S.C., and Vanderleyden, J. (2007). Adherence factors of *Lactobacillus* in the human gastrointestinal tract. FEMS Microbiol. Lett. *276*, 140–148.

Ventura, M., Canchaya, C., Bernini, V., Altermann, E., Barrangou, R., McGrath, S., Claesson, M.J., Li, Y., Leahy, S., Walker, C.D., *et al.* (2006). Comparative genomics and transcriptional analysis of prophages identified in the genomes of *Lactobacillus gasseri*, *Lactobacillus salivarius*, and *Lactobacillus casei*. Appl. Environ. Microbiol. *72*, 3130–3146.

Vidgren, G., Palva, I., Pakkanen, R., Lounatmaa, K., and Palva, A. (1992). S-layer protein gene of *Lactobacillus brevis*: cloning by polymerase chain reaction and determination of the nucleotide sequence. J. Bacteriol. *174*, 7419–7427.

Waligora, A.J., Hennequin, C., Mullany, P., Bourlioux, P., Collignon, A., and Karjalainen, T. (2001). Characterization of a cell surface protein of *Clostridium difficile* with adhesive properties. Infect. Immun. *69*, 2144–2153.

Walter, J. (2008). Ecological role of lactobacilli in the gastrointestinal tract: implications for fundamental and biomedical research. Appl. Environ. Microbiol. *74*, 4985–4996.

Wang, B., Wei, H., Yuan, J., Li, Q., Li, Y., Li, N., and Li, J. (2008). Identification of a surface protein from *Lactobacillus reuteri* JCM1081 that adheres to porcine gastric mucin and human enterocyte-like HT-29 cells. Curr. Microbiol. *57*, 33–38.

Welman, A.D., and Maddox, I.S. (2003). Exopolysaccharides from lactic acid bacteria: perspectives and challenges. Trends Biotechnol. *21*, 269–274.

West, A.P., Koblansky, A.A., and Ghosh, S. (2006). Recognition and signaling by toll-like receptors. Annu. Rev. Cell. Dev. Biol. *22*, 409–437.

Woese, C.R. (1987). Bacterial evolution. Microbiol. Rev. *51*, 221–271.

Wu, S.W., and De Lencastre, H. (1999). Mrp – a new auxiliary gene essential for optimal expression of methicillin resistance in *Staphylococcus aureus*. Microb. Drug. Resist. *5*, 9–18.

Xiaodong, W., Xuan, G., and Rakshit, S.K. (1997). Direct fermentative production of lactic acid on cassava and other starch substrates. Biotechnol. Lett. *19*, 841–843.

Yang, X., Izadi, H., Coleman, A.S., Wang, P., Ma, Y., Fikrig, E., Anguita, J., and Pal, U. (2008). *Borrelia burgdorferi* lipoprotein BmpA activates pro-inflammatory responses in human synovial cells through a protein moiety. Microbes Infect. *10*, 1300–1308.

Yarovinsky, F., Zhang, D., Andersen, J.F., Bannenberg, G.L., Serhan, C.N., Hayden, M.S., Hieny, S., Sutterwala, F.S., Flavell, R.A., Ghosh, S., *et al.* (2005). TLR11 activation of dendritic cells by a protozoan profilin-like protein. Science *308*, 1626–1629.

Yebra, M.J., Zuniga, M., Beaufils, S., Perez-Martinez, G., Deutscher, J., and Monedero, V. (2007). Identification of a gene cluster enabling *Lactobacillus casei* BL23 to utilize myo-inositol. Appl. Environ. Microbiol. *73*, 3850–3858.

Zhou, M., Boekhorst, J., Francke, C., and Siezen, R.J. (2008). LocateP: genome-scale subcellular-location predictor for bacterial proteins. BMC Bioinformatics *9*, 173.

Current Status of *Bifidobacterium* Gene Manipulation Technologies

2

Satoru Fukiya,* Tohru Suzuki, Yasunobu Kano and Atsushi Yokota

Abstract

Bifidobacteria are one of the beneficial intestinal bacteria that exert health-promoting effects in humans. As evidence for their beneficial effects has accumulated, the scientific and commercial importance of bifidobacteria is increasing. In contrast, the mechanisms of the beneficial effects of these bacteria have not been well-clarified primarily because the development of gene manipulation methodology in bifidobacteria is far behind that in other probiotic bacteria, such as lactobacilli. However, the recent accumulation of genomic sequence information has opened the way for the gene function analysis in this bacterium. This chapter will describe the current knowledge of bifidobacterial genetic materials, such as cryptic plasmids and transposable elements. Additionally, the current status and future perspectives on the development of gene manipulation tools in bifidobacteria will be discussed.

Introduction

Bifidobacteria are a major component of the intestinal microbiota in humans and other animals. Together with the lactobacilli, bifidobacteria are widely used as probiotics, defined as 'live microorganisms, which, when administered in adequate amounts, confer a health benefit on the host' (Food and Agriculture Organization/World Health Organization (FAO/WHO), 2001). It is well known that bifidobacteria exert health-promoting effects, such as prevention of diarrhoea and microbial infection, alleviation of lactose intolerance and inflammatory bowel disease, and modulation of immune function (Leahy *et al.*, 2005).

The mechanisms responsible for these effects must be investigated to understand fully the impact of probiotic administration on the gastrointestinal microbiota and to prevent the occurrence of undesirable sequelae. This is also necessary for understanding the role of bifidobacterial genes in the interaction between the host and potentially beneficial allochthonous microbes.

The availability of accurate bifidobacterial genome sequence information is a prerequisite for the investigation of the mechanisms of bacterial–host interactions. The genome sequences of several *Bifidobacterium* species, such as *Bifidobacterium longum* subsp. *longum* (Schell *et al.*, 2002; Lee *et al.*, 2008), *B. longum* subsp. *infantis* (Sela *et al.*, 2008), *Bifidobacterium breve* (van Sinderen *et al.*, unpublished), *Bifidobacterium adolescentis* (Suzuki *et al.*, 2006, Genbank accession no. AP009256), *Bifidobacterium animalis* subsp. *lactis* (Barrangou *et al.*, 2009; Kim *et al.*, 2009; Garrigues *et al.*, 2010), and *Bifidobacterium dentium* (Ventura *et al.*, 2009), are now available.

However, because of a lack of gene manipulation techniques, such as targeted gene disruption and insertional mutagenesis, progress in bifidobacterial gene function analysis lags far behind that of lactobacilli (Kullen and Klaenhammer, 2000). In this chapter, the current development status of gene manipulation techniques including plasmid transformation, vector development, gene disruption, and insertional mutagenesis, are reviewed. Additionally, issues relating to, and future perspectives for, various uncompleted technical developments are discussed.

*corresponding author

Cryptic plasmids and their application in bifidobacteria

Cryptic plasmids in bifidobacteria

Identification of bifidobacterial cryptic plasmids was initially conducted by screening plasmid-bearing strains and plasmid profile analyses. These studies indicated that bifidobacterial strains harbouring cryptic plasmids were both rare and limited to specific species: *B. longum*, *Bifidobacterium asteroides*, *Bifidobacterium globosum*, *Bifidobacterium indicum*, and *B. breve* (Sgorbati *et al.*, 1982; Sgorbati *et al.*, 1986a,b; Iwata and Morishita, 1989). Characterization of cryptic plasmids began after molecular cloning and nucleotide sequencing technology became widely available in the 1990s. The first molecular characterization of bifidobacterial plasmids was conducted by Matteuzzi and colleagues. Plasmid pMB1 was identified in *B. longum* B2577 and successfully cloned into an *Escherichia coli* vector (Matteuzzi *et al.*, 1990). Sequencing of pMB1 revealed a length of 1847 bp and the presence of two putative open reading frames (orf). The first, *orf1*, encoded a putative plasmid replication protein (Rep) that resembled *E. coli* ColE-type Rep proteins. Subcloning of pMB1 fragments and transformation into *B. longum* indicated that both *orf1* and *orf2* were necessary for plasmid replication (Rossi *et al.*, 1996).

About 30 bifidobacterial cryptic plasmids have been characterized (Table 2.1). Compared with those of *Lactobacillus*, bifidobacterial plasmids are relatively small (up to 10 kb). Despite extensive molecular investigation of bifidobacterial plasmids, their biological significance remains unknown. This is primarily because the functions of most of the orfs remain unclear, except for those involved in plasmid replication and propagation. Minimal replicon analysis of pBC1 from *Bifidobacterium catenulatum* L48 indicated that regions (other than *rep*) affect transformation efficiency, copy number, and stability in *Bifidobacterium* transformants (Álvarez-Martín *et al.*, 2007b). The transformation efficiencies of pBC1-derived plasmids tended to increase with the length of the cloned fragment (Álvarez-Martín *et al.*, 2007b).

Bifidobacterial plasmids possess two replication mechanisms: rolling circle (RC) and theta-type replication (Khan, 1997; del Solar *et al.*, 1998). Required features for RC replication are Rep protein, double-strand origin (DSO), and single-strand origin (SSO). RC replication starts from the binding of the Rep protein to the DSO that leads to the synthesis of single-strand (ss) DNA intermediates of leading strand. Subsequent lagging strand synthesis from the SSO leads to the double-strand (ds) DNA plasmid molecule. Several bifidobacterial plasmids have been revealed to possess RC replication mechanisms by S1 nuclease treatment, because RC replication requires ssDNA intermediates (O'Riordan and Fitzgerald, 1999; Park *et al.*, 1999, 2000; Corneau *et al.*, 2004; Guglielmetti *et al.*, 2007). Based on the similarity of the Rep protein amino acid sequence and the conservation of structures for replication initiation, around half of bifidobacterial plasmids undergo RC replication (Table 2.1).

Theta-type replication has been demonstrated in only two bifidobacterial plasmids. In addition to the Rep protein, theta-type replication requires the specific genetic features, such as thermodynamically unstable AT-rich region with DnaA box and short direct repeats for Rep protein binding site, called iteron. pDOJH10S (*B. longum* DJO10A) is known to undergo theta replication owing to the similarity of its Rep protein sequence to other bacterial theta-type replication plasmids and a lack of ssDNA intermediates (Lee and O'Sullivan, 2006). Plasmid pFI2576 (*B. longum*) undergoes theta-type replication; this was discovered by direct observation of plasmid molecules by atomic force microscopy (Moon *et al.*, 2009). Two other plasmids (pBC1 and pCIBAO89; Table 2.1) have been suggested to use a theta-replication mechanism, based on the similarity of their Rep protein sequence and structure of the region upstream of *rep* (Álvarez-Martín *et al.*, 2007a, 2008; Cronin *et al.*, 2007).

A key element for determining mechanism of plasmid replication is the Rep protein. Rep regulates replication initiation through formation of nucleoprotein complexes (del Solar *et al.*, 1998). A phylogenetic tree of bifidobacterial plasmids, based on multiple alignments of Rep amino acid sequences, is shown in Fig. 2.1. These plasmids are classified into five groups. This finding is similar to that of a previous study (Corneau *et al.*, 2004);

Table 2.1 Cryptic plasmids in bifidobacteria

Organism	Plasmid name	Accession no.	Length (bp)	Replication type[a]	Group[b]	Reference
B. asteroides DSM 20089	pCIBAO89	EU030683	2111	theta?	2	Cronin *et al.* (2007)
B asteroides DSM 20089	pAP1	Y11549	2140	unknown	2	Unpublished
B. bifidum B80	pB80	DQ305402	4898	RC?	1c	Shkoporov *et al.* (2008b)
B. breve B21a	pB21a	DQ497626	5206	RC?	4	Shkoporov *et al.* (2008b)
B. breve NCFB 2258	pCIBb1	AF 085719	5750	RC	4	O'Riordan and Fitzgerald (1999)
B. breve ATCC 15698	pNBb1	E17316	5600	unknown	4	Bourget *et al.* (1993)
B. catenulatum L48	pBC1	DQ011664	2540	unknown	3	Álvarez-Martín *et al.* (2007a)
B. longum DPC6043	p6043A	DQ458910	4896	unknown	1c	Unpublished
B. longum DPC6043	p604B	DQ458911	3680	unknown	1a	Unpublished
B. longum FI10564	pFI2576	DQ452864	2197	theta	3	Moon *et al.* (2009)
B. longum MG1	pMG1	AY210701	3682	RC	1a	Park *et al.* (2003)
B. longum BK51	pTB6	AB187597	3624	RC?	1a	Tanaka *et al.* (2005)
B. longum A24	pBIFA24	DQ286581	4892	RC	1c	Park Y.S. *et al.* (2008)
B. longum B2577	pMB1	X84655	1847	theta?	3	Rossi *et al.* (1996)
B. longum biovar *longum* NAL8	pNAL8L	AM183145	3489	RC	1b	Guglielmetti *et al.* (2007)
B. longum biovar *longum* NAL8	pNAL8M	AM183144	4910	unknown	1c	Guglielmetti *et al.* (2007)
B. longum biovar *longum* NAL8	pNAL8H	AM183142-3	10,000	unknown	unknown[c]	Guglielmetti *et al.* (2007)
B. longum DJO10A	pDOJH10L	AF538868	10,073	unknown	1c, 1a and <u>2</u>[d]	Lee and O'Sullivan (2006)
B. longum DJO10A	pDOJH10S	AF538869	3661	theta	3	Lee and O'Sullivan (2006)
B. longum KJ	pKJ36	AF139129	3625	RC	1a	Park *et al.* (2000)
B. longum KJ	pKJ50	U76614	4960	RC	1c	Park *et al.* (1999)
B. longum NCC2705	pBLO1	AF540971	3626	unknown	1a	Schell *et al.* (2002)
B. longum RW041	pNAC2	AY112723	3684	RC?	1a	Corneau *et al.* (2004)
B. longum RW041	pNAC3	AY112722	10,224	unknown	2	Corneau *et al.* (2004)
B. longum RW048	pNAC1	AY112724	3538	RC	1b	Corneau *et al.* (2004)
B. longum VMKB44	pB44	AY066026	3624	unknown	1a	Shkoporov *et al.* (2008b)
B. pseudocatenulatum VMKB4M	p4M	AF359574	4488	unknown	5	Gibbs *et al.* (2006)
B. pseudolongum subsp. *globosum*	pASV479	DQ103758	4815	RC?	4	Sangrador-Vegas *et al.* (2007)

[a]'?' indicates that the replication mechanism was predicted, but not demonstrated in each reference.

[b]Group was classified based on alignment of the Rep protein aa sequences indicated in Fig. 2.1.

[c]Not included because the entire sequence of this plasmid was not determined.

[d]This plasmid contains three Rep genes: *repA*, *repB*, and *repC*. These were classified into groups 1c, 1a, and 2, respectively. Because *repC* was revealed to be responsible for plasmid replication (Lee and O'Sullivan, 2006), the corresponding group number (2) is underlined.

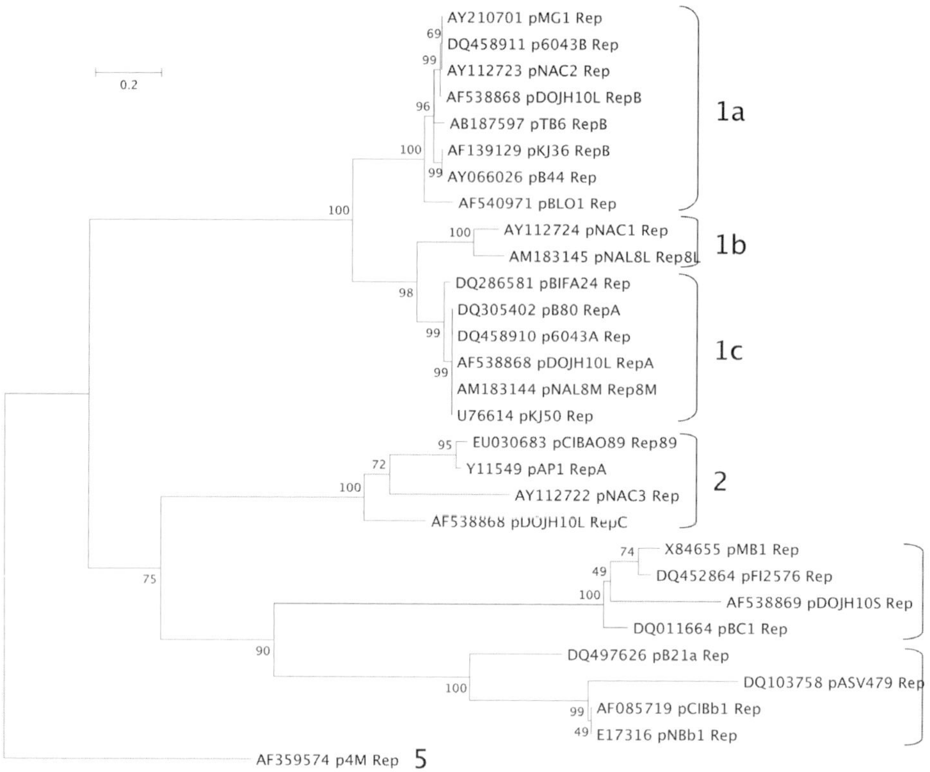

Figure 2.1 Phylogenetic relationships of bifidobacterial plasmid Rep proteins. Rep protein sequences were obtained from the Genbank/EMBL/DDBJ database. Database accession number is indicated in front of the name of the plasmid. Detailed information of plasmids is shown in Table 2.1. A sequence alignment was generated by the ClustalW program (DDBJ, available at http://clustalw.ddbj.nig.ac.jp/top-e.html), and the neighbour-joining method in the MEGA 4.1 software package (Tamura *et al.*, 2007) was used to construct the phylogenetic tree. The Rep protein of plasmid p4M was used as an outgroup. The scale bar represents 0.2 amino acid substitutions per site. A bootstrap analysis with 1000 replicates was conducted to determine the statistical stability of each node using the MEGA 4.1 software. Bootstrap values are indicated at the nodes. Names of groups/subgroups are indicated to the right of each cluster.

however, the largest group (Group 1) contained three subgroups (1a, 1b, and 1c).

Group 1 consists primarily of plasmids from *B. longum*. Most RC-replication Rep proteins are represented. Subgroups consist of a specific plasmid type: subgroup 1a, *B. longum* 3.6 kb plasmids; subgroup 1b, *B. longum* 3.5 kb plasmids; subgroup 1c, *B. longum* 5.0 kb plasmids (except for pB80 from *B. bifidum*). Group 2 consists of the plasmids whose replication mechanisms are unknown, whereas known theta-replicating plasmids are included in group 3. Group 4 includes *B. breve* and *B. pseudolongum* plasmids that are predicted to undergo RC replication. Group 5 consists of only p4M, the *rep* gene of which was probably generated by interspecies gene recombination (Gibbs *et al.,* 2006). Although the plasmid pMB1 is widely used for shuttle vector construction (see next section), its replication mechanism is not fully understood. Based on its phylogenetic classification (Group 3), pMB1 probably undergoes theta-replication (Fig. 2.1). Thus, the sequence similarity of Rep proteins renders the replication type of bifidobacterial plasmids predictable, to some extent.

Transformation with *E. coli*–*Bifidobacterium* shuttle vectors

Construction of *E. coli*–*Bifidobacterium* shuttle vectors using cryptic plasmids has progressed towards the development of a bifidobacterial host-vector system. The first attempt to establish a *Bifidobacterium* transformation system used the pMB1-based *E. coli–B. longum* shuttle vector pRM2 (Missich *et al.*, 1994). Transformation was conducted by electroporation, based on an established *Lactobacillus* protocol. Transformation of *B. longum* B2577 with pRM2 was successful, and its transformation efficiency was $3.4–3.8 \times 10^2$ CFU/μg DNA (Missich *et al.*, 1994). Several attempts at optimizing electroporation parameters were then conducted. Addition of high-osmotic solutes, such as sucrose and fructo-oligosaccharides, to the culture medium and wash buffer improved transformation efficiency, presumably because of the induction of limited autolysis that decreased the cell wall barrier effect (Argnani *et al.*, 1996; Matsumura *et al.*, 1997; Rossi *et al.*, 1997). Incubation of competent cells in transformation buffer at 4°C for several hours was also effective (Argnani *et al.*, 1996). Further studies suggested that a relatively high voltage (12–12.5 kV/Cm electrode) was optimal for high-efficiency transformation (Argnani *et al.*, 1996; Matsumura *et al.*, 1997).

After optimization, various *Bifidobacterium* species were used as host strains. Extant shuttle vectors and their transformation efficiencies are shown in Table 2.2. In addition to the major probiotic *Bifidobacterium* species, such as *B. longum*,

Table 2.2 Currently developed *E. coli–Bifidobacterium* shuttle vectors

Host species and strain	Vector	Length (kpb)	Transformation efficiency (CFU/μg DNA)	Resource plasmid	Reference
B. adolescentis	**Two vectors**				
ATCC 15703	pKKT427	3.9	9.0×10^4 to 4.0×10^5	pTB6	Yasuiet *et al.* (2009)
LMG 10502	pAM4	7.6	9.2×10^2	pBC1	Álvarez-Martín *et al.* (2008)
B. animalis	**15 vectors**				
–	pEBM3	9.6	2.0×10^3	pBL1	Argnani *et al.* (1996)
–	pECM2	10.3	2.0×10^3	pCV35	Argnani *et al.* (1996)
ATCC 27536	pDG7	7.3	5.0×10^3	pMB1	Argnani *et al.* (1996)
ATCC 27536	pNC2	7.1	1.0×10^4	pNAC1	Corneau *et al.* (2004)
MB209	pBKJ36F	6.8	1.2×10^1	pKJ36	Park *et al.* (2000)
MB209	pBKJ50	8.1	2.0×10^2	pKJ50	Park *et al.* (1999)
MB209	pDG7	7.3	3.7×10^3	pMB1	Rossi *et al.* (1996)
MB209	pNC7	4.9	Successful	pMB1	Rossi *et al.* (1996)
MB209	pKG7	7.1	Successful	pMB1	Rossi *et al.* (1996)
MB209	pNC7	4.9	3.0×10^4	pMB1	Rossi *et al.* (1997)
MB209	pCLJ15	5.9	Successful	pMB1	Rossi *et al.* (1998)
MB209	pLF5	5.7	Successful	pMB1	Rossi *et al.* (1998)
MB209	pSPEC1	5.9	Successful	pMB1	Rossi *et al.* (1998)
MB209	pTRE3	2.8	Successful	pMB1	Rossi *et al.* (1998)
ScMa7	pGOSBif33	6.7	5.0×10^1	pNAL8L	Guglielmetti *et al.* (2007)
subsp. *animalis* LMG 10508	pAM4	7.6	4.0×10^1	pBC1	Álvarez-Martín *et al.* (2008)
subsp. *lactis* Bb12	pAM4	7.6	1.6×10^2	pBC1	Álvarez-Martín *et al.* (2008)
subsp. *lactis*	pPKCm1	6.2	1.0×10^1	pCIBAO89	Cronin *et al.* (2007)
B. bifidum	**Nine vectors**				
U3	pDG7	7.3	3.0×10^2	pMB1	Argnani *et al.* (1996)

Table 2.2 continued

Host species and strain	Vector	Length (kpb)	Transformation efficiency (CFU/μg DNA)	Resource plasmid	Reference
ATCC 15696	pDG7	7.3	7.4×10^3	pMB1	Argnani *et al.* (1996)
ATCC 15696	pSUW1/114	7.8	Up to 1.0×10^1	pB44	Shkoporov *et al.* (2008)
ATCC 15696	pSUW53/114	8.2	Up to 1.0×10^1	pB44	Shkoporov *et al.* (2008)
ATCC 15696	pSUW64/123	4.9	Up to 1.0×10^1	pB44	Shkoporov *et al.* (2008)
MB254	pNC7	4.9	7.2×10^4	pMB1	Rossi *et al.* (1997)
MB254	pSPEC1	5.9	Successful	pMB1	Rossi *et al.* (1998)
MB254	pCLJ15	5.9	Successful	pMB1	Rossi *et al.* (1998)
MB254	pLF5	5.7	Successful	pMB1	Rossi *et al.* (1998)
B. breve	**15 vectors**				
4	pDG7	7.3	1.3×10^4	pMB1	Argnani *et al.* (1996)
AS	pDG7	7.3	2.0×10^2	pMB1	Argnani *et al.* (1996)
LMG 13208	pAM4	7.6	1.0×10^2	pBC1	Álvarez-Martín *et al.* (2008)
MB226	pNC7	4.9	6.6×10^4	pMB1	Rossi *et al.* (1997)
MB252	pNC7	4.9	2.3×10^4	pMB1	Rossi *et al.* (1997)
NCFB 2258	pASV480	9	up to 1.0×10^5	pASV479	Sangrador-Vegas *et al.* (2007)
NCIMB 8807	pASV480	9	up to 1.0×10^5	pASV479	Sangrador-Vegas *et al.* (2007)
UCC2003	pPKCm1	6.2	3.8×10^6	pCIBAO89	Cronin *et al.* (2007)
UCC2003	pSUW64/123	4.9	5.0×10^1 to 3.0×10^2	pB44	Shkoporov *et al.* (2008)
UCC2003	pEESH80	6.5	1.2×10^3 to 2.8×10^3	pB80	Shkoporov *et al.* (2008)
UCC2003	pCESH90	6.4	8.1×10^3 to 3.1×10^4	pB80	Shkoporov *et al.* (2008)
UCC2003	pAM1	6.1	2.3×10^2	pBC1	Álvarez-Martín *et al.* (2008)
UCC2003	pAM4	7.6	1.4×10^2	pBC1	Álvarez-Martín *et al.* (2008)
UCC2003	pAM5	7.7	1.6×10^3	pBC1	Álvarez-Martín *et al.* (2007b)
UCC2003	pBC1.2	8.7	1.2×10^3	pBC1	Álvarez-Martín *et al.* (2007b)
UCC2003	pBC1.3	8.4	8.5×10^2	pBC1	Álvarez-Martín *et al.* (2007b)
UCC2003	pBC1.4	8.3	2.5×10^2	pBC1	Álvarez-Martín *et al.* (2007b)
UCC2003	pBC1.5	8	1.0×10^1	pBC1	Álvarez-Martín *et al.* (2007b)
B. dentium	**Three vectors**				
F101	pAM4	7.6	9.5×10^1	pBC1	Álvarez-Martín *et al.* (2008)
MB269	pNC7	4.9	3.6×10^1	pMB1	Rossi *et al.* (1997)
NCFB 2843	pPKCm1	6.2	1.0×10^4	pCIBAO89	Cronin *et al.* (2007)

Table 2.2 continued

Host species and strain	Vector	Length (kpb)	Transformation efficiency (CFU/µg DNA)	Resource plasmid	Reference
B. longum	**18 vectors**				
105-A	pBLES100	9.1	1.6×10^4	pTB6	Matsumura *et al.* (1997)
105-A	pBRASTA100	5.6	5.9×10^4	pTB6	Tanaka *et al.* (2005)
105-A	pBRASTA101	5	2.5×10^6	pTB6	Tanaka *et al.* (2005)
105-A	pBRASTA102	5.6	7.5×10^4	pTB6	Tanaka *et al.* (2005)
105-A	pBRASTA103	6.2	7.3×10^2	pTB6	Tanaka *et al.* (2005)
105-A	pDSO44Sp	4.3	1.3×10^6	pTB6	Tanaka *et al.* (2005)
105-A	pKKT427	3.9	1.0×10^6 to 3.0×10^6	pTB6	Yasuiet *et al.* (2009)
ATCC 15708	pNC2	7.1	1.0×10^4	pNAC1	Corneau *et al.* (2004)
B2577	pRM2	6	3.4×10^2 to 3.8×10^2	pMB1	Missich *et al.* (1994)
L25	pAM4	7.6	6.6×10^1	pBC1	Álvarez-Martín *et al.* (2008)
MB219	pNC7	4.9	4.1×10^2	pMB1	Rossi *et al.* (1997)
MB219	pCLJ15	5.9	Successful	pMB1	Rossi *et al.* (1998)
MB219	pLF5	5.7	Successful	pMB1	Rossi *et al.* (1998)
MB219	pSPEC1	5.9	Successful	pMB1	Rossi *et al.* (1998)
MB219	pNC7	4.9	9.3×10^2	pMB1	Rossi *et al.* (1997)
MG1	pBES2	7.6	7.3×10^3	pMG1	Park *et al.* (2003)
NCC2705	GOSBif33	6.7	2.0×10^2	pNAL8L	Guglielmetti *et al.* (2007)
NCIMB 8809	pPKCm1	6.2	1.0×10^2	pCIBAO89	Cronin *et al.* (2007)
U2	pDG7	7.3	2.6×10^3	pMB1	Argnani *et al.* (1996)
VMK44	pDOJHR	8.6	2.0×10^1	pDOJHR10S	Lee and O'Sullivan (2006)
Wiesby 2	pDG7	7.3	7.0×10^4	pMB1	Argnani *et al.* (1996)
B. longum* subsp. *infantis	**Six vectors**				
ATCC 27920	pDG7	7.3	4.0×10^4	pMB1	Argnani *et al.* (1996)
ATCC 27920	pBKJ36F	6.8	2.6×10^2	pKJ36	Park *et al.* (2000)
MB208	pNC7	4.9	1.2×10^5	pMB1	Rossi *et al.* (1997)
MB231	pNC7	4.9	2.8×10^2	pMB1	Rossi *et al.* (1997)
MB263	pNC7	4.9	9.3×10^3	pMB1	Rossi *et al.* (1997)
MB263	pSPEC1	5.9	Successful	pMB1	Rossi *et al.* (1998)
MB263	pCLJ15	5.9	Successful	pMB1	Rossi *et al.* (1998)
MB263	pLF5	5.7	Successful	pMB1	Rossi *et al.* (1998)
U1	pDG7	7.3	2.5×10^2	pMB1	Argnani *et al.* (1996)
B. magnum	**Four vectors**				
MB267	pNC7	4.9	1.8×10^3	pMB1	Rossi *et al.* (1997)
MB267	pSPEC1	5.9	Successful	pMB1	Rossi *et al.* (1998)
MB267	pCLJ15	5.9	Successful	pMB1	Rossi *et al.* (1998)
MB267	pLF5	5.7	Successful	pMB1	Rossi *et al.* (1998)

Table 2.2 continued

Host species and strain	Vector	Length (kpb)	Transformation efficiency (CFU/μg DNA)	Resource plasmid	Reference
B. pseudocatenulatum	**Four vectors**				
LMG10505	pPKCmI	6.2	1.0×10^3	pCIBAO89	Cronin *et al.* (2007)
M115	pAM1	6.1	2.5×10^5	pBC1	Álvarez-Martín *et al.* (2008)
M115	pAM4	7.6	1.0×10^5	pBC1	Álvarez-Martín *et al.* (2008)
MB264	pNC7	4.9	5.0×10^1	pMB1	Rossi *et al.* (1997)
B. pseudolongum	**Two vectors**				
LMG11571	pAM4	7.6	6.3×10^1	pBC1	Álvarez-Martín *et al.* (2008)
NCIMB 2244	pPKCmI	6.2	1.0×10^2	pCIBAO89	Cronin *et al.* (2007)
subsp. *globosum* JCM 5820	pPKCmI	6.2	1.0×10^3	pCIBAO89	Cronin *et al.* (2007)
B. ruminale	**One vector**				
MB266	pNC7	4.9	7.2×10^2	pMB1	Rossi *et al.* (1997)
B. thermophilum	**One vector**				
LMG 11571	pAM4	7.6	4.6×10^1	pBC1	Álvarez-Martín *et al.* (2008)

B. breve, and *B. animalis* subsp. *lactis*, minor species such as *B. pseudocatenulatum*, *B. dentium*, *Bifidobacterium ruminale*, and *Bifidobacterium magnum* were also used as host strains. As shown in Table 2.2, these strains' transformation efficiencies varied greatly; the median transformation efficiency was ~10^3 transformants/μg plasmid DNA. This is far lower than that of *E. coli* K-12 or *Bacillus subtilis*, for which many molecular analysis tools have been developed. Transformation efficiency was dependent on bifidobacterial species, strains, and plasmids. In our opinion, efficiency is strain-, rather than species-dependent. In an early work, Rossi *et al.* reported that transformation efficiency was dependent on species, because different strains of the same species (*B. longum*, *B. breve*, and *B. infantis*) had similar transformation efficiencies (Rossi *et al.*, 1997). However, Matsumura *et al.* reported that the transformation efficiency of *B. longum* differed markedly from strain to strain; *B. longum* strain 105-A exhibited 1.6×10^4/μg pBLES100 (9.1 kb) DNA, whereas no transformants were obtained with the same vector in *B. longum* JCM1217 type strain (Matsumura *et al.*, 1997). A similar phenomenon has been reported elsewhere (Argnani *et al.*, 1996; Álvarez-Martín *et al.*, 2007b). This ambiguity seemed to be due to differences in both the strains used and in plasmid construction.

Improvement of transformation efficiency: PAM concept

The transformation efficiency of bifidobacteria is generally low. Possible limiting factors include (i) the cell-surface physicochemical barrier (e.g. peptidoglycan and polysaccharides), (ii) damage to cells (e.g. electric pulse and oxygen exposure) and (iii) degradation of exogenous DNA by host nucleases. Of these, the latter seems to be the most important because the restriction system in bifidobacteria effectively degrades the shuttle vector (foreign DNA), resulting in poor transformation efficiency. This effect of the bifidobacterial restriction system was apparent when transformation efficiency improved if the shuttle vector was prepared from previously transformed bifidobacterial strains (Rossi *et al.*, 1998; Shkoporov *et al.*, 2008b; Yasui *et al.*, 2009). This was used to improve transformation efficiency in bifidobacteria, resulting in a method termed 'plasmid artificial modification' (PAM) (Yasui *et al.*, 2009). Yasui and colleagues identified two putative modification genes that encode either DNA-methyltransferases homologous to *Sau*3AI, or *Kpn*2kI methyltransferase from *B. adolescentis* ATCC 15703 genome. These genes were cloned and simultaneously expressed in *E. coli*. The shuttle vector was transformed in this *E. coli* strain and modified using the cloned bifidobacterial modification genes. Because this modified shuttle vector was not restricted in the host *B. adolescentis* ATCC 15703, transformation efficiency of this strain was greatly improved: $1–3 \times 10^0$ transformants/μg plasmid DNA using purified vector from an *E. coli* host strain *versus* up to 4×10^5 transformants/μg plasmid DNA using the PAM-treated vector. The PAM concept can most probably be applied to other bifidobacteria, because it was also effective in *Lactococcus lactis* IO-1 (Yasui *et al.*, 2009).

Heterologous gene expression

Several vectors for heterologous gene expression have been developed. Most of these studies intended to use recombinant bifidobacteria therapeutically; for instance, as a delivery vector for enzyme/pro-drug therapy (Nakamura *et al.*, 2002; Hamaji *et al.*, 2007; Hidaka *et al.*, 2007), production of recombinant human proteins (Escogido *et al.*, 2007; Shkoporov *et al.*, 2008a), or vaccination against infectious disease (Takata *et al.*, 2006). Initial and extensive studies on enzyme/pro-drug cancer therapies using recombinant bifidobacteria that express *E. coli* cytosine deaminase (CD) have been conducted. This enzyme converts the pro-drug 5-fluorocytosine to the anti-cancer agent 5-fluorouracil (Nakamura *et al.*, 2002). In these studies, *E. coli* CD was expressed under the control of a *hup* promoter and terminator. The *hup* gene encodes a histone-like DNA-binding protein, HU, and is constitutively expressed in *B. longum* (Takeuchi *et al.*, 2002). The *hup* promoter is currently the first choice for recombinant protein expression in bifidobacteria (Nakamura *et al.*, 2002; Takata *et al.*, 2006; Escogido *et al.*, 2007; Hamaji *et al.*, 2007; Hidaka *et al.*, 2007; Shkoporov *et al.*, 2008a). Other bifidobacterial promoters, such as α-amylase (Rhim *et al.*, 2006), glyceraldehyde-3-phosphate dehydrogenase (Shkoporov *et al.*, 2008a), and 16S rRNA (Park, M.S. *et al.*, 2008) have also been used. Secretion of recombinant protein by *Bifidobacterium* was successfully achieved by fusion of a secretion signal peptide to the recombinant protein (Moon *et al.*, 2005; Shkoporov *et al.*, 2008a).

In some cases, expression constructs from bifidobacteria were recognized and recombinant proteins have been expressed in *E. coli* (Rhim *et al.*, 2006; Escogido *et al.*, 2007). This suggests that the structural consensus of promoters and translation-initiation sites is consistent between *Bifidobacterium* spp. and *E. coli*. Additionally, the *E. coli*-derived expression systems *araC*-P_{BAD} and the T5 phage promoter have been used for the expression of therapeutic and luciferase proteins, respectively (Guglielmetti *et al.*, 2008; Long *et al.*, 2010). The possibility of gene expression from *E. coli* promoters in *Bifidobacterium* spp. should be investigated further.

Reporter plasmids

Reporter genes are used for a variety of purposes: for example, the measurement of promoter and metabolic activities, detection of microorganisms *in vivo*, and identification of protein export signal sequences. Several types of reporter vectors in bifidobacteria have been developed recently. The first is a protein export-specific reporter plasmid, pFUN, which uses the staphylococcal nuclease gene, but lacks the export signal (MacConaill *et al.*, 2003). A genomic library was constructed by the insertion of bifidobacterial genomic DNA fragments into the upstream region of the nuclease gene in pFUN. Using this library, MacConaill and colleagues screened bifidobacterial protein export signal sequences using a two-stage system. The first screen was conducted using *L. lactis* subsp. *cremoris* MG1363 as a host strain, and positive clones were transferred to the second host strain, *B. breve* UCC2003. Successful export of the nuclease was detected in 10 clones. Seven of these exhibited strong nuclease activity in culture supernatants, indicating that the protein secretion signal sequence was present in the cloned DNA fragment (MacConaill *et al.*, 2003). In another study, the *E. coli* β-glucuronidase gene was used for the development of a system for measuring bifidobacterial promoter activity (Klijn *et al.*, 2006). Luciferase has also been used as a reporter system in bifidobacteria. The first example of this was the insect luciferase that requires ATP and oxygen to catalyse the conversion of exogenous D-luciferin to oxyluciferin, which produces light of wavelength 560 nm. This system was used for measurement of the metabolic activity of bifidobacterial cells under acidic stress by using the emitted light as an indicator of cellular ATP levels (Guglielmetti *et al.*, 2008). Another example is the bacterial luciferase (LuxABCDE), which does not require an exogenous substrate for light emission. Luciferase was used for detection of bioluminescence from *B. breve* cells administrated in the murine intestine (Cronin *et al.*, 2008).

Development of targeted gene inactivation systems

Targeted gene inactivation is a fundamental method for the investigation of microbial gene function. However, this method is currently under development in bifidobacteria. An exception is the report of the functional analysis of the amylopullulanase-encoding gene *apuB* in *B. breve* UCC2003 (O'Connell Motherway *et al.*, 2008). A plasmid insertion mutant of *apuB* was generated by introduction of the insertional gene inactivation system for *L. lactis* (Law *et al.*, 1995). In this system, an internal *apuB* fragment was cloned into the Rep-negative plasmid pORI19, and then the plasmid was introduced into *B. breve* UCC2003 harbouring the thermo-sensitive plasmid pTGB019 that produces Rep protein. Selection by growth at a non-permissive temperature (42°C, 50 generations) allowed loss of pTGB019 from *B. breve* cells because this plasmid has strong segregational instability as well as being thermo-sensitive. pORI19 cannot replicate in *B. breve* cells without a supply of Rep protein from pTGB019. Thus, insertion of pORI19 into *apuB* occurred by homologous recombination, and insertion mutants were isolated by antibiotic selection. The *apuB* mutant could grow well on glucose, but not on amylopectin, glycogen, pullulan, or starch as sole carbon sources. This report demonstrates the possibility of gene inactivation by homologous recombination in *B. breve* UCC2003 (O'Connell Motherway *et al.*, 2008).

Despite this success, development of gene-inactivation systems in bifidobacteria remains incomplete. Various issues must first be overcome: first, homologous recombination activity in bifidobacteria has been little investigated. *B. longum* NCC2705 lacks *recBCD* genes, suggesting the difficulty of homologous recombination in bifidobacteria (Schell *et al.*, 2002). In addition to *B. breve* UCC2003, homologous recombination activity should be assessed in other *Bifidobacterium* strains. Second, high transformation efficiency is generally required for successful gene inactivation by homologous recombination. As mentioned above, efficiency comparable to that of model bacteria, such as *E. coli* and *B. subtilis,* may be required for gene inactivation. Third, plasmid-inserted gene mutants, which are inactivated by a single-crossover reaction, are known to be susceptible to reversion to wild type, owing to excision of the inserted plasmid. Gene inactivation by double-crossover is commonly used in other bacteria.

Considering the problems of transformation efficiency and low homologous recombination activity, the probability of successful development of bifidobacterial gene inactivation by double-crossover is limited. Induction of the secondary recombination event and use of an efficient selection strategy are keys for the isolation of double-crossover mutants. One possibility is to develop a thermo-sensitive plasmid replicon that works in bifidobacteria. If this type of plasmid was available, transformation efficiency would cease to be a problem, because plasmid-inserted strains could be screened from single plasmid-bearing strain using a non-permissive temperature. A double-crossover mutant could also be screened from the plasmid-inserted strains at the permissive temperature. Another possibility would be to use an intrinsic bifidobacterial plasmid. A combination of Rep-positive and -negative plasmids, which possess the same *ori* region, could be applied to induce the double-crossover event. Introduction of a Rep-positive plasmid into a strain, into the genome of which has been inserted a Rep-negative plasmid, could induce excision and loss of the inserted plasmid owing to mutual incompatibility. Additionally, incorporation of a counter-selectable marker gene, such as *sacB*, which encodes levansucrase in *B. subtilis* (Gay *et al.*, 1983), into the chromosome-integrated plasmid could assist in isolation of gene-inactivated mutants. SacB induces a conditional lethal phenotype in various bacterial species in the presence of high sucrose concentrations. When an integrated plasmid containing the *sacB* marker is excised from the chromosome by a secondary crossover event, the bacterium regains the ability to replicate in the presence of sucrose. Thus, the *sacB* marker is frequently used for positive selection of strains in which a secondary crossover event has occurred (Steinmetz *et al.*, 1983; Jäger *et al.*, 1992). The double-crossover system described above is available not only in lactic acid bacteria, but also in high G+C content *Actinobacteria*, such as *Corynebacterium*, *Streptomyces*, and *Mycobacterium* spp. (Clark-Curtiss and Haydel, 2003; Gust *et al.*, 2004; Vertès *et al.*, 2005). Extensive efforts to develop the double-crossover system in bifidobacteria are required to allow a greater understanding of the probiotic effect of this bacterium.

Transposable elements and their application in bifidobacteria

Compared with the progress in plasmid research, investigation of bifidobacterial transposable elements has been slow. Alongside *Bifidobacterium* genome sequencing efforts, the presence of transposable elements, such as insertion sequences (IS) and mobile integrase cassettes (MIC), has been demonstrated. In this section, several aspects of these elements will be reviewed and the characteristics required for development of a random mutagenesis system in bifidobacteria will be discussed.

Mobile integrase cassettes

Mobile integrase cassettes (MIC) were first identified in the genome sequence of *B. longum* NCC2705 (Schell *et al.*, 2002) and published along with the *B. longum* DJO10A genome sequence (Lee *et al.*, 2008). The organization of this MIC is quite unique; three different integrase genes are repeated and are flanked by inverted repeats associated with a palindromic sequence in both termini. This MIC is thought to be a new type of transposable element owing to its unusual organization. Interestingly, with regard to these two genomes, MIC are always associated with IS3-like sequences, which are split by the MIC. It has been suggested that these elements are mobile, owing both to MIC organization and differences in the chromosomal location of the MICs in both *B. longum* genomes (Lee *et al.*, 2008). Although the function and distribution of these elements in bifidobacteria remain unclear, their distribution among *Bifidobacterium* spp. is probably limited because the complete set has been detected thus far in only two *B. longum* genomes (Lee *et al.*, 2008).

Insertion sequences

Insertion sequences (IS) are ubiquitous transposable elements found in both the *Bacteria* and *Archaea*. IS are small (<2.5 kb), cryptic segments of DNA with a compact genetic organization (transposase genes with recognition sequences) that are able to insert at multiple sites in a target molecule (Mahillon and Chandler, 1998). Putative IS elements, predicted from orfs annotated as transposases in completed *Bifidobacterium* genomes, are indicated in Table 2.3. The number and variety of IS elements in each genome varies markedly. *B. longum* subsp. *infantis* ATCC 15697 has the most abundant IS elements (58 elements). IS elements that belong to the IS3, IS*30*, IS*21*, and IS256 families predominate in the *B. longum* group. IS3 family elements are also dominant in *B. adolescentis* ATCC 15703 and *B. dentium* Bd1. In contrast, IS*L3* family elements predominate in *B. animalis* genomes.

Generally, transposons possess a repeat of an IS element at each terminus, and contain between them genes encoding some other function, such as antibiotic resistance. Antibiotic resistance genes in bifidobacteria [e.g. the tetracycline resistance gene *tet*(W) and erythromycin resistance gene *erm*(X)] were shown to have IS transposase orfs in their flanking regions (Kazimierczak *et al.*, 2006; Ammor *et al.*, 2008; van Hoek *et al.*, 2008). Indeed, two IS transposase orfs were identified in the flanking region of *erm*(X) in *Bifidobacterium thermophilum*, the organization of which was similar to that of Tn5432, described in other high G+C Gram-positive bacteria (van Hoek *et al.*, 2008). The *tet*(W) gene is also known to be a part of Tn*B1230* in *Butyrivibrio fibrisolvens* (Melville *et al.*, 2004). An IS*4*-family element was found in the region upstream of *tet*(W) in *B. longum* F8 (Kazimierczak *et al.*, 2006). The *tet*(W) gene has been identified in several *Bifidobacterium* strains, where it is associated with putative transposase genes that belong to the IS*3*, IS*30*, IS*256*, and IS*5* families (Ammor *et al.*, 2008).

IS transposase orfs were found in the prophage-like element Bbr-1 in *B. breve* UCC 2003 (Ventura *et al.*, 2005a). Analysis of Bbr-1 organization suggested that a putative composite mobile element was inserted into the prophage-like sequence. This composite element contains five transposase-like and one integrase-like orf. Two of the five are pseudogenes, and three of the five encode transposases of the IS*21*, IS*200*/IS*605*, and IS*256* families. The composite mobile element is assumed to have integrated anciently with the Bbr-1 prophage-like element and co-evolved with Bbr-1 (Ventura *et al.*, 2005a).

Many types of putative IS element have been identified, primarily by genomic sequence analysis. However, demonstrably active IS are rare in bifidobacteria. IS*1999* was first identified as an active IS element by transposition to the plasmid vector in *B. animalis* ATCC 27536 during plasmid stability analysis in continuous culture (González Vara *et al.*, 2003). Although IS*1999* has not been further characterized, this element is classified as a member of the IS*L3* family in the IS Finder database (Siguier *et al.*, 2006). Because the IS*L3* family elements predominated in the *B. animalis* genome, they may not have suffered from modification or inactivation in this species. The second active IS element is the IS*30* family in *B. longum* DJO10A. Comparison of the genomes of the original strain DJO10A and the laboratory-adopted strain DJO10A-JH1 indicated that only the IS*30* family element changed position in the genome. This suggests that the IS*30* family element was active in this strain (Lee *et al.*, 2008).

The transposed element of the IS*200*/IS*605* family has been identified (Fukiya *et al.*, 2010). IS*Blo15* was found in the cryptic plasmid pBK283 in *B. longum* BK28. This 6.5-kb plasmid has strong sequence similarity to the 5.0 kb *B. longum* plasmid pKJ50 (Park *et al.*, 1999), but is about 1.5 kb larger. Nucleotide sequence analysis indicated that IS*Blo15* was inserted into the pKJ50-like region of pBK283. IS*Blo15* possesses characteristics typical of the IS*200*/IS*605* family, such as terminal palindromic sequences and a transposase gene *orfB* that is conserved in this family (Fukiya *et al.*, 2010). Based on these reports, we expect some IS elements to be active in bifidobacteria.

Development of random mutagenesis system using transposable elements

Mutagenesis is a basic methodology for bacterial genetic analysis. Random mutagenesis systems are valuable for clarification of gene functions that contribute to the objective phenotypes in the target bacteria. To date, random mutagenesis systems have not been established in bifidobacteria, primarily because of the paucity of available gene manipulation tools. However, as discussed above, we now have both a transformation system and evidence for the activity of endogenous IS elements in bifidobacteria. Thus, it seems worthwhile to discuss issues important to the future development of bifidobacterial random mutagenesis systems.

Several strategies for the development of a random mutagenesis system in bifidobacteria are outlined below. These are based on use of (i) broad host-range transposons, (ii) conjugative transposons, (iii) *in vivo* transposomics, (iv) endogenous mobile elements and (v) heterologous mobile elements from related bacterial genera.

Broad host-range transposons

Generally, broad host-range transposons, such as Tn*5*, Tn*10*, and Tn*916*, are the first choice for the development of transposon mutagenesis systems in genetically uncharacterized bacterial species. This is because detailed information relating to randomness of insertion, activity and other characteristics of these transposons is available (Dyson, 1999). As noted in the previous section, the transformation efficiency of bifidobacteria seems strain dependent and generally low, primarily because of the strong effect of the restriction-modification (RM) system in this organism. Thus, use of broad host-range transposons for random mutagenesis

Table 2.3 Number of putative IS elements identified in the completed *Bifidobacterium* genomes[a]

	Genomic copy number						
IS family	*B. longum* NCC2705[b]	*B. longum* DJO10A	*B. longum* subsp. *infantis* ATCC 15697	*B. adolentis* ATCC 15703	*B. animalis* subsp. *lactis* DSM 10140/ BI-04	*B. animalis* subsp. *lactis* AD011	*B. dentium* Bd1
IS*110*			9				
IS*200*/IS*605*			6				
IS*21*	5	10	3	4			
IS*256*	5	3	9	1	3	2	
IS3	2	11	12	8		1	7
IS*30*	3	8	11		1		
IS5			3			1	
IS*607*	1	1	2				
IS*66*		2					
IS*L3*		4	3		5	6	
Total	**16**	**39**	**58**	**13**	**9**	**10**	**7**

[a]Except for *B. longum* NCC2705, IS families were identified by similarity analysis of the deduced aa sequences of annotated transposase orfs in each genome sequence. Similarity analysis was conducted using the BlastP program on accumulated IS sequences in IS finder database (http://www-is.biotoul.fr/is.html). The family to which each putative IS element belonged was determined, based on the family of the most similar IS sequence in the database.

[b]Twelve different IS sequences have already been identified in this strain according to the IS finder database (IS*Blo1*–IS*Blo12*). Because 4 of the 12 sequences were truncated or fragmented in this genome sequence, they are not included in this table.

requires both highly efficient transformable strains and circumvention of the RM system.

Conjugative transposons

This type of transposon is also widely used in other bacterial species. It requires the conjugative transfer of DNA from host bacteria to bifidobacteria, but conjugative transfer from *E. coli* to *Bifidobacterium* has seen only limited success (Kazimierczak *et al.*, 2006; Shkoporov *et al.*, 2008b). Establishment of a reliable conjugation system and investigation of the activity of the conjugative transposons of bifidobacteria are, therefore, required.

In vivo transposomics

This technology is based on *in vitro* electro-transformation with the stable transposome complex that consists of transposon DNA and transposase protein (EZ:TN™ Transposome™, Epicentre Biotechnologies). This system does not require specific host factors and may be the easiest way to acquire random gene mutants in bifidobacteria because this system is generally applicable, so long as the electro-transformation system has been established. Indeed, this system has been applied not only to Gram-negative, but also to the high-G+C Gram-positive bacteria, including *Corynebacterium glutamicum*, *Mycobacterium tuberculosis*, and *Rhodococcus erythropolis* (Tanaka *et al.*, 2002; Maus *et al.*, 2005; Kawaguchi *et al.*, 2006; Suzuki *et al.*, 2006). However, the effect of the bifidobacterial RM system and the appropriateness of intracellular physiological conditions for transposome activity may be problematic.

Endogenous mobile elements

As mentioned above, IS are the main endogenous mobile element in bifidobacteria and three are known to be active (González Vara *et al.*, 2003; Lee *et al.*, 2008; Fukiya *et al.*, 2010). Application of those active IS elements would be a straightforward way of developing a system. Endogenous IS elements are advantageous for expression of transposition activity because the structures necessary for transposase expression (promoter, ribosome binding site, codon usage) are, of course, more appropriate than those of heterologous transposable elements. It is also certain that compatibility between endogenous mobile elements and host proteins that involved in the transposition will be most appropriate.

Heterologous mobile elements from related bacterial genera

Bifidobacterium belongs to the high G+C Gram-positive bacteria. This group includes the genera *Mycobacterium, Corynebacterium, Rhodococcus,* and other *Actinobacteria.* Genetic engineering methods, including random mutagenesis, have been established within each of these (Camacho *et al.*, 1999; Sallam *et al.*, 2006; Suzuki *et al.*, 2006). Thus, application of those heterologous mobile elements may be an alternative for the development of a system. Activity of heterologous elements in bifidobacteria is expected because the structures necessary for transposase gene expression are similar to those of bifidobacteria. However, activity is not guaranteed, because host-derived proteins other than transposase are thought to be involved in transposition. Of the above possibilities, *in vivo* transposomics and endogenous mobile elements seem most appropriate for the construction of a random mutagenesis system in bifidobacteria.

In the practical stage of system development, other factors must be taken into consideration. The most important of these is random insertion of the mobile elements into the genome. Because this is primarily dependent on the nature of the mobile element, random insertion of candidate elements must be checked prior to development to ensure a legitimate system. Introduction of an element containing a selective marker gene to bifidobacterial cells and Southern hybridization analysis of insertion mutants is effective for an analysis of randomness. Sequencing analysis of the inserted element's flanking regions is the most reliable method.

The second important factor is transposase expression. Generally, transposition of transposable elements is strictly controlled, because it is disadvantageous rather than advantageous for the evolution of the relevant organism's genome: transposition might contribute to the acquisition of a new phenotypic trait, but excess transposition may result in the disruption of essential or important genes. Thus, expression of transposase in native IS elements is generally low. Transposase should be expressed under the control of a strong promoter and appropriate ribosome binding sequence. Unfortunately, only a small number of bifidobacterial promoters have been fully characterized. Bifidobacterial promoters of general stress response proteins (e.g. GroEL and DnaK) and constitutive DNA-binding proteins (e.g. Hup, a histone-like protein) will be the candidate promoters of transposase expression (Takeuchi *et al.*, 2002; Ventura *et al.*, 2004, 2005b).

The third important factor is stability of the transposed fragment in the genome. If the transposed sequence possesses a complete set of transposition elements (transposase gene and its recognition sequence for transposition, generally inverted repeats in many IS species), it is possible for secondary transposition from the primary insertion site to occur. Secondary transposition results in loss of mutants with an interesting phenotype and to ambiguous results in the phenotype analysis. This possibility can be reduced by separating the required transposition units; transposase should be expressed outside of the transposable unit that contains its recognition sequences.

The choice of vector also affects the stability of the inserted fragment. Vectors that can replicate in target bifidobacteria have an advantage in the continuous supply of transposase; however, this may induce secondary transposition. Vectors that cannot replicate in target bifidobacterial cells have little opportunity for secondary transposition, because the duration of transposase expression is limited. Strong expression of transposase is required to obtain an appropriate number of insertion mutants. To satisfy these seemingly incompatible requirements, thermo-sensitive (ts) plasmids have been used in other bacteria, including lactic acid bacteria (Maguin *et al.*, 1996; Serror *et al.*, 2003). The ts-plasmids can replicate at the permissive temperature and such replication is terminated when cultivation shifts to a non-permissive temperature. This has been cleverly applied to development of a random mutagenesis system: in the first round, plasmid-bearing cells are cultivated at the permissive temperature. After incubation for a time appropriate for transposition, shifting to a non-permissive temperature greatly reduces the opportunity for secondary transposition, because the replication-terminated plasmids are lost along with cell division. Until now, bifidobacterial ts-plasmids have not been reported and application of lactic acid bacteria plasmids resulted in only limited success (Kullen and Klaenhammer, 2000; O'Connell Motherway *et al.*, 2008). Further trials of modification of endogenous plasmids' replication function using thermo-sensitive replicons should thus be conducted.

The purpose of developing a random mutagenesis system is the acquisition of a number of random insertion mutants sufficient for genetic analysis of phenotypic traits and functions. Although we have discussed a number of possibilities for the development of such a system in bifidobacteria and the factors that we expect to impact the probability of success, one 'perfect' system is not necessary. A combination of several methodologies may well be the most practical way to accomplish the purpose.

Conclusions

In this chapter, we have discussed the genetic elements present in bifidobacteria, including cryptic plasmids and mobile genetic elements, and their application in the development of gene manipulation techniques. To date, numerous reports of the health-promoting effects of bifidobacteria have been published. However, the mechanisms underlying these effects remain to be elucidated. Various molecular biological analyses should be conducted to uncover the fundamental mechanisms underlying the interaction between bifidobacteria and their human host. Recent advances such as transcriptome and proteome analysis represent powerful tools for investigating these interactions. In combination with genomic, transcriptomic, and proteomic analysis, classical molecular genetic analyses (gene inactivation and random mutagenesis) will contribute to understanding the molecular mechanisms of the host–probiotic interaction. Because bifidobacteria possess homologous recombination activity (O'Connell Motherway *et al.*, 2008) and active IS elements (González Vara *et al.*, 2003; Lee *et al.*, 2008; Fukiya *et al.* 2010), systems for gene inactivation and random mutagenesis will probably be developed in the near future. These techniques will help clarify the mechanisms underlying the interaction between bifidobacteria and their host, which will in turn lead to more sophisticated, mechanism-based knowledge of the health-promoting effects of probiotic bifidobacteria.

References

Álvarez-Martín, P., Flórez, A.B., and Mayo, B. (2007a). Screening for plasmids among human bifidobacteria species: sequencing and analysis of pBC1 from *Bifidobacterium catenulatum* L48. Plasmid *57*, 165–174.

Álvarez-Martín, P., O'Connell-Motherway, M., van Sinderen, D., and Mayo, B. (2007b). Functional analysis of the pBC1 replicon from *Bifidobacterium catenulatum* L48. Appl. Microbiol. Biotechnol. *76*, 1395–1402.

Álvarez-Martín, P., Flórez, A.B., Margolles, A., del Solar, G., and Mayo, B. (2008). Improved cloning vectors for bifidobacteria, based on the *Bifidobacterium catenulatum* pBC1 replicon. Appl. Environ. Microbiol. *74*, 4656–4665.

Ammor, M.S., Flórez, A.B., Álvarez-Martín, P., Margolles, A., and Mayo, B. (2008). Analysis of tetracycline resistance *tet*(W) genes and their flanking sequences in intestinal *Bifidobacterium* species. J. Antimicrob. Chemother. *62*, 688–693.

Argnani, A., Leer, R.J., van Luijk, N., and Pouwels, P.H. (1996). A convenient and reproducible method to genetically transform bacteria of the genus *Bifidobacterium*. Microbiology *142*, 109–114.

Barrangou, R., Briczinski, E.P., Traeger, L.L., Loquasto, J.R., Richards, M., Horvath, P., Coûté-Monvoisin, A.C., Leyer, G., Rendulic, S., Steele, J.L., *et al.* (2009). Comparison of the complete genome sequences of *Bifidobacterium animalis* subsp. *lactis* DSM 10140 and Bl-04. J. Bacteriol. *191*, 4144–4151.

Bourget, N., Simonet, J.M., and Decaris, B. (1993). Analysis of the genome of the five *Bifidobacterium breve* strains: plasmid content, pulsed-field gel electrophoresis genome size estimation and *rrn* loci number. FEMS Microbiol. Lett. *110*, 11–20.

Camacho, L.R., Ensergueix, D., Perez, E., Gicquel, B., and Guilhot, C. (1999). Identification of a virulence gene cluster of *Mycobacterium tuberculosis* by signature-tagged transposon mutagenesis. Mol. Microbiol. *34*, 257–267.

Clark-Curtiss, J.E., and Haydel, S.E. (2003). Molecular genetics of *Mycobacterium tuberculosis* pathogenesis. Ann. Rev. Microbiol. *57*, 517–549.

Corneau, N., Émond, É., and LaPointe, G. (2004). Molecular characterization of three plasmids from *Bifidobacterium longum*. Plasmid *51*, 87–100.

Cronin, M., Knobel, M., O'Connell-Motherway, M., Fitzgerald, G.F., and van Sinderen, D. (2007). Molecular dissection of a bifidobacterial replicon. Appl. Environ. Microbiol. 73, 7858–7866.

Cronin, M., Sleator, R.D., Hill, C., Fitzgerald, G.F., and van Sinderen, D. (2008). Development of a luciferase-based reporter system to monitor *Bifidobacterium breve* UCC2003 persistence in mice. BMC Microbiol. *8*, 161.

Dyson, P.J. (1999). Isolation and development of transposons. In Genetic Methods for Diverse Prokaryotes, M.C.M. Smith, and R.E. Sockett, eds. (Academic Press: London, UK), pp. 133–167.

Escogido, M.L.R., Rodríguez, A.D.L., and de la Rosa, A.P.B. (2007). A novel binary expression vector for production of human IL-10 in *Escherichia coli* and *Bifidobacterium longum*. Biotechnol. Lett. *29*, 1249–1253.

FAO/WHO. (2001). Health and nutritional properties of probiotics in food including powder milk with live lactic acid bacteria (FAO/WHO Expert Consultation: Córdoba, Argentina). Available from: ftp://ftp.fao.org/es/esn/food/probio_report_en.pdf. Accessed 4 May 2010.

Fukiya, S., Sugiyama, T., Kano, Y., and Yokota, A. (2010). Characterization of an insertion sequence-like element, IS*Blo15*, identified in a size-increased cryptic plasmid pBK283 in *Bifidobacterium longum* BK28. J. Biosci. Bioeng. *110*, 141–146.

Garrigues, C., Johansen, E., and Pedersen, M.B. (2010). Complete genome sequence of *Bifidobacterium animalis* subsp. *lactis* BB-12, a widely consumed probiotic strain. J. Bacteriol. *192*, 2467–2468.

Gay, P., Lecoq, D., Steinmetz, M., Ferrari, E., and Hoch, J.A. (1983). Cloning structural gene *sacB*, which codes for exoenzyme levansucrase of *Bacillus subtilis*: expression of the gene in *Escherichia coli*. J. Bacteriol. *153*, 1424–1431.

Gibbs, M.J., Smeianov, V.V., Steele, J.L., Upcroft, P., and Efimov, B.A. (2006). Two families of Rep-like genes that probably originated by interspecies recombination are represented in viral, plasmid, bacterial, and parasitic protozoan genomes. Mol. Biol. Evol. *23*, 1097–1100.

González Vara, A., Rossi, M., Altomare, L., Eikmanns, B., and Matteuzzi, D. (2003). Stability of recombinant plasmids on the continuous culture of *Bifidobacterium animalis* ATCC 27536. Biotechnol. Bioeng. *84*, 145–150.

Guglielmetti, S., Karp, M., Mora, D., Tamagnini, I., and Parini, C. (2007). Molecular characterization of *Bifidobacterium longum* biovar *longum* NAL8 plasmids and construction of a novel replicon screening system. Appl. Microbiol. Biotechnol. *74*, 1053–1061.

Guglielmetti, S., Ciranna, A., Mora, D., Parini, C., and Karp, M. (2008). Construction, characterization and exemplificative application of bioluminescent *Bifidobacterium longum* biovar *longum*. Int. J. Food Microbiol. *124*, 285–290.

Gust, B., Chandra, G., Jakimowicz, D., Yuqing, T., Bruton, C.J., and Chater, K.F. (2004). λ red-mediated genetic manipulation of antibiotic-producing *Streptomyces*. In Advances in Applied Microbiology, 54, A.I. Laskin, J.W. Bennett, and G.M. Gadd, eds. (Elsevier Academic Press: San Diego, CA), pp. 107–128.

Hamaji, Y., Fujimori, M., Sasaki, T., Matsuhashi, H., Matsui-Seki, K., Shimatani-Shibata, Y., Kano, Y., Amano, J., and Taniguchi, S. (2007). Strong enhancement of recombinant cytosine deaminase activity in *Bifidobacterium longum* for tumor-targeting enzyme/prodrug therapy. Biosci. Biotechnol. Biochem. *71*, 874–883.

Hidaka, A., Hamaji, Y., Sasaki, T., Taniguchi, S., and Fujimori, M. (2007). Exogeneous cytosine deaminase gene expression in *Bifidobacterium breve* I-53–8w for tumor-targeting enzyme/prodrug therapy. Biosci. Biotechnol. Biochem. *71*, 2921–2926.

van Hoek, A.H.A.M., Mayrhofer, S., Domig, K.J., and Aarts, H.J.M. (2008). Resistance determinant *erm*(X) is borne by transposon Tn*5432* in *Bifidobacterium thermophilum* and *Bifidobacterium animalis* subsp. *lactis*. Int. J. Antimicrob. Agents *31*, 544–548.

Iwata, M., and Morishita, T. (1989). The presence of plasmids in *Bifidobacterium breve*. Lett. Appl. Microbiol. *9*, 165–168.

Jäger, W., Schäfer, A., Pühler, A., Labes, G., and Wohlleben, W. (1992). Expression of the *Bacillus subtilis sacB* gene leads to sucrose sensitivity in the Gram-positive bacterium *Corynebacterium glutamicum* but not in *Streptomyces lividans*. J. Bacteriol. *174*, 5462–5465.

Kawaguchi, H., Vertès, A.A., Okino, S., Inui, M., and Yukawa, H. (2006). Engineering of a xylose metabolic pathway in *Corynebacterium glutamicum*. Appl. Environ. Microbiol. 72, 3418–3428.

Kazimierczak, K.A., Flint, H.J., and Scott, K.P. (2006). Comparative analysis of sequences flanking *tet*(W) resistance genes in multiple species of gut bacteria. Antimicrob. Agents Chemother. *50*, 2632–2639.

Khan, S.A. (1997). Rolling-circle replication of bacterial plasmids. Microbiol. Mol. Biol. Rev. *61*, 442–455.

Kim, J.F., Jeong, H., Yu, D.S., Choi, S.H., Hur, C.G., Park, M.S., Yoon, S.H., Kim, D.W., Ji, G.E., Park, H.S., *et al.* (2009). Genome sequence of the probiotic bacterium *Bifidobacterium animalis* subsp. *lactis* AD011. J. Bacteriol. *191*, 678–679.

Klijn, A., Moine, D., Delley, M., Mercenier, A., Arigoni, F., and Pridmore, R.D. (2006). Construction of a reporter vector for the analysis of *Bifidobacterium longum* promoters. Appl. Environ. Microbiol. 72, 7401–7405.

Kullen, M.J., and Klaenhammer, T.R. (2000). Genetic modification of intestinal lactobacilli and bifidobacteria. Curr. Issues Mol. Biol. *2*, 41–50.

Law, J., Buist, G., Haandrikman, A., Kok, J., Venema, G., and Leenhouts, K. (1995). A system to generate chromosomal mutations in *Lactococcus lactis* which allows fast analysis of targeted genes. J. Bacteriol. *177*, 7011–7018.

Leahy, S.C., Higgins, D.G., Fitzgerald, G.F., and van Sinderen, D. (2005). Getting better with bifidobacteria. J. Appl. Microbiol. *98*, 1303–1315.

Lee, J.H., and O'Sullivan, D.J. (2006). Sequence analysis of two cryptic plasmids from *Bifidobacterium longum* DJO10A and construction of a shuttle cloning vector. Appl. Environ. Microbiol. *72*, 527–535.

Lee, J.H., Karamychev, V.N., Kozyavkin, S.A., Mills, D., Pavlov, A.R., Pavlova, N.V., Polouchine, N.N., Richardson, P.M., Shakhova, V.V., Slesarev, A.I., *et al.* (2008). Comparative genomic analysis of the gut bacterium *Bifidobacterium longum* reveals loci susceptible to deletion during pure culture growth. BMC Genomics *9*, 247.

Long, R.T., Zeng, W.S., Chen, L.Y., Guo, J., Lin, Y.Z., Huang, Q.S., and Luo, S.Q. (2010). *Bifidobacterium* as an oral delivery carrier of oxyntomodulin for obesity therapy: inhibitory effects on food intake and body weight in overweight mice. Int. J. Obes. *34*, 712–719.

MacConaill, L.E., Fitzgerald, G.F., and van Sinderen, D. (2003). Investigation of protein export in *Bifidobacterium breve* UCC2003. Appl. Environ. Microbiol. *69*, 6994–7001.

Maguin, E., Prévost, H., Ehrlich, S., and Gruss, A. (1996). Efficient insertional mutagenesis in lactococci and other Gram-positive bacteria. J. Bacteriol. *178*, 931–935.

Mahillon, J., and Chandler, M. (1998). Insertion sequences. Microbiol. Mol. Biol. Rev. *62*, 725–774.

Matsumura, H., Takeuchi, A., and Kano, Y. (1997). Construction of *Escherichia coli* – *Bifidobacterium longum* shuttle vector transforming *B. longum* 105-A and 108-A. Biosci. Biotechnol. Biochem. *61*, 1211–1212.

Matteuzzi, D., Brigidi, P., Rossi, M., and Di, D. (1990). Characterization and molecular cloning of *Bifidobacterium longum* cryptic plasmid pMB1. Lett. Appl. Microbiol. *11*, 220–223.

Maus, C.E., Plikaytis, B.B., and Shinnick, T.M. (2005). Mutation of *tlyA* confers capreomycin resistance in *Mycobacterium tuberculosis*. Antimicrob. Agents Chemother. *49*, 571–577.

Melville, C.M., Brunel, R., Flint, H.J., and Scott, K.P. (2004). The *Butyrivibrio fibrisolvens tet*(W) gene is carried on the novel conjugative transposon Tn*B1230*, which contains duplicated nitroreductase coding sequences. J. Bacteriol. *186*, 3656–3659.

Missich, R., Sgorbati, B., and LeBlanc, D.J. (1994). Transformation of *Bifidobacterium longum* with pRM2, a constructed *Escherichia coli–B. longum* shuttle vector. Plasmid *32*, 208–211.

Moon, G.S., Pyun, Y.R., Park, M.S., Ji, G.E., and Kim, W.J. (2005). Secretion of recombinant pediocin PA-1 by *Bifidobacterium longum*, using the signal sequence for bifidobacterial α-amylase. Appl. Environ. Microbiol. *71*, 5630–5632.

Moon, G.S., Wegmann, U., Gunning, A.P., Gasson, M.J., and Narbad, A. (2009). Isolation and characterization of a theta-type cryptic plasmid from *Bifidobacterium longum* FI10564. J. Microbiol. Biotechnol. *19*, 403–408.

Nakamura, T., Sasaki, T., Fujimori, M., Yazawa, K., Kano, Y., Amano, J., and Taniguchi, S. (2002). Cloned cytosine deaminase gene expression of *Bifidobacterium longum* and application to enzyme/pro-drug therapy of hypoxic solid tumors. Biosci. Biotechnol. Biochem. *66*, 2362–2366.

O'Connell Motherway, M., Fitzgerald, G.F., Neirynck, S., Ryan, S., Steidler, L., and van Sinderen, D. (2008). Characterization of ApuB, an extracellular type II amylopullulanase from *Bifidobacterium breve* UCC2003. Appl. Environ. Microbiol. *74*, 6271–6279.

O'Riordan, K., and Fitzgerald, G.F. (1999). Molecular characterisation of a 5.75-kb cryptic plasmid from *Bifidobacterium breve* NCFB 2258 and determination of mode of replication. FEMS Microbiol. Lett. *174*, 285–294.

Park, M.S., Shin, D.W., Lee, K.H., and Ji, G.E. (1999). Sequence analysis of plasmid pKJ50 from *Bifidobacterium longum*. Microbiology *145*, 585–592.

Park, M.S., Shin, D.W., Lee, K.H., and Ji, G.E. (2000). Characterization of plasmid pKJ36 from *Bifidobacterium longum* and construction of an *E. coli-Bifidobacterium* shuttle vector. J. Microbiol. Biotechnol. *10*, 312–320.

Park, M.S., Moon, H.W., and Ji, G.E. (2003). Molecular characterization of plasmid from *Bifidobacterium longum*. J. Microbiol. Biotechnol. *13*, 457–462.

Park, Y.S., Kim, K. H., Park, J.H., Oh, I.K., and Yoon, S.S. (2008). Isolation and molecular characterization of a cryptic plasmid from *Bifidobacterium longum*. Biotechnol. Lett. *30*, 145–151.

Park, M.S., Kwon, B., Shim, J.J., Huh, C.S., and Ji, G.E. (2008). Heterologous expression of cholesterol oxidase in *Bifidobacterium longum* under the control of 16S rRNA gene promoter of bifidobacteria. Biotechnol. Lett. *30*, 165–172.

Rhim, S.L., Park, M.S., and Ji, G.E. (2006). Expression and secretion of *Bifidobacterium adolescentis* amylase by *Bifidobacterium longum*. Biotechnol. Lett. *28*, 163–168.

Rossi, M., Brigidi, P., Gonzalez Vara y Rodriguez, A., and Matteuzzi, D. (1996). Characterization of the plasmid pMB1 from *Bifidobacterium longum* and its use for shuttle vector construction. Res. Microbiol. *147*, 133–143.

Rossi, M., Brigidi, P., and Matteuzzi, D. (1997). An efficient transformation system for *Bifidobacterium* spp. Lett. Appl. Microbiol. *24*, 33–36.

Rossi, M., Brigidi, P., and Matteuzzi, D. (1998). Improved cloning vectors for *Bifidobacterium* spp. Lett. Appl. Microbiol. *26*, 101–104.

Sallam, K.I., Mitani, Y., and Tamura, T. (2006). Construction of random transposition mutagenesis system in *Rhodococcus erythropolis* using IS*1415*. J. Biotechnol. *121*, 13–22.

Sangrador-Vegas, A., Stanton, C., van Sinderen, D., Fitzgerald, G.F., and Ross, R.P. (2007). Characterization of plasmid pASV479 from *Bifidobacterium pseudolongum* subsp. *globosum* and its use for expression vector construction. Plasmid *58*, 140–147.

Schell, M.A., Karmirantzou, M., Snel, B., Vilanova, D., Berger, B., Pessi, G., Zwahlen, M.C., Desiere, F., Bork, P., Delley, M., *et al.* (2002). The genome sequence of *Bifidobacterium longum* reflects its adaptation to the human gastrointestinal tract. Proc. Natl. Acad. Sci. U.S.A. *99*, 14422–14427.

Sela, D.A., Chapman, J., Adeuya, A., Kim, J.H., Chen, F., Whitehead, T.R., Lapidus, A., Rokhsar, D.S., Lebrilla, C.B., German, J.B., *et al.* (2008). The genome sequence of *Bifidobacterium longum* subsp. *infantis* reveals adaptations for milk utilization within the infant microbiome. Proc. Natl. Acad. Sci. U.S.A. *105*, 18964–18969.

Serror, P., Ilami, G., Chouayekh, H., Ehrlich, S.D., and Maguin, E. (2003). Transposition in *Lactobacillus delbrueckii* subsp. *bulgaricus*: identification of two thermosensitive replicons and two functional insertion sequences. Microbiology *149*, 1503–1511.

Sgorbati, B., Scardovi, V., and Leblanc, D.J. (1982). Plasmids in the genus *Bifidobacterium*. J. Gen. Microbiol. *128*, 2121–2131.

Sgorbati, B., Scardovi, V., and Leblanc, D.J. (1986a). Related structures in the plasmid profiles of *Bifidobacterium asteroides, B. indicum* and *B. globosum*. Microbiologica *9*, 443–454.

Sgorbati, B., Scardovi, V., and Leblanc, D.J. (1986b). Related structures in the plasmid profiles of *Bifidobacterium longum*. Microbiologica *9*, 415–422.

Shkoporov, A.N., Efimov, B.A., Khokhlova, E.V., Kafarskaia, L.I., and Smeianov, V.V. (2008a). Production of human basic fibroblast growth factor (FGF-2) in *Bifidobacterium breve* using a series of novel expression/secretion vectors. Biotechnol. Lett. *30*, 1983–1988.

Shkoporov, A.N., Efimov, B.A., Khokhlova, E.V., Steele, J.L., Kafarskaia, L.I., and Smeianov, V.V. (2008b). Characterization of plasmids from human infant *Bifidobacterium* strains: Sequence analysis and construction of *E. coli-Bifidobacterium* shuttle vectors. Plasmid *60*, 136–148.

Siguier, P., Perochon, J., Lestrade, L., Mahillon, J., and Chandler, M. (2006). ISfinder: the reference centre for bacterial insertion sequences. Nucleic Acids Res. *34*, D32-D36.

del Solar, G., Giraldo, R., Ruiz-Echevarría, M.J., Espinosa, M., and Díaz-Orejas, R. (1998). Replication and control of circular bacterial plasmids. Microbiol. Mol. Biol. Rev. *62*, 434–464.

Steinmetz, M., Lecoq, D., Djemia, H.B., and Gay, P. (1983). Genetic analysis of *sacB*, the structural gene of a secreted enzyme, levansucrase of *Bacillus subtilis* Marburg. Mol. Gen. Genet. *191*, 138–144 (in French).

Suzuki, N., Okai, N., Nonaka, H., Tsuge, Y., Inui, M., and Yukawa, H. (2006). High-throughput transposon mutagenesis of *Corynebacterium glutamicum* and construction of a single-gene disruptant mutant library. Appl. Environ. Microbiol. *72*, 3750–3755.

Takata, T., Shirakawa, T., Kawasaki, Y., Kinoshita, S., Gotoh, A., Kano, Y., and Kawabata, M. (2006). Genetically engineered *Bifidobacterium animalis* expressing the *Salmonella* flagellin gene for the mucosal immunization in a mouse model. J. Gene. Med. *8*, 1341–1346.

Takeuchi, A., Matsumura, H., and Kano, Y. (2002). Cloning and expression in *Escherichia coli* of a gene, *hup*, encoding the histone-like protein HU of *Bifidobacterium longum*. Biosci. Biotechnol. Biochem. *66*, 598–603.

Tamura, K., Dudley, J., Nei, M., and Kumar, S. (2007). MEGA4: Molecular evolutionary genetics analysis (MEGA) software version 4.0. Mol. Biol. Evol. *24*, 1596–1599.

Tanaka, Y., Yoshikawa, O., Maruhashi, K., and Kurane, R. (2002). The *cbs* mutant strain of *Rhodococcus erythropolis* KA2–5–1 expresses high levels of Dsz enzymes in the presence of sulfate. Arch. Microbiol. *178*, 351–357.

Tanaka, K., Samura, K., and Kano, Y. (2005). Structural and functional analysis of pTB6 from *Bifidobacterium longum*. Biosci. Biotechnol. Biochem. *69*, 422–425.

Ventura, M., Canchaya, C., Zink, R., Fitzgerald, G.F., and van Sinderen, D. (2004). Characterization of the *groEL* and *groES* loci in *Bifidobacterium breve* UCC 2003: Genetic, transcriptional, and phylogenetic analyses. Appl. Environ. Microbiol. *70*, 6197–6209.

Ventura, M., Lee, J.H., Canchaya, C., Zink, R., Leahy, S., Moreno-Munoz, J.A., O'Connell-Motherway, M., Higgins, D., Fitzgerald, G.F., O'Sullivan, D.J., *et al.* (2005a). Prophage-like elements in bifidobacteria: insights from genomics, transcription, integration, distribution, and phylogenetic analysis. Appl. Environ. Microbiol. *71*, 8692–8705.

Ventura, M., Zink, R., Fitzgerald, G.F., and van Sinderen, D. (2005b). Gene structure and transcriptional organization of the *dnaK* operon of *Bifidobacterium breve* UCC 2003 and application of the operon in bifidobacterial tracing. Appl. Environ. Microbiol. *71*, 487–500.

Ventura, M., Turroni, F., Zomer, A., Foroni, E., Giubellini, V., Bottacini, F., Canchaya, C., Claesson, M.J., He, F., Mantzourani, M., *et al.* (2009). The *Bifidobacterium dentium* Bd1 genome sequence reflects its genetic adaptation to the human oral cavity. PLoS Genet. *5*, e1000785.

Vertès, A.A., Inui, M., and Yukawa, H. (2005). Manipulating corynebacteria, from individual genes to chromosomes. Appl. Environ. Microbiol. *71*, 7633–7642.

Yasui, K., Kano, Y., Tanaka, K., Watanabe, K., Shimizu-Kadota, M., Yoshikawa, H., and Suzuki, T. (2009). Improvement of bacterial transformation efficiency using plasmid artificial modification. Nucleic Acids Res. *37*, e3.

Metabolic Pathway of Human Milk Oligosaccharides in Bifidobacteria

3

Motomitsu Kitaoka,* Takane Katayama and Kenji Yamamoto

Abstract

It has long been considered that human milk oligosaccharides (HMOs) are the growth-promoting factors for bifidobacteria, which makes bifidobacteria the predominant intestinal microbiota in breast-fed infants. However, the mechanism responsible for the selective growth of bifidobacteria has not yet been identified because of the difficulty caused by the complicated content of HMOs, which includes more than 100 kinds of molecules. The lacto-*N*-biose I (LNB) hypothesis made it possible to investigate the bifidobacterial metabolism of HMOs systematically. As predicted by the LNB hypothesis, several extracellular enzymes that liberate LNB from HMOs have been identified from *Bifidobacterium bifidum* JCM1254. The intracellular metabolic pathway of bifidobacteria specific to galacto-*N*-biose and LNB, led by β-1,3-galactosyl-*N*-acetylhexosamine phosphorylase, has been fully characterized. Meanwhile, a practical method to produce LNB on a large scale has been established. The long-standing enigma of the bifidus factor present in human milk might be clarified in the near future.

Background

It has been known that within a week after birth, breast-fed infants form intestinal microbiota consisting predominantly of bifidobacteria, which accounts for 95–99.9% of the intestinal microorganisms (Rotimi and Duerden, 1981; Benno and Mitsuoka, 1986). Intestinal colonization by bifidobacteria seems to prevent diarrhoea and infection by some pathogenic bacteria (Bezkorovainy, 1989). Conversely, bifidobacteria do not readily colonize in the intestine of bottle-fed infants, and early in the 20th century they often suffered infections of pathogenic bacteria. To improve the growth of intestinal bifidobacteria, infant formula is now supplemented with prebiotic oligosaccharides, resulting in healthier bottle-fed infants. However, the intestinal microbiota of bottle-fed infants is still different from that of breast-fed infants; in particular an increased population of Enterobacteriaceae is observed (Benno *et al.*, 1984).

The growth-stimulating factors for bifidobacteria present in human milk have been investigated for a long time. Initially, a nitrogen-containing sugar was assumed to be the bifidus factor, but subsequent studies revealed that the originators of this idea were misled because of the use of a particular GlcNAc-auxotrophic strain of *Bifidobacterium bifidum* var. *pennsylvanicus* (Gyorgy *et al.*, 1954; Veerkamp, 1969). Further studies revealed that oligosaccharides in human milk (human milk oligosaccharides, HMOs) are promising candidates as real bifidus factors (Bezkorovainy, 1989), but the mechanism underlying their effects remains obscure.

Cow's milk contains lactose as the sole saccharide, at a concentration of around 50 g/l (Newburg and Neubauer, 1995; Kunz *et al.*, 2000). The major carbohydrate present in human milk is also lactose (60 g/l), but it contains HMOs as well, at a concentration of 22–24 g/l in colostrum and 12–13 g/l in mature milk (Newburg and Neubauer, 1995). HMOs are composed of one of the 12 core structures (Fig. 3.1), and these core structures are often

*corresponding author

1. lactose — Galβ1→4Glc

2. **lacto-*N*-tetraose** — Galβ1→3GlcNAcβ1→3Galβ1→4Glc

3. lacto-*N*-neotetraose — Galβ1→4GlcNAcβ1→3Galβ1→4Glc

4. **lacto-*N*-hexaose**

   Galβ1→4GlcNAcβ1
↓
6
Galβ1→3GlcNAcβ1→3Galβ1→4Glc

5. lacto-*N*-neohexaose

   Galβ1→4GlcNAcβ1
↓
6
Galβ1→4GlcNAcβ1→3Galβ1→4Glc

6. ***para*-lacto-*N*-hexaose** — Galβ1→3GlcNAcβ1→3Galβ1→4GlcNAcβ1→3Galβ1→4Glc

7. *para*-lacto-*N*-neohexaose — Galβ1→4GlcNAcβ1→3Galβ1→4GlcNAcβ1→3Galβ1→4Glc

8. **lacto-*N*-octaose**

   Galβ1→4GlcNAcβ1→3Galβ1→4GlcNAcβ1
↓
6
Galβ1→3GlcNAcβ1→3Galβ1→4Glc

9. **lacto-*N*-neooctaose**

   Galβ1→3GlcNAcβ1→3Galβ1→4GlcNAcβ1
↓
6
Galβ1→4GlcNAcβ1→3Galβ1→4Glc

10. ***iso*-lacto-*N*-octaose**

   Galβ1→3GlcNAcβ1→3Galβ1→4GlcNAcβ1
↓
6
Galβ1→3GlcNAcβ1→3Galβ1→4Glc

11. ***para*-lacto-*N*-octaose**

   Galβ1→3GlcNAcβ1→3Galβ1→4GlcNAcβ1→3Galβ1→4GlcNAcβ1→3Galβ1→4Glc

12. **lacto-*N*-decaose**

   Galβ1→4GlcNAcβ1
↓
6
Galβ1→3GlcNAcβ1→3Galβ1→4GlcNAcβ1
↓
6
Galβ1→3GlcNAcβ1→3Galβ1→4Glc

Figure 3.1 Core structures found in HMOs. Bold, containing LNB structure (type I); underline, containing LacNAc structure (type II).

modified with α-fucosyl and/or α-sialyl residues (Haeuw-Fievre *et al.*, 1993). At least 93 kinds of molecules have been identified so far. Core structures of HMOs other than lactose are classified into type I and II based on the disaccharide unit at the non-reducing end. Type I and II sugars contain lacto-*N*-biose I (LNB, Galβ1→3GlcNAc) and *N*-acetyllactosamine (LacNAc, Galβ1→4GlcNAc) structures at the non-reducing end, respectively. Type I oligosaccharides are predominant in HMOs and type II oligosaccharides are minor components. For example, 2′-fucosyllactose (Fucα1→2Galβ1→4Glc, 2′FL), lacto-*N*-tetraose (Galβ1→3GlcNAcβ1→3Galβ1→4Glc, LNT), and lacto-*N*-fucopentaose I (Fucα1→2Galβ1→3GlcNAcβ1→3Galβ1→4Glc, LNFPI) are the most abundant components in HMOs (Asakuma *et al.*, 2008), and the latter two are type I oligosaccharides. The type I dominant composition of milk oligosaccharides is a specific feature of human milk, and not of other mammals, including anthropoids (Urashima *et al.*, 2009).

The mechanism whereby HMOs promote the selective growth of bifidobacteria has long been unsolved because of the difficulty caused by the complicated content of HMOs. In this chapter, recent progress in our understanding of the metabolic pathway of HMOs in bifidobacteria is described.

The lacto-*N*-biose I hypothesis

In 1999, β-1,3-galactosyl-*N*-acetylhexosamine phosphorylase (GalHexNAcP, EC 2.4.1.211) was found in a cell-free extract of *B. bifidum* DSM20082, which reversibly phosphorolysed LNB and galacto-*N*-biose (GNB, Galβ1→3GalNAc) to α-D-galactose 1-phosphate (Gal1*P*) and the corresponding *N*-acetylhexosamine (Derensy-Dron *et al.*, 1999). Kitaoka *et al.* (2005) purified GalHexNAcP from *B. bifidum* JCM1254 (=DSM20082) to homogeneity and found that its N-terminal and internal amino acid sequences showed high degrees of similarity to a hypothetical protein encoded by the BL1641 gene of *Bifidobacterium longum* subsp. *longum* NCC2705, the complete genome sequence of which is available (Schell *et al.*, 2002). They cloned and expressed the corresponding gene, *lnpA*, from *B. longum* subsp. *longum* JCM1217^T, the type strain, in *Escherichia coli* and confirmed that *lnpA* encoded GalHexNAcP (Kitaoka *et al.*, 2005).

The GalHexNAcP gene, BL1641, is located in a putative operon in the *B. longum* subsp. *longum* NCC2705 genome (Fig. 3.2). In the operon, BL1638–1640 genes are annotated as component proteins of ATP-binding cassette (ABC)-type sugar transporters. BL1642, BL1643, and BL1644 are annotated as mucin desulfatase, UTP-hexose-1-phosphate uridylyltransferase (GalT2, EC 2.7.7.10), and UDP-glucose 4-epimerase (GalE, EC 5.1.3.2), respectively. Considering the annotated functions, this putative operon appeared to be involved in the metabolism of GNB and LNB.

GNB is abundantly present in mucin glycoproteins as *O*-linked core 1 sugar. The extracellular enzyme that liberates GNB, endo-α-*N*-acetylgalactosaminidase, has already been isolated from *B. longum* subsp. *longum* JCM1217^T (Fujita *et al.*, 2005; Katayama *et al.*, 2005). On the other hand, the most important source of LNB available in the human intestine is HMOs. Based on the facts that type I oligosaccharides containing the LNB structure are predominant in HMOs and that bifidobacteria possess GalHexNAcP, Kitaoka *et al.* (2005) proposed the LNB hypothesis, in which the LNB structures in HMOs acted as growth-stimulating factors for bifidobacteria. These authors predicted the occurrence of enzymes that are involved in liberating LNB from type I HMOs, and then transporting it in bifidobacteria (Fig. 3.3). They later succeeded in isolating these genes as described below.

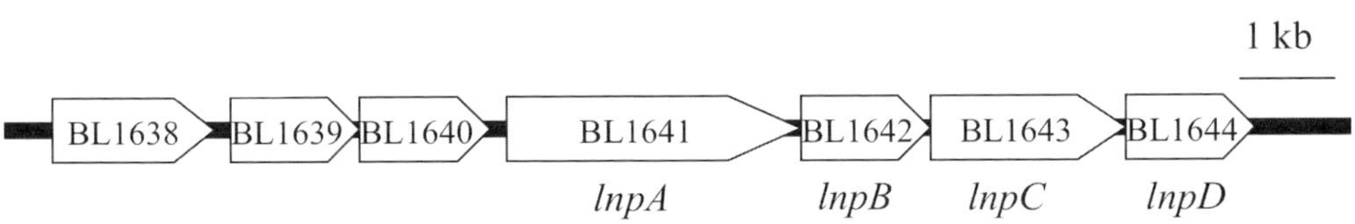

	annotated	identified
BL1638	solute binding protein	GNB/LNB binding protein (GL-BP)
BL1639	permease	
BL1640	permease	
BL1641	hypothetical protein	GalHexNAcP
BL1642	desulfatase for mucin	*N*-acetylhexosamine 1-kinase (NahK)
BL1643	EC 2.7.1.10	EC 2.7.1.12 active on GlcNAc1*P*/GalNAc1*P*
BL1644	EC 5.1.3.2	EC 5.1.3.2 active on UDP-GlcNAc

Figure 3.2 Gene cluster of *B. longum* subsp. *longum* NCC2705 containing the GalHexNAc gene.

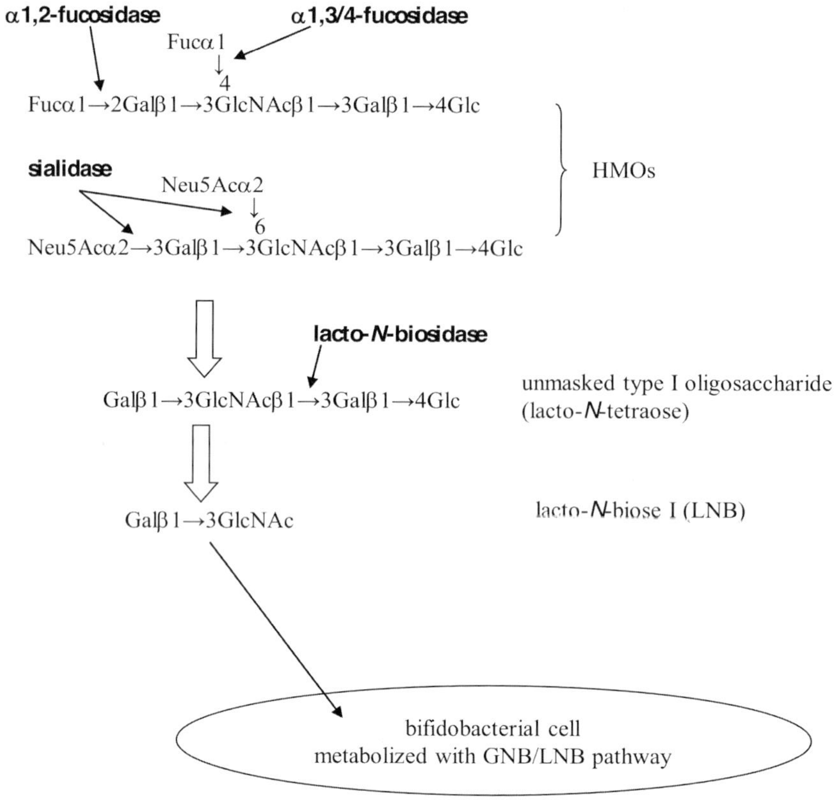

Figure 3.3 Schematic representation of the lacto-*N*-biose I hypothesis.

Extracellular enzymes involved in liberating LNB from HMOs

Enzymes involved in liberating core structures from HMOs

Core structures of HMOs (Fig. 3.1) are often modified by α-fucosylation and/or α-sialylation. Since lacto-*N*-biosidase (see section 'Lacto-*N*-biosidase') is quite specific to unmodified type 1 chains, α-fucosidases and α-sial idases are required for bifidobacteria to utilize HMOs.

Katayama *et al.* (2004) discovered α-1,2-fucosidase activity in various strains of bifidobacteria, and cloned the gene (*afcA*) by screening the genomic library of *B. bifidum* JCM1254 for the ability to hydrolyse 2′FL (Katayama *et al.*, 2004). The enzyme (AcfA) was a huge protein with 1959 amino acid residues, possessing a secreting signal sequence at the N-terminal and a membrane-anchoring motif at the C-terminal. The truncated recombinant protein selectively hydrolysed α-1,2-linked L-fucosyl residues at the non-reducing end. It effectively liberated L-fucose from glycoproteins such as porcine stomach mucin, as well as HMOs and blood-type antigens. This was the first report on the molecular cloning of α-1,2-linkage-specific α-L-fucosidase, and at that time the gene (*afcA*) and its deduced amino acid sequence were completely different from the other α-L-fucosidases previously identified. In the Carbohydrate-Active Enzyme (CAZy) database, AfcA was classified in the glycoside hydrolase family 95, a new family created for AfcA. Recently, the crystal structure of AfcA (Nagae *et al.*, 2007) and the synthesis of 2′FL using mutated AfcA (Wada *et al.*, 2008b) were reported.

α-1,2-L-Fucoside-hydrolysing activity was observed in some species of bifidobacteria such as *B. bifidum* (JCM1254, JCM7004, and JCM29251) and *Bifidobacterium longum* subsp. *infantis* JCM1222^{T}, but not in *Bifidobacterium breve* (203 and JCM1192^{T}) and *B. longum* subsp.

longum (33R and JCM1217^T). Homologous genes are found in genomic sequences of some intestinal bacteria, such as *Bacteroides*, *Clostridium*, and *Ruminococcus*, and in some other bacteria, such as *Streptomyces*, *Streptococcus*, and *Xanthomonas*. Among them, α-1,2-fucosidase has already been purified from a culture filtrate of *Clostridium perfringens* (Aminoff and Furukawa, 1970).

Other side-chain cleaving enzymes from HMOs include cell surface-bound α-1,3/4-fucosidase (Ashida *et al.*, 2009) and α-2,3/6-sialidases (Kiyohara *et al.*, 2011), which were isolated and molecularly cloned from *B. bifidum* JCM1254. Since major modifications in the core structures of HMOs are α-1,2/3/4-fucosylation and α-2,3/6-sialylation, the presence of these enzymes is beneficial for bifidobacteria to unmask the core structure of HMOs and to further degrade them (see below).

Lacto-*N*-biosidase

The hydrolytic activity to liberate LNB from LNT was observed in several strains of *B. bifidum* (JCM1254, JCM1255^T, and JCM7004) and *B. longum* subsp. *longum* (JCM1217^T and JCM7054) (Wada *et al.*, 2008a). In contrast, most of the other intestinal bacteria, including other bifidobacterial strains such as *B. breve* JCM1192^T and *Bifidobacterium adolescentis* (JCM1275^T and JCM7046), did not hydrolyse LNT at all. Wada *et al.* (2008a) cloned the gene for lacto-*N*-biosidase (*lnbB*) from *B. bifidum* JCM1254 based on the amino acid sequence of the *Streptomyces* sp. strain 142 lacto-*N*-biosidase (Sano *et al.*, 1992, 1993), which was the only sequence of the enzyme available at that time. The gene encoded a protein of 1112 amino acid residues with a predicted molecular mass of 120 kDa. The deduced amino acid sequence contained a signal peptide and a membrane anchor at the N-terminal and C-terminal ends, respectively, a structure similar to that of AfcA. The LnbB protein exhibited 38% amino acid identity to the lacto-*N*-biosidase protein from *Streptomyces* sp.

β-*N*-Acetylhexosaminidases and lacto-*N*-biosidases are members of GH family 20, and accordingly, the catalytically important residues identified in β-*N*-acetylhexosaminidases (Williams *et al.*, 2002; Mark *et al.*, 2003) are well conserved in sequences of LnbB. However, the residues involved in the recognition of O-3 and O-6 of *N*-acetylhexosamine in β-*N*-acetylhexosaminidases are not conserved in LnbB, which might reflect the difference in substrate specificity between the two enzymes. No homologues with significant similarity were found in the genomic sequences of *B. longum* subsp. *longum* (NCC2705 and DJO10A) and *B. adolescentis* strains (ATCC 15703 and L2–32). It should be noted that a partial contig sequence showing a 87% identity with *lnbB* was discovered in an infant gut metagenome (Kurokawa *et al.*, 2007), suggesting the presence of LnbB-positive bacteria in the intestinal environment.

A truncated form of LnbB was expressed in *E. coli* and purified. The recombinant LnbB hydrolysed LNT into LNB and lactose. It hydrolysed *p*NP-β-LNB, and also hydrolysed *p*NP-β-GNB at 30% of the rate at which it hydrolysed *p*NP-β-LNB. The LnbB protein did not act on any β-linked *p*NP monosaccharides, including *p*NP-β-GlcNAc and *p*NP-β-GalNAc. The enzyme did not hydrolyse type II oligosaccharides such as lacto-*N*-neotetraose, (Galβ1→4GlcNAcβ1→3Galβ1→4Glc) nor did it hydrolyse the fucosylated forms of type I oligosaccharides such as LNFPI and lacto-*N*-fucopentaose II (Galβ1→3(Fucα1→4)GlcNAcβ1→3Galβ1→4Glc), indicating that LnbB was specific to the unmodified type I core structures.

The GNB/LNB transporter

The presence of extracellular LNB- and GNB-liberating enzymes (lacto-*N*-biosidase and endo-α-*N*-acetylgalactosaminidase) and an intracellular GalHexNAcP suggested that bifidobacteria possessed a specific transport system to incorporate GNB and LNB. The ABC transporter genes located upstream of the GalHexNAcP gene of *B. longum* subsp. *longum* JCM1217^T (Fig. 3.2) were cloned and introduced into *B. longum* subsp. *longum* 105-A with a multi-copy plasmid. The recombinant *B. longum* subsp. *longum* 105-A exhibited significantly higher LNB uptake activity than the control strain (T. Katayama *et al.*, unpublished result).

The solute-binding protein of the transporter from *B. longum* subsp. *longum* JCM1217^T was produced in *E. coli* and purified (Wada *et al.*, 2007). The protein (named GNB/LNB-binding protein, GL-BP) specifically binds to GNB (K_d = 10 nM) and LNB (K_d = 87 nM) in an enthalpy-driven manner, but not to monosaccharides and other disaccharides, including LacNAc (Suzuki *et al.*, 2008). It possesses the capacity to bind with LNT with a much higher K_d value (11 μM), indicating that LNT is a poor ligand of GL-BP. These results indicate that several bifidobacterial strains can liberate LNB from type I HMOs through the action of lacto-*N*-biosidase and then import it through the GNB/LNB-specific transporter.

The residues involved in binding GNB and LNB were identified through structural analysis of GL-BP (Suzuki *et al.*, 2008). Two tryptophans, Trp^{278} and Trp^{257}, take part in aromatic stacking interactions with the sugar rings of Gal and GlcNAc/GalNAc, respectively. As suggested by the results of thermodynamic analysis, hydrogen bonds play a dominant role in the binding of ligands to GL-BP. The hydroxyl groups of ligands form direct hydrogen bonds with Arg^{49}, Gln^{105}, Asp^{154}, Glu^{203}, and Ser^{316}.

Proteins sharing more than 30% identity with GL-BP are not found in genomic sequences of intestinal bacteria other than bifidobacteria. Apart from the intestinal bacteria, homologous genes are found in species of *Propionibacterium*, *Arthrobacter*, and *Streptomyces*. The amino acid residues involved in binding with GNB and LNB are well conserved in these homologous proteins.

The intracellular GNB/LNB pathway

GalHexNAcP

Nishimoto and Kitaoka (2007a) cloned the genes of the galactose metabolism operon from *B. longum* subsp. *longum* JCM1217^T, which corresponds to BL1641–BL1644 of *B. longum* subsp. *longum* NCC2705, and named them *lnpA*, *lnpB*, *lnpC*, and *lnpD*, respectively (Fig. 3.2). The product coded by *lnpA* is GalHexNAcP. LnpA phosphorolyses GNB and LNB but cannot phosphorolyse LNT, indicating that HMOs must be digested to LNB before the phosphorolytic cleavage by LnpA.

Homologous genes were found in *B. longum* subsp. *infantis* ATCC15697^T (Sela *et al.*, 2008) and *B. bifidum* JCM1254 (Nishimoto and Kitaoka, 2007b), but not in the genomic sequence of *B. adolescentis* ATCC15703^T and *B. animalis* subsp. *lactis* AD011 (Kim *et al.*, 2009). Homologous genes were also found in the genomic sequence of *C. perfringens* 13 (Shimizu *et al.*, 2002), *Propionibacterium acnes* KPA171202 (Bruggemann *et al.*, 2004), and *Vibrio vulnificus* (CMCP6 and YJ016). Since all the strains are pathogenic or commensal in humans, the enzyme might be related to their inhabitation in human tissues. Characterization of the enzymes from *B. longum* subsp. *longum* JCM1217^T (Kitaoka *et al.*, 2005), *B. bifidum* JCM1254 (Nishimoto and Kitaoka, 2007b), *C. perfringens* ATCC13124^T (Nakajima *et al.*, 2008), *P. acnes* JCM6425 (Nakajima *et al.*, 2009), and *V. vulnificus* CMCP6 (Nakajima and Kitaoka, 2008) revealed that GalHexNAcPs can be separated into three classes based on their substrate preference. GalHexNAcPs from bifidobacteria phosphorolyses both GNB and LNB equally and are classified as galacto-*N*-biose/lacto-*N*-biose I phosphorylase (GLNBP). Those from *C. perfringens* and *P. acnes* phosphorolyse GNB much faster than LNB and are classified as galacto-*N*-biose phosphorylase (GNBP). The *V. vulnificus* enzyme is lacto-*N*-biose I phosphorylase (LNBP), that phosphorolyses LNB much faster than GNB. It should be noted that GalHexNAcP from *C. perfringens*, an intestinal bacterium, is GNBP and is not considered to be related to the metabolism of HMOs. The absence of lacto-*N*-biosidase activity in this bacterium (Wada *et al.*, 2008a) supports this notion.

The 3D structure of GalHexNAcP from *B. longum* subsp. *longum* JCM1217^T resembles that of GH42 β-galactosidase, suggesting that GalHexNAcP is related to glycoside hydrolases rather than glycosyl transferases (Hidaka *et al.*, 2009). On the basis of this finding, GalHexNAcPs are grouped in a new family, GH112.

N-Acetylhexosamine 1-kinase

BL1642 was annotated as a mucin desulfatase, but LnpB from *B. longum* subsp. *longum* JCM1217^T did not exhibit any sulfatase activity. Since a BLAST search showed that BL1642 was moderately similar to some kinases, the kinase activity of LnpB was examined using GlcNAc as the phosphate acceptor in the presence of ATP. It produced α-GlcNAc 1-phosphate (GlcNAc1*P*), and not GlcNAc 6-phosphate, indicating that the protein encoded by the *lnpB* gene has a novel kinase activity (Nishimoto and Kitaoka, 2007a). This protein had similar activity on GalNAc, yielding α-GalNAc 1-phosphate (GalNAc1*P*), and weak activities on several monosaccharides. The enzyme was named as *N*-acetylhexosamine 1-kinase (NahK), with a new EC number, EC 2.7.1.162. NahK does not have a signal peptide sequence and is considered to be an intracellular enzyme. Like other kinases, NahK requires magnesium ions for activity. It was found to act by a sequential bi bi (two substrates-two products) mechanism, the reaction occurring after the binding of both ATP and *N*-acetylhexosamine.

Generally, most kinases for hexoses phosphorylate their sugar substrates at the six position. Three enzymes, galactokinase (EC 2.7.1.6), *N*-acetylgalactosamine kinase (EC 2.7.1.157), and fucokinase (EC 2.7.1.52), are known to phosphorylate the anomeric hydroxyl group of their sugar substrates. Therefore, NahK was the first enzyme to be discovered that recognizes gluco-type substrates and phosphorylates the anomeric hydroxyl group.

$GalT_{Bl}$ and $GalE_{Bl}$

BL1643 was annotated as GalT2 (EC 2.7.7.10), the enzyme transferring the UMP unit of UTP to Gal1*P*. Although the protein encoded by the *lnpC* gene from *B. longum* subsp. *longum* JCM1217^T did not have GalT2 activity, it showed UDP-glucose-hexose 1-phosphate uridylyltransferase (GalT, EC 2.7.7.12) activity, by transferring the UMP unit from UDP-Glc to Gal1*P*. Unusually, the enzyme also transferred the UMP unit to GlcNAc1*P* and GalNAc1*P* (Nishimoto and Kitaoka, 2007a). The UMP-transferring activities from UDP-Glc to GlcNAc1*P* (2.24 U/mg protein) and GalNAc1*P* (2.30 U/mg protein) were found to be 2-fold higher than that to Gal1*P* (1.08 U/mg protein). LnpC is a novel type of GalT, having wide substrate specificity, as no other GalT has been found to recognize GlcNAc1*P* or GalNAc1*P*.

LnpD from *B. longum* subsp. *longum* JCM1217^T has GalE activity as annotated. It can also epimerize UDP-GalNAc to UDP-GlcNAc, with an activity of 147 U/mg protein, the same level of activity as that for the conversion of UDP-Gal to UDP-Glc (158 U/mg) (Nishimoto and Kitaoka, 2007a). The catalytic activities of the LnpC and LnpD proteins were named $GalT_{Bl}$ and $GalE_{Bl}$ (GalT and GalE of *B. longum*), respectively. Neither LnpC nor LnpD has a signal peptide sequence.

Comprehensive understanding of the GNB/LNB pathway

The enzyme activities of LnpA, LnpB, LnpC, and LnpD suggest that these genes constitute an operon to metabolize whole molecules of LNB and GNB (Fig. 3.4) (Nishimoto and Kitaoka, 2007a). LNB from HMOs and GNB from mucin are thought to be transported into the cell by a specific GNB/LNB transporter encoded by the genes located upstream of this cluster (the BL1638–BL1640 genes). Subsequently, GNB/LNB is phosphorolysed by LnpA to form Gal1*P* and GlcNAc/GalNAc. Gal1*P* is converted to Glc1*P* by the concerted actions of $GalT_{Bl}$ and $GalE_{Bl}$, and then joins the glycolytic pathway. On the other hand, GalNAc/GlcNAc is phosphorylated by NahK to form GalNAc1*P*/GlcNAc1*P*. GalNAc1*P* is converted into GlcNAc1*P* by the concerted actions of $GalT_{Bl}$ and $GalE_{Bl}$. Finally, GlcNAc1*P* enters the aminosugar metabolic pathway. The operon encodes sufficient enzymes to fully metabolize GNB and LNB molecules to substrates of a subsequent central metabolic pathway. The metabolic pathway encoded by this operon was named the GNB/LNB pathway.

In most organisms, Gal is metabolized to Glc1*P* by the Leloir pathway (Leloir, 1951), which consists of four enzymes: galactose mutarotase (GalM), galactokinase (GalK), GalT, and GalE (Fig. 3.5). Although the GNB/LNB pathway resembles the Leloir pathway, the two are different. In the Leloir pathway, GalK phosphorylates galactose with the consumption of ATP, whereas in the GNB/LNB pathway LnpA produces Gal1*P* without consuming ATP by the phosphorolysis of the galactosides. The utilization of *N*-acetylhexosamine, the other component of GNB/LNB, in the LNB/GNB pathway also differs from that in the Leloir pathway.

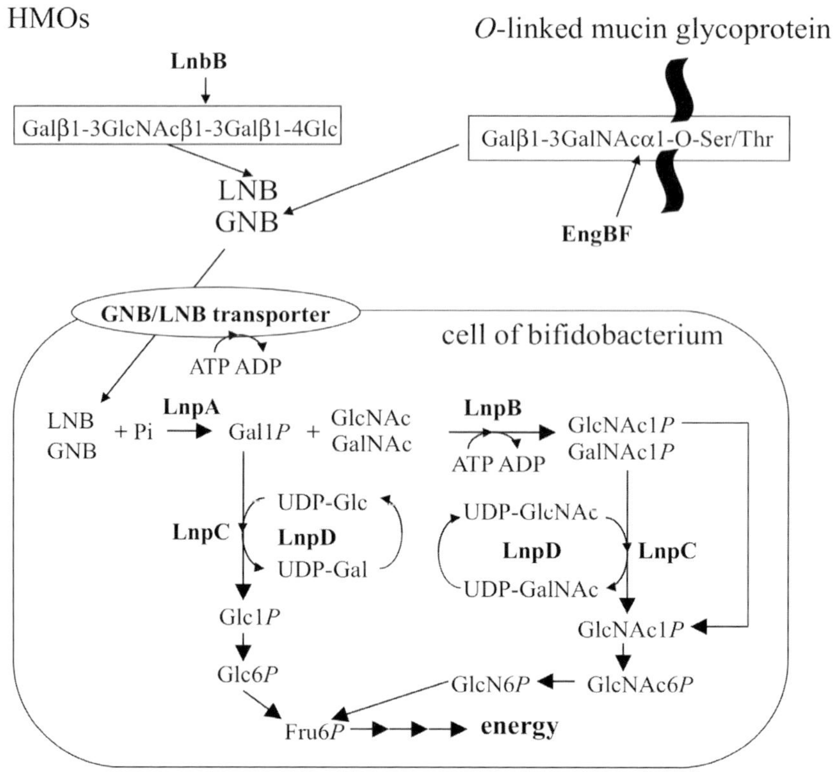

Figure 3.4 Schematic representation of the bifidobacterial intracellular GNB/LNB pathway.

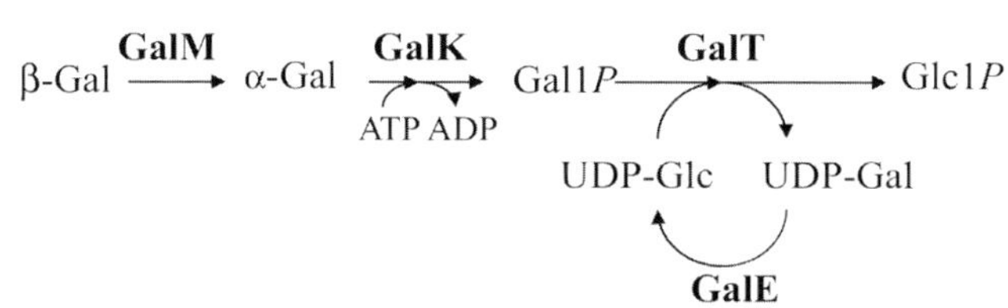

Figure 3.5 The Leloir pathway of galactose metabolism.

No gene cluster encoding the full set of enzymes involved in the Leloir pathway was found in *B. longum* NCC2705. Other than the *lnpABCD* cluster, a single gene cluster containing genes for two of the Leloir enzymes, *galK* (BL1210) and *galT* (BL1211), was found in the genome. In contrast, genes encoding the Leloir enzymes for galactose metabolism, *galM, galK, galT*, and *galE*, are often found in operons in other species. These results suggest that the GNB/LNB pathway is the main metabolic pathway for galactose as an energy source in *B. longum*.

Practical production of LNB

According to the LNB hypothesis, LNB is expected to act as the promoting factor for the growth of bifidobacteria. Although some enzymatic methods of producing LNB by using β-galactosidase have been reported (Murata *et al.*, 1996), they are not suitable to be used on a large scale for several reasons such as the high cost of the substrate and the requirement for chromatography. Thus, a practical method for preparation of LNB was needed. Nishimoto and Kitaoka

(2007c) reported one-pot enzymatic production method for LNB that was suitable for scaling up. The scheme of their method is summarized in Fig. 3.6. In this method, sucrose and GlcNAc are converted into LNB and fructose by the concerted reactions of four enzymes, sucrose phosphorylase (SP, EC 2.4.1.7), GalT, GalE, and GalHexNAcP, in the presence of catalytic amounts of UDP-glucose and inorganic phosphate. Sucrose is phosphorolysed to Glc1*P* and fructose by SP. Glc1*P* is converted to UDP-Glc concomitantly with the conversion of UDP-Gal into Gal1*P* by GalT. UDP-Glc is converted to UDP-Gal by GalE, and the resulting UDP-Gal is consumed in the reaction of GalT. Finally, LNB is synthesized from Gal1*P* and GlcNAc by GalHexNAcP.

LNB was produced at 30°C in 10 l of reaction mixture consisting of 660 mM sucrose, 600 mM GlcNAc, 1 mM UDP-Glc, 30 mM phosphate buffer (pH 7.0), 10 mM $MgCl_2$, 26 mg/l (1716 U/l) SP, 22 mg/l (3080 U/l) GalE, 17 mg/l (3570 U/l) GalT, and 43 mg/l (150 U/l) GalHexNAcP. The LNB concentration reached 500 mM (190 g/l) after 600 hours of reaction. The yield was 83% based on the amount of GlcNAc used. Although this method requires expensive UDP-Glc, the amount of UDP-Glc used was 1/600 of GlcNAc, indicating that UDP-Glc acted as a catalyst with a turnover number of 500. The reaction mixture was treated with baker's yeast to remove fructose and sucrose, and was concentrated to obtain 1.8 kg of crystalline LNB (95% purity). Recrystallization of LNB yielded 1.4 kg of crystalline LNB (99.6% purity).

This method is carried out with unit processes that are ready for scaling up. The reaction mixture containing all the substrates and enzymes with cofactors is kept at a controlled temperature to produce LNB. The process for the isolation of LNB does not require any chromatography, which is often difficult to scale up.

Growth-promoting activity of LNB

The practical production of LNB has made experiments that consume large amounts of LNB feasible. Kiyohara *et al.* (2009) examined the *in vitro* growth-promoting activity of LNB in various intestinal bacteria.

LNB showed growth-promoting activity in several bifidobacteria, including *B. bifidum* JCM1254, *B. breve* JCM1192^T, *B. longum* subsp. *longum* JCM1217^T and JCM7054, *B. longum* subsp. *infanits* JCM1210 and JCM1222^T, and *Bifidobacterium scardovii* JCM12489^T. It was not

A

Suc + Pi	$\rightleftarrows$ Glc1*P* + Fru	(EC 2.4.1.7, SP)
Glc1*P* + UDP-Gal	$\rightleftarrows$ UDP-Glc + Gal1*P*	(EC 2.7.1.12, GalT)
UDP-Glc	$\rightleftarrows$ UDP-Gal	(EC 5.1.3.2, GalE)
GlcNAc + Gal1*P*	$\rightleftarrows$ LNB + Pi	(EC 2.4.1.211, GalHexNAcP)

Overall reaction
Suc + GlcNAc $\rightleftarrows$ LNB + Fru (Catalysts: 4 enzymes, UDP-Glc, Pi)

B

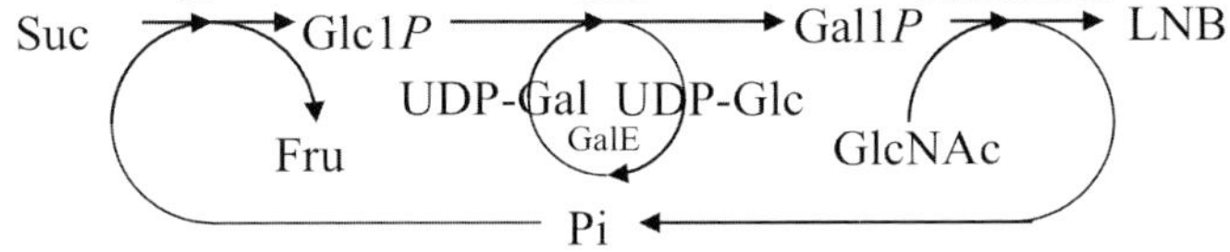

Figure 3.6 One-pot enzymatic production of LNB. (A) Principle of the reaction. Four enzymatic reactions are carried out in one pot, and the compounds appearing on both sides are recycled in the overall reaction. Thus, the overall reaction is to produce LNB and fructose from sucrose and GlcNAc in the presence of four enzymes and catalytic amounts of UDP-Glc and Pi. (B) Schematic description of the reaction.

effective in other strains of bifidobacteria, including *B. adolescentis* JCM1275^T and JCM7046, *Bifidobacterium angulatum* JCM7096^T, *B. animalis* subsp. *lactis* JCM10602^T, *Bifidobacterium catenulatum* JCM1194^T, *Bifidobacterium dentium* JCM1195^T, *Bifidobacterium gallicum* JCM8224^T, *Bifidobacterium pseudocatenulatum* JCM1200^T, and *Bifidobacterium pseudolongum* subsp. *pseudolongum* JCM1205^T. It should be noted that strains capable of growing on LNB are often isolated from infant faeces. LNB promoted the growth of bifidobacterial strains possessing the GNB/LNB pathway (*B. bifidum* JCM1254, *B. longum* subsp. *longum* JCM1217^T, and *B. longum* subsp. *infanits* JCM1222^T) but was not effective in strains lacking this pathway (*B. adolescentis* JCM1275^T and *B. animalis* subsp. *lactis* JCM10602^T). These results suggest that the GNB/LNB pathway might be essential for the metabolism of LNB.

LNB did not show a significant growth-promoting effect in any other bacteria examined such as other intestinal bacteria (including *Bacteroides ovatus* JCM5824^T, *Bacteroides thetaiotaomicron* JCM5827^T, *Clostridium celatum* JCM1394^T, *Clostridium hiranonis* JCM10541^T, *Clostridium hylemonae* JCM10539^T, *Clostridium scidens* JCM6567^T, *Lactobacillus johnsonii* JCM1101^T, *Lactobacillus reuteri* JCM1112^T, *Lactobacillus rhamnosus* JCM1136^T, and *Ruminінococcus productus* JCM1471^T), lactic acid bacteria from sources other than intestine (including *Lactobacillus plantarum* subsp. *plantarum* JCM1149^T, *Lactobacillus brevis* JCM1059^T, *Lactobacillus casei* JCM1134^T, *Lactobacillus delbrueckii* subsp. *delbrueckii* JCM1012^T, *Lactococcus lactis* subsp. *lactis* JCM5805^T), and a commensal of human skin (*P. acnes* JCM6425^T). LNB did not promote the growth of several strains of *C. perfringens* and *P. acnes* that possessed GalHexNAcP. This might be related to the fact that their GalHexNAcPs are classified as GNBP (Nakajima *et al.*, 2008, 2009) and that they do not recognize LNB effectively.

Metabolism of HMOs in *B. longum* subsp. *infantis*

Recently, the genomic sequence of *B. longum* subsp. *infantis* ATCC15697^T, a predominant bifidobacterial species in the infant gut that is known to metabolize HMOs (LoCascio *et al.*, 2007; Ward *et al.*, 2006, 2007), has been published (Sela *et al.*, 2008). The organism seems to possess the intracellular GNB/LNB pathway, including the GNB/LNB transporter, but its metabolic pathway of HMOs could be different from that of *B. bifidum* JCM1254.

The genome of *B. longum* subsp. *infantis* ATCC15697^T contains a 43-kb gene cluster specific for HMO degradation that is not found in other bifidobacterial species. In the cluster, homologous genes of α-1,2-fucosidase, α-1,3/4-fucosidase, and α-2,3/6-sialidase are present in addition to the genes for β-galactosidase and β-*N*-acetlyhexosaminidase, but they are considered to be intracellular proteins because of the lack of signal peptides. Furthermore, no homologous gene for lacto-*N*-biosidase is found. Since LNB must be liberated from HMOs before entering the GNB/LNB pathway, it is not clear from the genome information as to whether the strain metabolizes HMOs through this pathway. It should also be noted that most bifidobacterial β-galactosidases do not hydrolyse the Galβ1→3GlcNAc linkage (Goulas *et al.*, 2009), making it difficult to understand the metabolic pathway of type I oligosaccharides, the major components of HMOs, in this strain. In order to understand how *B. longum* subsp. *infantis* ATCC15697^T metabolizes HMOs, further studies are required to determine whether it possesses a lacto-*N*-biosidase or a β-galactosidase that hydrolyses the Galβ1→3GlcNAc linkage.

Genomic sequences of several bifidobacterial strains are becoming available. Such genomic information together with the characterization of the related enzymes will enhance the understanding of the metabolic pathway of HMOs for each bifidobacterial strain.

Conclusion

The LNB hypothesis opens up the possibility to systematically investigate HMO metabolism in bifidobacteria. *B. bifidum* JCM1254, one of the bifidobacterial species often found in infant faeces, has both the extracellular hydrolytic enzymes required to liberate LNB from complex HMOs and the intracellular GNB/LNB pathway.

On the other hand, the two genome-sequenced strains of *B. longum* subsp. *longum* (NCC2705 and DJO10A) possess the intracellular GNB/LNB pathway but do not have α-fucosidase, α-sialidase, or lacto-*N*-biosidase, indicating that it can metabolize LNB but not HMOs.

LNB is expected to function as a specific prebiotic for certain strains of bifidobacteria that are found in infant faeces. The production of LNB on an industrial scale might allow its use as a functional food ingredient, and hopefully as a supplement to formula milk.

References

Ashida, H., Miyake, A., Kiyohara, M., Wada, J., Yoshida, E., Kumagai, H., Katayama, T., and Yamamoto, K. (2009). Two distinct α-L-fucosidases from *Bifidobacterium bifidum* are essential for the utilization of fucosylated milk oligosaccharides and glycoconjugates. Glycobiology *19*, 1010–1017.

Aminoff, D., and Furukawa, K. (1970). Enzymes that destroy blood group specificity. I. Purification and properties of α-L-fucosidase from *Clostridium perfringens*. J. Biol. Chem. *245*, 1659–1669.

Asakuma, S., Urashima, T., Akahori, M., Obayashi, H., Nakamura, T., Kimura, K., Watanabe, Y., Arai, I., and Sanai, Y. (2008). Variation of major neutral oligosaccharides levels in human colostrum. Eur. J. Clin. Nutr. *62*, 488–494.

Benno, Y., and Mitsuoka, T. (1986). The development of gastrointestinal micro-flora in humans and animals. Bifidobacteria Microfl. *5*, 13–25.

Benno, Y., Sawada, K., and Mitsuoka, T. (1984). The intestinal microflora of infants: composition of fecal flora in breast-fed and bottle-fed infants. Microbiol. Immunol. *28*, 975–986.

Bezkorovainy, A. (1989). Ecology of bifidobacteria. In Biochemistry and Physiology of Bifidobacteria, Bezkorovainy, A., and Miller-Catchpole, R., eds. (CRC Press, Cleveland).

Bruggemann, H., Henne, A., Hoster, F., Liesegang, H., Wiezer, A., Strittmatter, A., Hujer, S., Durre, P., and Gottschalk, G. (2004). The complete genome sequence of *Propionibacterium acnes*, a commensal of human skin. Science *305*, 671–673.

Derensy-Dron, D., Krzewinski, F., Brassart, C., and Bouquelet, S. (1999). β–1,3-Galactosyl-N-acetylhexosamine phosphorylase from *Bifidobacterium bifidum* DSM 20082: characterization, partial purification and relation to mucin degradation. Biotechnol. Appl. Biochem. *29*, 3–10.

Fujita, K., Oura, F., Nagamine, N., Katayama, T., Hiratake, J., Sakata, K., Kumagai, H., and Yamamoto, K. (2005). Identification and molecular cloning of a novel glycoside hydrolase family of core 1 type *O*-glycan-specific endo-α-*N*-acetylgalactosaminidase from *Bifidobacterium longum*. J. Biol. Chem. *280*, 37415–37422.

Goulas, T., Goulas, A, Tzortzis, G., and Gibson, G.R. (2009). Comparative analysis of four β-galactosidases from *Bifidobacterium bifidum* NCIMB41171: purification and biochemical charcterisation. Appl. Microbiol. Biotechnol. *82*, 1079–1088.

Gyorgy, P., Rose, C.S., and Springer, G.F. (1954). Enzymatic inactivation of bifidus factor and blood group substances. J. Lab. Clin. Med. *43*, 543–552.

Haeuw-Fievre, S., Wieruszeski, J.M., Plancke, Y., Michalski, J.C., Montreuil, Jr., and Strecker, G. (1993). Primary structure of human milk octa-, dodeca- and tridecasaccharides determined by a combination of ^{1}H-NMR spectroscopy and fast-atom-bombardment mass spectrometry. Evidence for a new core structure, the para-lacto-*N*-octaose. Eur. J. Biochem. *215*, 361–371.

Hidaka, M., Nishimoto, M., Kitaoka, M., Wakagi, T., Shoun, H., and Fushinobu, S. (2009). The crystal structure of galacto-*N*-biose/lacto-*N*-biose I phosphorylase: a large deformation of a TIM barrel scaffold. J. Biol. Chem. *284*, 7273–7283.

Katayama, T., Sakuma, A., Kimura, T., Makimura, Y., Hiratake, J., Sakata, K., Yamanoi, T., Kumagai, H., and Yamamoto, K. (2004). Molecular cloning and characterization of *Bifidobacterium bifidum* 1,2-α-L-fucosidase (AfcA), a novel inverting glycosidase (glycoside hydrolase family 95). J. Bacteriol. *186*, 4885–4893.

Katayama, T., Fujita, K., and Yamamoto, K. (2005). Novel bifidobacterial glycosidases acting on sugar chains of mucin glycoproteins. J. Biosci. Biotechnol. *99*, 457–465.

Kim, J.F., Jeong, H., Yu, D.S., Choi, S.H., Hur, C.G., Park, M.S., Yoon, S.H., Kim, D.W., Ji, G.E., Park, H.S., and Oh, T.K. (2009). Genome sequence of the probiotic bacterium *Bifidobacterium animalis* subsp. *lactis* AD011. J. Bacteriol. *191*, 678–679.

Kitaoka, M., Tian, J., and Nishimoto, M. (2005). Novel putative galactose operon involving lacto-*N*-biose phosphorylase in *Bifidobacterium longum*. Appl. Environ. Microbiol. *71*, 3158–3162.

Kiyohara, M., Tachizawa, A., Nishimoto, M., Kitaoka, M., Ashida, H., and Yamamoto, K. (2009). Prebiotic effect of lacto-*N*-biose I on bifidobacterial growth. Biosci. Biotechnol. Biochem. *73*, 1175–1179.

Kiyohara, M., Tachizawa, A., Chaiwangsri, T., and Yamamoto, T. (2011). An exo-x-sialidase from bifidobacteria involved in the degradation of sialyloligosaccharides in human milk and intestinal glycoconjugates. Glycobiology (in press). doi: 10.1093/glycob/cwq175.

Kunz, C., Rudloff, S., Baier, W., Klein, N., and Strobel, S. (2000). Oligosaccharides in human milk: structural, functional, and metabolic aspects. In Annual Review of Nutrition, McCormick, D.B., Bier, D.M., and Cousins, R.J., eds. (Annual Reviews, Palo Alto), pp. 699–722.

Kurokawa, K., Itoh, T., Kuwahara, T., Oshima, K., Toh, H., Toyoda, A., Takami, H., Morita, H., Sharma, V.K., Srivastava, T.P., *et al.* (2007). Comparative metagenomics revealed commonly enriched gene sets in human gut microbiomes. DNA Res. *14*, 169–181.

Leloir, L.F. (1951). The enzymatic transformation of uridine diphosphate glucose into a galactose derivative. Arch. Biochem. *33*, 186–190.

LoCascio, R.G., Ninonuevo, M.R., Freeman, S.L., Sela, D.A., Grimm, R., Lebrilla, C.B., Mills, D.A., and German, J.B. (2007). Glycoprofiling of bifidobacterial consumption of human milk oligosaccharides demonstrates strain specific, preferential consumption of small chain glycans secreted in early human lactation. J. Agric. Food. Chem. *55*, 8914–8919.

Mark, B.L., Mahuran, D.J., Cherney, M.M., Zhao, D., Knapp, S., and James, M.N. (2003). Crystal structure of human beta-hexosaminidase B: understanding the molecular basis of Sandhoff and Tay-Sachs disease. J. Mol. Biol. *327*, 1093–1109.

Murata, T., Itoh, T., Hayakawa, Y., and Usui, T. (1996). Convenient synthesis of β-(1→3)-galactosyl disaccharide α-glycoside and its analogs as mimic units of mucin-type carbohydrate. J. Biochem. *120*, 851–855.

Nagae, M., Tsuchiya, A., Katayama, T., Yamamoto, K., Wakatsuki, S., and Kato, R. (2007). Structural basis of the catalytic reaction mechanism of novel 1,2-α-L-fucosidase from *Bifidobacterium bifidum*. J. Biol. Chem. *282*, 18497–18509.

Nakajima, M., and Kitaoka, M. (2008). Identification of lacto-*N*-biose I phosphorylase from *Vibrio vulnificus* CMCP6. Appl. Environ. Microbiol. *74*, 6333–6337.

Nakajima, M., Nihira, T., Nishimoto, M., and Kitaoka, M. (2008). Identification of galacto-*N*-biose phosphorylase from *Clostridium perfringens* ATCC13124. Appl. Microbiol. Biotechnol. *78*, 465–471.

Nakajima, M., Nishimoto, M., and Kitaoka, M. (2009). Characterization of β-1,3-galactosyl-*N*-acetylhexosamine phosphorylase from *Propionibacterium acnes*. Appl. Microbiol. Biotechnol. *83*, 109–115.

Newburg, D.S., and Neubauer, S.H. (1995). Carbohydrate in milks: analysis, quantities, and significance. In Handbook of Milk Composition, Jensen, R.G., ed. (Academic Press, San Diego), pp. 273–349.

Nishimoto, M., and Kitaoka, M. (2007a). Identification of *N*-acetylhexosamine 1-kinase in the complete lacto-*N*-biose I/galacto-*N*-biose metabolic pathway in *Bifidobacterium longum*. Appl. Environ. Microbiol. *73*, 6444–6449.

Nishimoto, M., and Kitaoka, M. (2007b). Identification of the putative proton donor residue of lacto-*N*-biose phosphorylase (EC 2.4.1.211). Biosci. Biotechnol. Biochem. *71*, 1587–1591.

Nishimoto, M., and Kitaoka, M. (2007c). Practical preparation of lacto-*N*-biose I, a candidate for the bifidus factor in human milk. Biosci. Biotechnol. Biochem. *71*, 2101–2104.

Rotimi, V.O., and Duerden, B.I. (1981). The development of the bacterial flora in normal neonates. J. Med. Microbiol. *14*, 51–58.

Sano, M., Hayakawa, K., and Kato, I. (1992). An enzyme releasing lacto-*N*-biose from oligosaccharides. Proc. Natl. Acad. Sci. U.S.A. *89*, 8512–8516.

Sano, M., Hayakawa, K., and Kato, I. (1993). Purification and characterization of an enzyme releasing lacto-*N*-biose from oligosaccharides with type 1 chain. J. Biol. Chem. *268*, 18560–18566.

Schell, M.A., Karmirantzou, M., Snel, B., Vilanova, D., Berger, B., Pessi, G., Zwahlen, M.C., Desiere, F., Bork, P., Delley, M., *et al.* (2002). The genome sequence of *Bifidobacterium longum* reflects its adaptation to the human gastrointestinal tract. Proc. Natl. Acad. Sci. U.S.A. *99*, 14422–14427.

Sela, D.A., Chapman, J., Adeuya, A., Kim, J.H., Chen, F., Whitehead, T.R., Lapidus, A., Rokhsar, D.S., Lebrilla, C.B., German, J.B., *et al.* (2008). The genome sequence of *Bifidobacterium longum* subsp. *infantis* reveals adaptations for milk utilization within the infant microbiome. Proc. Natl. Acad. Sci. U.S.A. *105*, 18964–18969.

Shimizu, T., Ohtani, K., Hirakawa, H., Ohshima, K., Yamashita, A., Shiba, T., Ogasawara, N., Hattori, M., Kuhara, S., and Hayashi, H. (2002). Complete genome sequence of *Clostridium perfringens*, an anaerobic flesh-eater. Proc. Natl. Acad. Sci. U.S.A. *99*, 996–1001.

Suzuki, R., Wada, J., Katayama, T., Fushinobu, S., Wakagi, T., Shoun, H., Sugimoto, H., Tanaka, A., Kumagai, H., Ashida, H., *et al.* (2008). Structural and thermodynamic analyses of solute-binding protein from *Bifidobacterium longum* specific for core 1 disaccharide and lacto-*N*-biose I. J. Biol. Chem. *283*, 13165–13173.

Urashima, T., Odaka, G., Asakuma, S., Uemura, Y., Goto, K., Senda, A., Saito, T., Fukuda, K., Messer, M., and Oftedal, O. T. (2009). Chemical characterization of oligosaccharides in chimpanzee, bonobo, gorilla, orangutan, and siamang milk or colostrum. Glycobiology *19*, 499–508.

Veerkamp, J.H. (1969). Uptake and metabolism of determinatives of 2-deoxy-2-amino-D-glucose in *Bifidobacterium bifidum* var. *pennsylvanicus*. Arch. Biochem. Biophys. *129*, 248–256.

Wada, J., Suzuki, R., Fushinobu, S., Kitaoka, M., Wakagi, T., Shoun, H., Ashida, H., Kumagai, H., Katayama, T., and Yamamoto, K. (2007). Purification, crystallization and preliminary X-ray analysis of the galacto-*N*-biose-/lacto-*N*-biose I-binding protein (GL-BP) of the ABC transporter from *Bifidobacterium longum* JCM1217. Acta Crystallogr. Sect. F Struct.Biol. Cryst. Commun. *63*, 751–753.

Wada, J., Ando, T., Kiyohara, M., Ashida, H., Kitaoka, M., Yamaguchi, M., Kumagai, H., Katayama, T., and Yamamoto, K. (2008a). *Bifidobacterium bifidum* lacto-*N*-biosidase, a critical enzyme for the degradation of human milk oligosaccharides with a type 1 structure. Appl. Environ. Microbiol. *74*, 3996–4004.

Wada, J., Honda, Y., Nagae, M., Kato, R., Wakatsuki, S., Katayama, T., Taniguchi, H., Kumagai, H., Kitaoka, M., and Yamamoto, K. (2008b). 1,2-α-L-Fucosynthase: a glycosynthase derived from an inverting α-glycosidase with an unusual reaction mechanism. FEBS Lett. *582*, 3739–3743.

Ward, R.E., Ninonuevo, M., Mills, D.A., Lebrilla, C.B., and German, J.B. (2006). *In vitro* fermentation of breast milk oligosaccharides by *Bifidobacterium infantis* and *Lactobacillus gasseri*. Appl. Environ. Microbiol. *72*, 4497–4499.

Ward, R.E., Ninonuevo, M., Mills, D.A., Lebrilla, C.B., and German, J.B. (2007). *In vitro* fermentability of human milk oligosaccharides by several strains of bifidobacteria. Mol. Nutr. Food. Res. *51*, 1398–1405.

Williams, S.J., Mark, B.L., Vocadlo, D.J., James, M.N., and Withers, S.G. (2002). Aspartate 313 in the *Streptomyces plicatus* hexosaminidase plays a critical role in substrate-assisted catalysis by orienting the 2-acetamido group and stabilizing the transition state. J. Biol. Chem. *277*, 40055–40065.

Energy Generation Coupled with Decarboxylation Reactions in Lactic Acid Bacteria

4

Kei Nanatani and Keietsu Abe*

Abstract

In bacteria, many biological reactions are sustained by metabolic energy present in the form of phosphoester bonds, in compounds such as adenosine triphosphate (ATP) and phosphoenolpyruvate (PEP), or in the form of ion gradients, such as the proton motive force (PMF) and the sodium motive force. The two forms of metabolic energy can be interconverted by FoF1-ATPases, which catalyse the translocation of H^+ (or Na^+) concomitant with either the hydrolysis or synthesis of ATP. Nutrient transport in bacteria is usually thought to consume metabolic energy; however, over the last two decades, a new class of nutrient transport reactions has been identified, in which substrate transport generates rather than consumes energy. The reaction consists of two steps: (1) electrogenic exchange of a precursor (amino acid or other organic acid) with its intracellular metabolic product produced by decarboxylation, and (2) intracellular decarboxylation of the transported precursor. The precursor–product exchange causes a net charge movement, which generates a membrane potential of physiological polarity, and the intracellular decarboxylation consumes cytoplasmic protons to generate both a pH gradient of physiological polarity and an outward concentration gradient of the end-product, which drives precursor uptake. The combined activities constitute a metabolically driven proton pump (proton-motive metabolic cycle), which provides sufficient energy to generate ATP in a process called 'decarboxylative phosphorylation'. Thus, the proton-motive metabolic cycle can be recognized as a new class of ATP generation system that is distinct from substrate-level phosphorylation, oxidative phosphorylation, and photo-phosphorylation. We consider that the proton-motive metabolic cycle could be made available as an artificial energy supply system in various industrial fermentation organisms with the use of recombinant DNA technology in the future.

Introduction

Microorganisms are used in industry to produce primary metabolites, including organic acids, amino acids, and enzymes, as well as secondary metabolites such as antibiotics, in a process called 'fermentation'. To maximize the productivity of each target in fermentation, the metabolic pathways in the microorganism are usually engineered, which leads to problems in controlling the energy-supply systems and a decreased level of cellular ATP. Production of cellular ATP can become a limiting factor when attempts are made to improve the productivity of target compounds. Thus, modern industrial fermentation requires a novel energy supply system, which is not negatively affected by other metabolic pathways and does not perturb the metabolic pathways that control the biosynthesis of the targets. Ideally, the new energy supply systems should be capable of being easily introduced into industrial strains by DNA recombination techniques. In this chapter, we discuss bacterial energy generation systems coupled with decarboxylation of amino acids or organic acids, as examples of energy supply systems suitable for use in fermentation.

*corresponding author

Decarboxylation reactions, with amino acid or organic acid precursors, have been well known in bacteria for a long time; examples are listed in Table 4.1 (Maloney *et al.*, 1994; Lolkema *et al.*, 1996). Some Gram-positive facultative anaerobes including lactic acid bacteria are capable of this type of reaction.

Decarboxylation of amino acids and organic acids in bacteria impacts on humanity with unfavourable or beneficial consequences depending on the circumstances. Carbon dioxide gas produced by bacterial decarboxylation reactions in packaged foods causes spoilage of the foods during distribution. During food manufacture, microbial production of amines by decarboxylation of amino acids has to be controlled to maintain food safety. In contrast, L-lactic acid production by decarboxylation of malic acid (malo-lactic fermentation) is used to improve the quality of wine (Salema *et al.*, 1994), and in soy sauce fermentation, L-aspartate, which has a strong sour taste, produces L-alanine and CO_2 by decarboxylation, resulting in a sauce with a milder, more acceptable taste (Uchida, 1988).

Research into the physiological meaning and mechanism of decarboxylation reactions began in the 1990s (Maloney *et al*,. 1994; Lolkema *et al.*, 1996). Understanding the molecular mechanism is clearly important from the industrial standpoint. Here, we overview biological systems involved in energy generation and ATP synthesis, and discuss the physiological meaning of energy generation coupled with decarboxylation reactions.

Table 4.1 Proton-motive metabolic cycles

Precursor/product[1]	Organism
Oxalate/formate	*Oxalobacter formigenes*
Malate/lactate	*Oenococcus oeni*
	Lactococcus lactis
	Lactobacillus plantarum
Lysine/cadaverine	*Escherichia coli*
Glutamate/GABA	*Lactobacillus* sp.
	Escherichia coli
	Listeria monocytogenes
Histidine/histamine	*Lactobacillus buchneri*
Phenylalanine/phenylethyl amine	*Lactobacillus buchneri*
Tyrosine/tyramine	*Lactobacillus buchneri*
Aspartate/alanine	*Lactobacillus* sp.
	Tetragenococcus halophilus
Citrate/lactate	*Oenococcus oeni*
	Leuconostoc mesenteroides
	Lactococcus lactis
Arginine/agmatine	*Escherichia coli*

[1]Conversion of precursor into product requires carrier-mediated entry of precursor into the cell, followed by an internal decarboxylation and export of product by way of the same or a different carrier. The net exchange of precursor and product brings in a single negative charge; a single cytosolic proton is consumed during decarboxylation as CO_2 is generated. The organic product need not be excreted, but may be used biosynthetically; in which case the requirements of a proton pump may be satisfied by other restraints. For example, in the case of divalent anion entry (oxalate^{2-}), a proton pump is preserved if precursor entry is accompanied by hydroxyl export (or proton entry) via by the same or different carrier (see Anantharam *et al.*, 1989; Ruan *et al.*, 1992). (Modified from Maloney, 1994.)

Bioenergetics

Types of energy in biological organisms

The types of energy in biological organisms are generally classified into four forms (Harold, 1986).

1 'Chemical energy' is stored in high energy phosphorylated compounds such as ATP, phosphoenolpyruvate, and acetyl-phosphate.
2 'Chemiosmotic energy' consists of electrochemical gradients of solutes such as protons (H^+) and sodium ions (Na^+) across biomembranes. The electrochemical gradient of protons across a membrane is called a proton motive force (PMF). A PMF is generated by respiratory chains, which export protons in combination with oxidation of respiratory substrates, and by FoF1-ATPase (F-ATPase), which excludes protons by coupling with hydrolysis of ATP.
3 'Mechanical energy' in living cells is represented by the rotation of flagella coupled with an ion gradient across membranes, and by muscle contraction.
4 'Photo energy' is converted to a PMF in chloroplasts, and the PMF is then used to synthesize ATP by chloroplast ATP synthase (cF-ATPase).

The four forms of biological energy are interconverted at biomembranes. F-type ATPases, which are found in the mitochondria of eukaryotes and the cytoplasmic membranes of bacteria, catalyse ATP synthesis (chemical energy) coupled with a PMF (chemiosmotic energy) concomitant with the rotation of F1 against Fo (mechanical energy). Moreover, F-ATPase catalyses the reverse reaction to generates a PMF by ATP hydrolysis.

Mechanisms of ATP generation

ATP is often described as representing a form of biological energy, and the mechanisms of ATP generation can be classified into four categories including substrate-level phosphorylation, oxidative phosphorylation, photo-phosphorylation, and decarboxylative phosphorylation, based on energy transduction (Harold, 1986; Maloney, 1994).

Substrate-level phosphorylation

Intermediates of glycolysis such as phosphoenolpyruvate and 1,3-bisphosphoglycerate (chemical energy) directly transfer their high-energy phosphate to ADP through an enzymatic reaction, which results in the synthesis of ATP (chemical energy).

Oxidative phosphorylation

Respiratory chains in mitochondrial inner membranes, or bacterial cytoplasmic membranes, result in oxidation of respiratory substrates (chemical energy), which drives proton pumps and generates a PMF (chemiosmotic energy). The PMF is then used to synthesize ATP (chemical energy) by F-type ATPases in the membranes.

Photo-phosphorylation

Photo-energy initiates a redox reaction or reduction-oxidation reaction (chemical energy) in electron-transfer systems in chloroplasts, and the redox reaction generates a PMF (chemiosmotic energy) that can be converted to ATP (chemical energy) by cF-ATPase.

Decarboxylative phosphorylation

More recently, it has been discovered that a PMF can also be generated in bacteria by decarboxylation reactions with amino acids or organic acids as the precursor. The precursors are transported across the cell membrane and intracellularly decarboxylated (chemical energy). A cytoplasmic proton is chemically connected to the decarboxylated position in the product, and the subsequent export of the products from the cell generates an internally alkaline proton gradient across the membrane (chemiosmotic energy). The exchange of the precursor $R\text{-}(COO^-)_n$ with the product $R\text{-}(COO^-)_{n-1}$ across the membranes also generates an internally negative membrane potential (chemiosmotic energy). Overall, the combination of the precursor–product exchange reaction with intracellular decarboxylation generates a PMF (chemiosmotic energy). The resultant PMF can be used to synthesize ATP (chemical energy) by F-ATPase.

Decarboxylative phosphorylation

Discovery of decarboxylative phosphorylation reactions

Nutrient transport by bacteria is usually thought to consume metabolic energy because this step is typically driven by an ion-motive gradient (e.g. a proton or sodium ion gradient) or by hydrolysis of a phosphoester bond (e.g. ATP or PEP) (Maloney, 1987; Postma, 1987). However, a new class of nutrient transport reactions has been identified in which substrate transport is used to generate energy. The first discovered example of such a reaction occurs in *Oxalobacter formigenes*, a Gram-negative, obligate anaerobe that exploits the decarboxylation of the carboxylic acid oxalate to support transmembrane ion-motive gradients (Anantharam *et al.*, 1989; Ruan *et al.*, 1992). This bacterium mediates the exchange of divalent oxalate (oxalate^{2-}) with the product of its intracellular decarboxylation, monovalent formate (formate^{1-}) (Baetz, 1989), using a membrane transporter named OxlT (or oxalate transporter) (Ruan *et al.*, 1992). The one-for-one exchange of oxalate^{2-} and formate^{1-} polarizes the membrane (electrically negative inside), while the decarboxylation reaction consumes a single cytoplasmic proton and hence generates an internal alkalinity. As a result, the metabolic sequence of oxalate^{2-} entry, oxalate^{2-} decarboxylation, and formate^{1-} exit, acts as a proton pump (Anantharam *et al.*, 1989) or proton-motive metabolic cycle (Maloney, 1994). Similarly, some strains of the Gram-positive facultative anaerobe lactobacilli catalyse the decarboxylation of either L-aspartate^{1-} or L-glutamate^{1-} with a near-stoichiometric release of the products, L-alanine^{0} or γ-aminobutyrate^{0}, respectively, plus the release of CO_2 (Abe *et al.*, 1996; Higuchi *et al.*, 1997). These decarboxylation reactions support ATP synthesis in a manner consistent with the idea that processing of these anions involves a proton-motive metabolic cycle. As their central element, proton-motive metabolic cycles have a vectorial component(s) that mediates the electrogenic exchange of precursor and product. The membranes of *Lactobacillis* sp. M3 display an electrogenic L-aspartate^{1-}:L-alanine^{0} exchange of the sort required by a proton-motive metabolic cycle (Abe *et al.*, 1996). This precursor-product antiport is catalysed by a single protein, termed AspT (or aspartate transporter), which catalyseds both the homologous self-exchange of L-aspartate^{1-} and the heterologous antiport of L-aspartate^{1-} and L-alanine^{0}. Additionally, in other bacteria, the transport (vectorial) and decarboxylation (scalar) reactions associated with the conversion of L-malate^{1-} to L-lactate^{0} (Kunkee, 1991; Olsen *et al.*, 1991; Poolman *et al.*, 1991; Salema *et al.*, 1994), L-glutamate^{1-} to γ-aminobutyrate^{0} (Smith *et al.*, 1992; Higuchi *et al.*, 1997; Sanders *et al.*, 1998), or L-histidine to histamine (Molenaar *et al.*, 1993) have been shown to act as proton-motive metabolic cycles. These discoveries suggest a new way of interpreting the relationship between anion (except for L-histidine) transport and decarboxylation reactions in microorganisms.

ATP generation coupled with decarboxylation

Lactobacillus sp. M3 was isolated as a contaminant from soy sauce with low salt content (<8% wt/vol), in which CO_2 gas was formed by the decarboxylation of L-aspartate. The growth rate of *Lactobacillus* sp. M3 approximately doubles in the presence of L-aspartate in MRS-glucose medium (Fig. 4.1). *Lactobacillus* sp. M3 catalyses decarboxylation of L-aspartate with nearly stoichiometric release of L-alanine and CO_2; the β-carboxyl group in the side chain of L-aspartate is eliminated by the decarboxylation reaction. These observations lead to the intriguing question as to why the growth of *Lactobacillus* sp. M3 is promoted by such a simple decarboxylation reaction. The dependence of the increased growth rate on the decarboxylation of L-aspartate implies that energy generation is coupled with the decarboxylation reaction. Experiments in intact *Lactobacillus* sp. M3 cells have demonstrated that the intracellular levels of ATP increase following decarboxylation of L-aspartate to form L-alanine (Table 4.2) (Abe *et al.*, 1996). As described above, the mechanisms of ATP synthesis are categorized into two major groups: substrate-level phosphorylation that produces ATP directly by metabolic processes, and ATP synthesis by PMF-driven F-type ATPases. To determine whether the ATP

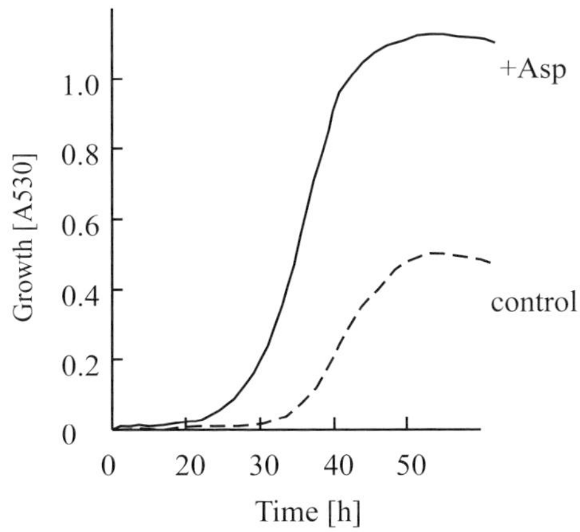

Figure 4.1 Effect of L-aspartate on the growth curve of the aspartate decarboxylative lactic acid bacterium *Lactobacillus* sp. M3. Cells were grown by static culture in MRS broth (5% NaCl) at 30°C with or without 30 mM L-aspartate.

Table 4.2 Aspartate-dependent ATP production

Additions	Cellular ATP (nmol/mg dry weight)
None	0.3
Aspartate (10 mM)	25.3
Aspartate + DCCD (1 mM)	8.7
Aspartate + nigricine (2 μM)	
+ Valinomycin (2 μM)	0.3
Aspartate + CCCP (2 μM)	0.3

Cells were washed in 100 mM 2-(*N*-morpholino)ethanesulfonic acid (MES)/KOH (pH 6), 0.67 M KCl (40 mg dry weight/ml). Then the washed cells were preincubated for 10 minutes at 37°C with DCCD (an inhibitor of F-ATPase), CCCP (a protonophore that collapses PMFs), nigericin plus valinomycin, or carrier ethanol, before 10-fold dilution into the same buffer containing 10 mM potassium aspartate. After 6 minutes, ATP levels were determined in boiled cell extracts. Reprinted from Abe *et al.* (1996).

synthesis is directly coupled with the decarboxylation of L-aspartate, studies have been carried out in the presence of a protonophore CCCP, which collapses the PMF across membranes, and an F-ATPase inhibitor DCCD (Abe *et al.*, 1996). These studies demonstrate that both CCCP and DCCD inhibit production of ATP, but not decarboxylation of L-aspartate, in intact cells of *Lactobacillus* sp. M3, suggesting that conversion of L-aspartate to L-alanine generates a PMF that subsequently drives F-ATPase for ATP synthesis. Since decarboxylation of L-aspartate, but not ATP production, was observed in cell-free extracts of *Lactobacillus* sp. M3, the ATP synthesis requires membranes. Furthermore, a high extracellular concentration of L-alanine (>300 mM) inhibits both L-alanine production from L-aspartate and ATP generation coupled with decarboxylation, suggesting that such high levels of extracellular L-alanine inhibit generation of the PMF that is required for ATP synthesis by F-ATPase (Abe *et al.*, 1996).

Proton-motive metabolic cycle

Aspartate:alanine exchange reaction

Proteoliposomes are an ideal model system for studying membrane transport processes because the composition of proteins in the reconstituted membrane can be dictated and the salt composition both within the aqueous cavity and outside the proteoliposome can be controlled. Thus, use of proteoliposomes allows the artificial conditions required for analysis of transport processes to be established. The making of proteoliposomes refers to the reconstitution of membrane proteins into liposomes and is achieved by the following protocol. Membrane proteins are solubilized from cell membranes with mild detergents, and excessive amounts of phospholipids (prepared from *Escherichia coli* or soybeans) are added to the solubilized proteins in the presence of detergent. The detergent is then removed either by dilution with a large amount of buffer, or by gel filtration or ion-exchange chromatography, thereby inducing the formation of proteoliposomes composed of a lipid bilayer. Membrane proteins that have hydrophobic segments are usually automatically inserted into the lipid bilayer. To study the function of transporters, the components of the aqueous cavity of the proteoliposome can be controlled by adding materials such as substrates to the buffers used for proteoliposome formation. The proteoliposomes are then collected by ultracentrifugation (>100,000 g), washed, and resuspended in buffer. It is possible to control the buffer components outside the proteoliposome by using several kinds of buffers for wash and resuspension. To analyse the activities of transporters, radioactive or fluorescent substrates are used to detect movements of substrates across the lipid bilayer.

To determine whether a transporter catalyses the import or export reactions of a particular solute across the membrane, a high concentration of a non-labelled substrate (10–200 mM) is loaded into proteoliposomes and a low concentration of the radioactive substrate is added outside the proteoliposomes. When a transporter is present and active, the radioactive solute is transported by a self-exchange reaction via the transporter, in a process called 'counter flow'. After the short-term incorporation of radioactive substrates into the proteoliposomes, the import reaction reaches a steady-state phase. At this phase, addition of a non-radioactive solute (10–50 mM) outside the proteoliposomes induces the export of the imported radioactive solute from the proteoliposome, in a process called an 'exchange reaction'. Thus, the detection of counter flow and exchange reactions can be used to demonstrate the presence of a transporter. Fig. 4.2 shows the L-aspartate transport reaction in proteoliposomes reconstituted with the membrane proteins of *Lactobacillus* sp. M3 (Abe *et al.*, 1996). These proteoliposomes import radioactive L-aspartate until a steady state is reached. In contrast, liposomes, which lack membrane proteins, do not show an influx of radioactive L-aspartate.

The addition of a high concentration of nonradioactive L-alanine to the buffers outside

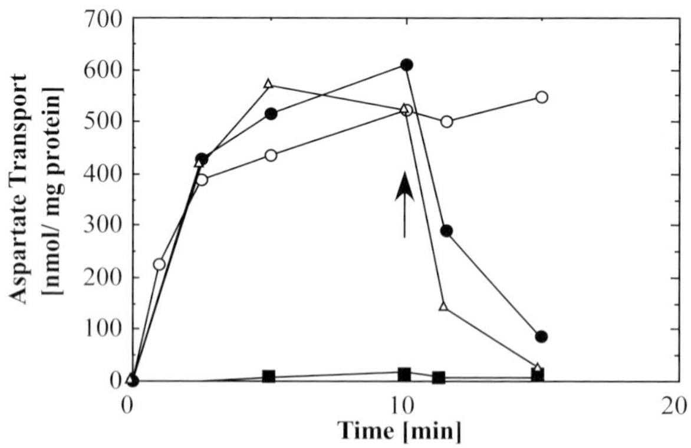

Figure 4.2 Reconstitution of L-aspartate self-exchange and L-aspartate:L-alanine exchange. Proteoliposomes and liposomes were loaded with 100 mM potassium L-aspartate plus 100 mM potassium phosphate (pH 7), and washed and resuspended as described in Abe *et al.* 1996. To start the experiment, proteoliposomes (○, ● and △) were placed in 100 mM K_2SO_4 plus 100 mM potassium phosphate (pH 7) at 3.3 µg of protein/ml, and 100 μM [^{3}H]L-aspartate was added to the external media. An equivalent volume of liposomes (■) was treated in the same way. To estimate substrate transport, aliquots were taken at the times indicated for filtration and washing. The arrow denotes the addition of 15 mM unlabelled L-aspartate (△) or 30 mM unlabelled L-alanine (●). The addition of neither amino acid is denoted by (○). With permission from Abe *et al.* (1996).

proteoliposomes (Fig. 4.2) induces the rapid efflux of imported radioactive L-aspartate from the proteoliposomes suggesting that an L-aspartate:L-alanine exchange reaction occurs in this model system (Abe *et al.*, 1996). This result raises the possibility that the carrier protein AspT, which is present in the membranes of *Lactobacillus* sp. M3 from which the proteoliposomes are derived, carries out the heterologous exchange of L-aspartate and L-alanine. Because conversion of L-aspartate to L-alanine is very nearly stoichiometric in *Lactobacillus* sp. M3, the L-aspartate:L-alanine exchange ratio is expected to be a one-for-one antiport. Furthermore, the reaction is predicted to be electrogenic because at near physiological pH (pH 7) the net charge on L-aspartate is −1 [$pK_1(COOH) = 2.09$, $pK_2(NH_3) = 9.82$, $pK_3(COOH) = 3.86$], whereas the net charge on L-alanine is 0 [$pK_1(COOH) = 2.34$, $pK_2(NH_3) = 9.69$]. These predictions were tested in a study of L-alanine-loaded proteoliposomes using valinomycin that is a potassium-selective ionophore and high concentrations of either external or internal potassium to generate a membrane potential that would be expected to accelerate or retard such an electrogenic reaction (Abe *et al.*, 1996). As shown in the schematic model (Fig. 4.3), if the L-aspartate^{1-}:L-alanine^0 exchange is electrogenic, the artificially generated inside negative membrane potential strongly inhibits the L-aspartate^{1-}:L-alanine^0 exchange reaction and the artificially generated inside positive membrane potential promotes the same exchange reaction. The electrogenic property of L-aspartate^{1-}:L-alanine^0 exchange reaction was examined as follows.

In the experiments, [^{3}H]L-aspartate was added to the external media, and the movement of [^{3}H]L-aspartate to the inside of the proteoliposomes was assayed under various conditions (Fig. 4.4). Proteoliposomes that were prepared and assayed with N-methyl-glucamine (NMG) instead of potassium as both the internal and external cation took up [^{3}H]L-aspartate from the external media, and the level of uptake was largely unaffected by valinomycin. Proteoliposomes that were prepared with potassium as the internal cation, and assayed with NMG as the external cation, established an

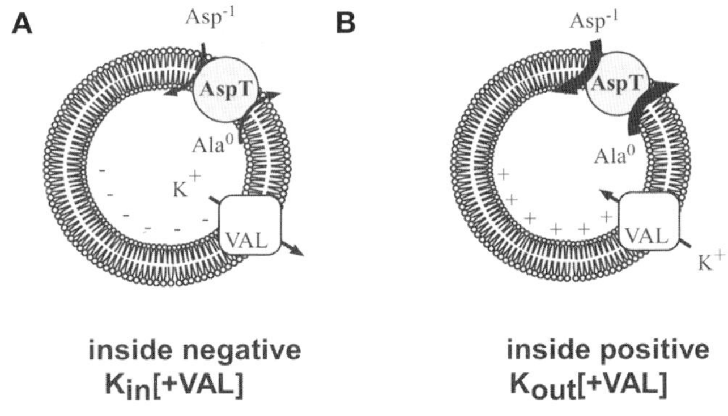

Figure 4.3 A schematic model showing the electrogenic nature of the L-aspartate^{1-}:L-alanine^0 exchange reaction. The process is demonstrated in AspT-reconstituted proteoliposomes, which are electrically polarized by the movement of potassium ions in the presence of valinomycin. AspT, L-aspartate:L-alanine antiporter; Val, valinomycin; K^+, potassium ion; Asp^{1-}, L-aspartate with net charge −1 at pH 7; Ala^0, L-alanine with net charge 0 at pH 7. The width of the curved arrows reflects the level of L-aspartate:L-alanine exchange. When AspT-reconstituted proteoliposomes are loaded with K^+-based buffer containing L-alanine^0 and are subsequently diluted into K^+-free buffer containing L-aspartate^{1-} in the presence of valinomycin, the outward K^+-diffusion potential generates a negative membrane potential that inhibits inward movement of the negatively charged L-aspartate^{1-} coupled with outward movement of L-alanine^0 (A). When AspT-reconstituted proteoliposomes are loaded with K^+-free buffer containing L-alanine^0 and are subsequently diluted into K^+-based buffer containing L-aspartate^{1-} in the presence of valinomycin, the inward K^+-diffusion potential generates a positive membrane potential that accelerates inward movement of negatively charged L-aspartate^{1-} (electrophoretic) coupled with outward movement of L-alanine^0 (B).

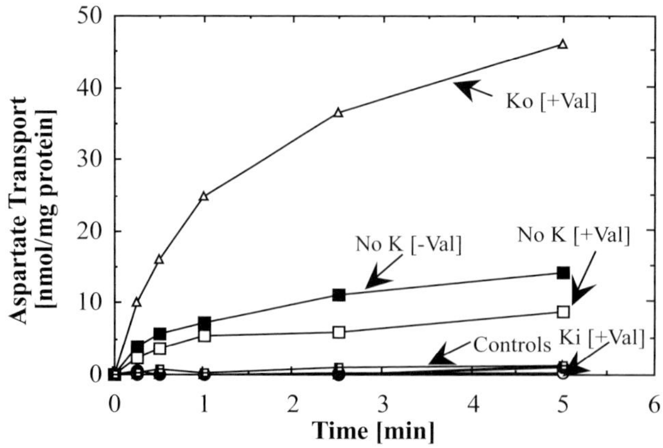

Figure 4.4 The electrogenic nature of L-aspartate^{1-}:L-alanine^0 exchange. Proteoliposomes or liposomes were loaded with 100 mM L-alanine^0 plus 100 mM phosphate (pH 7) as either potassium (○) or NMG (△, □, and ■) salts. Proteoliposomes (or liposomes), were suspended at 1.9 mg of protein/ml in their respective loading buffers. To start the experiment, the proteoliposomes (or liposomes) were diluted 133-fold into assay media containing 100 μM [^{3}H]L-aspartate^{1-}, 100 mM SO_4, plus 100 mM phosphate as potassium (△) or NMG (○, □, ■) salts. With one exception (■), 1 μM valinomycin (Val) was present in the media. Control liposomes and proteoliposomes were assayed in potassium-based medium in the presence of valinomycin. The controls included liposomes loaded with either potassium or NMG salts of L-alanine, and proteoliposomes loaded with 75 mM NMG sulfate rather than 100 mM NMG L-alanine. Because these controls showed no significant L-aspartate transport, and for reasons of clarity, they are not individually represented in the figure. The level of L-aspartate transport was determined for the various combinations of potassium or NMG L-alanine-loaded proteoliposomes using potassium- or NMG-based assay media, with and without valinomycin. 'K_i' and 'K_o' denote the location of the potassium, inside or outside the proteoliposome, respectively. 'No K' denotes the absence of potassium. With permission from Abe *et al.* (1996).

internal negative membrane potential in the presence of valinomycin, and a profound inhibition of [^{3}H]L-aspartate transport occurred. In the reciprocal trial, with NMG as the internal cation and potassium as the external cation, an internal positive membrane potential was established, and [^{3}H]L-aspartate movement was stimulated. From these results, the electrogenic nature of L-aspartate^{1-}:L-alanine^0 exchange was confirmed (Fig. 4.3).

Mechanism of proton-motive metabolic cycle

The L-aspartate:L-alanine exchange reaction can be used to illustrate the mechanism of a proton-motive metabolic cycle (Fig. 4.5). The electrogenic L-aspartate^{1-}:L-alanine^0 exchange reaction catalysed by AspT generates an internally negative membrane potential in *Lactobacillus* sp. M3 cells. In addition, the intracellular decarboxylation reaction of L-aspartate to L-alanine consumes cytoplasmic scalar protons, and thus generates a pH gradient of physiological polarity across the membrane. The internally negative membrane potential and the pH gradient (inside alkaline) form a PMF across the membrane. If the PMF were to become too large, the overpolarized membrane would retard the L-aspartate:L-alanine exchange reaction, even if the cytoplasmic L-alanine concentration were sufficiently high. However, in bacterial cells, F-ATPase in the membrane catalyses ATP synthesis by consuming the generated PMF, which consequently circumvents the over-polarization. Therefore, the L-aspartate^{1-}:L-alanine^0 exchange reaction coupled with cytoplasmic L-aspartate^{1-} decarboxylation is continuously maintained until L-aspartate is completely depleted in the medium. In *Oxalobacter formigenes*, the intracellular decarboxylation of oxalate^{2-}, and the antiport reaction of oxalate^{2-} and formate^{1-}, combine to generate a PMF with almost the same mechanism as the L-aspartate^{1-} decarboxylation in *Lactobacillus* sp. M3 described above (Ruan *et al.*, 1992).

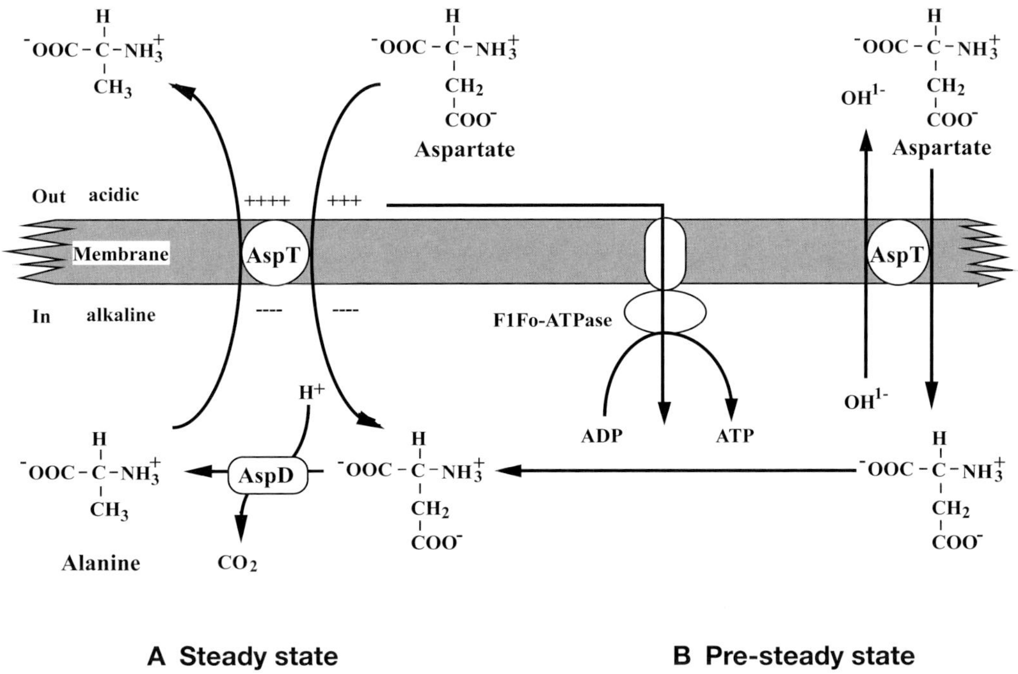

Figure 4.5 The proton-motive metabolic cycle associated with L-aspartate decarboxylation. Two possible scenarios are shown. Part A (left) outlines the *steady state* operation of a proton-motive metabolic cycle based on the function of AspT. In this case, inward transport of L-aspartate^{1-} is followed by its decarboxylation and the outward movement of uncharged L-alanine^{0}. Because entry of a single negative charge (on aspartate) is stoichiometric with consumption of a single internal proton (during decarboxylation), these reactions comprise a thermodynamic proton pump. The generated electrical potential ($\Delta\psi$) and proton gradient (ΔpH) are components of a PMF (= $\Delta\psi - Z\Delta$pH), which supplies energy for ATP synthesis by F-ATPase. Part B (right) illustrates a suggested alternative *pre-steady state* function of AspT. Here, an electrically neutral exchange of L-aspartate^{1-} with OH^{1-} (or the electrically neutral symport of L-aspartate^{1-} and H^+), allows the initial internalization of substrate, after which continued decarboxylation can expand the internal L-alanine pool to a suitable size. With permission from Abe *et al.* (1996).

This reaction can be summarized as a three-step process involving (1) import of the precursor (electric charge $-n$), (2) intracellular decarboxylation of the precursor, and (3) export of the decarboxylated product (electric charge $-[n-1]$). A negative membrane potential ($\Delta\psi$) is generated by the precursor^{-n}:product$^{-(n-1)}$ exchange reaction. During substrate decarboxylation one cytoplasmic scalar proton is consumed as it forms a covalent bond with the carbon at the position from which the carboxyl group is released, and this proton is exported from the cell as part of the decarboxylation product. The intracellular pH then becomes alkaline and a proton gradient is generated across the membrane. The generated electrical potential ($\Delta\psi$) and proton gradient (ΔpH) are components of a PMF (= $\Delta\psi - Z\Delta$pH), which supplies energy for ATP synthesis by F-ATPase. Because the motive force of precursor import is an outward gradient of its decarboxylated product [$(\text{product})_{\text{in}} > (\text{product})_{\text{out}}$], this reaction is not energy consuming. The outward diffusion of CO_2 gas across the membrane keeps the decarboxylation reaction irreversible. The series of reactions described above is the so called proton-motive metabolic cycle, and the resultant ATP synthesis is called decarboxylative phosphorylation, as described previously (Maloney *et al.,* 1994).

Other proton-motive metabolic cycles

Variation among proton-motive metabolic cycles

The decarboxylation reactions listed in Table 4.1 contribute to proton-motive metabolic cycles that are dependent on similar mechanisms to those described in the section above. Most of these reactions rely upon pyridoxal-5′-phosphate (PLP)-dependent decarboxylases and antiporters for the electrogenic exchange of precursors and decarboxylated products, and for the generation of a PMF. In *O. formigenes*, decarboxylation of oxalate^{2-} to produce formate^{1-} requires two enzymes: oxalyl-CoA decarboxylase (OxC) and formyl-CoA transferase (FrC) (Sidhu *et al.*, 1997). In *Leuconostoc mesenteroides*, conversion of citrate^{2-} to lactate^{1-} is catalysed by three enzymatic reactions in the pathway citrate^{2-} → oxaloacetate → pyruvate → lactate^{1-} (Marty-Teysset *et al.*, 1995; Lolkema *et al.*, 1996; Bandell *et al.*, 1998).

Lactobacillus buchneri generates metabolic energy from the decarboxylation of L-histidine and the electrogenic antiport of L-histidine and histamine (Molenaar *et al.*, 1993). The *L. buchneri* strain ST2A was originally isolated from Swiss cheese that was implicated in an outbreak of food poisoning (Sumner *et al.*, 1985). In food products, the presence of histamine or other biogenic amines such as cadaverine, putrescine, or tyramine, can cause food poisoning (Rice *et al.*, 1976; Sumner *et al.*, 1985; ten Brink *et al.*, 1990). *L. buchneri* ST2A decarboxylates L-histidine and produces histamine, which is then excreted into the medium. Cells grown in the presence of L-histidine generate both an internally alkaline pH gradient and a negative membrane potential. The mechanism of L-histidine uptake and histamine excretion has been investigated in membrane vesicles and proteoliposomes that are devoid of cytosolic L-histidine decarboxylase activity (Molenaar *et al.*, 1993). The results of this study demonstrate that L-histidine uptake, histamine efflux, and L-histidine:histamine exchange are electrogenic processes. L-Histidine:histamine exchange is much faster than the unidirectional flux of these substrates. The exchange is inhibited by an inside negative membrane potential and is stimulated by an inside positive membrane potential. These findings suggest that the generation of metabolic energy coupled with L-histidine decarboxylation originates from electrogenic L-histidine:histamine exchange and indirect proton extrusion due to the combined action of the decarboxylase and carrier-mediated exchange (Molenaar *et al.*, 1993). The abundance of amino acid decarboxylation reactions among bacteria suggests that this mechanism of metabolic energy generation and/or pH regulation is widespread.

When the intracellular metabolism of a substrate in some proton-motive metabolic cycle is catalysed by multi-step enzyme reactions, the uptake of the substrate and the extrusion of its metabolic product are sometime catalysed by independent transporters (Lolkema *et al.*, 1996). Systems in which a single antiporter catalyses both entry of the precursor and export of its product, appear to be more robust against perturbation by other metabolic pathways, and can more readily sustain the driving force required for substrate entry.

Use of various proton-motive metabolic cycles according to environmental pH

Since the movement of electric charge across membranes generates chemiosmotic energy, the charge of the imported substrate or exported product is important for energy generation. Because the electric charges of imported substrates or exported products are affected by environmental pH, levels of generated chemiosmotic energy are predicted to depend on the environmental pH as described below. In the L-aspartate^{1-}:L-alanine^{0} exchange, dissociation of the β-carboxyl group in L-aspartate to form COO^- and H^+ depends on the external pH. Therefore, the external pH significantly affects the level of PMF generated. At the near physiological pH (pH 7), owing to the level of dissociation of the carboxyl and amino groups, the net charge on L-aspartate is −1 [$pK_1(COOH) = 2.09$, $pK_2(NH_3^+) = 9.82$, $pK_3(COOH) = 3.86$] while that on L-alanine is 0 [$pK_1(COOH) = 2.34$, $pK_2(NH_3^+) = 9.69$]. In contrast, at around pH 3 there is very little dissociation of the β-carboxyl [$pK_3(COOH) = 3.86$]

group of L-aspartate, the L-aspartate:L-alanine exchange is mostly electroneutral, and the exchange reaction generates little or no membrane potential. When protonated L-aspartate is imported at pH 3, the net proton extrusion in the reaction cycle is zero, and only a small pH gradient (inside alkaline) is generated. The optimum external pH for the L-aspartate to L-alanine conversion in *Lactobacillus* sp. M3 cells is approximately pH 6. The reaction cycle generates a PMF when the pH is greater than five but not when the pH is less than 4, implying that the major substrate for AspT is L-aspartate^{1-}.

In *Lactobacillus* sp. EI, L-glutamate (glutamate) is converted to γ-aminobutylate (GABA), when the α-carboxyl group [pK_1(COOH) = 2.19] of L-glutamate is eliminated by decarboxylation (Higuchi *et al.*, 1997). In intact cells, the optimum pH for the glutamate to GABA conversion is around pH 5. At this pH, the α-carboxyl group of L-glutamate is negatively charged, suggesting that its elimination contributes to the generation of a membrane potential. Some strains of *Lactococcus lactis* show L-malate to L-lactate conversion, and in intact cells the elimination of the carboxyl group [pK_1(COOH) = 3.40] in L-malate has an optimum pH range of pH 3–4 (Poolman *et al.*, 1991). The major proton pumps in lactic acid bacteria are F-ATPases and their optimum pH is generally around pH 7. The optimum pH values of F-type ATPase and proton-motive metabolic cycles are shown in Fig. 4.6. Bacterial cells appear to employ various proton-motive metabolic cycles as a proton pump to cover different pH ranges (Table 4.1).

Genes and proteins associated with proton motive metabolic cycles

Aspartate:alanine exchange reaction

Asp operon

The Gram-positive lactic acid bacterium *Tetragenococcus halophilus* is used to ferment soy sauce, which contains large amounts of amino acids, such as L-aspartate, as well as high concentrations of sodium chloride (17%, wt/vol). Under high salt conditions, the growth of *T. halophilus* changes the conditions from a neutral pH (pH 7) towards an acidic pH (pH 4–5) owing to the production of lactic acid (Uchida, 1988). Higuchi *et al.* (1997) discovered that a 25-kb plasmid is responsible for the trait of L-aspartate decarboxylation in this bacterium, and proceeded to clone and sequence the *asp* operon within this plasmid. The operon consists of two genes designated as $aspD_{Th}$ and $aspT_{Th}$, which encode aspartate decarboxylase ($AspD_{Th}$) and aspartate:alanine antiporter ($AspT_{Th}$), respectively (Abe *et al.*, 2002). The results of sequence

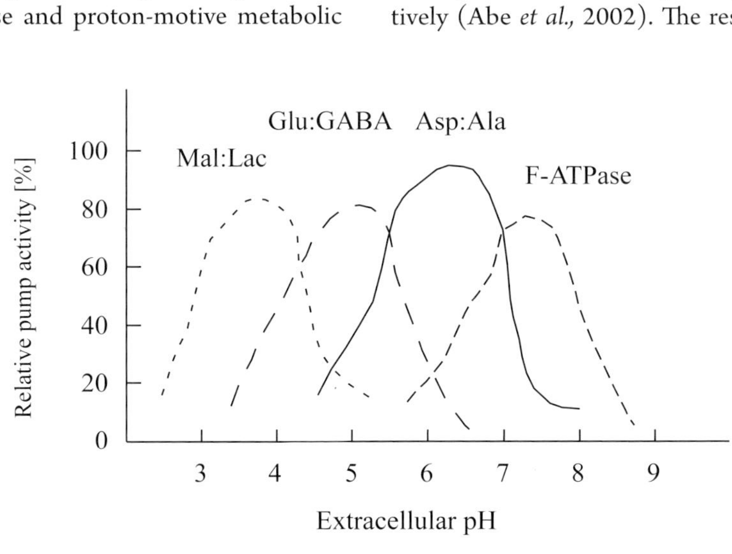

Figure 4.6 Schematic image of pH dependence of the relative activities of proton pumps found in lactic acid bacteria. F-ATPase, FoF1-ATPase; Asp:Ala, L-aspartate:L-alanine exchange reaction; Glu:GABA, L-glutamate:GABA exchange reaction; Mal:Lac, L-malate:L-lactate exchange reaction.

analysis of the intergenic region between $aspD_{Th}$ and $aspT_{Th}$ suggest that this region does not form a secondary structure, and that the organization of the *asp* operon is promoter → $aspD_{Th}$ → $aspT_{Th}$, with the predicted upstream promoter co-transcribing the two genes. These predictions are supported by the nearly stoichiometric conversion of L-aspartate to L-alanine observed in *E. coli* cells, in which the *asp* operon has been artificially expressed, and by the finding that purified histidine-tagged $AspD_{Th}$ shows L-aspartate-4-decarboxylase activity. Evidence of the electrogenic nature of recombinant $AspT_{Th}$, in a proteoliposome reconstitution system, further supports the proposal that the gene products of the *asp* operon, $AspD_{Th}$ and $AspT_{Th}$, coordinate a proton motive metabolic cycle (Abe *et al.*, 2002).

Some strains of a Gram-negative bacterium *Comamonas testosteroni* show a high level of L-aspartate decarboxylation activity, and have thus been used for industrial production of L-alanine. The *C. testosteroni aspD* gene ($aspD_{Ct}$) was isolated by Rozzell (1991). A search of the DNA sequence surrounding $aspD_{Ct}$, revealed that part of an open reading frame (ORF) encoding a membrane protein is located upstream of $aspD_{Ct}$ (Fujiki *et al.*, 2007). Determination of the full nucleotide sequence of this region revealed that a gene encoding a putative membrane protein, with multiple transmembrane (TM) segments, is located 47-bp upstream of the 5′ end of $aspD_{Ct}$, and that this gene, together with $aspD_{Ct}$, probably comprise the *asp* operon in *C. testosterone* (Fujiki *et al.*, 2007). The gene encoding this membrane protein is designated $aspT_{Ct}$. There is no potential ORF, more than 40 amino acids in length, immediately upstream of $aspT_{Ct}$ or downstream of $aspD_{Ct}$. The organization of the *asp* operon in *C. testosteroni* ($aspT_{Ct}$ → $aspD_{Ct}$) is the inverse of that of the *asp* operon in *T. halophilus* ($aspD_{Th}$ → $aspT_{Th}$) (Fig. 4.7) (Fujiki *et al.*, 2007).

Aspartate-4–decarboxylase (AspD) (EC 4.1.1.12)

L-Aspartate-4-decarboxylase is a PLP-dependent enzyme that catalyses the β-decarboxylation of L-aspartate to produce L-alanine and CO_2. This prokaryotic enzyme is known for its catalytic versatility (Tate *et al.*, 1971; Wang *et al.*, 2006), its activation by α-keto acids, (Nishimura *et al.*, 1962; Novogrodsky *et al.*, 1963; Chibata *et al.*, 1968; Tate *et al.*, 1971) and its use in the industrial-scale biocatalytic synthesis of L-alanine by immobilized *Pseudomonas dacunhae* (*C. testosteroni*) and other organisms (Yamakatsu *et al.*, 1983; Fusee *et al.*, 1984). AspDs are found in several bacteria, and their enzymology has been well studied. Since the first discovery of AspD in *Pseudomonas mycobacterium* (Mardashev *et al.*, 1948), AspDs have been found in *Clostridium perfringens* (Meister *et al.*, 1951; Nishimura *et al.*, 1962), *Desulfovibrio desulfurican* (Senez *et al.*, 1956; Cattaneo-Lacombe *et al.*, 1958), *Nocardia globerula* (Crawford, 1958), a strain of *Acetobacter* (Cooksey, 1962),

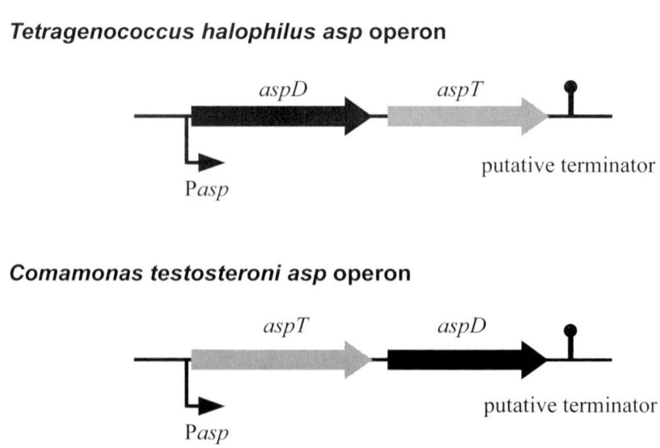

Figure 4.7 Gene organization in *asp* operons from *T. halophilus* D10 and *C. testosteroni*. Shaded arrows indicate the L-aspartate:L-alanine antiporter (AspT) gene and black arrows indicate the L-aspartate-4-decarboxylase (AspD) gene. *Pasp* denotes the promoter.

Achromobacter d-15 (Wilson, 1963; Wilson *et al.*, 1963), and *Alcaligenes faecalis* (Novogrodsky *et al.*, 1963, 1964). *In vitro* studies have shown that AspD from *Pseudomonas* sp. ATCC 19121 forms homododecameric particles, which are most active at pH 5.5, and that it partially dissociates into dimers with lower catalytic activity as the pH increases (Wang *et al.*, 2006).

The deduced amino acid sequence of the AspD from *T. halophilus* ($AspD_{Th}$) is only moderately related to those of the AspD proteins from *P. dacunhae* and *A. faecalis* (37–38% identity); however, the AspD proteins from *P. dacunhae* and *A. faecalis* are highly related to each other (93% identity). Although $AspD_{Th}$ appears to be evolutionally distant from the two other decarboxylases, the PLP attachment sites are well conserved among the three decarboxylases.

AspD proteins derived from *A. faecalis* CCRC 11585 and *Pseudomonas* sp. ATCC 19121 were recently crystallized, and their three-dimensional (3D) atomic structures were determined (Chen *et al.*, 2009; Lima *et al.*, 2009). These AspD proteins form a novel dodecameric assembly with truncated tetrahedron symmetry. On the basis of the crystal structure, a number of mutants at the subunit interface and active site have been designed and characterized, to help understand the molecular assembly and catalytic mechanism (Chen *et al.*, 2009).

Aspartate:alanine antiporter (AspT)

AspT belongs to the newly classified aspartate–alanine exchanger (AAE) family (TC# 2.A.81) in the transporter classification system developed by the Saier Lab Bioinformatics Group (http://www.tcdb.org/index.php). The results of a BLAST (http://blast.ncbi.nlm.nih.gov/Blast.cgi; 7) search of the $aspT_{Th}$ nucleotide sequence and the $AspT_{Th}$ amino acid sequence against current nucleotide and protein databases suggest that AAE family transporters are conserved among many bacterial species (Barabote *et al.*, 2006; Fujiki *et al.*, 2007).

Because the conversion of L-aspartate to L-alanine was very nearly stoichiometric in *E. coli* that harboured pTrcAsp, a plasmid that encodes $AspD_{Th}$ and $AspT_{Th}$, Abe *et al.* (2002) expected that the heterologous L-aspartate:L-alanine exchange would be a one-for-one antiport. They also anticipated that the reaction would be an electrogenic exchange, since L-aspartate is monovalent negative at physiological pH. This view was initially supported by the results of studies that measured the uptake and release of [^{3}H] L-aspartate by proteoliposomes derived from *E. coli* that harbours pTrcAsp with K^+-selective ionophore valinomycin as mentioned above (Abe *et al.*, 2002). Abe *et al.* (2002) concluded that the L-aspartate^{1-}:L-alanine^0 exchange reaction observed in the proteoliposomes derived from *E. coli* containing pTrcAsp is attributable to the function of AspT. The pattern of responses indicates that the L-aspartate^{1-}:L-alanine^0 exchange is electrogenic, with a negative charge moving in parallel with L-aspartate (Abe *et al.*, 2002).

Structure and substrate transport mechanism of AspT

Secondary structure of AspT in the membrane

$AspT_{Th}$ is a membrane protein containing 543 amino acids (57.2 kDa). The membrane topology of $AspT_{Th}$ has been studied by means of alkaline phosphatase and β-lactamase fusion methods (Nanatani *et al.*, 2005), and more recently by means of the substituted-cysteine accessibility method (SCAM) (Nanatani *et al.*, 2007). SCAM uses an impermeant, fluorescent thiol-specific probe, Oregon Green 488 maleimide (OGM), and an impermeant, nonfluorescent thiol-specific probe, [2-(trimethylammonium)ethyl]methanethiosulfonate bromide (MTSET) (Akabas *et al.*, 1992), and has been used to elucidate the membrane topologies of other proteins such as OxlT (Fu *et al.*, 1998, 2001; Ye *et al.*, 2001, 2002), the tetracycline:H^+ antiporter (Tamura *et al.*, 2003), the Na^+:H^+ antiporter (Kuwabara *et al.*, 2004), lactose permease (Kaback *et al.*, 2001), the multidrug transporter P-glycoprotein (Loo *et al.*, 1995), and the glutamate transporter GLT-1 (Grunewald *et al.*, 1998).

To establish the membrane topology of $AspT_{Th}$, Nanatani *et al.* (2007) used a cysteine-less $AspT_{Th}$ template to generate 23 variants, each containing a single cysteine, which was placed either at the end of a putative TM segment, or in a loop that connects TM segments. Each of the variants retained normal or near-normal (≥20%) specific activity after reconstitution in proteoliposomes, indicating that the introduced cysteines did not significantly perturb the structure of AspT. Thus, the topology of each of these variants is considered to reflect the location of the original residues in the cysteine-less AspT template, and in wild-type AspT. The location of the introduced cysteines, relative to the membrane was then assayed in intact cells. If a cysteine in a variant is modified by exposure to OGM in intact cells, this provides evidence that the targeted position is exposed to the periplasm. In contrast, if cysteine modification occurs only after cell lysis, the targeted residue is deemed to be located in the cytoplasm. These analyses revealed that AspT has a unique topology: the protein has 10 TM helices, a large hydrophilic cytoplasmic loop (about 180 amino acids) between TM5 and TM6, and N- and C-termini located in the periplasm (Fig. 4.8A).

Structure of the large hydrophilic loop region

Ninety *aspT* orthologues, such as *yidE* and *ybjL* of *E. coli*, have been discovered in 60 bacterial or archaebacterial genomes (Fujiki *et al.*, 2007). All the putative AspT orthologues possess typical AAE domains, which are listed as PF06826 Asp-Ala_Ex (http://www.sanger.ac.uk/cgi-bin/Pfam/getacc?PF06826) in the Pfam database (http://www.sanger.ac.uk/Software/Pfam/), or IPR006512 YidE/YbjL duplication (http://www.ebi.ac.uk/interpro/IEntry?ac=IPR006512) in the InterPro database (http://www.ebi.ac.uk/interpro/index.html). Analyses of hydropathy and sequence alignment of the AspT orthologues suggest that most of them have 9–11 TM helices and a large hydrophilic loop centred between these TM helices. The large hydrophilic loops of the AspT orthologues are computationally predicted to possess one or two TrkA_C domains [PF02080 TrkA_C domain (http://www.sanger.ac.uk//cgi-bin/Pfam/getacc?PF02080), IPR006037 Regulator of K^+ conductance, C-terminal (http://www.ebi.ac.uk/interpro/IEntry?ac=IPR006037); Barabote *et al.*, 2006], which were originally found in the regulatory subunit TrkA of the bacterial potassium uptake system (Anantharaman *et al.*, 2001).

Nanatani *et al.* (2007) performed homology modelling of the 3D structure of the large hydrophilic loop region of $AspT_{Th}$ using FAMS software (Ogata *et al.*, 2000; Iwadate *et al.*, 2001). The crystal structures of the putative potassium channel-related proteins from *Methanothermobacter thermautotrophicus* [Protein Data Bank (PDB) code 1LNQ] (Jiang *et al.*, 2003) and *Pyrococcus horikoshii* (PDB code 1VCT) were selected as templates. The large hydrophilic loop possessed two alpha–beta complex structures (domain 1, valine 208 to phenylalanine 277; and domain 2, threonine 283 to glutamine 352) (Fig. 4.8B and C). Nanatani *et al.* (2007) then produced and purified a recombinant protein of the hydrophilic loop region (arginine 178 to asparagine 361). The circular dichroism spectrum of the recombinant protein was analysed, and the results confirmed that the hydrophilic loop has secondary structures such as α-helices and β-structures. Interestingly, the two alpha–beta complex structures in the hydrophilic loop have similar 3D structures when modelled by FAMS; the root mean square deviation of the structures is 1.29Å, and the Z-score is 5.2 (Fig. 4.8D). The similarity in 3D structures occurs despite the amino acid sequence identity of the structures being very low (10.3%). An analysis of the amino acid homology between AspT and known protein domains by means of InterProScan Sequence Search (http://www.ebi.ac.uk/InterProScan/; Apweiler *et al.*, 2001) suggests that $AspT_{Th}$ has only one TrkA_C domain, in the N-terminal half of the large hydrophilic loop (domain 1). In contrast, the results of the homology modelling by FAMS discussed above suggest that $AspT_{Th}$ has a TrkA_C-like domain in the C-terminal half of the loop region (domain 2), which has the specific 3D structure of a TrkA_C domain.

TrkA_C domains have also been identified in members of the divalent anion:Na^+ symporter family (TC# 2.A.47) (Barabote *et al.*, 2006). Even though TrkA_C domains are broadly distributed in membrane transport proteins of archaebacteria, bacteria, and eukaryotes, the functions of these domains are not well defined (Anantharaman *et al.*, 2001; Barabote *et al.*, 2006).

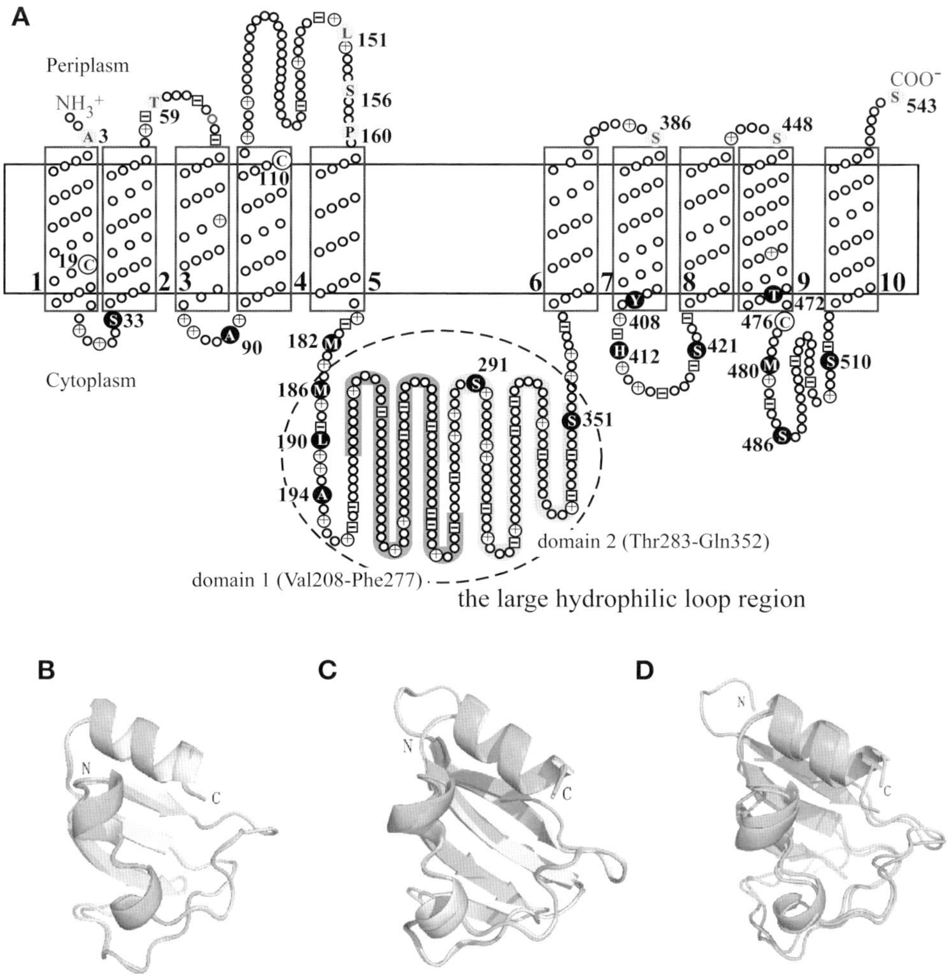

Figure 4.8 Membrane topology and modelling structure of the large hydrophilic loop region of AspT. (A) The predicted topology of AspT is shown together with topological assignments determined by site-directed fluorescence labelling of 23 single-cysteine derivatives. The residues replaced by a cysteine residue are indicated by letters in circles, and the numbers indicate the positions. Single-cysteine variants labelled by addition of OGM to the intact cells are indicated by grey circles. Single-cysteine variants labelled by addition of OGM to membrane ghosts prepared after pretreatment of intact cells with MTSET are indicated by black circles. Positively charged residues and negatively charged residues are indicated by open circles and open boxes, respectively. The three cysteine residues (cysteine 19, cysteine 110, and cysteine 476) that were replaced by alanine are indicated by letters in open circles. The cysteine-free variant was used as a template of the 23 single-cysteine variants. The large hydrophilic loop region is surrounded by a dashed circle. The alpha–beta complex structures (domain 1, valine 208 to phenylalanine 277; domain 2, threonine 283 to glutamine 352) that were predicted by means of homology modelling of the 3D structure of the large hydrophilic loop region, using FAMS software, are indicated by dark grey and light grey shading (Nanatani *et al.*, 2007). (B–D) Modelling of the large hydrophilic loop region of AspT. (B) Superposition of the model of AspT loop domain 1 (valine 208 to phenylalanine 277) (dark grey) and the structure of the MthK TrkA_C domain (methionine 264 to isoleucine 3340) (light grey). The root mean square deviation (RMSD) of these structures is 0.38Å. (C) Superposition of the model of AspT loop domain 2 (threonine 283 to glutamine 3520) (dark grey) and the structure of the protein PH0236 TrkA_C domain (isoleucine 122 to isoleucine 192) (light grey). The RMSD of these structures is 0.54 Å. (D) Superposition of the model of AspT loop domain 1 (dark grey) and the model of AspT loop domain 2 (light grey). The structures of domain 1 and domain 2 were superposed by means of the combinatorial extension algorithm (Shindyalov *et al.*, 1998). The RMSD of these structures is 1.29Å. (Modified from Nanatani *et al.*, 2007.)

Substrate transport mechanism of AspT

Multiple alignment of the amino acid sequences of 10 members of the AAE family of transporters (Nanatani *et al.,* 2009) revealed that TM3 has many amino acid residues that are conserved among AAE family members (Fig. 4.9A). The topological analysis described previously (Nanatani *et al.,* 2007) indicated that there are isolated charged residues (aspartate 67 and arginine 76) in the middle of TM3 (Fig. 4.9A). For transporters of polar molecules such as L-aspartate^{1-}, the substrate transport pathway, which includes the substrate binding site, can be expected to be enriched with residues that are more hydrophilic than residues found elsewhere in TM segments. To clarify the role of TM3 in L-aspartate^{1-} transport, Nanatani *et al.* (2009) performed a complete cysteine-scanning mutagenesis of TM3. In the first step, each amino acid from threonine 59 to phenylalanine 88, including TM3 (isoleucine 64 to methionine 85), was individually replaced with cysteine in a functional AspT molecule devoid of all three native cysteine residues. The membrane-topology of each TM3 cysteine-variant was then studied by assessing the accessibility of the introduced cysteines to OGM labelling in membrane ghosts (Fig. 4.9B) (Nanatani *et al.,* 2009). Some positions in TM3 show variable accessibility to OGM dependent on whether or not L-aspartate is present. This phenomenon has been explained by an L-aspartate-induced conformational alteration that makes these positions accessible or inaccessible to the aqueous phase, and leads to L-aspartate binding at, or close to, these residues. In the case of the cysteine introduced at proline 79 (P79C), which is in the aqueous cavity, the accessibility of the cysteine was markedly increased in the presence of L-aspartate. In contrast, in the case of the cysteine introduced at glycine 62 (G62C), which is in a periplasmic loop, the presence of L-aspartate decreased the accessibility of the cysteine. It is possible that the L-aspartate-dependent protection of G62C against OGM labelling is achieved by direct steric hindrance caused by L-aspartate binding to positions close to glycine 62, but the same reasoning can not be used to explain the stimulatory effect of L-aspartate on the OGM labelling of P79C. A second possibility is that there is an L-aspartate-induced conformational alteration that buries or exposes the residues within the protein. In brief, in the absence of substrate, glycine 62 is exposed to the periplasmic surface, and proline 79 is located at the cytoplasmic end of the hydrophobic core. Substrate binding induces a structural conformation change, and glycine 62 move to the periplasmic end of the hydrophobic core and proline 79 moves to the cytoplasmic cavity. This hypothesis explains both the protective effect and stimulatory effect of substrate binding on OGM labelling.

Glutamate:GABA exchange reaction

The combination of amino acid antiport and amino acid decarboxylation can be seen as a general system for pH homeostasis, since an amino acid is exchanged for a more basic amine, and a proton is consumed intracellularly by the decarboxylation reaction (Molenaar *et al.,* 1993). In *E. coli* and *Shigella flexneri,* putative glutamate:GABA antiporters confer acid resistance to cells grown in L-glutamate-supplemented media (Hersh *et al.,* 1996; Waterman *et al.,* 1996). The genes for these antiporters are linked to genes encoding L-glutamate decarboxylases. The same type of acid resistance mechanism has been found and characterized in *L. lactis.* The acid resistance in *L. lactis* is attributable to a putative glutamate:GABA antiporter GadC, and an L-glutamate decarboxylase, GadB, which are both encoded in the *gad* operon (Nomura *et al.* 1999). Waterman *et al.* (1996) proposed a model in which GadC is involved in the antiport of L-glutamate and GABA, while GadB converts internalized glutamate to GABA with the simultaneous consumption of a proton and production of one molecule of CO_2. According to this model, the net result of the combined action of GadC and GadB is the removal of a proton from the cytosol, with a concomitant increase in the internal pH. In addition, extracellular L-glutamate^{-1} is exchanged for the more alkaline GABA^0. An alternative function for *gadCB* may be the generation of metabolic energy. ATP is generated by L-glutamate metabolism in resting *Lactobacillus* sp. strain E1 cells, and it has been proposed that this ATP synthesis is coupled with a PMF

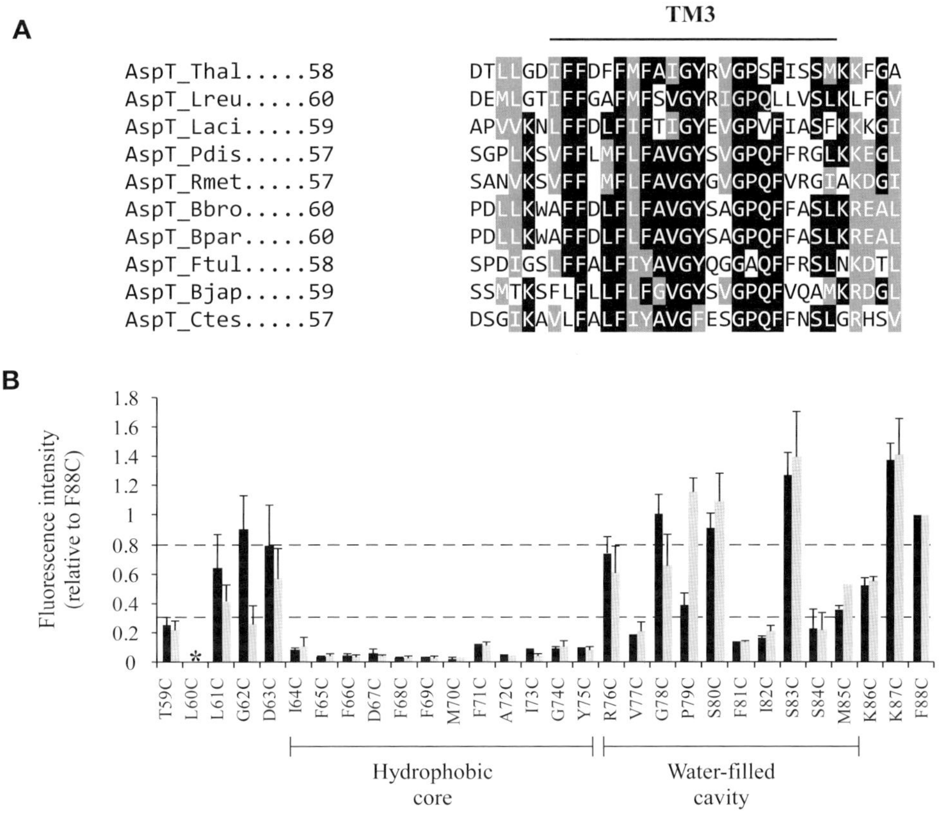

Figure 4.9 Analysis of the AspT TM3 region: protein sequence alignment and affect of aspartate on membrane topology. (A) Multiple sequence alignment of members of the AAE family of transporters. A multiple alignment was generated by ClustalW2 (version 1.83) (http://www.ebi.ac.uk/Tools/clustalw2/index.html) with sequences obtained from a BLASTP search (http://www.ncbi.nlm.nih.gov/blast/Blast.cgi) against AspT of *T. halophilus* (UniProtKB/Swiss-Prot Q8L3K8) (Pearson, 1988; Pearson, 1990). The BLOSUM matrix was chosen, and the penalties were set at 10 (gap open), 0.05 (gap extension), and 10 (gap separation distance). The resulting alignment was optically improved by the Boxshade 3.31 program (http://bioweb.pasteur.fr/seqanal/interfaces/boxshade.html); conserved residues are highlighted in black (identical to consensus) or grey (similar to consensus). The term AspT refers to the homology with the AspT of *T. halophilus* (AspT_Thal). The bacterial host strains and UniProtKB/Swiss-Prot accession numbers of the AspT homologs are as follows: AspT_Lreu, *Lactobacillus reuteri*, ABQ82551; AspT_Laci, *Lactobacillus acidophilus*, AAV43508; AspT_Pdis, Parabacteroides distasonis, ABR42837; AspT_Rmet, *Ralstonia metallidurans*, ABF08864; AspT_Bbro, *Bordetella bronchiseptica*, CAE32035; AspT_Bpar, *Bordetella parapertussis*, CAE37441; AspT_Ftul, *Francisella tularensis*, ABK89250; AspT_Bjap, *Bradyrhizobium japonicum*, BAC48740; and AspT_Ctes, *Comamonas testosteroni*, BAC65228. Experimental evidence for aspartate:alanine antiporter activity exists only for *T. halophilus* and *C. testosteroni*. Numbers indicate the position in the primary sequence and refer to the first amino acid of the alignment. (B) Membrane topology of single-cysteine variants spanning the entire TM3 region. Membrane ghosts derived from each variant were exposed to OGM in the absence or presence of L-aspartate, and the proteins were solubilized and purified under denaturing conditions. The proteins were then separated on non-reducing sodium dodecyl sulfate-polyacrylamide gels. The intensity of fluorescence and Coomassie Brilliant Blue staining were normalized to the corresponding levels in the F88C variant, which has a cysteine introduced outside the TM3 region in an intracellular hydrophilic loop. The fluorescent intensity of cysteine in each variant is shown as the ratio of the normalized levels of fluorescence intensity and Coomassie Brilliant Blue staining. A high fluorescent intensity indicates that the cysteine was accessible to OGM in the aqueous cavity. The black bars represent absence of L-aspartate and the grey bars represent presence of L-aspartate. The asterisk indicates that L60C showed strong disulfide dimer formation; the amount of monomer was too small to normalize the strength of the fluorescence. (Modified from Nanatani *et al.*, 2009.)

generated by the electrogenic glutamate:GABA antiport and the intracellular L-glutamate decarboxylation reactions (Higuchi *et al.*,1997).

Arginine:agmatine exchange reaction

Decarboxylation reactions are observed in other bacteria besides lactic acid bacteria. For instance, the glutamate:GABA exchange reaction causes acid resistance in *E. coli* and *S. flexneri* and the L-arginine:agmatine exchange is thought to be essential for acid tolerance in *E. coli*. Acid tolerance is required for the survival of enteric bacteria under low pH conditions (pH 1.5–4) in infectious processes through gastric tracts of mammals. The L-arginine to agmatine conversion catalysed by the arginine:agmatine antiporter (AdiC) and intracellular L-arginine decarboxylase (AdiA) is thought to be essential for acid tolerance in *E. coli* (Iyer *et al.*, 2003). L-Arginine imported by AdiC is decarboxylated by cytoplasmic AdiA and converted to agmatine, which is then exported by AdiC as a one for one exchange reaction. This process consumes a cytoplasmic proton by decarboxylation and generates a membrane potential by the exchange reaction, and thus generates a PMF. A recent crystallography study of AdiC (Fang *et al.*, 2009), suggests that AdiC possesses 12 TM helices and that TM1–TM5 of the N-terminal half and TM6–TM10 of the C-terminal half are localized as pseudo 2-fold axis. Although AdiC forms a dimer in membranes and under soluble conditions, an AdiC protomer shows transport activity (Fang *et al.*, 2009). AdiC shows a structural similarity to Na^+-coupled solute transporters that are not functionally related to AdiC. This similarity may reflect the structural evolution of AdiC. Crystallography of AdiC has given the first 3D structure of a precursor–product antiporter that is coupled with intracellular decarboxylation. Future studies that use this 3D structure to determine how AdiC selectively recognizes the precursor L-arginine and the decarboxylated product L-agamatine will be of great interest.

Conclusion

Decarboxylation reactions of amino acid or organic acid substrates have been well known in bacteria for a long time, and have impacted both favourably and deleteriously on humanity and industry. From the viewpoints of bacterial physiology, the decarboxylation reactions contribute generation of chemiosmotic energy and acid tolerance via proton motive metabolic cycles. The proton motive metabolic cycles driven by the decarboxylation reactions consist of at least two components such as (1) precursor:decarboxylated product antiporters and (2) intracellular decarboxylases. The liberation of CO_2 gas keeps the reactions thermodynamically irreversible. For these reasons, the proton motive metabolic cycles are scarcely affected by other metabolic pathways, and work as robust closed circuits. This robustness seems to be beneficial for bacterial cells, in particular lactic acid bacteria, which naturally grow from a neutral pH (pH 7) towards an acidic pH (pH 4–5) owing to lactic acid production. The acidic growing conditions of lactic acid bacteria force their cells to pump out protons. Because the natural habitats of lactic acid bacteria are generally enriched with amino acids or organic acids, proton motive metabolic cycles using such substrates are thought to be evolutionally diverged in lactic acid bacteria.

Acknowledgements

We thank Dr Peter C. Maloney (Department of Physiology, Johns Hopkins School of Medicine) for his valuable collaboration and kind support throughout our research on AspT. We deeply appreciate Dr Ichiro Yamato, Dr Kinji Uchida, Dr Chiyuke Kanbe and Dr Takeshi Higuchi for their helpful discussion, and Kuniko Shiraishi for excellent technical assistance in the cloning of the *asp* operon. We also thank the kikkoman Corporation for its gift of the *T. halophilus asp* operon.

References

Abe, K., Hayashi, H., and Maloney, P.C. (1996). Exchange of aspartate and alanine. Mechanism for development of a proton-motive force in bacteria. J. Biol. Chem. *271*, 3079–3084.

Abe, K., Ohnishi, F., Yagi, K., Nakajima, T., Higuchi, T., Sano, M., Machida, M., Sarker, R.I., and Maloney, P.C. (2002). Plasmid-encoded asp operon confers a proton motive metabolic cycle catalysed by an aspartate-alanine exchange reaction. J. Bacteriol. *184*, 2906–2913.

Akabas, M.H., Stauffer, D.A., Xu, M., and Karlin, A. (1992). Acetylcholine receptor channel structure probed in cysteine-substitution mutants. Science *258*, 307–310.

Anantharam, V., Allison, M.J., and Maloney, P.C. (1989). Oxalate:formate exchange. The basis for energy coupling in *Oxalobacter*. J. Biol. Chem. *264*, 7244–7250.

Anantharaman, V., Koonin, E.V., and Aravind, L. (2001). Regulatory potential, phyletic distribution and evolution of ancient, intracellular small-molecule-binding domains. J. Mol. Biol. *307*, 1271–1292.

Apweiler, R., Attwood, T.K. Bairoch, A., Bateman, A., Birney, E., Biswas, M., Bucher, P., Cerutti, L., Corpet, F., Croning, M.D., *et al.* (2001). The InterPro database, an integrated documentation resource for protein families, domains and functional sites. Nucleic Acids Res. *29*, 37–40.

Baetz, A.L., and Allison, M.J. (1989). Purification and characterization of oxalyl-coenzyme A decarboxylase from *Oxalobacter formigenes*. J. Bacteriol. *171*, 2605–2608.

Bandell, M., Lhotte, M.E., Marty-Teysset, C., Veyrat, A., Prévost, H., Dartois, V., Diviès, C., Konings, W.N., and Lolkema, J.S. (1998). Mechanism of the citrate transporters in carbohydrate and citrate cometabolism in *Lactococcus* and *Leuconostoc* species. Appl. Environ. Microbiol. *64*, 1594–1600.

Barabote, R.D., Tamang, D.G., Abeywardena, S.N., Fallah, N.S., Fu, J.Y., Lio, J.K., Mirhosseini, P., Pezeshk, R., Podell, S., Salampessy, M.L., *et al.* (2006). Extra domains in secondary transport carriers and channel proteins. Biochim. Biophys. Acta *1758*. 1557–1579.

ten Brink, B., Damink, C., Joosten, H.M., and Huis in 't Veld, J.H. (1990). Occurrence and formation of biologically active amines in foods. Int. J. Food Microbiol. *11*, 73–84.

Cattaneo-Lacombe, J., Senez, J.C., and Beaumont, P. (1958). Purification of 4-aspartic acid decarboxylase of *Desulfovibrio desulfuricans*. Biochim. Biophys. Acta *30*, 458–465.

Chen, C.C., Chou, T.L., and Lee, C.Y. (2000). Cloning, expression and characterization of L-aspartate ß-decarboxylase gene from *Alcaligenes faecalis* CCRC 11585. J. Ind. Microbiol. Biotechnol. *25*, 132–140.

Chen, H.J., Ko, T.P., Lee, C.Y., Wang, N.C., and Wang, A.H. (2009). Structure, assembly, and mechanism of a PLP-dependent dodecameric L-aspartate beta-decarboxylase. Structure *17*, 517–529.

Chibata, I., Kakimoto, T., Kato, J., Shibatani, T., and Nishimura, N. (1968). On the activation mechanism of L-aspartate beta-decarboxylase from Pseudomonas dacunhae by alpha-ketoglutarate. Biochem. Biophys. Res. Commun. *32*, 375–379.

Cooksey, K.E., and Rainbow, C. (1962). Metabolic patterns in acetic acid bacteria. J. Gen. Microbiol. *27*, 135–142.

Crawford, L.V. (1958). Studies on the aspartic decarboxylase of *Nocardia globerula*. Biochem. J. *68*, 221–225.

Fang, Y., Jayaram, H., Shane, T., Kolmakova-Partensky, L., Wu, F., Williams, C., Xiong, Y., and Miller, C. (2009). Structure of a prokaryotic virtual proton pump at 3.2 A resolution. Nature *460*, 1040–1043.

Fu, D., and Maloney, P.C. (1998). Structure–function relationships in OxlT, the oxalate/formate transporter of *Oxalobacter formigenes*. Topological features of transmembrane helix 11 as visualised by site-directed fluorescent labeling. J. Biol. Chem. *273*, 17962–17967.

Fu, D., Sarker, R.I. Abe, K., Bolton, E., and Maloney, P.C. (2001). Structure/function relationships in OxlT, the oxalate-formate transporter of *Oxalobacter formigenes*. Assignment of transmembrane helix 11 to the translocation pathway. J. Biol. Chem. *276*, 8753–8760.

Fujiki, T., Nanatani, K., Nishitani, K., Yagi, K., Ohnishi, F., Yoneyama, H., Uchida, T., Nakajima, T., and Abe, K. (2007). Membrane topology of aspartate:alanine antiporter AspT from *Comamonas testosteroni*. J. Biochem. *141*, 85–91.

Furui, M., and Yamashita, K. (1983). Pressurized reaction method for continuous production of L-alanine by immobilized *Pseudomonas dacunhae* cells. J. Ferment. Technol. *61*, 587–591.

Fusee, M.C., and Weber, J.E. (1984). Immobilization by polyurethane of *Pseudomonas dacunhae* cells containing l-aspartate beta-decarboxylase activity and application to l-alanine production. Appl. Environ. Microbiol. *48*, 694–698.

Grunewald, M., Bendahan, A., and Kanner, B.I. (1998). Biotinylation of single cysteine mutants of the glutamate transporter GLT-1 from rat brain reveals its unusual topology. Neuron *21*, 623–632.

Harold, F.M. (1986). The vital force: a study of bioenergetics (W.H. Freeman and Company, New York).

Hersh, B.M., Farooq, F.T., Barstad, D.N., Blankenhorn, D.L., and Slonczewski J.L. (1996). A glutamate-dependent acid resistance gene in *Escherichia coli*. J. Bacteriol. *178*, 3978–3981.

Higuchi, T., Hayashi, H., and Abe, K. (1997). Exchange of glutamate and gamma-aminobutyrate in a *Lactobacillus* strain. J. Bacteriol. *179*, 3362–3364.

Iyer, R., Williams, C., and Miller, C. (2003). Arginine-agmatine antiporter in extreme acid resistance in *Escherichia coli*. J. Bacteriol. *185*, 6556–6561.

Iwadate, M., Ebisawa, K., and Umeyama, H. (2001). Comparative modeling of CAFASP2 competition. Chem. Bio. Informatics J. *1*, 136–148.

Jiang, Y., Ruta, V., Chen, J., Lee, A., and MacKinnon, R. (2003). The principle of gating charge movement in a voltage-dependent K^+ channel. Nature *423*, 42–48.

Kaback, H.R., Sahin-Toth, M., and Weinglass, A.B. (2001). The kamikaze approach to membrane transport. Nat. Rev. Mol. Cell Biol. *2*, 610–620.

Kunkee, R.E. (1991). Some roles of malic acid in the malolactic fermentation in wine making. FEMS Microbiol. Lett. *88*, 55–72.

Kuwabara, N., Inoue, H., Tsuboi, Y., Nakamura, N., and Kanazawa, H. (2004). The fourth transmembrane domain of the *Helicobacter pylori* Na^+/H^+ antiporter

NhaA faces a water-filled channel required for ion transport. J. Biol. Chem. *279*, 40567–40575.

Lima, S., Sundararaju, B., Huang C., Khristoforov, R., Momany, C., and Phillips, R.S. (2009). The crystal structure of the *Pseudomonas dacunhae* aspartate-beta-decarboxylase dodecamer reveals an unknown oligomeric assembly for a pyridoxal-5′-phosphate-dependent enzyme. J. Mol. Biol. *388*, 98–108.

Lolkema, J.S., Poolman, B., and Konings, W.N. (1996). Secondary transporters and metabolic energy generation in bacteria. Transport processes in eukaryotic and prokaryotic organisms. Handbook of Biological Physics (Elesevier, Amsterdam), pp. 229–260.

Loo, T.W., and Clarke, D.M. (1995). Membrane topology of a cysteine-less mutant of human P-glycoprotein. J. Biol. Chem. *270*, 843–848.

Maloney, P.C. (1987). in *Escherichia coli* and *Salmonella typhimurium*. Cell. Mol. Biol., *1*, 222–243.

Maloney, P.C. (1994). Bacterial transporters. Curr. Opin. Cell. Biol. *6*, 571–582.

Maloney, P.C., Yan, R.T., and Abe, K. (1994). Bacterial anion exchange: reductionist and integrative approaches to membrane biology. J. Exp. Biol. *196*, 471–482.

Mardashev, S.R., and Gladkova V.N. (1948). Determination of L-aspartic acid with bacterial decarboxylase. Biokhimiya *13*, 315.

Marty-Teysset, C., Lolkema, J.S., Schmitt, P., Divies, C., and Konings, W.N. (1995). Membrane potential-generating transport of citrate and malate catalysed by CitP of *Leuconostoc mesenteroides*. J. Biol. Chem. *270*, 25370–25376.

Meister, A., Sober, H.A., and Tice, S.V. (1951). Determination of aspartic and glutamic acids by enzymatic decarboxylation. J. Biol. Chem. *189*, 591–595.

Molenaar, D., Bosscher, J.S., ten Brink, B., Driessen, A.J., and Konings, W.N. (1993). Generation of a proton motive force by histidine decarboxylation and electrogenic histidine/histamine antiport in *Lactobacillus buchneri*. J. Bacteriol. *175*, 2864–2870.

Nanatani, K., Ohonishi, F., Yoneyama, H., Nakajima, T., and Abe, K. (2005). Membrane topology of the electrogenic aspartate-alanine antiporter AspT of *Tetragenococcus halophilus*. Biochem. Biophys. Res. Commun. *328*, 20–26.

Nanatani, K., Fujiki, T., Kanou, K.M., Takeda-Shitaka, M., Umeyama, H., Ye, L., Wang, X., Nakajima, T., Uchida, T., Maloney, P.C., and Abe, K. (2007). Topology of AspT, the aspartate:alanine antiporter of *Tetragenococcus halophilus*, determined by site-directed fluorescence labeling. J. Bacteriol. *189*, 7089–7097.

Nanatani, K., Maloney, P.C., and Abe, K. (2009). Structural and functional importance of transmembrane domain 3 (TM3) in the aspartate:alanine antiporter AspT: topology and function of the residues of TM3 and oligomerization of AspT. J. Bacteriol. *191*, 2122–2132.

Nishimura, J.S., Manning, J.M., and Meister, A. (1962). Studies on the mechanism of activation acid beta-decarboxylase by alpha-keto acids and pyridoxal 5′-phosphate. Biochemistry *1*, 442–447.

Nomura, M., Nakajima, I., Fujita, Y., Kobayashi, M., Kimoto, H., Suzuki, I., and Aso, H. (1999). *Lactococcus lactis* contains only one glutamate decarboxylase gene. Microbiology *145*, 1375–1380.

Novogrodsky, A., and Meister A. (1964). Control of aspartate beta-decarboxylase activity by transamination. J. Biol. Chem. *239*, 879–888.

Novogrodsky, A., Nishimura, J.S., and Meister, A. (1963). Transamination and beta-decarboxylation of aspartate catalysed by the same pyridoxal phosphate-enzyme. J. Biol. Chem. *238*, 1903–1905.

Ogata, K., and Umeyama H. (2000). An automatic homology modeling method consisting of database searches and simulated annealing. J. Mol. Graph Model *18*, 258–272.

Olsen, E.B., Russell, J.B., and Henick-Kling, T. (1991). Electrogenic L-malate transport by *Lactobacillus plantarum*: a basis for energy derivation from malolactic fermentation. J. Bacteriol. *173*, 6199–6206.

Pearson, W.R. (1990). Rapid and sensitive sequence comparison with FASTP and FASTA. Methods Enzymol. *183*, 63–98.

Poolman, B., Molenaar, D., Smid, E.J., Ubbink, T., Abee, T., Renault, P.P., and Konings, W.N. (1991). Malolactic fermentation: electrogenic malate uptake and malate/lactate antiport generate metabolic energy. J. Bacteriol. *173*, 6030–6037.

Postma, P.W. (1987). Phosphotransferase system for glucose and other sugars. In *Escherichia coli* and *Salmonella typhimurium*: Cellular and Molecular Biology, Neidhart, F.C., Ingraham, J.L., Low, K.B., Magasarik, B., Schaechter, M., and Umbargen, H.E., eds (ASM Press, Washington, D.C.).

Rice, S.L., Eitenmiller, R.R., and Koehler, P.E. (1976). Biologically active amines in food. J. Milk Food Technol. *39*, 353–358.

Rozzell, J.D. (1991). Method and compositions for the production of L-alanine and derivatives thereof. U.S. Patent 5,019,509.

Ruan, Z.S., Anantharam, V., Crawford, I.T., Ambudkar, S.V., Rhee, S.Y., Allison, M.J., and Maloney, P.C. (1992). Identification, purification, and reconstitution of OxlT, the oxalate: formate antiport protein of *Oxalobacter formigenes*. J. Biol. Chem. *267*, 10537–10543.

Salema, M., Poolman, B., Lolkema, J.S., Dias, M.C., and Konings, W.N. (1994). Uniport of monoanionic L-malate in membrane vesicles from *Leuconostoc oenos*. Eur. J. Biochem. *225*, 289–295.

Sanders, J.W., Venema, G., Kok, J., and Leenhouts, K. (1998). Identification of a sodium chloride-regulated promoter in *Lactococcus lactis* by single-copy chromosomal fusion with a reporter gene. Mol. Gen. Genet. *257*, 681–685.

Senez, J.C., and Cattaneo-Lacombe J. (1956). Transformation de l'acide L-asparatique en alpha-alanine par des extraits de *Desulforibrio desulfuricans*. C. R. Hebd. Seances. Acad. Sci. *242*, 941–943.

Shindyalov, I.N., and Bourne, P.E. (1998). Protein structure alignment by incremental combinatorial extension (CE) of the optimal path. Protein Eng. *11*, 739–747.

Sidhu, H., Ogden, S.D., Lung, H.Y., Luttge, B.G., Baetz, A.L., and Peck, A.B. (1997). DNA sequencing and expression of the formyl coenzyme A transferase gene, frc, from *Oxalobacter formigenes*. J. Bacteriol. *179*, 3378–3381.

Smith, D.K., Kassam, T., Singh, B., and Elliott, J.F. (1992). Escherichia coli has two homologous glutamate decarboxylase genes that map to distinct loci. J. Bacteriol. *174*, 5820–5826.

Sumner, S.S., Speckhard, M.W., Somers, E.B., and Taylor, S.L. (1985). Isolation of histamine-producing *Lactobacillus buchneri* from Swiss cheese implicated in a food poisoning outbreak. Appl. Environ. Microbiol. *50*, 1094–1096.

Tamura, N., Konishi, S., and Yamaguchi, A. (2003). Mechanisms of drug/H+ antiport: complete cysteine-scanning mutagenesis and the protein engineering approach. Curr. Opin. Chem. Biol. *7*, 570–579.

Tate, S.S., and Meister, A. (1971). L-aspartate-beta-decarboxylase: structure, catalytic activities, and allosteric regulation. Adv. Enzymol. Relat. Areas Mol. Biol. *35*, 503–543.

Uchida, K. (1988). Trends in preparation and uses of fermented and acid-hydrolysed soy sauce. (Proceeding of the World Congress on Vegetable Protein Utilization in Human Foods and Animal Feedstuffs, Singapore).

Wang, N.C., and Lee, C.Y. (2006). Molecular cloning of the aspartate 4-decarboxylase gene from *Pseudomonas* sp. ATCC 19121 and characterization of the bifunctional recombinant enzyme. Appl. Microbiol. Biotechnol. *73*, 339–348.

Waterman, S.R., and Small, P.L. (1996). Identification of sigma S-dependent genes associated with the stationary-phase acid-resistance phenotype of *Shigella flexneri*. Mol. Microbiol. *21*, 925–940.

Wilson, E.M. (1963). Crystalline L-aspartate 4-carboxy-lyase. Biochim. Biophys. Acta. *67*, 345–348.

Wilson, E.M., and Kornberg, H.L. (1963). Properties of Crystalline L-aspartate 4-carboxy-lyase from *Achromobacter* sp. Biochem. J. *88*, 578–587.

Ye, L., and Maloney, P.C. (2002). Structure/function relationships in OxlT, the oxalate/formate antiporter of *Oxalobacter formigenes*: assignment of transmembrane helix 2 to the translocation pathway. J. Biol. Chem. *277*, 20372–20378.

Ye, L., Jia, Z., Jung, T., and Maloney, P.C. (2001). Topology of OxlT, the oxalate transporter of *Oxalobacter formigenes*, determined by site-directed fluorescence labeling. J. Bacteriol. *183*, 2490–2496.

Part II

Stress Response

Oxidative Stress and Oxygen Metabolism in Lactic Acid Bacteria

5

Yuji Yamamoto,* Philippe Gaudu and Alexandra Gruss

Abstract

Lactic acid bacteria (LAB) are used industrially for their fermentation properties, and as such, have been generally regarded as anaerobic bacteria that contain neither a respiratory chain nor a catalase. However, most LAB can grow under aerobic conditions and consume molecular oxygen by unique flavoprotein oxidases. Some lactic acid bacteria, such as *Enterococcus faecalis* and *Lactococcus lactis,* construct a functional respiratory chain when a source of haem, an essential cofactor of cytochrome oxidases, is provided. Others, such as *Lactobacillus plantarum* and *Streptococcus agalactiae,* undergo respiration metabolism when both haem and a quinone are provided. These unique characteristics of LAB have been studied by biochemical approaches and more recently by molecular and genetic approaches. In this chapter, we present an overview of the unique features of oxygen metabolism and mechanisms of oxygen tolerance in LAB.

Toxicity of oxygen and reactive oxygen species

Oxygen is a powerful oxidant and therefore a good electron acceptor for respiration metabolism. However oxygen can also be toxic to all living organisms. Incomplete reduction of oxygen generates reactive oxygen species (ROS) such as superoxide, hydrogen peroxide (H_2O_2), and hydroxyl radicals, which can have deleterious effects on biomolecules including DNA, RNA, membrane lipid, and protein (Fig. 5.1). The respiratory chain is considered to be a major source of ROS in aerobic organisms such as *Escherichia coli.* ROS can be also generated by radiation, UV and X-ray exposure, and catalysis of transition metals (Imlay, 2003, 2008). Therefore, organisms living in air cannot avoid encountering ROS.

Oxygen itself also has toxic effects. Although reactivity of oxygen is lower than ROS, oxygen selectively inactivates certain cellular molecules. Pyruvate formate lyase, which is required for anaerobic acetyl-CoA synthesis of many bacteria, is extremely sensitive to oxygen, and the enzyme is rapidly inactivated if bacteria are exposed to air (Sawers and Watson, 1998). It is also speculated that the defective growth of an obligate anaerobe in air may partly be due to the inactivation of pyruvate:ferredoxin oxidoreductase, an essential metabolic enzyme, by oxygen (Pan and Imlay, 2001).

Organisms developed various defence systems to overcome the toxic effects of oxygen and ROS. ROS scavenging enzymes, such as superoxide dismutase (SOD), catalase, and peroxidases, are ubiquitous in aerobic organisms. Enzymes that

$$O_2 \xrightarrow{e^-} O_2^{\bullet -} \xrightarrow{e^-} H_2O_2 \xrightarrow{e^-} {}^{\bullet}OH$$

Figure 5.1 Oxygen and reactive oxygen species. Incomplete reduction of oxygen generates reactive oxygen species such as superoxide ($O_2^{\bullet -}$), hydrogen peroxide (H_2O_2), and hydroxyl radical (${}^{\bullet}OH$).

*corresponding author

consume molecular oxygen may reduce oxygen toxicity by eliminating environmental oxygen itself. Defence strategies also include enzymes that repair oxidative lesions in proteins and nucleic acids, regulation of cellular iron metabolism to limit the generation of ROS, and regulatory systems that recognize and respond to oxygen and ROS (Imlay, 2003, 2008).

Effects of oxygen on lactic acid bacteria

The LAB are a group of bacteria that produce lactic acid as a major fermentation product. Members of this class of bacteria lack the ability to synthesize haem, which is a cofactor of haem catalases and cytochrome oxidases. Accordingly, energy generation of LAB basically depends on anaerobic fermentation. But unlike other anaerobic bacteria, most LAB are not sensitive to oxygen and grow under aerobic conditions. The majority of LAB lack catalases, and most, except for Streptococcaceae, also lack SOD. However, in view of a generally good tolerance of LAB to oxygen, the lack of these known antioxidative enzymes may suggest the presence of unique oxygen tolerance mechanisms in LAB.

Respiration metabolism has an energetic advantage compared with fermentation metabolism. For example, the growth yield of *E. coli* in air is much higher than that under anaerobic conditions. In contrast, LAB that grow via fermentation metabolism in the presence of oxygen generally end up with a similar or a worse growth yield in this condition (note however, that oxygen is required for assimilation of several non-glucose carbon sources, and also affects carbon metabolism of LAB; Pugh and Knowles, 1982; Higuchi *et al.*, 1999). Several LAB can perform aerobic respiration metabolism by utilizing exogenous respiratory components such as haem and quinone (Winstedt *et al.*, 2000; Duwat *et al.*, 2001; Yamamoto *et al.*, 2005; Brooijmans *et al.*, 2009).

In the first part of this chapter we introduce unique ROS detoxifying factors and possible mechanisms of oxygen tolerance in LAB. In the later part, enzymes and factors involved in oxygen utilization in LAB will be described.

ROS scavenging systems

LAB are among the few known organisms that are able to grow in air but lack one or both ROS scavenging enzymes. In their place, other H_2O_2-scavenging enzymes that may compensate for the lack of catalase have been characterized: NADH peroxidase is one of the H_2O_2-degrading enzymes whose activities are widely demonstrated in LAB (Sakamoto and Komagata, 1996). To date, details of molecular characteristics of two types of NADH peroxidases have been reported: one is *Enterococcus faecalis* NADH peroxidase, a flavoprotein that forms a homotetramer of a 49.5-kDa subunit. The active enzyme site requires the cysteine residue located in the N-terminal part of the protein. The other enzyme is a bi-component peroxidase, consisting of AhpC and Nox-1, of *Streptococcus mutans* (Poole *et al.*, 2000). Nox-1, a reductive subunit of a two-component peroxidase, was originally purified as an H_2O_2-forming NADH oxidase, and was subsequently revealed to act as a bi-component peroxidase together with AhpC, the other subunit of peroxidase (Higuchi *et al.*, 1993; Poole *et al.*, 2000). Nox-1 is a 55-kDa flavoprotein that transfers electrons from NADH to AhpC. AhpC is a cysteine peroxidase with two essential cysteine residues, positioned near the N- and C-termini of the protein. Reduction of H_2O_2 or peroxide by AhpC converts reduced AhpC to the oxidized form. Then, Nox-1 reduces oxidized AhpC at the expense of NADH (Poole *et al.*, 2000). It is now known that AhpC belongs to the class of peroxidase family called 'peroxiredoxin' that is present in almost all living cells from prokaryotes to higher eukaryotes (Dubbs and Mongkolsuk, 2007). Therefore it is not surprising that AhpC homologues are also widely distributed in LAB. The genetic studies of AhpC homologues and other peroxiredoxins in *Streptococcus pyogenes, Streptococcus pneumoniae,* and *E. faecalis* suggest that these enzymes play an important role in oxidative stress (Wan *et al.*, 1997; Brenot *et al.*, 2005; La Carbona *et al.*, 2007).

The other H_2O_2-degrading enzyme is manganese catalase of *Lactobacillus plantarum* (Kono and Fridovich, 1983b). Typical catalase has a haem prosthetic group in the active site and forms homotetramer. In contrast, manganese catalase contains a dinuclear manganese active site in place of haem, and forms a hexameric structure (Kono and Fridovich, 1983b; Barynin *et al.*, 2001). Analysis of the catalytic mechanism of manganese catalase indicated that the dinuclear manganese active site performs two electron cycles, as does the haem counterpart. Manganese catalase activity has been reported in only some strains of *L. plantarum* (Kono and Fridovich, 1983a). Recent genome analysis of LAB indicated the presence of manganese catalase homologues in *Pediococcus pentosaceus* and *Enterococcus faecium*, although it is not known whether these homologues are functional. *E. faecalis, Lactobacillus sakei,* and *L. plantarum* synthesize haem-catalases when an exogenous haem source is provided (Knauf *et al.*, 1992; Frankenberg *et al.*, 2002; Abriouel *et al.*, 2004). Frankenberg *et al.* (2002) purified and characterized haem catalase from *E. faecalis*, and found that it is a typical monofunctional catalase that contains one haem *b* per subunit as a prosthetic group.

Manganese SOD (Mn-SOD) is a superoxide-scavenging enzyme present in numerous LAB, including streptococci, lactococci, enterococci, and several lactobacilli such as *L. sakei* and *Lactobacillus sanfrancisco* (Vance *et al.*, 1972; Britton *et al.*, 1978; Nakayama, 1992; Gibson and Caparon, 1996; Chang and Hassan, 1997; De Angelis and Gobbetti, 1999; Yesilkaya *et al.*, 2000; Amanatidou *et al.*, 2001; Poyart *et al.*, 2001; Andrus *et al.*, 2003). Mn-SOD forms a homodimer that contains one manganese ion per subunit. Martin *et al.* purified SOD of *S. mutans* and showed that it contained either manganese or iron as a cofactor (Martin *et al.*, 1986). The physiological roles of SOD have been studied by constructing SOD-deficient mutants of different LAB. In *L. lactis*, for example, a *sodA* mutant is thermosensitive; it displays good growth, but poor survival in aerobic conditions (Sanders *et al.*, 1995; Duwat *et al.*, 2001). SODs of other Streptococcaceae are reportedly not always essential for bacterial growth in air, although this may depend on how tests were performed.

Manganese ions have been reported as being effective superoxide scavengers in some LAB (Archibald and Fridovich, 1981; Archibald, 1986). *L. plantarum* is a LAB that grows well under aerobic conditions despite its lack of the SOD enzyme. Archibald *et al.* found that *L. plantarum* can accumulate up to 30 mM intracellular manganese, and its aerobic growth is positively affected by the availability of manganese ions in the medium (Archibald and Fridovich, 1981). Measurement of superoxide scavenging activity of *L. plantarum* crude extracts showed that it was resistant to heat treatment, and sensitive to dialysis and metal chelation (Archibald and Fridovich, 1981). These observations indicated that *intracellular* manganese ions might scavenge superoxide. Manganese ions catalyse the dismutation of superoxide *in vitro* with a rate constant in the range of 10^6/M/s, supporting this hypothesis (Imlay, 2008). A high affinity manganese ion transporter, MntA, was also identified in *L. plantarum* (Hao *et al.*, 1999). Media used to cultivate LAB, such as MRS and API, contain relatively high concentrations of manganese ions (331 μM and 710 μM for MRS and API, respectively), which may favour its assimilation. The above evidence suggests that manganese ion activity as a superoxide scavenger may participate in aerobic survival of various LAB.

Other factors contribute to oxygen tolerance

The unique ROS-scavenging systems present in LAB, summarized in Table 5.1, do not always explain the behaviour of LAB in air: Many LAB secrete and accumulate H_2O_2 in the culture medium without diminishing their growth (Marty-Teysset *et al.*, 2000; Seki *et al.*, 2004; Pridmore *et al.*, 2008; Taniai *et al.*, 2008). The H_2O_2 concentration accumulated in cultures may reach the order of several millimolar, which is high enough to kill other catalase-positive bacteria such as *Staphylococcus aureus* and *Salmonella enterica* (Seki *et al.*, 2004; Otero and Nader-Macias, 2006; Pridmore *et al.*, 2008). In an ecological niche, H_2O_2 produced by LAB is believed to be a weapon to eliminate other bacteria. These observations raise the question of how LAB manage to survive in such high H_2O_2 concentrations.

A hypothesis that was proposed to explain this unique property of LAB concerns the relationship with cellular iron metabolism of these bacteria (Archibald, 1986). The toxic properties attributed to iron in the presence of oxygen are due to the generation of reactive oxygen species. H_2O_2

Table 5.1 Enzymes and antioxidants that react with molecular oxygen or ROS in lactic acid bacteria

Antioxidants and oxidases	Function or reaction catalysed
NADH peroxidase AhpC, Nox-1	H_2O_2 + NADH + H^+ → $2H_2O$ + NAD^+ ROOH + NADH + H^+ → ROH + H_2O + NAD^+
Manganase catalase Haem catalase	$2H_2O_2 \rightarrow O_2 + 2H_2O$
SOD Manganese ions	$2O_2^- + 2H^+ \rightarrow O_2 + H_2O_2$
Dpr	Ferritin-like iron-binding protein. Inhibits Fenton reaction ($Fe^{2+} + H_2O_2 \rightarrow Fe^{3+} + {}^{\bullet}OH + OH^-$)
NADH oxidase	$NADH + H^+ + O_2 \rightarrow NAD^+ + H_2O_2$ (H_2O_2-forming oxidase) $2NADH + 2H^+ + O_2 \rightarrow 2NAD^+ + 2H_2O$ (H_2O -forming oxidase)
Pyruvate oxidase	Pyruvate + O_2 + phosphate → acetylphosphate + CO_2 + H_2O_2
Lactate oxidase	Lactate + O_2 → pyruvate + H_2O_2
α-Glycerophosphate oxidase	α-Glycerophosphate + O_2 → dihydroxyacetone phosphate + H_2O_2
L-amino acid oxidase	L-amino acid + H_2O + O_2 → 2-oxo acid + NH_3 + H_2O_2

can be converted to hydroxyl radicals, which are among the most reactive ROS, in the presence of ferrous ions (Fenton reaction) while H_2O_2 itself is a moderate oxidant and selectively oxidizes certain cellular molecules (Imlay, 2003, 2008). Thus H_2O_2 is less toxic in the absence, or in limited concentrations of iron ions. In *E. coli*, iron–sulfur cluster enzymes such as aconitase, fumarase, and dehydratases are reportedly sensitive to superoxide and/or H_2O_2 (Imlay, 2003, 2008). For example, submicromolar concentrations of H_2O_2 are reported to be high enough to interfere with growth of *E. coli* Hpx^- mutant (*katG, katE, ahp*) by inactivating isopropylmalate isomerase, a [4Fe–4S] dehydratase, required for leucine biosynthesis (Jang and Imlay, 2007). In contrast, LAB require little or no iron for growth, probably because they lack enzymes that require iron as the essential cofactor, such as those in the TCA cycle (Archibald, 1986). Limited cellular iron stocks may benefit LAB by limiting toxicity of H_2O_2, and thereby contribute to their oxygen tolerance.

In addition to the physiological characteristics of LAB mentioned above, a novel type of antioxidant that regulates cellular iron status was found in LAB. Dpr (dps-like peroxide resistance) was originally found as a peroxide resistance gene from a genomic library of *S. mutans*, based on the observation that a peroxidase deficient mutant still showed the same level of peroxide and oxygen tolerance as did the wild type strain (Yamamoto *et al.*, 2000a,b). The construction of a series of *dpr* deficient mutants of *S. mutans* revealed that *dpr* was vital for aerobic growth (Yamamoto *et al.*, 2000b). Further studies verified that Dpr is a ferritin-like iron-binding protein that forms a spherical dodecamer and incorporates up to 480 iron atoms per dodecamer (Yamamoto *et al.*, 2002). Since the function of Dpr is to bind iron ions, the mechanism of oxygen and peroxide tolerance by Dpr is proposed to prevent hydroxyl radical formation by limiting intracellular free iron ions. Consistent with this hypothesis, Dpr protein was shown to inhibit iron dependent hydroxyl radical formation *in vitro* (Yamamoto *et al.*, 2002). Moreover, Dpr also reduced toxicity of oxygen *in vivo* by lowering the intracellular free iron ion concentration (Yamamoto *et al.*, 2004). These results support the hypothesis that Dpr may confer oxygen tolerance by limiting intracellular free iron ions. *E. coli* Dps, which was originally identified as a non-specific DNA-binding protein, has similar phenotypes and functions (Zhao *et al.*, 2002). Dpr homologues are predicted by genomic sequences of most LAB, and demonstrated in several LAB species (Pericone *et al.*, 2003; Nicodeme *et al.*, 2004). Several laboratories have constructed or tried to construct *dpr* deficient mutants; results show that *dpr* is responsible for H_2O_2 resistance of *Streptococcus suis* and *S. pyogenes*, and is probably essential in *S. pneumoniae* (as deduced from unsuccessful attempts to isolate a *dpr* mutant) (Pericone *et al.*, 2003; Pulliainen *et al.*, 2003; Ulijasz *et al.*, 2004; Tsou *et al.*, 2008). Although the role of Dpr in LAB other than Streptococci is not known, a mechanism of oxygen tolerance by Dpr via iron scavenging may explain why LAB can grow in the presence of H_2O_2.

Recently an interesting contribution of peroxidases to the growth of H_2O_2-producing enterococci has been reported (La Carbona *et al.*, 2007). *E. faecalis* is a typical H_2O_2-producing LAB that accumulates H_2O_2 in the hundred micromolar range into their growth medium. La Carbona *et al.* constructed three *E. faecalis* mutants defective in peroxide degradation enzymes alkyl hydroperoxide reductase, NADH peroxidase, and thiol peroxidase. These peroxidases were found to be required for the full protection of the bacterium from self producing H_2O_2 (La Carbona *et al.*, 2007). This finding is interesting because functional peroxidases protect the bacterium without eliminating H_2O_2. Possibly, these peroxidases contribute to the elimination of more toxic forms of peroxides such as lipid peroxides, thereby allowing growth of the bacterium in the presence of H_2O_2.

E. faecalis also has the unusual property of generating very high amounts of extracellular superoxide (production was estimated at 17 nmol per minute per 10^9 cells). Genetic studies associated this property with demethylmenaquinones present on the membrane outer surface (Huycke *et al.*, 2001), which may reduce oxygen to superoxide when not involved in respiration metabolism (i.e. when haem is absent).

Thiol metabolism, repair system and other antioxidative factor(s)

Cytoplasmic disulfide bond reducing systems contribute to maintaining the redox status, and are used to deliver electrons to various essential enzymes such as the ribonucleotide reductases. Thioredoxin/thioredoxin reductase and glutaredoxin/glutathione/glutathione reductase systems are major disulfide reducing systems in organisms. These systems also deliver electrons to peroxidases such as peroxiredoxin and glutathione peroxidase (Ritz and Beckwith, 2001).

Generally, Gram-positive bacteria including LAB use thioredoxin as a primary cytoplasmic disulfide reducing system. Because thioredoxin is essential for bacterial growth, few studies have succeeded in isolating thioredoxin reductase defective mutants (Scharf *et al.*, 1998; Uziel *et al.*, 2004). In LAB, Vido *et al.* isolated a thioredoxin reductase (*trxB1*) mutant of *L. lactis* and found that the gene is almost essential for aerobic growth. The *trxB1* mutant had a very long lag phase, but reached the same final cell density as the wild-type strain after prolonged microaerobic incubation; *trxB1* aerobic growth was normal in the presence of dithiothreitol (Vido *et al.*, 2005).

Glutathione, L-γ-glutamyl-L-cysteinylglycine, or GSH, is a low-molecular-weight thiol commonly found in Gram-negative bacteria, but is absent in numerous Gram-positive bacteria including LAB (Fahey *et al.*, 1978; Kim *et al.*, 2008). Although the presence of GSH in some LAB has been reported, it is not always clear how it is acquired. GSH is synthesized by the action of two enzymes, γ-glutamyl-cysteine synthetase (γ-GCS) and glutathione synthetase (GS), which are encoded by *gshA* and *gshB*, respectively. γ-GCS catalyses the synthesis of γ-glutamyl-cysteine (γ-GC) from L-cysteine and L-glutamine. GS catalyses the final step of GSH synthesis (Ritz and Beckwith, 2001). Copley and Dhillon (2002) searched for the presence of γ-GCS and GS homologues in bacteria and found that low-GC Gram-positive bacteria appeared to have γ-GCS but not GS homologues. Studies of *S. mutans* revealed that this bacterium assimilates GSH from the medium instead of synthesizing it (Sherrill and Fahey, 1998). These observations led to the assumption that availability of GSH in LAB might depend on the import from their growth medium. However, Janowiak and Griffith (2005) have found a novel GSH synthetase from *S. agalactiae*. Interestingly, the enzyme they purified was found to catalyse both the ATP-dependent synthesis of γ-GC from L-glutamate and L-cysteine and the ATP-dependent synthesis of GSH from γ-GC and glycine. They designated this novel bifunctional enzyme as γ-GCS-GS (encoded by the *gshF* gene). γ-GCS-GS homologues are present in the genomes of several LAB, including *Streptococcus thermophilus, S. suis, S. mutans, E. faecalis, and E. faecium* (Janowiak and Griffith, 2005), indicating the potential ability of these bacteria to synthesize GSH. On the other hand, Kim *et al.* (2008) reported that *Leuconostoc kimchii* and *Leuconostoc mesenteroides* use γ-GC as the major low-molecular-weight thiol instead of GSH, which accumulates intracellularly.

Protein and DNA repair systems constitute part of the antioxidative response that reportedly function in LAB. Methionine sulfoxide reductase (Msr) is a cysteine enzyme that reduces methionine sulfoxide protein residues or free methionine sulfoxide. Interestingly *msrB* of *Lactobacillus reuteri* has been identified as a gene required for colonization of the bacterium in the mouse gut (Walter *et al.*, 2005). As the gut is generally considered as being mainly anaerobic (however, see below for evidence of oxygen in the gut), it was unexpected to find a requirement for this antioxidant enzyme for colonization. Walter *et al.* (2005) speculated that MsrB may be necessary to repair protein oxidative damage caused by nitric oxide produced by gut epithelial cells through the action of nitric oxide synthase. Nevertheless, the gut is likely to be microaerobic: first, dissolved oxygen in human colon gas was estimated to be ~30 torr (Kirk, 1949), and similar values have been reported for colon gas from dogs and germ-free, gnotobiotic and conventional rats (Steggerda, 1968; Bornside *et al.*, 1976; Huycke and Moore, 2002). In addition, the microaerobic environment of the gut is suggested by colonization studies of *E. coli* respiration mutants of *E. coli* (Jones *et al.*, 2007) and *S. agalactiae* (YY, unpublished data).

RecA, which is required for homologous recombination and DNA repair, was shown in *L. lactis* to be involved in multiple stress resistance such as heat, salt, and oxidative stresses. An *L. lactis recA* mutant is sensitive to H_2O_2 compared with the wild type strain, indicating that H_2O_2 stress might cause critical DNA damage (Duwat *et al.*, 1995). Carotenoid pigment produced by *S. agalactiae* has been proposed as a non-protein antioxidative factor. Liu *et al.* (2004) isolated a carotenoid pigment defective mutant of *S. agalactiae* and demonstrated that the mutant is sensitive to superoxide, singlet oxygen, and H_2O_2 compared with the wild-type strain. The existence of other non-protein antioxidative species has been reported in LAB (Kaizu *et al.*, 1993; Terahara *et al.*, 2001; Saide and Gilliland, 2005), although the nature of their activities remains to be explored.

Oxygen metabolism by flavoprotein oxidases

LAB do not synthesize haem, and only some LAB encode cytochrome oxidases, which are essential components for energy-linked aerobic metabolism. However, LAB do consume molecular oxygen through the action of unique flavoprotein oxidases, including NADH oxidase, pyruvate oxidase, α-glycerophosphate oxidase, L-amino acid oxidase, and lactate oxidase (Koditschek and Umbreit, 1969; Carlsson *et al.*, 1987; Higuchi *et al.*, 1993, 1999; Lorquet *et al.*, 2004; Seki *et al.*, 2004; Yamamoto *et al.*, 2006; Barre *et al.*, 2007; Tong *et al.*, 2008). Reactions catalysed by each enzyme are shown in Table 5.1.

LAB NADH oxidases, which have been well studied, catalyse two- or four-electron reduction of molecular oxygen to form H_2O_2 (H_2O_2-forming NADH oxidase) or H_2O (H_2O-forming NADH oxidases), respectively. Genetic studies aimed at analysing the role of NADH oxidases of *S. mutans*, *S. pneumoniae*, *S. pyogenes*, *S. agalactiae*, *E. faecium*, and *L. lactis*, reveal that NADH oxidase inactivation affects multiple cellular processes including aerobic growth, aerobic carbon metabolism, fatty acid synthesis, development of competence, H_2O_2 production, and virulence (Higuchi *et al.*, 1999; Gibson *et al.*, 2000; Marty-Teysset *et al.*, 2000; Echenique and Trombe, 2001; Yu *et al.*, 2001; Moy *et al.*, 2004; Yamamoto *et al.*, 2006). A major role of NADH oxidases is presumably to provide NAD^+ required for smooth operation of glycolysis. Lactate dehydrogenase is an enzyme that regenerates NAD^+ from NADH coupled with the production of lactate. In contrast, the use of NADH oxidase allows LAB to use oxygen as an electron acceptor and regenerate NAD^+ without producing lactate. Consistent with this hypothesis, deletion of NADH oxidase activity increased lactate production and reduced acetate and acetoin, while overexpression of NADH oxidase in *L. lactis* abolished lactate production and markedly increased acetate and acetoin production (Lopez de Felipe *et al.*, 1998; Higuchi *et al.*, 1999; Yamamoto *et al.*, 2006). Pyruvate oxidase is reportedly involved in the production of acetate under aerobic conditions. Acetyl phosphate, the principal product of pyruvate oxidase, is converted to acetate through the action of acetate kinase with accompanying production of ATP (Goffin *et al.*, 2006; Taniai *et al.*, 2008). Higher intracellular levels of acetyl phosphate due to pyruvate oxidase activity are proposed to be involved in phosphorylation-dependent signal transduction, and in the synthesis of acetyl-CoA, an essential cofactor for synthesis of cellular components (Spellerberg *et al.*, 1996; Pericone *et al.*, 2003). α-Glycerophosphate oxidase, L-amino acid oxidase, and lactate oxidase are involved in the utilization of glycerol, L-amino acid, and lactic acid (Pugh and Knowles, 1982; Seki *et al.*, 2004; Tong *et al.*, 2008).

Unlike membrane-associated cytochrome oxidases, most LAB flavoprotein oxidases are reportedly localized in the cytoplasm, and are not always coupled with energy production. The presence of these flavoprotein oxidases expands the metabolic capacity of LAB by allowing them to utilize oxygen as an electron acceptor. Overall, the flavoprotein oxidases may contribute to oxygen tolerance of LAB by eliminating environmental oxygen. On the other hand, these enzymes can be a source of H_2O_2 in LAB, because they all generate H_2O_2, with the exception of the H_2O-forming NADH oxidase. Genetic studies have shown that pyruvate oxidase, lactate oxidase, L-amino acid oxidase, and NADH oxidase are involved in H_2O_2 production of LAB (Gibson *et al.*, 2000; Moy *et al.*, 2004; Pericone *et al.*, 2003; Taniai *et al.*, 2008; Tong *et al.*, 2008).

Respiration metabolism

A functional respiratory chain consists of (i) NADH dehydrogenases, serving as electron donors, (ii) quinones, which deliver electrons to the terminal oxidoreductases, and (iii) terminal oxidoreductases, which contain haem as an essential cofactor and use oxygen as final electron acceptor (Poole and Cook, 2000). Bacteria that use respiration generally synthesize all the necessary components (Poole and Cook, 2000). Owing to the lack of a complete haem biosynthesis pathway, LAB cannot undergo autonomous respiration metabolism. However, several LAB such as *L. lactis, E. faecalis,* and *L. mesenteroides* undergo respiration metabolism when these bacteria are aerobically cultured in the presence of a haem source (Ritchey and Seely, 1976) (Fig. 5.2). These observations suggested that these LAB have all the necessary respiratory components except haem synthesis, as is now confirmed experimentally and by genomic sequence data. More recent studies of *L. lactis* and *E. faecalis* respiration metabolism have shown that these

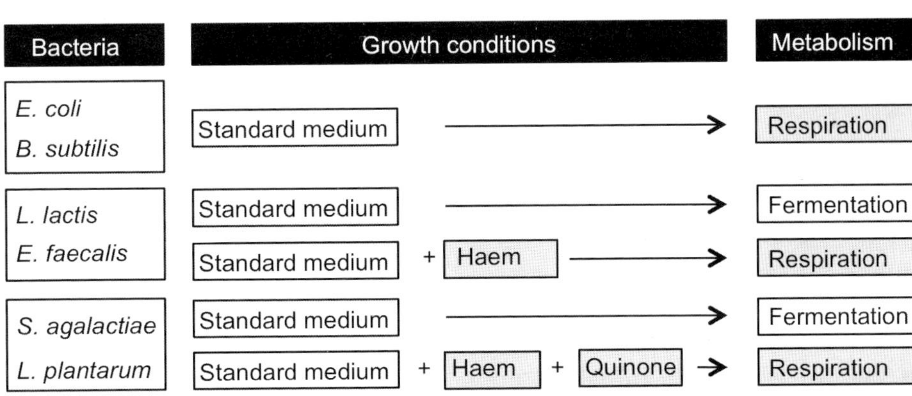

Figure 5.2 Distinct class of respiring bacteria. Standard respiring bacteria, such as *E. coli* and *B. subtilis*, synthesize all respiratory components. On the other hand, *L. lactis* and *E. faecalis* require exogenous haem sources for respiration. *S. agalactiae* and *L. plantarum* need both exogenous haem and quinone sources to activate respiration metabolism.

bacteria use cytochrome *bd* quinol oxidase as a terminal oxidase and synthesize menaquinones or demethylmenaquinones (Winstedt *et al.*, 2000; Duwat *et al.*, 2001; Huycke *et al.*, 2001; Rezaiki *et al.*, 2008). Intriguingly, *S. agalactiae* was found to lack both haem and quinone synthetic pathways, but encoded the cytochrome *bd* quinol oxidase. This bacterium was demonstrated to perform respiration metabolism when both respiratory components, haem and menaquinone, were provided (Yamamoto *et al.*, 2005). Similarly, the addition of both haem and menaquinone to the growth medium activates respiratory metabolism of *L. plantarum* (Brooijmans *et al.*, 2009) (Fig. 5.2).

Interestingly, the shift from fermentation to respiration metabolism dramatically changes the cellular physiology of these bacteria. In the case of *L. lactis*, major positive impacts are increase of biomass and long-term survival (Duwat *et al.*, 2001; Rezaiki *et al.*, 2004). The growth yield of *L. lactis* is approximately doubled by respiration compared with fermentation growth. Reduced production of lactate and an energetic advantage of respiration metabolism contribute largely to the improved features of respiration growth (Duwat *et al.*, 2001). Surprisingly, long-term survival of respiration-grown cells of *L. lactis* is extended from a few days to several months at 4°C (Duwat *et al.*, 2001; Rezaiki *et al.*, 2004). This is attributed to elimination of oxygen that limits the life span of bacteria via oxidative stress, and to the increased pH of cultures as a result of the metabolic shift (Rezaiki *et al.*, 2004). Higher stability and yield of respiration-grown compared with fermentation-grown cells are attractive features for industrial production of LAB. This idea was realized to application by a Danish company, i.e. respiration growth was developed for large-scale production of lactococcal starter cultures (Pedersen *et al.*, 2005).

S. agalactiae, a leading cause of neonatal infection, showed stimulated growth in the presence of haem and menaquinone. The presence of these factors in animal hosts suggested a possible role of respiration in virulence. This hypothesis was confronted in virulence studies of *S. agalactiae* wild type compared with a respiration-negative mutant in a neonatal rat model (Yamamoto *et al.*, 2005). Attenuated virulence of the respiration-negative mutant indicated that *S. agalactiae* may use respiration metabolism by acquiring haem and menaquinone in the host environment and that respiration metabolism may contribute to *S. agalactiae* virulence (Yamamoto *et al.*, 2005). These findings might be applicable to several lactobacilli that, like *S. agalactiae*, are able to use respiration metabolism under specific conditions, where haem, menaquinone, and oxygen are present.

Regulation of oxidative stress responses

Information on oxygen- or ROS- mediated regulation of metabolic pathways and antioxidants in LAB remains limited. Several reports and genome information of LAB indicate the presence of some key regulatory elements in LAB. PerR, a repressor protein widely conserved among Gram-positive bacteria including LAB, functions as an H_2O_2 sensor (Lee and Helmann, 2006). PerR has a non-haem iron centre that reacts with H_2O_2, probably forming hydroxyl radicals. Radical formation by H_2O_2 inactivates PerR, thus allowing expression of the genes in the regulon (Lee and Helmann, 2006). Genetic studies of *S. pyogenes* demonstrated the involvement of *perR* in its oxidative stress responses (King *et al.*, 2000). FLP is an Fnr-like regulatory protein reported in *L. casei* and *L. lactis* (Gostick *et al.*, 1998; Scott *et al.*, 2000). Flp of *L. lactis* is proposed to have oxygen-sensitive iron-sulphur clusters, whose state is suggested to have a sensing role (Scott *et al.*, 2000). HypR of *E. faecalis* has been identified as a regulator required for H_2O_2 resistance of the bacterium, though it is not known whether HypR has a sensing function (Verneuil *et al.*, 2004). Another candidate redox-responsive regulator is Spx, whose role was most extensively characterized in *Bacillus subtilis* (Nakano *et al.*, 2003). Spx interacts with the α-subunit factor of the RNA polymerase. An N-terminal CxxC motif (and sometimes CxxS) is conserved in the family of related proteins. Oxidation of Spx cysteine residues is proposed to affect the global transcriptional profile by changing promoter specificity of RNA polymerase. *spx* positively regulates *B. subtilis* thioredoxin and thioredoxin reductase expression (Nakano *et al.*, 2003). *L. lactis* encodes seven Spx-like proteins: *spxA* was initially identified as a suppressor mutation (*trmA*) for multiple stress resistance, and as a *clpP* mutant suppressor (Duwat *et al.*, 1999; Frees *et al.*, 2001). A second *spx* homologue, *spxB*, responds to cell wall stress by activating a cascade, leading to increased cell wall acetylation that might prevent further cell wall damage (Veiga *et al.*, 2007). Spx homologues are widely found in LAB genomes, suggesting that Spx may be a common redox responsive regulator in LAB.

References

Abriouel, H., Herrmann, A., Starke, J., Yousif, N.M., Wijaya, A., Tauscher, B., Holzapfel, W., and Franz, C.M. (2004). Cloning and heterologous expression of hematin-dependent catalase produced by *Lactobacillus plantarum* CNRZ 1228. Appl. Environ. Microbiol. *70*, 603–606.

Amanatidou, A., Bennik, M.H., Gorris, L.G., and Smid, E.J. (2001). Superoxide dismutase plays an important role in the survival of *Lactobacillus* sake upon exposure to elevated oxygen. Arch. Microbiol. *176*, 79–88.

Andrus, J.M., Bowen, S.W., Klaenhammer, T.R., and Hassan, H.M. (2003). Molecular characterization and functional analysis of the manganese-containing superoxide dismutase gene (*sodA*) from *Streptococcus thermophilus* AO54. Arch. Biochem. Biophys. *420*, 103–113.

Archibald, F. (1986). Manganese: its acquisition by and function in the lactic acid bacteria. Crit. Rev. Microbiol. *13*, 63–109.

Archibald, F.S., and Fridovich, I. (1981). Manganese and defenses against oxygen toxicity in *Lactobacillus plantarum*. J. Bacteriol. *145*, 442–451.

Barre, O., Mourlane, F., and Solioz, M. (2007). Copper induction of lactate oxidase of *Lactococcus lactis*: a novel metal stress response. J. Bacteriol. *189*, 5947–5954.

Barynin, V.V., Whittaker, M.M., Antonyuk, S.V., Lamzin, V.S., Harrison, P.M., Artymiuk, P.J., and Whittaker, J.W. (2001). Crystal structure of manganese catalase from *Lactobacillus plantarum*. Structure *9*, 725–738.

Bornside, G.H., Donovan, W.E., and Myers, M.B. (1976). Intracolonic tensions of oxygen and carbon dioxide in germfree, conventional, and gnotobiotic rats. Proc. Soc. Exp. Biol. Med. *151*, 437–441.

Brenot, A., King, K.Y., and Caparon, M.G. (2005). The PerR regulon in peroxide resistance and virulence of *Streptococcus pyogenes*. Mol. Microbiol. *55*, 221–234.

Britton, L., Malinowski, D.P., and Fridovich, I. (1978). Superoxide dismutase and oxygen metabolism in *Streptococcus faecalis* and comparisons with other organisms. J. Bacteriol. *134*, 229–236.

Brooijmans, R.J., de Vos, W.M., and Hugenholtz, J. (2009). The electron transport chains of *Lactobacillus plantarum* WCFS1. Appl. Environ. Microbiol. *75*, 3580–3585.

Carlsson, J., Edlund, M.B., and Lundmark, S.K. (1987). Characteristics of a hydrogen peroxide-forming pyruvate oxidase from *Streptococcus sanguis*. Oral Microbiol. Immunol. *2*, 15–20.

Chang, S.K., and Hassan, H.M. (1997). Characterization of superoxide dismutase in *Streptococcus thermophilus*. Appl. Environ. Microbiol. *63*, 3732–3735.

Copley, S.D., and Dhillon, J.K. (2002). Lateral gene transfer and parallel evolution in the history of glutathione biosynthesis genes. Genome Biol. *3*, research 0025.

De Angelis, M., and Gobbetti, M. (1999). *Lactobacillus sanfranciscensis* CB1: manganese, oxygen, superoxide dismutase and metabolism. Appl. Microbiol. Biotechnol. *51*, 358–363.

Dubbs, J.M., and Mongkolsuk, S. (2007). Peroxiredoxins in bacterial antioxidant defense. Subcell. Biochem. *44*, 143–193.

Duwat, P., Sourice, S., Ehrlich, S.D., and Gruss, A. (1995). *recA* gene involvement in oxidative and thermal stress in *Lactococcus lactis*. Dev. Biol. Stand. *85*, 455–467.

Duwat, P., Ehrlich, S.D., and Gruss, A. (1999). Effects of metabolic flux on stress response pathways in *Lactococcus lactis*. Mol. Microbiol. *31*, 845–858.

Duwat, P., Sourice, S., Cesselin, B., Lamberet, G., Vido, K., Gaudu, P., Le Loir, Y., Violet, F., Loubiere, P., and Gruss, A. (2001). Respiration capacity of the fermenting bacterium *Lactococcus lactis* and its positive effects on growth and survival. J. Bacteriol. *183*, 4509–4516.

Echenique, J.R., and Trombe, M.C. (2001). Competence modulation by the NADH oxidase of *Streptococcus pneumoniae* involves signal transduction. J. Bacteriol. *183*, 768–772.

Fahey, R.C., Brown, W.C., Adams, W.B., and Worsham, M.B. (1978). Occurrence of glutathione in bacteria. J. Bacteriol. *133*, 1126–1129.

Frankenberg, L., Brugna, M., and Hederstedt, L. (2002). *Enterococcus faecalis* heme-dependent catalase. J. Bacteriol. *184*, 6351–6356.

Frees, D., Varmanen, P., and Ingmer, H. (2001). Inactivation of a gene that is highly conserved in Gram-positive bacteria stimulates degradation of non-native proteins and concomitantly increases stress tolerance in *Lactococcus lactis*. Mol. Microbiol. *41*, 93–103.

Gibson, C.M., and Caparon, M.G. (1996). Insertional inactivation of *Streptococcus pyogenes sod* suggests that *prtF* is regulated in response to a superoxide signal. J. Bacteriol. *178*, 4688–4695.

Gibson, C.M., Mallett, T.C., Claiborne, A., and Caparon, M.G. (2000). Contribution of NADH oxidase to aerobic metabolism of *Streptococcus pyogenes*. J. Bacteriol. *182*, 448–455.

Goffin, P., Muscariello, L., Lorquet, F., Stukkens, A., Prozzi, D., Sacco, M., Kleerebezem, M., and Hols, P. (2006). Involvement of pyruvate oxidase activity and acetate production in the survival of *Lactobacillus plantarum* during the stationary phase of aerobic growth. Appl. Environ. Microbiol. *72*, 7933–7940.

Gostick, D.O., Green, J., Irvine, A.S., Gasson, M.J., and Guest, J.R. (1998). A novel regulatory switch mediated by the FNR-like protein of *Lactobacillus casei*. Microbiology *144*, 705–717.

Hao, Z., Chen, S., and Wilson, D.B. (1999). Cloning, expression, and characterization of cadmium and manganese uptake genes from *Lactobacillus plantarum*. Appl. Environ. Microbiol. *65*, 4746–4752.

Higuchi, M., Shimada, M., Yamamoto, Y., Hayashi, T., Koga, T., and Kamio, Y. (1993). Identification of two distinct NADH oxidases corresponding to H_2O_2-forming oxidase and H_2O-forming oxidase induced in *Streptococcus* mutans. J. Gen. Microbiol. *139*, 2343–2351.

Higuchi, M., Yamamoto, Y., Poole, L.B., Shimada, M., Sato, Y., Takahashi, N., and Kamio, Y. (1999). Functions of two types of NADH oxidases in energy metabolism and oxidative stress of *Streptococcus mutans*. J. Bacteriol. *181*, 5940–5947.

Huycke, M.M., and Moore, D.R. (2002). *In vivo* production of hydroxyl radical by *Enterococcus faecalis* colonizing the intestinal tract using aromatic hydroxylation. Free Radic. Biol. Med. *33*, 818–826.

Huycke, M.M., Moore, D., Joyce, W., Wise, P., Shepard, L., Kotake, Y., and Gilmore, M.S. (2001). Extracellular superoxide production by *Enterococcus faecalis* requires demethylmenaquinone and is attenuated by functional terminal quinol oxidases. Mol. Microbiol. *42*, 729–740.

Imlay, J.A. (2003). Pathways of oxidative damage. Annu. Rev. Microbiol. *57*, 395–418.

Imlay, J.A. (2008). Cellular defenses against superoxide and hydrogen peroxide. Annu. Rev. Biochem. *77*, 755–776.

Jang, S., and Imlay, J.A. (2007). Micromolar intracellular hydrogen peroxide disrupts metabolism by damaging iron–sulfur enzymes. J. Biol. Chem. *282*, 929–937.

Janowiak, B.E., and Griffith, O.W. (2005). Glutathione synthesis in *Streptococcus agalactiae*. One protein accounts for gamma-glutamylcysteine synthetase and glutathione synthetase activities. J. Biol. Chem. *280*, 11829–11839.

Jones, S.A., Chowdhury, F.Z., Fabich, A.J., Anderson, A., Schreiner, D.M., House, A.L., Autieri, S.M., Leatham, M.P., Lins, J.J., Jorgensen, M., *et al.* (2007). Respiration of *Escherichia coli* in the mouse intestine. Infect. Immun. *75*, 4891–4899.

Kaizu, H., Sasaki, M., Nakajima, H., and Suzuki, Y. (1993). Effect of antioxidative lactic acid bacteria on rats fed a diet deficient in vitamin E. J. Dairy Sci. *76*, 2493–2499.

Kim, E.K., Cha, C.J., Cho, Y.J., Cho, Y.B., and Roe, J.H. (2008). Synthesis of gamma-glutamylcysteine as a major low-molecular-weight thiol in lactic acid bacteria *Leuconostoc* spp. Biochem. Biophys. Res. Commun. *369*, 1047–1051.

King, K.Y., Horenstein, J.A., and Caparon, M.G. (2000). Aerotolerance and peroxide resistance in peroxidase and PerR mutants of *Streptococcus pyogenes*. J. Bacteriol. *182*, 5290–5299.

Kirk, E. (1949). The quantity and composition of human colonic flatus. Gastroenterology *12*, 782–794.

Knauf, H.J., Vogel, R.F., and Hammes, W.P. (1992). Cloning, sequence, and phenotypic expression of *katA*, which encodes the catalase of *Lactobacillus sake* LTH677. Appl. Environ. Microbiol. *58*, 832–839.

Koditschek, L.K., and Umbreit, W.W. (1969). Alpha-glycerophosphate oxidase in *Streptococcus faecium* F 24. J. Bacteriol. *98*, 1063–1068.

Kono, Y., and Fridovich, I. (1983a). Functional significance of manganese catalase in *Lactobacillus plantarum*. J. Bacteriol. *155*, 742–746.

Kono, Y., and Fridovich, I. (1983b). Isolation and characterization of the pseudocatalase of *Lactobacillus plantarum*. J. Biol. Chem. *258*, 6015–6019.

La Carbona, S., Sauvageot, N., Giard, J.C., Benachour, A., Posteraro, B., Auffray, Y., Sanguinetti, M., and Hartke, A. (2007). Comparative study of the physiological roles of three peroxidases (NADH peroxidase, Alkyl hydroperoxide reductase and Thiol peroxidase) in oxidative stress response, survival inside macrophages and virulence of *Enterococcus faecalis*. Mol. Microbiol. *66*, 1148–1163.

Lee, J.W., and Helmann, J.D. (2006). The PerR transcription factor senses H_2O_2 by metal-catalysed histidine oxidation. Nature *440*, 363–367.

Liu, G.Y., Doran, K.S., Lawrence, T., Turkson, N., Puliti, M., Tissi, L., and Nizet, V. (2004). Sword and shield: linked group B streptococcal beta-hemolysin/cytolysin and carotenoid pigment function to subvert host phagocyte defense. Proc. Natl. Acad. Sci. U.S.A. *101*, 14491–14496.

Lopez de Felipe, F., Kleerebezem, M., de Vos, W.M., and Hugenholtz, J. (1998). Cofactor engineering: a novel approach to metabolic engineering in *Lactococcus lactis* by controlled expression of NADH oxidase. J. Bacteriol. *180*, 3804–3808.

Lorquet, F., Goffin, P., Muscariello, L., Baudry, J.B., Ladero, V., Sacco, M., Kleerebezem, M., and Hols, P. (2004). Characterization and functional analysis of the *poxB* gene, which encodes pyruvate oxidase in *Lactobacillus plantarum*. J. Bacteriol. *186*, 3749–3759.

Martin, M.E., Byers, B.R., Olson, M.O., Salin, M.L., Arceneaux, J.E., and Tolbert, C. (1986). A *Streptococcus mutans* superoxide dismutase that is active with either manganese or iron as a cofactor. J. Biol. Chem. *261*, 9361–9367.

Marty-Teysset, C., de la Torre, F., and Garel, J. (2000). Increased production of hydrogen peroxide by *Lactobacillus delbrueckii* subsp. *bulgaricus* upon aeration: involvement of an NADH oxidase in oxidative stress. Appl. Environ. Microbiol. *66*, 262–267.

Moy, T.I., Mylonakis, E., Calderwood, S.B., and Ausubel, F.M. (2004). Cytotoxicity of hydrogen peroxide produced by *Enterococcus faecium*. Infect. Immun. *72*, 4512–4520.

Nakano, S., Kuster-Schock, E., Grossman, A.D., and Zuber, P. (2003). Spx-dependent global transcriptional control is induced by thiol-specific oxidative stress in *Bacillus subtilis*. Proc. Natl. Acad. Sci. U.S.A. *100*, 13603–13608.

Nakayama, K. (1992). Nucleotide sequence of *Streptococcus mutans* superoxide dismutase gene and isolation of insertion mutants. J. Bacteriol. *174*, 4928–4934.

Nicodeme, M., Perrin, C., Hols, P., Bracquart, P., and Gaillard, J.L. (2004). Identification of an iron-binding protein of the Dps family expressed by *Streptococcus thermophilus*. Curr. Microbiol. *48*, 51–56.

Otero, M.C., and Nader-Macias, M.E. (2006). Inhibition of *Staphylococcus aureus* by H_2O_2-producing *Lactobacillus gasseri* isolated from the vaginal tract of cattle. Anim. Reprod. Sci. *96*, 35–46.

Pan, N., and Imlay, J.A. (2001). How does oxygen inhibit central metabolism in the obligate anaerobe *Bacteroides thetaiotaomicron*. Mol. Microbiol. *39*, 1562–1571.

Pedersen, M.B., Iversen, S.L., Sorensen, K.I., and Johansen, E. (2005). The long and winding road from the research laboratory to industrial applications of lactic acid bacteria. FEMS Microbiol. Rev. *29*, 611–624.

Pericone, C.D., Park, S., Imlay, J.A., and Weiser, J.N. (2003). Factors contributing to hydrogen peroxide resistance in *Streptococcus pneumoniae* include pyruvate oxidase (SpxB) and avoidance of the toxic effects of the fenton reaction. J. Bacteriol. *185*, 6815–6825.

Poole, L.B., Higuchi, M., Shimada, M., Calzi, M.L., and Kamio, Y. (2000). *Streptococcus mutans* H_2O_2-forming NADH oxidase is an alkyl hydroperoxide reductase protein. Free Radic. Biol. Med. *28*, 108–120.

Poole, R.K., and Cook, G.M. (2000). Redundancy of aerobic respiratory chains in bacteria? Routes, reasons and regulation. Adv. Microb. Physiol. *43*, 165–224.

Poyart, C., Pellegrini, E., Gaillot, O., Boumaila, C., Baptista, M., and Trieu-Cuot, P. (2001). Contribution of Mn-cofactored superoxide dismutase (SodA) to the virulence of *Streptococcus agalactiae*. Infect. Immun. *69*, 5098–5106.

Pridmore, R.D., Pittet, A.C., Praplan, F., and Cavadini, C. (2008). Hydrogen peroxide production by *Lactobacillus johnsonii* NCC 533 and its role in anti-*Salmonella* activity. FEMS Microbiol. Lett. *283*, 210–215.

Pugh, S.Y., and Knowles, C.J. (1982). Growth of *Streptococcus faecalis* var. *zymogenes* on glycerol: the effect of aerobic and anaerobic growth in the presence and absence of haematin on enzyme synthesis. J. Gen. Microbiol. *128*, 1009–1017.

Pulliainen, A.T., Haataja, S., Kahkonen, S., and Finne, J. (2003). Molecular basis of H_2O_2 resistance mediated by streptococcal *dpr*. Demonstration of the functional involvement of the putative ferroxidase center by site-directed mutagenesis in *Streptococcus suis*. J. Biol. Chem. *278*, 7996–8005.

Rezaiki, L., Cesselin, B., Yamamoto, Y., Vido, K., van West, E., Gaudu, P., and Gruss, A. (2004). Respiration metabolism reduces oxidative and acid stress to improve long-term survival of *Lactococcus lactis*. Mol. Microbiol. *53*, 1331–1342.

Rezaiki, L., Lamberet, G., Derre, A., Gruss, A., and Gaudu, P. (2008). *Lactococcus lactis* produces short-chain quinones that cross-feed Group B *Streptococcus* to activate respiration growth. Mol. Microbiol. *67*, 947–957.

Ritchey, T.W., and Seely, H.W., Jr. (1976). Distribution of cytochrome-like respiration in streptococci. J. Gen. Microbiol. *93*, 195–203.

Ritz, D., and Beckwith, J. (2001). Roles of thiol-redox pathways in bacteria. Annu. Rev. Microbiol. *55*, 21–48.

Saide, J.A., and Gilliland, S.E. (2005). Antioxidative activity of lactobacilli measured by oxygen radical absorbance capacity. J. Dairy Sci. *88*, 1352–1357.

Sakamoto, M., and Komagata, K. (1996). Aerobic growth of and activities of NADH oxidase and NADH peroxidase in lactic acid bacteria. J. Ferment. Bioeng. *82*, 210–216.

Sanders, J.W., Leenhouts, K.J., Haandrikman, A.J., Venema, G., and Kok, J. (1995). Stress response *in Lactococcus lactis*: cloning, expression analysis, and mutation of the lactococcal superoxide dismutase gene. J. Bacteriol. *177*, 5254–5260.

Sawers, G., and Watson, G. (1998). A glycyl radical solution: oxygen-dependent interconversion of pyruvate formate-lyase. Mol. Microbiol. *29*, 945–954.

Scharf, C., Riethdorf, S., Ernst, H., Engelmann, S., Volker, U., and Hecker, M. (1998). Thioredoxin is an essential protein induced by multiple stresses in *Bacillus subtilis*. J. Bacteriol. *180*, 1869–1877.

Scott, C., Guest, J.R., and Green, J. (2000). Characterization of the *Lactococcus lactis* transcription factor FlpA and demonstration of an *in vitro* switch. Mol. Microbiol. *35*, 1383–1393.

Seki, M., Iida, K., Saito, M., Nakayama, H., and Yoshida, S. (2004). Hydrogen peroxide production in *Streptococcus pyogenes*: involvement of lactate oxidase and coupling with aerobic utilization of lactate. J. Bacteriol. *186*, 2046–2051.

Sherrill, C., and Fahey, R.C. (1998). Import and metabolism of glutathione by *Streptococcus mutans*. J. Bacteriol. *180*, 1454–1459.

Spellerberg, B., Cundell, D.R., Sandros, J., Pearce, B.J., Idanpaan-Heikkila, I., Rosenow, C., and Masure, H.R. (1996). Pyruvate oxidase, as a determinant of virulence in *Streptococcus pneumoniae*. Mol. Microbiol. *19*, 803–813.

Steggerda, F.R. (1968). Gastrointestinal gas following food consumption. Ann. N.Y. Acad. Sci. *150*, 57–66.

Taniai, H., Iida, K., Seki, M., Saito, M., Shiota, S., Nakayama, H., and Yoshida, S. (2008). Concerted action of lactate oxidase and pyruvate oxidase in aerobic growth of *Streptococcus pneumoniae*: role of lactate as an energy source. J. Bacteriol. *190*, 3572–3579.

Terahara, M., Kurama, S., and Takemoto, N. (2001). Prevention by lactic acid bacteria of the oxidation of human LDL. Biosci. Biotechnol. Biochem. *65*, 1864–1868.

Tong, H., Chen, W., Shi, W., Qi, F., and Dong, X. (2008). SO-LAAO, a novel L-amino acid oxidase that enables *Streptococcus oligofermentans* to outcompete *Streptococcus mutans* by generating H_2O_2 from peptone. J. Bacteriol. *190*, 4716–4721.

Tsou, C.C., Chiang-Ni, C., Lin, Y.S., Chuang, W.J., Lin, M.T., Liu, C.C., and Wu, J.J. (2008). An iron-binding protein, Dpr, decreases hydrogen peroxide stress and protects *Streptococcus pyogenes* against multiple stresses. Infect. Immun. *76*, 4038–4045.

Ulijasz, A.T., Andes, D.R., Glasner, J.D., and Weisblum, B. (2004). Regulation of iron transport in *Streptococcus pneumoniae* by RitR, an orphan response regulator. J. Bacteriol. *186*, 8123–8136.

Uziel, O., Borovok, I., Schreiber, R., Cohen, G., and Aharonowitz, Y. (2004). Transcriptional regulation of the *Staphylococcus aureus* thioredoxin and thioredoxin reductase genes in response to oxygen and disulfide stress. J. Bacteriol. *186*, 326–334.

Vance, P.G., Keele, B.B., Jr., and Rajagopalan, K.V. (1972). Superoxide dismutase from *Streptococcus mutans*. Isolation and characterization of two forms of the enzyme. J. Biol. Chem. *247*, 4782–4786.

Veiga, P., Bulbarela-Sampieri, C., Furlan, S., Maisons, A., Chapot-Chartier, M.P., Erkelenz, M., Mervelet, P., Noirot, P., Frees, D., Kuipers, O.P., *et al.* (2007). SpxB regulates O-acetylation-dependent resistance of *Lactococcus lactis* peptidoglycan to hydrolysis. J. Biol. Chem. *282*, 19342–19354.

Verneuil, N., Sanguinetti, M., Le Breton, Y., Posteraro, B., Fadda, G., Auffray, Y., Hartke, A., and Giard, J.C. (2004). Effects of the *Enterococcus faecalis hypR* gene encoding a new transcriptional regulator on oxidative stress response and intracellular survival within macrophages. Infect. Immun. *72*, 4424–4431.

Vido, K., Diemer, H., Van Dorsselaer, A., Leize, E., Juillard, V., Gruss, A., and Gaudu, P. (2005). Roles of thioredoxin reductase during the aerobic life of *Lactococcus lactis*. J. Bacteriol. *187*, 601–610.

Walter, J., Chagnaud, P., Tannock, G.W., Loach, D.M., Dal Bello, F., Jenkinson, H.F., Hammes, W.P., and Hertel, C. (2005). A high-molecular-mass surface protein (Lsp) and methionine sulfoxide reductase B (MsrB) contribute to the ecological performance of *Lactobacillus reuteri* in the murine gut. Appl. Environ. Microbiol. *71*, 979–986.

Wan, X.Y., Zhou, Y., Yan, Z.Y., Wang, H.L., Hou, Y.D., and Jin, D.Y. (1997). Scavengase p20: a novel family of bacterial antioxidant enzymes. FEBS Lett. *407*, 32–36.

Winstedt, L., Frankenberg, L., Hederstedt, L., and von Wachenfeldt, C. (2000). *Enterococcus faecalis* V583 contains a cytochrome bd-type respiratory oxidase. J. Bacteriol. *182*, 3863–3866.

Yamamoto, Y., Higuchi, M., Poole, L.B., and Kamio, Y. (2000a). Identification of a new gene responsible for the oxygen tolerance in aerobic life of *Streptococcus mutans*. Biosci. Biotechnol. Biochem. *64*, 1106–1109.

Yamamoto, Y., Higuchi, M., Poole, L.B., and Kamio, Y. (2000b). Role of the *dpr* product in oxygen tolerance in *Streptococcus mutans*. J. Bacteriol. *182*, 3740–3747.

Yamamoto, Y., Poole, L.B., Hantgan, R.R., and Kamio, Y. (2002). An iron-binding protein, Dpr, from *Streptococcus mutans* prevents iron-dependent hydroxyl radical formation *in vitro*. J. Bacteriol. *184*, 2931–2939.

Yamamoto, Y., Fukui, K., Koujin, N., Ohya, H., Kimura, K., and Kamio, Y. (2004). Regulation of the intracellular free iron pool by Dpr provides oxygen tolerance to *Streptococcus mutans*. J. Bacteriol. *186*, 5997–6002.

Yamamoto, Y., Poyart, C., Trieu-Cuot, P., Lamberet, G., Gruss, A., and Gaudu, P. (2005). Respiration metabolism of Group B *Streptococcus* is activated by environmental haem and quinone and contributes to virulence. Mol. Microbiol. *56*, 525–534.

Yamamoto, Y., Pargade, V., Lamberet, G., Gaudu, P., Thomas, F., Texereau, J., Gruss, A., Trieu-Cuot, P., and Poyart, C. (2006). The Group B *Streptococcus* NADH oxidase Nox-2 is involved in fatty acid biosynthesis during aerobic growth and contributes to virulence. Mol. Microbiol. *62*, 772–785.

Yesilkaya, H., Kadioglu, A., Gingles, N., Alexander, J.E., Mitchell, T.J., and Andrew, P.W. (2000). Role of manganese-containing superoxide dismutase in oxidative stress and virulence of *Streptococcus pneumoniae*. Infect. Immun. *68*, 2819–2826.

Yu, J., Bryant, A.P., Marra, A., Lonetto, M.A., Ingraham, K.A., Chalker, A.F., Holmes, D.J., Holden, D., Rosenberg, M., and McDevitt, D. (2001). Characterization of the *Streptococcus pneumoniae* NADH oxidase that is required for infection. Microbiology *147*, 431–438.

Zhao, G., Ceci, P., Ilari, A., Giangiacomo, L., Laue, T.M., Chiancone, E., and Chasteen, N.D. (2002). Iron and hydrogen peroxide detoxification properties of DNA-binding protein from starved cells. A ferritin-like DNA-binding protein of *Escherichia coli*. J. Biol. Chem. *277*, 27689–27696.

Response of *Bifidobacterium* Species to Oxygen

6

Shinji Kawasaki

Abstract

Members of the genus *Bifidobacterium* are classified as anaerobes that are known to be beneficial to human health. This group is expected to prove highly valuable for use in milk products, health promotion, and treatment of intestinal disorders. However, their sensitivity to O_2 may limit their probiotic activity to solely anaerobic habitats. Recent studies have reported that the *Bifidobacterium* strains exhibit various types of oxic growth. Similar to the effectof CO_2, low concentrations of O_2 can have a stimulatory effect on the growth of *Bifidobacterium* strains. Based on the growth profiles under different O_2 concentrations, the *Bifidobacterium* species were classified into four classes: O_2-hypersensitive, O_2-sensitive, O_2-tolerant, and microaerophilic. The primary factor responsible for aerobic growth inhibition is proposed to be the production of H_2O_2 in the growth medium. An H_2O_2-forming NADH oxidase was purified from O_2-sensitive *Bifidobacterium bifidum* and was identified as a *b*-type dihydroorotate dehydrogenase. Kinetic parameters suggested that the enzyme could be involved in H_2O_2 production in highly aerated environments.

Oxygen and anaerobes

Anaerobes such as sulfate-reducing bacteria; methanogens; and *Bacteroides, Clostridium,* and *Bifidobacterium* species show inhibited growth under aerobic conditions. Over the last half-century, several approaches have been used to determine the mechanisms underlying aerobic growth inhibition in anaerobes. By using these approaches, several theories have been proposed to account for this phenomenon, including the lack of reactive oxygen species (ROS) scavenging enzymes such as catalase and superoxide dismutase, molecular O_2 toxicity, and imbalances in the oxidative/redox potential (McCord *et al.,* 1971; O'Brien and Morris, 1971; Woods and Jones, 1986; Storz *et al.,* 1990). Recent research has revealed that these anaerobes exhibit various types of oxic growth over an O_2 concentration range of 0.1–21% along with the expression of O_2-responsive enzymes (Rocha *et al.,* 1996; Fareleira *et al.* 2003; Kawasaki *et al.,* 2004a,b; Baughn and Malamy, 2004; Dolla *et al.* 2006; Wildschut *et al.,* 2006; Lobo *et al.,* 2007; Mukhopadhyay *et al.,* 2007). For example, sulfate-reducing bacteria and *Clostridium* species, both of which are classified as obligate anaerobes, express unique enzyme complexes that efficiently scavenge O_2 and ROS when they are grown in oxic environments (Wildschut *et al.,* 2006; Kawasaki *et al.,* 2009b). These findings indicate that anaerobes have developed systems in order to survive in growth habitats that are commonly contaminated with O_2.

History of research on the bifidobacterial oxidative stress response

Since the isolation of *Bif. bifidum* from human faeces in 1900, *Bifidobacterium* strains have been isolated using anaerobic techniques. de Vries and Stouthamer (1969) classified *Bifidobacterium* species into three categories according to their O_2 sensitivity. They proposed that O_2-sensitive species produce H_2O_2 owing to the activity of NADH oxidase that was detected in cell extracts. Since then, several different approaches have been used to determine the correlation between oxidative growth inhibition and NADH oxidase activity. Shimamura *et al.* (1990) reported the presence of O_2 uptake activity that was suggested to be catalysed by NADH oxidase. It was also demonstrated that three *Bifidobacterium* species – *Bifidobacterium infantis* ATCC15697, *Bifidobacterium breve* ATCC15700, and *Bifidobacterium longum* ATCC15707 – possess NADH oxidase and NADH peroxidase activity, and that higher activities of these enzymes were detected in these species as compared with the O_2-sensitive species *Bifidobacterium adolescentis* ATCC15703 (Shimamura *et al.*, 1992). Although these species (*Bif. infantis*, *Bif. breve* and *Bif. longum*) were reported to be O_2 tolerant, their growth under semi-aerobic culture conditions (partial shaking for 10 seconds once every 30 minutes) was inhibited by half as compared with that under anaerobic conditions. Ahn *et al.* (2001) isolated O_2-tolerant *Bif. longum* strains from human faeces, which have been reported to grow under partial aerobic culture conditions (shaking at 40 strokes per minute), and their growth was inhibited by approximately one-third as compared with that under anoxic conditions. In a recent report by Gonzalez *et al.* (2004), *Bif. infantis* ATCC17930 was shown to grow well in 50–80% air-saturated liquid medium, and the growth profile thus obtained was similar to that of obtained under anaerobic growth. A metabolic shift in this strain was observed when the growth conditions changed from anoxic to oxic in that the production of lactic acid and its molar ratio relative to acetate increased in the presence of O_2. A highly O_2-tolerant species, *Bifidobacterium psychraerophilum* LMG21775^T, was isolated from porcine caecum by Simpson *et al.* (2004). This strain was reported to form colonies (~1 mm) after a 3-day incubation period under aerobic conditions, whereas colony size was 3 mm after a 3-day incubation under anaerobic

Table 6.1 Reports concerning the physiology of *Bifidobacterium* species in the presence of O_2

Authors	Year	Culture methods	Specific comments
Scardovi *et al.*	1960–70	Slant or stab culture under air	Identification of strains that can develop colonies under air with 10% CO_2 enrichment
de Vries *et al.*	1969	Liquid shaking culture under air (300 rpm at OD=0.3)	NADH oxidase is suggested to be a key enzyme in H_2O_2 production
Shimamura *et al.*	1991	Liquid shaking culture under air (partial aerobic: shaking every 30 minutes; aerobic: shaking at 45 rpm using L-tube)	O_2-sensitive species possess weak NADH peroxidase activity
Ahn *et al.*	2001	Liquid shaking culture under air of isolated *B.longum* strains (40 rpm)	Change in lipid composition under aerobic condition
Talwalkar *et al.*	2003	Static culture with agitation using a magnetic stirrer under 0%, 5%, 10%, 15%, and 21% O_2	Change in carbohydrate metabolism under different O_2 conditions
Gonzalez *et al.*	2004	Air-flow culture (50–80% air flow) of *B. infantis*	Change in carbohydrate metabolism under aerobic conditions
Simpson *et al.*	2005	Liquid shaking culture under air (150 rpm) using 500 ml conical flask containing 100 ml medium	Determination of intrinsic tolerance of *Bifidobacterium* species to O_2
Kawasaki *et al.*	2006	Liquid shaking culture under 0%, 5%, 10%, or 20% O_2 (130 rpm) using 150 ml serum bottles containing 30 ml medium	Identification of microaerophilic species whose metabolism does not change in the presence of O_2.
Kawasaki *et al.*	2007	Plate culture under 0%, 5%, 10%, or 20% O_2 in the presence or absence of CO_2	CO_2 is essential for colony development. O_2 can replace the effect of CO_2

conditions. Simpson *et al.* (2005) investigated the intrinsic tolerance of *Bifidobacterium* species to O_2 and found that *Bifidobacterium boum* LMG10736^T, *Bifidobacterium minimum* DSMZ10105^T, and *Bif. pyschraerophilum* LMG21775^T were highly tolerant to O_2, exhibiting an aerobic growth rate that was more than 30% as compared with the anaerobic growth. A summary of the research undertaken in the past half-century on the effects of O_2 on *Bifidobacterium* species physiology is shown in Table 6.1.

Physiological responses of *Bifidobacterium* species under different O_2 concentrations

To investigate the physiological characteristics of *Bifidobacterium* species towards O_2, selected species were grown under different O_2 concentrations (0%, 5%, 10%, and 20%) in liquid shaking cultures at 130 rpm (Kawasaki *et al.*, 2006). The results of the representative species are shown in Fig. 6.1A. Most of the selected species, including *Bif. bifidum* JCM1255^T, a type species of the genus *Bifidobacterium*, grew well in the presence of 5% O_2 (without growth inhibition), but stopped growing under 10% O_2. *Bif. longum* JCM1217^T stopped growing at 10% O_2, but the growth in the presence of 5% O_2 was increased in comparison to anoxic growth. Some of the species tested, such as *Bif. minimum* DSM20102^T and *Bifidobacterium asteroids* JCM8230^T, showed an O_2-tolerant profile in which growth was slightly inhibited in the presence of O_2. *Bif. boum* JCM1211^T and *Bifidobacterium thermophilum* JCM1207^T grew well in the presence of 20% O_2 with the growth being stimulated in comparison to that under anoxic conditions. Although the above tested strains all belong to the genus *Bifidobacterium*, the growth profiles of these species vary under different O_2 concentrations. We classified the *Bifidobacterium* species into four classes: O_2-hypersensitive, O_2-sensitive, O_2-tolerant, and microaerophilic (Fig. 6.1B). The species classified as O_2-tolerant and microaerophilic are all isolates from nonhuman sources: sewage (*Bif. minimum*), bovine

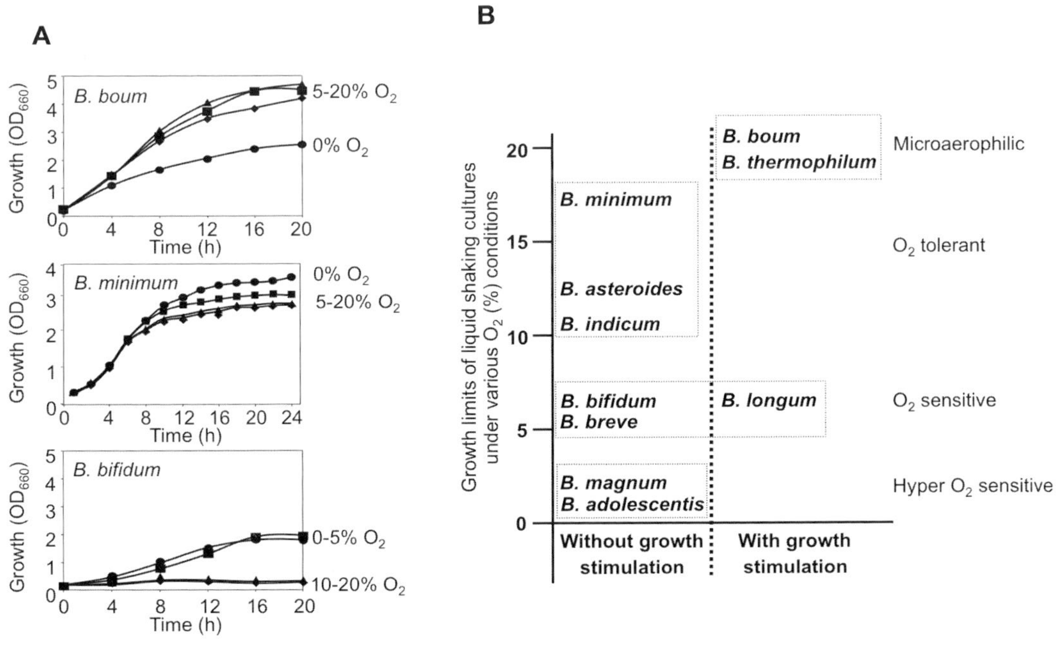

Figure 6.1 (A) Growth of the selected *Bifidobacterium* strains under various O_2 conditions in liquid shaking cultures. Circles, 100% N_2; squares, 5% O_2–95% N_2; triangles, 10% O_2–90% N_2; diamonds, 20% O_2–80% N_2. (B) Classification of strains according to O_2 tolerance. Tested strains were the type strain of species obtained from DSM or JCM.

rumen (*Bif. boum* and *Bif. thermophilum*), and honey-bee hindgut (*Bif. asteroides* and *Bifidobacterium indicum*).

Talwalkar and Kailasapathy (2003) reported a metabolic shift in some *Bifidobacterium* species under aerobic conditions. They tested the growth of several *Bifidobacterium* species (*Bif. infantis* CSIRO1912, *Bifidobacterium lactis* CSIRO1941, *Bifidobacterium pseudolongum* CSIRO1944 and *Bif. longum* ATCC55815) by agitating the static culture medium using a magnetic stirrer under several O_2 concentrations (0%, 5%, 10%, 15%, or 21%), and all of the strains except *Bif. infantis* CSIRO1912 were reported to grow in 21% O_2 condition (final growth yields were not presented in the text). Further, decreases and increases in lactate and acetate levels respectively were reported when these strains were grown aerobically. A similar fermentation shift was observed for *Bif. longum* JCM1217^T under 5% O_2 (Kawasaki *et al*, 2006). This aerobic metabolic shift was expected for the microaerophilic species *Bif. boum* JCM1211^T and *Bif. thermophilum* JCM1207^T, because of the drainage of reducing power to O_2, which has been speculated in some reports on lactic acid bacteria and *Bifidobacterium* species (Condon, 1987; Smart and Thomas, 1987; Sakamoto and Komagata, 1996; Talwalkar and Kailasapathy, 2003), but no significant metabolic shifts were observed in these two species (Kawasaki *et al*, 2006). In the case of *Bif. boum*, a slight decrease in carbon recovery from glucose to acetate and lactate production was observed under oxic growth conditions.

Effect of CO_2 on the growth of *Bifidobacterium* species

According to the description of *Bifidobacterium* species in Bergy's manual of systematic bacteriology (Jones and Collins, 1986), CO_2 does not affect anaerobic growth of most species except for *Bifidobacterium angulatum* and *Bif. asteroids*. It has also been demonstrated that some *Bifidobacterium* species such as *Bif. indicum*, *Bif. asteroids*, *Bifidobacterium globosum*, *Bif. boum*, and *Bif. thermophilum* grow in the presence of air particularly if the air is enriched with 10% CO_2 (Scardovi *et al.*, 1969a,b, 1979). Although CO_2 has conventionally been used in *Bifidobacterium* species culture atmospheres, the molecular mechanisms underlying the effect of CO_2 on the growth of *Bifidobacterium* species under anoxic and oxic conditions remain unknown.

To observe the effect of CO_2 on colony formation, four *Bifidobacterium* species, *Bif. bifidum* JCM1255^T, *Bif. longum* JCM1217^T, *Bif. boum* JCM1211^T, and *Bif. thermophilum* JCM1207^T were selected (Kawasaki *et al.*, 2007). Colony formation by these species was difficult under 100% N_2, but stimulation of colony development was observed in the presence of CO_2, and the stimulatory effects were similar under conditions of 1% CO_2_99% N_2 (Fig. 6.2 for *Bif. boum*) or 20% CO_2_80% N_2. In the absence of CO_2, *Bif. boum* efficiently developed colonies under 5% O_2_95% N_2 (Fig. 6.2); *Bif. longum* also exhibited a similar growth profile when grown under 0.1% O_2_99.9% N_2 (data not shown). These results indicate that CO_2 is not an essential contributing factor for colony development in *Bif. boum* and *Bif. longum* if

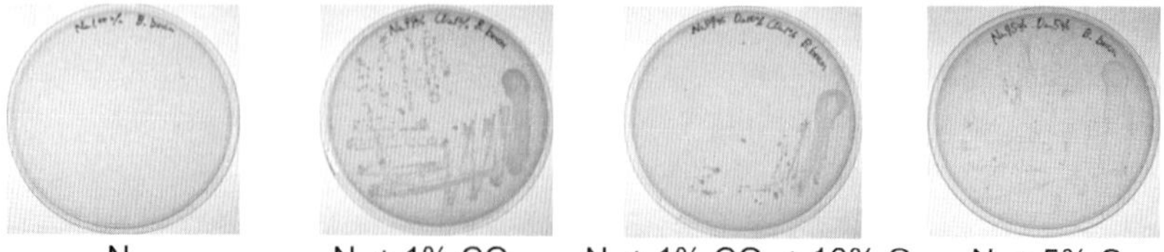

Figure 6.2 The effect of CO_2 and O_2 on colony development of *Bifidobacterium boum* JCM1211^T. *Bif. boum* showed stimulated colony development in the presence of 1% CO_2. This growth profile is similar to that of *Bif. bifidum* JCM1255^T, *Bif. longum* JCM1217^T, and *Bif. thermophilum* JCM1207^T (Kawasaki *et al.*, 2007). High O_2 concentrations inhibited colony development, and a low O_2 concentration could replace the effect of CO_2.

an appropriate concentration of O_2 is present as an alternative to CO_2. Although O_2 can at times serve as a growth stimulator, its presence appears to be intrinsically toxic for the growth of *Bifidobacterium* species (Kawasaki *et al.*, 2007). For example, *Bif. bifidum* and *Bif. longum* did not form colonies in an atmosphere containing 1% CO_2_5% O_2_94% N_2. Although *Bif. boum* and *Bif. thermophilum* developed colonies in an atmosphere containing N_2 including 1% CO_2 and 5% to 20% O_2 (Fig. 6.2 for *Bif. boum*), colony development was inhibited as the O_2 concentration increased (visible colonies were detected under 15% and 20% O_2, but the colony diameters were less than 0.1 mm). These results indicate that CO_2 is not so much a factor involved in improved O_2 tolerance, as it is a growth stimulator. Furthermore, the mechanism underlying the stimulatory effect of O_2 on *Bif. boum* colony development may in fact be similar to that underlying the effect of CO_2.

Determination of the factor involved in the inhibition of growth by O_2 in O_2-sensitive *Bifidobacterium* species

Although the growth response of *Bifidobacterium* species to O_2 has been investigated, molecular-based approaches have rarely been used to determine the O_2 sensitivity of these species. Growth profiles and the physiology of *Bifidobacterium* species under different O_2 conditions indicated that H_2O_2 production is the primary factor involved in the aerobic growth inhibition of *Bifidobacterium* species. In particular, the growth of O_2-sensitive species in the presence of O_2 recovered when catalase was exogenously added (de Vries and Stouthamer, 1969; Shimamura *et al.*, 1992; Kawasaki *et al.*, 2006). Two main theories have been proposed for the sensitivity of *Bifidobacterium* species to O_2; (a) the existence of H_2O_2 production systems and (b) the lack of H_2O_2 detoxification systems. In terms of systems that reduce O_2 to H_2O_2, only NADH oxidase activity has been detected in the cell extracts of all tested bifidobacterial species. Although the total NADH-dependent oxidase activity in cell extracts is similar among species, the type of NADH oxidase activity differs between O_2-sensitive and microaerophilic species – O_2-sensitive species exhibit H_2O_2-forming activity, whereas microaerophilic species exhibit H_2O-forming activity (Kawasaki *et al.*, 2006).

To investigate the enzyme that produces H_2O_2, H_2O_2-forming type NADH oxidase was purified and characterized from *Bif. bifidum* (Kawasaki

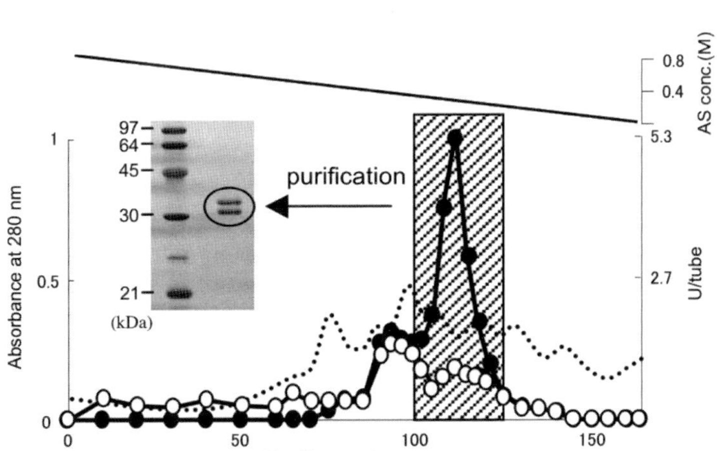

Figure 6.3 Fractionation of NADH oxidase activities in *Bifidobacterium bifidum* JCM1255^T. *B*-type dihydroorotate dehydrogenase was purified as a H_2O_2-forming type NADH oxidase. Active fractions (in the shaded region) were obtained using hydrophobic column chromatography from O_2-sensitive *Bif. bifidum*. *Dashed line*, protein absorption at 280 nm; *black circles*, NADH oxidase activity; *white circles*, NADPH oxidase activity; *solid line*, ammonium sulfate (AS) concentration. Fraction volume, 9 ml.

et al., 2009a). During purification, a dominant active fraction and a minor active fraction were detected (Fig. 6.3). The purified protein from the major active fraction is a homologue of *b*-type dihydroorotate dehydrogenase (DHOD; EC 1.3.1.14, EC 1.3.3.1, EC 1.3.99.11) that is composed of PyrK (31 kDa) and PyrDb (34 kDa), a flavoprotein enzyme that catalyses the oxidation of dihydroorotate to orotate in pyrimidine biosynthesis. Many enzymes that function as NADH oxidase *in vivo* have already been purified and characterized. However, to our knowledge, no DHOD has been purified as a main source of NADH oxidase activity in cell extracts. The kinetic parameters indicated that the enzyme that catalyses the DHOD reaction also could be involved in H_2O_2 production in highly aerated environments (e.g. K_m for $O_2 = 779 \pm 82$ μM) (Kawasaki *et al.,* 2009a). The O_2-dependent growth of *Bif. bifidum* (Fig. 6.1) is closely correlated with the reactivity of *Bif. bifidum* DHOD to O_2. The distribution of H_2O_2-forming NADH oxidase activity and the kinetic differences in DHOD among *Bifidobacterium* species are currently being investigated.

Conclusion

Bifidobacterium strains exhibit various species-dependent types of oxic growth. If O_2-sensitive species were unviable in oxic environments, it is likely that these species would not have survived into the earth's present environment. This suggests that *Bifidobacterium* species possess a novel survival system under oxic conditions. Further investigation of the O_2 metabolism and resistance to O_2 in *Bifidobacterium* species will help clarify mechanisms underlying oxygen adaptation.

References

Ahn, J.B., Hwang, H.J., and Park, J.H. (2001). Physiological responses of oxygen-tolerant anaerobic *Bifidobacterium longum* under oxygen. J. Microbiol. Biotechnol. *11,* 443–451.

Baughn, A.D., and Malamy M.H. (2004). The strict anaerobe *Bacteroides fragilis* grows in and benefits from nanomolar concentrations of oxygen. Nature *42,* 441–444.

Condon, S. (1987). Responses of lactic acid bacteria to oxygen. FEMS Microbiol. Rev. *46,* 269–280.

Dolla, A., Fournier, M., and Dermoun, Z. (2006). Oxygen defense in sulfate-reducing bacteria. J. Biotechnol. *126,* 87–100.

Fareleira, P., Santos, B.S., António, C., Moradas-Ferreira, P., LeGall, J., Xavier, A. V., and Santos, H. (2003). Response of a strict anaerobe to oxygen: survival strategies in *Desulfovibrio gigas.* Microbiology *149,* 1513–1522.

Gonzalez, R., Blancas, A., Santillana, R., Azaola, A., and Wacher, C. (2004). Growth and final product formation by *Bifidobacterium infantis* in aerated fermentations. Appl. Microbiol. Biotechnol. *65,* 606–610.

Jones, D., and Collins, M.D. (1986). Irregular, nonsporing Gram-positive rods. In Bergey's Manual of Systematic Bacteriology, Sneath, P.H.A., Mair, N.S., Sharpe M.E., and Holt, J.G., eds. (Williams and Wilkins Co., Baltimore), pp. 1261–1434.

Kawasaki, S., Ishikura, J., Chiba, D., Nishino, T., and Niimura, Y. (2004a). Purification and characterization of an H_2O-forming NADH oxidase from *Clostridium aminovalericum*: existence of an oxygen-detoxifying enzyme in an obligate anaerobic bacteria. Arch. Microbiol. *181,* 324–330.

Kawasaki, S., Ishikura, J., Watamura, Y., and Niimura, Y. (2004b). Identification of O_2-induced peptides in an obligatory anaerobe, *Clostridium acetobutylicum.* FEBS Lett. *571,* 21–25.

Kawasaki, S., Mimura, T., Satoh, T., Takeda, K., and Niimura, Y. (2006). Response of microaerophilic *Bifidobacterium boum* and *Bifidobacterium thermophilum* to oxygen. Appl. Environ. Microbiol. *72,* 6854–6858.

Kawasaki, S., Nagasaku, M., Mimura, T., Katashima, H., Ijuin, S., Satoh, T., and Niimura, Y. (2007). Requirement of CO_2 on the colony development of *Bifidobacterium* species. Appl. Environ. Microbiol. *73,* 7796–7798.

Kawasaki, S., Satoh, T., Todoroki, M., and Niimura Y. (2009a). *b*-type dihydroorotate dehydrogenase is purified as a H_2O_2-forming NADH oxidase from *Bifidobacterium bifidum.* Appl. Environ. Microbiol. *75,* 629–636.

Kawasaki, S., Sakai, Y., Takahashi, T., Suzuki, I., and Niimura Y. (2009b). O_2 and reactive oxygen species detoxification complex composed of O_2-responsive NADH:rubredoxin oxidoreductase – flavoportein A2 – desulfoferrodoxin operon enzymes, rubperoxin, and rubredoxin in *Clostridium acetobutylicum.* Appl. Environ. Microbiol. *75,* 1021–1029.

Lobo, S.A., Melo, A. M., Carita, J.N., Teixeira, M., and Saraiva, L.M. (2007). The anaerobe *Desulfovibrio desulfuricans* ATCC 27774 grows at nearly atmospheric oxygen levels. FEBS Lett. *581,* 433–436.

McCord, J.M., Keele, B.B., Jr, and Fridovich, I. (1971). An enzyme-based theory of obligate anaerobiosis: the physiological function of superoxide dismutase. Proc. Natl. Acad. Sci. U.S.A. *68,* 1024–1027.

Mukhopadhyay, A., Redding A.M., Joachimiak M.P., Arkin A.P., Borglin S.E., Dehal, P.S., Chakraborty, R., Geller, J.T., Hazen, T.C., He, Q., *et al.* (2007). Cell-wide responses to low-oxygen exposure in *Desulfovibrio vulgaris* Hildenborough. J. Bacteriol. *189,* 5996–6010.

O'Brien, R.W., and Morris J.G. (1971). Oxygen and growth and metabolism of *Clostridium acetobutylicum*. J. Gen. Microbiol. *68*, 307–318.

Rocha, E.R., Selby, T., Coleman, J.P., and Smith C.J. (1996). Oxidative stress response in an anaerobe, *Bacteroides fragilis*: a role for catalase in protection against hydrogen peroxide. J. Bacteriol. *178*, 6895–6903.

Scardovi, V., and Crociani, F. (1974). *Bifidobacterium catenulatum*, *Bifidobacterium dentium*, and *Bifidobacterium angulatum*: three new species and their deoxyribonucleic acid homology relationships. Int. J. Syst. Bacteriol. *24*, 6–20.

Scardovi, V., and Trovatelli, L. D. (1969). New species of bifid bacteria from *Apis mellifica* L., and *Apis indica* F. A contribution to the taxonomy and biochemistry of the genus *Bifidobacterium*. Zentralbl. Bakteriol. Parasitenkd. Infektionskr. Hyg. Zweite Abt. *123*, 64–88.

Scardovi, V., Trovatelli, L.D., Crociani, F., and Sgorbati, B. (1969). Bifid bacteria in bovine rumen. New species of the genus *Bifidobacterium*: *B. globosum* n.sp. and *B. ruminale* n.sp. Arch. Microbiol. *68*, 278–294.

Scardovi, V., Trovatelli, L.D., Biavati, B., and Zani, G. (1979). *Bifidobacterium cuniculi*, *Bifidobacterium choerinum*, *Bifidobacterium boum*, and *Bifidobacterium pseudocatenulatum*: Four new species and their deoxyribonucleic acid homology relationships. Int. J. Syst. Bacteriol. *29*, 291–311.

Shimamura, S., Abe, F., Ishibashi, N., Miyakawa, H., Yaeshima, T., and Tomita, M. (1990). Endogenous oxygen uptake and polysaccharide accumulation in *Bifidobacterium*. Agric. Biol. Chem. *54*, 2869–2874.

Shimamura, S., Abe, F., Ishibashi, N., Miyakawa, H., Yaeshima, T., Araya, T., and Tomita, M. (1992). Relationship between oxygen sensitivity and oxygen metabolism of *Bifidobacterium* species. J. Dairy. Sci. *75*, 3296–3306.

Simpson, P.J., Ross, R.P., Fitzgerald, G.F., and Stanton, C. (2004). *Bifidobacterium psychraerophilum* sp. nov. and *Aeriscardo*via *aeriphila* gen. nov., sp. nov., isolated from a porcine caecum. Int. J. Syst. Evol. Microbiol. *54*, 401–406.

Simpson, P.J., Stanton, C., Fitzgerald, G.F., and Ross, R.P. (2005). Intrinsic tolerance of *Bifidobacterium* species to heat and oxygen and survival following spray drying and storage. J. Appl. Microbiol. *99*, 493–501.

Smart, J.B., and Thomas D.T. (1987). Effect of oxygen on lactose metabolism in lactic streptococci. Appl. Environ. Microbiol. *53*, 533–541.

Storz, G., Tartaglia, L.A., Farr, S.B., and Ames, B.N. (1990). Bacterial defenses against oxidative stress. Trends Genet. *6*, 363–368.

Talwalkar, A., and Kailasapathy, K. (2003). Metabolic and biochemical responses of probiotic bacteria to oxygen. J. Dairy. Sci. *86*, 2537–2546.

de Vries, W., and Stouthamer, A.H. (1969). Factors determining the degree of anaerobiosis of *Bifidobacterium* strains. Arch. Microbiol. *65*, 275–287.

Wildschut, J.D., Lang R.M., Voordouw J.K., and Voordouw, G. (2006). Rubredoxin:oxygen oxidoreductase enhances survival of *Desulfovibrio vulgaris* Hildenborough under microaerophilic conditions. J. Bacteriol. *188*, 6253–6260.

Woods, D.R., and Jones D.T. (1986). Physiological responses of *Bacteroides* and *Clostridium* strains to environmental stress factors. Adv. Microb. Physiol. *28*, 1–64.

Bile Acid Stress in Lactic Acid Bacteria and Bifidobacteria

7

Abelardo Margolles and Atsushi Yokota*

Abstract

Bile is mainly composed of bile acids, detergent-like biological substances synthesized from cholesterol in the liver. They are involved in the generation of bile flow, and their major physiological function is to facilitate the absorption of lipophilic compounds from the diet, including vitamins and lipids. Furthermore, bile also plays a crucial role in the establishment of the intestinal microbiota in humans, since bile salts are potent antimicrobial agents and only microbial populations able to cope with physiological concentrations of bile are able to survive in the gut. Bile disturbs the cell membrane functionality in lactic acid bacteria (LAB) and bifidobacteria owing to its amphipatic property. Studies on the bile stress response in LAB and bifidobacteria revealed that bile resistance is a result of integration of multilateral responses, which include those that protect cell membrane from direct attack by bile acids, those that restore or degrade damaged proteins, those that eliminate oxidative stress and facilitate DNA repair, and those that enhance energy generation through up-regulated sugar metabolism. This chapter summarizes the current knowledge of the molecular mechanisms used by LAB and bifidobacteria to tolerate the deleterious action of bile salts.

Introduction

Bile is a heterogeneous mixture of electrolytes, bile salts, phospholipids, cholesterol and bilirrubin that is secreted into the duodenum during digestion to facilitate emulsification and absorption of liposoluble nutrients. Bile salts are organic compounds with strong antimicrobial activity. Their amphipatic property allows them to have a detergent action, which induces membrane damage in intestinal bacteria and severely impairs membrane functionality. Furthermore, bile salts are assumed to cause oxidative stress, DNA damage and protein misfolding, and trigger a series of molecular mechanisms directed to fight against the harmful action of these compounds. Thus, the ability of intestinal bacteria to tolerate bile salts is critical for their colonization/persistence in the human host.

The functionality of probiotics depends on their ability to survive and temporarily persist in the human host. Therefore, bile resistance of probiotics such as lactic acid bacteria (LAB) and bifidobacteria has been for a long time the focus of numerous works, and bile tolerance has now become an *in vitro* selection criterion for probiotic bacteria. Studies of the response to bile salts challenge in LAB and bifidobacteria for the past 20 years revealed the presence of integrated and multilateral strategies to tolerate physiological bile salts concentrations during passage through the gut. This chapter gives an overview of the molecular mechanisms involved in bile tolerance in LAB and bifidobacteria. Furthermore, we aim to facilitate the understanding of the potential benefits of bile tolerance in probiotic bacteria for the reader.

Bile: physiology and functions in humans

Synthesis, enterohepatic circulation and functions

Bile is a complex mixture of organic and inorganic compounds in an aqueous solution. It is isotonic with plasma, mainly owing to the presence of

*corresponding author

sodium, potassium and chloride, which are the main inorganic ions. Among the organic compounds, the major constituents include bile salts, cholesterol, phospholipids, and proteins, although many endogenous (bilirrubin, vitamins, steroids) and exogenous (antibiotics and drugs) substances can also be detected (Begley *et al.*, 2005).

In humans, synthesis of bile is regulated by many factors including nutrients, hormones, and bile acids (Fuchs, 2003). Bile is synthesized by the hepatocytes in the liver and secreted via the common hepatic duct, which joints the cystic duct from the gallbladder to form the bile duct. The gallbladder is a small pear-like organ in which bile is concentrated. It is not essential for bile secretion but facilitates its storage in preparation for fat digestion (Begley *et al.*, 2005; Fig. 7.1).

During digestion, bile is poured into the duodenum by the muscles sphincter Oddii, in a process that is dependent on enteric hormones (Calhoun and Hanks, 1984). Once in the intestine bile carries out its main physiological function, that is the emulsification of liposoluble substances to facilitate their absorption.

Bile is primarily composed of bile acids (Begley *et al.*, 2005). Bile acids that are synthesized from cholesterol in the hepatocyte are termed primary bile acids. They are C_{24} cyclopentanephenanthrene sterols, and the two main primary bile acids are cholic acid (CA) and chenodeoxycholic acid (CDCA) (Fig. 7.1; Fuchs, 2003; Ridlon *et al.*, 2006). Before secretion, all bile acids are conjugated as N-acyl amidates with either glycine or taurine. The ratio of glyco- to tauro-conjugates in

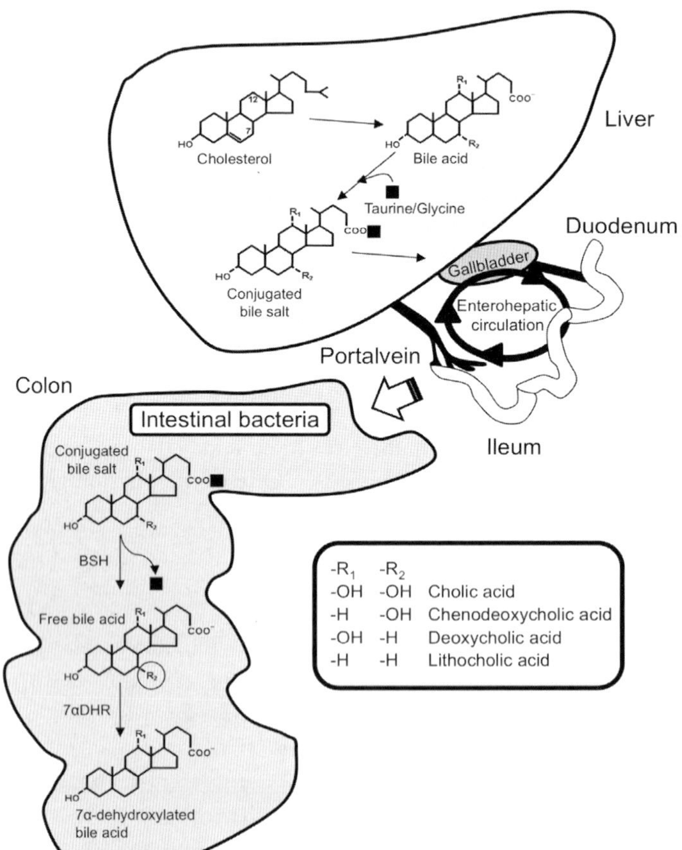

Figure 7.1 Synthesis, metabolism and enterohepatic circulation of bile salts. See text for details. BSH, bile salt hydrolase; 7αDHR, 7α-dehydroxylation reaction. For bile acid transformation in colon, only 7α-dehydroxylation reaction is illustrated as the representative reaction. ■, taurine or glycine.

human bile is usually around 3:1, although it can vary depending on specific individual populations or diet composition (Begley *et al.,* 2005). The conjugation decreases the pK_a of the acid and, at physiological pH, conjugated bile acids are almost fully ionized (in contrast to the unconjugated ones) and are named bile salts. Also, this conjugation has a deep impact on the solubility of the compounds, making them freely soluble over a wide range of ionic strengths and pH values (Begley *et al.,* 2005). Bile acids and bile salts are very often used synonymously and for practical reasons are used as such in this chapter.

After the pouring of bile into the small intestine, where fat digestion and absorption take place, bile salts are absorbed through active transport in the distal ileum, and returned to the liver through the portal vein. This process is known as the enterohepatic circulation of bile salts (Fig. 7.1). However, during their passage through the small intestine about 5% of the total bile salts escape active transport in the distal ileum and pass to the colon, in which they exert a strong selective pressure on the local microbiota. Therefore, along the intestinal tract there is a gradient of bile salts concentrations that goes from 40 mM (human hepatic bile) to less than 1 mM in the distal part of the large bowel (Ejderhamn *et al.,* 1991; Thomas *et al.,* 2001; Begley *et al.,* 2005; Hamilton *et al.,* 2007).

Metabolism of bile salts by gut microbiota

It is remarkable to point out the strong influence of the gut microbiota on the modifications undergone by bile salts during intestinal transit. These modifications are mainly carried out by fermentative bacteria in the large intestine, the most prevalent of these being deconjugation at the C-24 amide bond (Fig. 7.1). Deconjugation is carried out by bile salt hydrolases (BSH), enzymes common to many of the members of the intestinal microbiota, including *Lactobacillus* and *Bifidobacterium,* which release glycine or taurine from the steroid nuclei, rendering the free (unconjugated) bile acid (Begley *et al.,* 2006; Jones *et al.,* 2008). These reactions seem to be substrate limiting and go to completion in the large bowel (Ridlon *et al.,* 2006). When the structure of steroid nuclei of primary bile acids is modified by bacterial transformation, the resulting bile acids are termed secondary bile acids. The representative reactions are dehydroxylation and dehydrogenation of the hydroxyl groups at C-7. 7α-Dehydroxylation is the multi-step reaction and the most quantitatively important bacterial bile salt biotransformation in the human colon (Ridlon *et al.,* 2006). It is mainly carried out by members of the *Clostridium* group (Kitahara *et al.,* 2000, 2001; Wells *et al.,* 2003), and it seems to be absent in *Lactobacillus* and *Bifidobacterium* (Takahashi and Morotomi, 1994). This activity is restricted to free bile acids. Thus, a prerequisite for 7α-dehydroxylation is the previous removal of glycine/taurine by BSH (Batta *et al.,* 1990). By this reaction CA is converted to deoxycholic acid (DCA), and CDCA to lithocholic acid (LCA). Another reaction is 7α-dehydrogenation, which is an oxidative reaction catalysed by 7α-hydroxysteroid dehydrogenase (7α-HSDH), yielding 7-oxo derivatives of CA and CDCA. This enzyme has been detected in several intestinal bacteria, including *Clostridium scindens* (Baron *et al.,* 1991), several species of *Bacteroides* (Macdonald *et al.,* 1975; Sherrod and Hylemon, 1977; Bennett *et al.,* 2003; Fukiya *et al.,* 2009) and *Escherichia coli* (Yoshimoto *et al.,* 1991). Moreover, the resulting 7-oxo derivatives are further metabolized by other intestinal bacteria having 7β-hydroxysteroid dehydrogenase (7β-HSDH) to their 7β-OH derivatives (Masuda *et al.,* 1983). For example, the successive reactions by 7α-HSDH and 7β-HSDH yield ursodeoxycholic acid (UDCA) from CDCA.

One of the physiological consequences of the bile salt biotransformations of human intestinal bacteria is that faecal bile acid composition is significantly different from biliary bile salt composition. Comparison of bile acid composition in the gallbladder and faeces showed that primary bile acids (CA and CDCA) are predominant in the billiary bile (about 70% of the total bile acid pool). DCA is also accumulated in the billiary bile acid pool as a result of passive absorption, and the inability of the liver to hydroxylate DCA to CA. LCA, on the contrary, is hepatotoxic and poorly reabsorbed from the gut, thus it is hardly present in the bile. The secondary bile acids DCA and LCA are produced solely by microbial metabolism in the large intestine, and they predominate in human faeces (more than 60% of the total bile acid pool). CA and CDCA thus constitute less than 5% of total bile acids at this location (Ridlon *et al.*, 2006).

Physicochemical properties of bile salts

Understanding the physicochemical characteristics of bile salts is very relevant to interpreting their physiological function in humans and its antimicrobial activity. Bile acids are amphipatic molecules with distinctive detergent properties. This property is directly related to their capacity to emulsify and solubilize fats. Also the hydrophobicity-hydrophilicity character and detergent-like effect are related to their antimicrobial activities. A summary of the important physicochemical characteristics of bile salts is plotted in Table 7.1.

First of all, conjugation and deconjugation of the primary bile salts, CA and CDCA, exert an important influence on their physicochemical characteristics. Bile salts conjugation in the liver to either taurine or glycine decreases the pK_a and the hydrophobicity, although the conjugation with taurine confers a lower pK_a than that with glycine owing to the extreme acidity of the sulfonate group. On the other hand, deconjugation with BSH by intestinal bacteria increases pK_a and hydrophobicity (Carey, 1984; Begley *et al.*, 2005). Secondly, the number of functional hydroxyl groups also affects hydrophobicity of the bile salts. For example, bile acids with two hydroxyl groups, DCA (3α, 12α) and CDCA (3α, 7α), are much more hydrophobic than CA with three hydroxyl groups (3α, 7α, 12α). This holds true also with conjugated bile acids (Heuman, 1989; Table 7.1).

Bile acids self associate above a certain concentration (critical micellar concentration or CMC), in both aqueous and organic solvents, to form micelles. In aqueous solutions, bile salts form large micelles with non-polar hydrophobic interiors and polar hydrophilic surfaces facing the aqueous environment. The number of the bile acid molecules forming the micelles is dependent on bile salt structure, but usually each micelle contains between 4 and 50 molecules (Begley *et al.*, 2005). In physiological conditions bile salts can

Table 7.1 Physicochemical characteristics of main human bile salts

Bile salt	pK_a[1] Methanol/ water	pK_a[1] water	CMC[2] water	CMC[2] 0.15 mM NaCl	Hydrophobicity[3]
Cholic acid	6.58	5.05	13	11	0.13
Chenodeoxycholic acid	6.59	5.06	9	4	0.59
Deoxycholic acid	6.45	4.92	10	3	0.72
Lithocholic acid	–	~5	0.9	0.5	–
Ursodeoxycholic acid	6.57	5.04	19	7	–
Glycocholic acid	5.87	4.34	12	10	0.07
Glycochenodeoxycholic acid	5.80	4.27	6	2	0.51
Glycodeoxycholic acid	5.81	4.28	6	2	0.65
Glycolithocholic acid	–	–	–	–	1.05
Glycoursodeoxycholic acid	6.00	4.47	12	4	–
Taurocholic acid	2.15	0.74	10	6	0.00
Taurochenodeoxycholic acid	2.35	0.94	–	–	0.46
Taurodeoxycholic acid	2.34	1.02	–	–	0.59
Taurolithocholic acid	–	–	–	–	1.00
Tauroursodeoxycholic acid	2.35	0.94	–	–	–

[1]pK_a according to references (Carey, 1984; Cantafora and Di Biase, 1987).

[2]Critical micellar concentration reported in references (Roda *et al.*, 1983, 1990; Hofmann and Roda, 1984).

[3]Hydrophobicity is given as hydrophobicity indices of ionized forms based on reversed phase HPLC capacity factors (Heuman, 1989). –, not reported.

be present in the form of mixed micelles together with phospholipids, lecithin, and cholesterol, thus preventing the destruction of the cell membranes by monomeric bile acids. Also, these micelles serve to transport cholesterol from the liver to the intestine, thus enabling cholesterol to be excreted from the body.

Antimicrobial activity of bile acids in LAB and bifidobacteria: physiological studies

In order to understand antimicrobial activity of bile acids, it is appropriate to outline targets of bile acid attack and the mode of expression of resistance to bile acid in LAB and bifidobacteria.

Cell membrane: the main target of bile salts/acids attack

Electron microscopic analyses have revealed damaging effects of bile salts/acids on the cell surface of LAB and bifidobacteria (Bron *et al.* 2004b; Taranto *et al.*, 2006; Ruiz *et al.*, 2007). However, at this moment, these events have not been understood at molecular level. On the other hand, by far the most important target of bile salts/acids attack is undoubtedly the bacterial cell membrane. The property of bile salts/acids molecules (Table 7.1) interacting with the bacterial membrane is directly correlated to their cytotoxicity. Hydrophobic bile salts/acids are more toxic than hydrophilic ones, since their affinity for the membrane is higher. Conjugated bile acids are strong acids (e.g. tauroconjugate, Table 7.1) and are ionized at physiological pH values and usually unable to cross the membrane through passive diffusion (unless a transporter internalizes them). Free bile acids, on the contrary, are weak acids (Table 7.1) and their protonated form can pass across the lipid bilayer, without the need for a specific transporter, to enter the cell (Kurdi *et al.*, 2000, 2003). As a consequence, CA and DCA are much more inhibitory for cells than the corresponding taurine- or glycine-conjugated forms.

Minimal inhibitory concentrations (MICs) of bile have been described for *Lactobacillus* and *Bifidobacterium* species, and they usually range from 0.1% to 10% depending on the bile origin (porcine, bovine, or human) and the type of experiment performed (growth conditions, plate or liquid media, kind of media, etc.) (Dunne *et al.*, 2001; Margolles *et al.*, 2003; Delgado *et al.*, 2007, 2008; Ryan *et al.*, 2008). Although these values are important to consider bile tolerance of probiotics during passage in upper intestine, MICs for individual free bile acids are more important to evaluate their antimicrobial activities. MICs of free bile acids, CA and DCA, for many LAB and bifidobacteria have been investigated (Kurdi *et al.*, 2006). The MICs for CA ranged from 3 mM to 13 mM, while those of DCA appeared to be from 0.3 mM to 0.8 mM, suggesting higher cytotoxicity of DCA than CA (about 10 times). Since DCA with two hydroxyl groups is much more hydrophobic than CA with three hydroxyl groups (Table 7.1), DCA can interact with biological membrane better than CA. In fact, it has been reported that trans-bilayer movement (flip-flop) of DCA is at least 10 times faster than CA (Kamp and Hamilton, 1993). This means 10 times more DCA molecules than CA molecules permeate or cross the membrane within the same time period, thereby more severely disturbing the function of cell membrane than CA.

So, in which way can we correlate membrane disturbance by bile acids with antimicrobial activity? Detailed studies of interaction between bile acids and LAB/bifidobacteria at their different concentrations revealed that at around 0.1 mM CA, which is lower than physiological concentrations in human small intestine, no growth inhibition was observed and CA apparently does not disturb cell membrane. Under these conditions, CA is passively accumulated in the cell as a weak acid driven by the transmembrane proton gradient (ΔpH) (Kurdi *et al.*, 2000, 2003). At physiological concentrations of CA in human small intestine (2–6 mM) and DCA in human large intestine (0.2–0.6 mM), where growth inhibition was manifested, these bile acids lowered the internal pH levels and, at certain concentrations, dissipated both ΔpH and membrane potential (i.e. dissipation of proton-motive force). It was also shown that CA and DCA were able to disturb membrane integrity and that this effect can lead not only to the collapse of concentration gradient of ions (e.g. proton and potassium) but also to leakage of other cellular components (e.g. metabolic intermediates), and eventually to cell death. At extremely higher CA concentrations (15 mM), leakage of proteins was observed (Kurdi *et al.*, 2006). A series of these events is illustrated in Fig. 7.2. These results clearly demonstrated that the primary mechanism of growth inhibition by free bile acids *in vivo* is the membrane damage leading to the collapse of the barrier function, which resulted not only in the dissipation of proton-motive force (energy) but also in the loss of important solutes from the cytoplasm. These results also implied that protein leakage is physiologically not a direct mechanism of antimicrobial effects of bile acids.

Permeability increase owing to the treatment of bacterial cells with oxgall (bovine bile containing both free and conjugated bile acids) leading to the increased β-galactosidase activity in *Lactobacillus acidophilus* (Noh and Gilliland, 1993) and *Lactobacillus reuteri* (Valdez *et al.*, 1997) has been reported. Similarly, increased susceptibilities to aminoglycoside antibiotics in lactobacilli (Elkins and Mullis, 2004) and bifidobacteria (Kheadr *et al.*, 2007) have been reported, supporting the mode of antimicrobial action elucidated in LAB and bifidobacteria. Permeability increase owing to the treatment of lipid vesicles by CA has also been described (Treyer *et al.*, 2002).

Known mechanisms to counteract membrane damage by bile salts/acids attack

Alterations of fatty acid composition have been reported in response to bile attack in LAB and bifidobacteria (Gómez-Zavaglia *et al.*, 2002; Taranto *et al.*, 2003, 2006; Ruiz *et al.*, 2007). Since fatty acid composition has a deep impact on the physicochemical properties of biological membrane, these alterations must have biological importance as the stress response against bile attack. However, it seems difficult to draw consistent interpretation from the reported alterations as the defence strategies against bile challenge.

In relation to this, it has been demonstrated that adaptation to CA (see section 'Mode of expression of bile acid resistance' for definition) in a strain of *Lactobacillus gasseri* was correlated with alterations in the lipid composition of the cell membrane. During adaptation, significant alterations were detected in glycolipid and

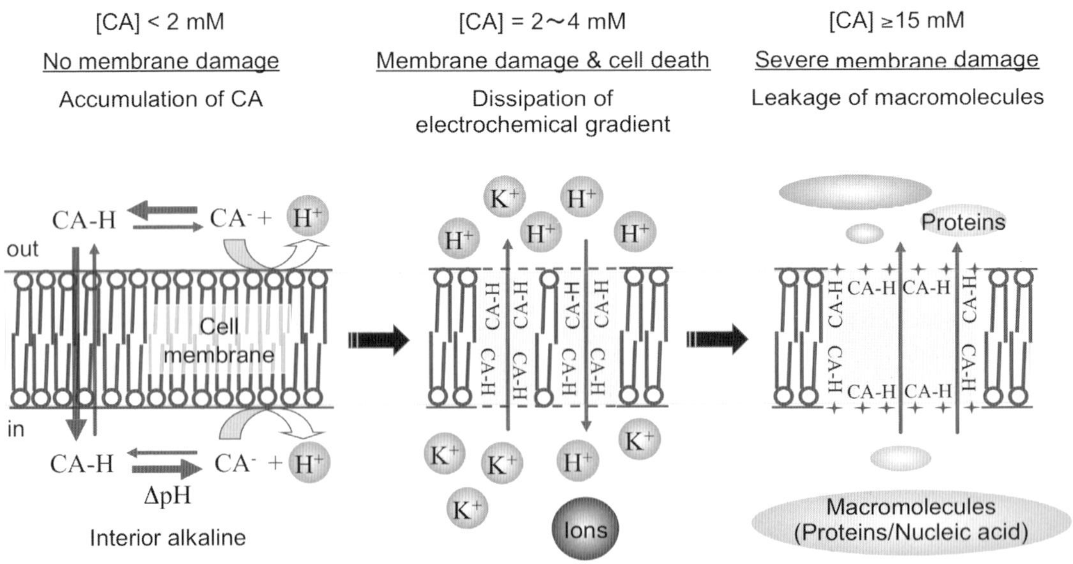

Figure 7.2 Interaction of cholic acid with bacterial cell membrane. See text for details. The thick arrows in the left illustration ([CA] < 2 mM) indicate a shift of the equilibration leading to accumulation of CA in the cell, which is driven by ΔpH. CA-H, protonated form of cholic acid; CA^-, dissociated form of cholic acid. Adopted from the original drawing appeared in Yokota, A. (2010). Bile acid (in Japanese). In Science of Lactic Acid Bacteria and Bifidobacteria. (Kyoto University Press, Kyoto, Japan), pp. 246–256.

phospholipids composition, with an increase in the length of the sugar chain in the glycolipid fraction and increase of cardiolipin (CL) content with concomitant decrease in phosphatidylglycerol (PG) in the phospholipid fraction. In fact, it was found that resistance of membrane vesicle against solubilization by CA increases with increasing CL/PG ratios. Although physiological importance of glycolipid alteration is not yet clear, the results highlight the previously unreported role of CL in conferring bile resistance to biological membrane (Yokota, *et al.* Ninth Symposium on Lactic Acid Bacteria, poster F024, 2008).

LAB and bifidobacteria display other mechanisms to counteract bile toxicity, such as active extrusion of bile acids through multidrug resistance (MDR) transporters that eliminate bile acids from cytoplasmic membrane (Yokota *et al.*, 2000; Price *et al.*, 2006; Zaidi *et al.*, 2008; Gueimonde *et al.*, 2009). Details will be described later in the corresponding sections.

One of the common physiological responses to bile in lactobacilli and bifidobacteria is the production of exocellular polymers. Bile induced the production of exopolysaccharide (EPS) in *Bifidobacterium animalis* subsp. *lactis* IPLA4549 and in its bile-resistant mutant 4549dOx (see section 'Physiological studies of the bile acid-resistant mutants') by overexpression of *gtf01207* that encodes a putative priming glycosyltransferase involved in EPS synthesis (Table 7.2) (Ruas-Madiedo *et al.*, 2009). By using cryo-scanning electron microscopy and quantification with lectin-conjugated probes, the increased production of EPS was demonstrated, probably as a mechanism of protection against a toxic action of bile (Ruas-Madiedo *et al.*, 2009). Bile strongly modulates the *in vitro* biofilm formation by *Lactobacillus rhamnosus* GG (Lebeer *et al.*, 2007). Thus, bile might therefore be a signal for *Lb. rhamnosus* GG to form biofilms in the small intestine (Lebeer *et al.*, 2007). In relation to this, a deletion mutant of *Lactobacillus johnsonii* NCC533 unable to synthesize exopolysaccharide displayed a significant alteration in residence time of the strain in the gastrointestinal tract of mice (Denou *et al.*, 2008). Enhanced biofilm formation by bile salt has also been reported in *Vibrio cholerae* (Hung *et al.*, 2006) and in *Bacteroides fragilis* (Pumbwe *et al.*, 2007), both of which have intestine as an environmental niche. These data suggest that bile could alter the biofilm and EPS production phenotypes in probiotics, probably facilitating the residence of these bacteria under a harmful environment containing bile, like the human intestine.

Other possible targets of bile acid attack: macromolecules

In mammalian cells, DCA and LCA have been shown to cause DNA damage and now these secondary bile acids are regarded as carcinogens. The mechanism is probably indirect, involving the induction of oxidative/nitrosative stress leading to the generation of reactive oxygen/nitrogen species, which then attack DNA (Bernstein *et al.*, 2005). Many of the genes that are consistently induced in LAB/bifidobacteria and other bacterial species by bile salts are responsive to oxidative stress/DNA repair (Bernstein *et al.*, 1999; Leverrier *et al.*, 2003; Bron *et al.*, 2006; Sánchez *et al.*, 2007b, 2008; Lee *et al.*, 2008). These results indirectly suggest that bile salts/acids induce oxidative stress/DNA damage also in bacterial cells, as in mammalian cells.

It has also been described in LAB and bifidobacteria that bile salts/acids stress induces molecular chaperons that promote proper protein folding and specific proteases that degrade irreparably damaged proteins, a common response to some stress stimuli (Flahaut *et al.*, 1996bc; Schmidt and Zink, 2000; Sánchez *et al.*, 2005, 2007b; Savijoki *et al.*, 2005; De Dea Lindner *et al.*, 2007). These results support protein denaturation as being one of the consequences encountered by bile-treated cells.

Mode of expression of bile acid resistance

There are three distinct modes of expression in the resistance to bile acids in LAB and bifidobacteria (van de Guchte *et al.*, 2002). (i) When exponentially growing cells are exposed to a sublethal concentration of a bile acid, these cells acquire increased resistance to otherwise lethal concentrations of the same bile acid. This phenomenon is called 'bile adaptation'. Not restricted to bile acid challenge, adaptation to environmental stresses is a commonly

Table 7.2 Main protein or gene players involved in bile response in *Lactococcus*, *Lactobacillus* and *Bifidobacterium*

Protein or gene	Putative function	Physiological relevance	Reference
L. lactis			
LmrCD↑	MDR transporter (ABC family)	Extrusion-based mechanism of bile salts	Lubelski *et al.* (2006), Zaidi *et al.* (2008)
LmrR	Transcriptional regulator	Negative regulator of *lmrCD*expression	Lubelski *et al.* (2006), Agustiandari *et al.* (2008)
groES+	GroES/Hsp10 chaperone	ND	Zaidi *et al.* (2008)
nah+	Na^+/H^+ antiporter	ND	Zaidi *et al.* (2008)
htrA+	Housekeeping protease	ND	Zaidi *et al.* (2008)
groEL➦	GroEL/Hsp60 chaperone	ND	Zaidi *et al.* (2008)
hslO+	Heat shock protein; 33-kDa chaperone	ND	Zaidi *et al.* (2008)
clpE×	ATP-dependent Clp protease ATP-binding subunit E	ND	Zaidi *et al.* (2008)
cdsA+	Phosphatidate cytidylyltransferase	ND	Zaidi *et al.* (2008)
murC+	UDP-*N*-acetyl muramate-alanine ligase	ND	Zaidi *et al.* (2008)
rgpE+	Glycosyltransferase RgpE	ND	Zaidi *et al.* (2008)
cluA+	Cell-surface antigen I/II precursor	ND	Zaidi *et al.* (2008)
Lb. plantarum			
lp_0085 ↑	Cation efflux protein	ND	Bron *et al.* (2004b)
lp_0237 ↑	Integral membrane protein	Highly induced in mouse duodenum	Bron *et al.* (2004a,b)
lp_0775 ↑	Argininosuccinate synthase	Highly induced in mouse duodenum	Bron *et al.* (2004a,b)
lp_0858 ↑	Redox protein, regulation of disulfide bond formation	ND	Bron *et al.* (2004b)
lp_1158 ↑	Lysozyme	ND	Bron *et al.* (2004b)
lp_1435 ↑	Integral membrane protein	ND	Bron *et al.* (2004b)
lp_2484 ↑	Transcriptional regulator (MarR family)	ND	Bron *et al.* (2004b)
lp_2564 ↑	Integral membrane protein	ND	Bron *et al.* (2004b)
lp_3145 ↑	Oxidoreductase, N-terminal fragment	ND	Bron *et al.* (2004b)
lp_3155 ↑	Cell surface protein, ErfK family	ND	Bron *et al.* (2004b)
lp_3154 ↑	Muramidase, C-terminal fragment	ND	Bron *et al.* (2004b)
lp_3153 ↑	Muramidase, middle fragment	ND	Bron *et al.* (2004b)
lp_3158 ↑	Oxidoreductase	ND	Bron *et al.* (2004b)
lp_3160↑	Multidrug transport protein	ND	Bron *et al.* (2004b)
lp_3175 ↑	Integral membrane protein	ND	Bron *et al.* (2004b)
lp_3344 ↑	Transcription regulator (MarR family)	ND	Bron *et al.* (2004b)
lp_3489↑	Oxidoreductase	ND	Bron *et al.* (2004b)
lp_3626 ↑	Sugar transport protein	ND	Bron *et al.* (2006)
lp_0254 ↑	Serine O-acetyltransferase (CysE)	ND	Bron *et al.* (2006)
lp_0255 ↑	Cystathionine beta-lyase (MetC)		Bron *et al.* (2006)
lp_0256 ↑	Cysteine synthase (CysK)	ND	Bron *et al.* (2006)
lp_0547 ↑	Cell division protein FtsH, ATP-dependent zinc metallopeptidase	ND	Bron *et al.* (2006)

Table 7.2 continued

Protein or gene	Putative function	Physiological relevance	Reference
lp_1253 ↑	Glutathione reductase	ND	Bron *et al.* (2006)
lp_2019 ↑	D-Alanine activating enzyme DltA	ND	Bron *et al.* (2006)
lp_2020 ↑	D-Ala-teichoic acid biosynthesis protein	ND	Bron *et al.* (2006)
lp_3687 ↑	Cell division protein	ND	Bron *et al.* (2006)
Lb. reuteri			
lr0004↑	Clp chaperone (ClpE)	Disruptant: no defect in bile shock	Whitehead *et al.* (2008)
lr0085↑	Hypothetical protein	Disruptant: sensitive to bile shock	Whitehead *et al.* (2008)
lr0597↑	Thioredoxin domain-containing protein	ND	Whitehead *et al.* (2008)
lr1291↑	Metalloproteinase	Disruptant: no defect in bile shock	Whitehead *et al.* (2008)
lr1351↑	Conserved membrane protein	Disruptant: no defect in bile shock	Whitehead *et al.* (2008)
lr1516↑	Putative esterase	Disruptant: sensitive to bile shock	Whitehead *et al.* (2008)
lr1584↑	MDR protein (major facilitator superfamily)	Disruptant: impaired growth in the presence of bile	Whitehead *et al.* (2008)
lr1706↑	DNA binding protein from starved cells (Dps)	Disruptant: no change in bile sensitivity	Whitehead *et al.* (2008)
lr1864↑	Clp chaperone (ClpL)	Disruptant: sensitive to bile shock	Whitehead *et al.* (2008)
G6PDH↑	Glucose-6-phosphate dehydrogenase	ND	Lee *et al.* (2008)
PGI ↑	Glucose-6-phosphate isomerase	ND	Lee *et al.* (2008)
LdhA↑	L-Lactate dehydrogenase	ND	Lee *et al.* (2008)
RPI↑	Ribose-5-phosphate isomerase	ND	Lee *et al.* (2008)
RpoD↑	RNA polymerase sigma factor RpoD	ND	Lee *et al.* (2008)
MAT ↑	S-Adenosylmethionine synthetase	ND	Lee *et al.* (2008)
RecR ↑	DNA-binding protein	ND	Lee *et al.* (2008)
RecN ↑	DNA-binding and repair protein	ND	Lee *et al.* (2008)
OCT↑	Ornithine carbamoyltransferase	ND	Lee *et al.* (2008)
MDH ↑	NAD-dependent oxidoreductase (mannitol 2-dehydrogenase)	ND	Lee *et al.* (2008)
EzrA ↑	Septation ring formation regulator	ND	Lee *et al.* (2008)
Lb. acidophilus			
LBA0083 ↑	Protease HtrA	ND	Pfeiler *et al.* (2007)
LBA0405 ↑	Cochaperonin GroES	ND	Pfeiler *et al.* (2007)
LBA0406 ↑	Chaperonin GroEL	ND	Pfeiler *et al.* (2007)
LBA0638 ↑	ATP-dependent Clp protease ClpE	ND	Pfeiler *et al.* (2007)
LBA1247 ↑	Heat shock protein DnaK	ND	Pfeiler *et al.* (2007)
LBA1248 ↑	Cochaperonin; GrpE; Hsp70 cofactor	ND	Pfeiler *et al.* (2007)

Table 7.2 continued

Protein or gene	Putative function	Physiological relevance	Reference
LBA1427 ↑	Oxidoreductase	Increased survival of deletion mutant in bile	Pfeiler *et al.* (2007)
LBA1428 ↑	Hypothetical proteins	Increased survival of deletion mutant in bile	Pfeiler *et al.* (2007)
LBA1429 ↑	Transporter	Reduced survival of deletion mutant in cholic acid-based bile	Pfeiler *et al.* (2007)
LBA1430 ↑	Histidine protein kinase (HPK)	Reduced survival of deletion mutant in bile	Pfeiler *et al.* (2007)
LBA1431 ↑	Response regulator (RR)	Reduced survival of deletion mutant in bile	Pfeiler *et al.* (2007)
LBA1432 ↑	Hypothetical protein	Reduced survival of deletion mutant in bile	Pfeiler *et al.* (2007)
LBA1446 ↑	MDR transporter	ND	Pfeiler *et al.* (2007)
LBA1458↑	Galactose-1-phosphate uridylyltransferase	ND	Pfeiler *et al.* (2007)
LBA1459↑	Galactokinase	ND	Pfeiler *et al.* (2007)
LBA1460 ↑	Mucus-binding protein precursor	ND	Pfeiler *et al.* (2007)
LBA1462↑	β-Galactosidase	ND	Pfeiler *et al.* (2007)
LBA1463↑	Lactose permease	ND	Pfeiler *et al.* (2007)
LBA1467↑	β-Galactosidase large subunit	ND	Pfeiler *et al.* (2007)
LBA1468↑	β-Galactosidase small subunit	ND	Pfeiler *et al.* (2007)
LBA1496 ↑	Fibrinogen-binding proteins	ND	Pfeiler *et al.* (2007)
LBA1652 ↑	Mucus-binding protein precursor	ND	Pfeiler *et al.* (2007)
Bif. breve			
Bbr_0838↑	Multidrug resistance protein	ND	Gueimonde *et al.* (2009)
Bif. longum			
Ctr	Sodium-dependent bile acid transporter	Cholic acid efflux	Price *et al.* (2006)
BL0920 ↑	Bile efflux transporter	Ursodeoxycholic acid efflux	Gueimonde *et al.* (2009)
HtrA ↑	Protease	ND	Savijoki *et al.* (2005)
DnaK ↑	Molecular chaperone	ND	Savijoki *et al.* (2005), Sánchez *et al.* (2005)
GroEL ↑	Chaperonin	ND	Savijoki *et al.* (2005)
Xfp ↑	Fructose-6-P phosphoketolase	Increased glucose consumption	Sánchez *et al.* (2005)
Tkt ↑	Transketolase	Increased glucose consumption	Sánchez *et al.* (2005)
Pyk ↑	Pyruvate kinase	Increased glucose consumption	Sánchez *et al.* (2005)
Pgk ↑	Phosphoglycerate kinase	Increased glucose consumption	Sánchez *et al.* (2005)
Gap ↑	Glyceraldehyde-3-phosphate Dehydrogenase C	Increased glucose consumption	Sánchez *et al.* (2005)
Ldh2 ↑	Lactate dehydrogenase	Increased glucose consumption	Sánchez *et al.* (2005)
Gpm ↑	Phosphoglycerate mutase	Increased glucose consumption	Sánchez *et al.* (2005)

Table 7.2 continued

Protein or gene	Putative function	Physiological relevance	Reference
GalE1 ↑	UDP-glucose 4-epimerase	ND	Sánchez *et al.* (2005)
BL1000 ↑a	Response regulator of two-component system	ND	Sánchez *et al.* (2005)
OppA ↑	Oligopeptide-binding protein of ABC transporter	Tripeptide uptake	Ruiz *et al.* (2009)
DppD ↑	ATP-binding protein of peptide ABC transporter	Tripeptide uptake	Ruiz *et al.* (2009)
Eno ↑	Enolase	ND, present in membrane fraction	Ruiz *et al.* (2009)
Gap ↑	Glyceraldehyde-3-phosphate dehydrogenase	ND, present in cell envelope fraction	Ruiz *et al.* (2009)
PbpA ↑	Cell division protein FtsI	ND, present in membrane fraction	Ruiz *et al.* (2009
RpsC ↑	Ribosomal protein S3	ND, present in membrane fraction	Ruiz *et al.* (2009)
RplD ↑	Ribosomal protein L4	ND, present in membrane fraction	Ruiz *et al.* (2009)
RpsE ↑	Ribosomal protein S5	ND, present in membrane fraction	Ruiz *et al.* (2009)
RplF ↑	Ribosomal protein L6P/L9E	ND, present in membrane fraction	Ruiz *et al.* (2009)
RplE ↑	Ribosomal protein L5	ND, present in membrane fraction	Ruiz *et al.* (2009)
Bif. animalis			
gtf01207↑	Priming glycosyltransferase	EPS production	Ruas-Madiedo *et al.* (2009)
AtpD+↑	H^+-ATPase, β-subunit	Regulation of intracellular pH	Sánchezet *al.* (2006)
BA10+	α-Glucosidase	ND	Sánchezet *al.* (2007b)
BA15+	α-Galactosidase	ND	Sánchezet *al.* (2007b)
BA28+	UDP-glucose 4-epimerase	ND	Sánchezet *al.* (2007b)
BA50+ ↑*	Glucose-6-phosphate 1-dehydrogenase	Rerouting of bifidobacterial shunt	Sánchezet *al.* (2007b)
BA14 ↑	6-Phosphogluconate dehydrogenase	Rerouting of bifidobacterial shunt	Sánchezet *al.* (2007b)
BA25 ↑*	Glucose-6-phosphate isomerase	Rerouting of bifidobacterial shunt	Sánchezet *al.* (2007b)
BA31 ↑*a BA41 ↑ BA35+	Xylulose-5-phosphate/fructose-6-phosphate phosphoketolase	Rerouting of bifidobacterial shunt	Sánchez *et al.* (2007b)
BA36+ BA38+↓ BA53×↑	Glyceraldehyde-3-phosphate dehydrogenase	Rerouting of bifidobacterial shunt	Sánchez *et al.* (2007b)
BA29+↑	Formyl-CoA transferase	Rerouting of bifidobacterial shunt	Sánchezet *al.* (2007b)
BA13 ↑	Formyl-CoA transferase	Rerouting of bifidobacterial shunt	Sánchezet *al.* (2007b)
BA32 ↑	Pyruvate formate-lyase	Rerouting of bifidobacterial shunt	Sánchezet *al.* (2007b)
BA5+↓	Methionine synthase	ND	Sánchezet *al.* (2007b)
BA40+↓**	O-Acetylhomoserine sulfhydrolase	ND	Sánchez *et al.* (2007b)

Table 7.2 continued

Protein or gene	Putative function	Physiological relevance	Reference
BA55×↓	Long-chain-fatty-acid CoA ligase	Shift on fatty acid composition	Sánchez *et al.* (2007b)
BA8+↑*	Heat shock protein ClpB	ND	Sánchez *et al.* (2007b)
BA1 ↑*a	Trypsin-like serine protease homologue to HtrA	ND	Sánchez *et al.* (2007b)
BA22 ↑*	Thioredoxin-dependent thiol peroxidase	ND	Sánchez *et al.* (2007b)
BA23 ↑	GrpE	ND	Sánchez *et al.* (2007b)
BA26 ↑	Chaperone protein GroES	ND	Sánchez *et al.* (2007b)
BA30 ↑	Chaperone protein GroEL	ND	Sánchez *et al.* (2007b)
BA12 ↑*	Chaperone protein DnaK	ND	Sánchez *et al.* (2007b)
BA21+↑*	DNA protection during starvation protein Dps	ND	Sánchez *et al.* (2007b)
BA9+	Cell division initiation protein	ND	Sánchez *et al.* (2007b)
BA2+	Chromosome segregating ATPase	ND	Sánchez *et al.* (2007b)

↑, induced in the presence of bile; ↓, repressed in the presence of bile; ↑*, induced in the presence of bile only in the wild type; ↓**,repressed in the presence of bile only in the mutant; +, constitutively (in the absence of bile) upregulated expression in the bile-resistant mutant; ×, constitutively (in the absence of bile) downregulated expression in the bile-resistant mutant; a, only detected in the presence of bile; ND, not determined. In the case of *Bif. animalis* subsp. *lactis*, ↑ and ↓ are applicable to both wild type and mutant.

displayed ability in bacteria. Sometimes the adapted cells show 'cross protection' or 'cross resistance' to a challenge with a different stress factor than the original one. (ii) Another mode of bile resistance is brought about by general stress response (GSR) expressed in stationary-phase cells under various stress environments. Bacterial cells at stationary phase normally induce many general stress proteins (GSPs), which takes place independently from bile acid stress and make the stationary-phase cells more resistant to many environmental stresses than the exponential-phase cells. Bile adaptation and GSR are sometimes overlapping, because some GSPs, as represented by molecular chaperons/housekeeping proteases (see sections 'Bile adaptation and cross resistance' and 'Functional genomics of bile tolerance'), are induced in the exponentially growing cells that acquired bile adaptation status. (iii) Independent from bile adaptation and GSR, bacteria can develop bile acid resistance as a consequence of acquiring mutations. In the bile acid-resistant mutants, the phenotypes are stable and can be transmitted to the next generations.

Bile adaptation and cross-resistance

Bile adaptation has been intensively investigated in a type strain of *Enterococcus faecalis* subsp. *faecalis* ATCC19433, a member of the human and animal gastrointestinal microbiota (Rince *et al.*, 2003). Exposure to sublethal concentrations of bile acid (0.08%, 1:1 mixture of CA and DCA both as sodium salt for 30 minutes) led to an acquired tolerance to otherwise lethal concentrations of bile acid (0.3% of the same mixture) (Flahaut *et al.*, 1996b). The adapted cells showed cross protection against sodium dodecyl sulfate (SDS) (Flahaut *et al.*, 1996b) and heat (Flahaut *et al.*, 1996c), but not to acid pH (Flahaut *et al.*, 1996c). Heterologous adaptations to SDS, heat, alkaline pH, NaCl but not acidic pH conferred cross protection to bile acid (Flahaut *et al.*, 1996a,b,c, 1997b; Pichereau *et al.*, 1999). Proteomic approach revealed consistent induction of molecular chaperons, GroEL and DnaK proteins, in response to these homologous and heterologous stresses including bile acid, heat, and alkaline pH (Flahaut *et al.*, 1996b,c, 1997a,b; Rince *et al.*, 2000), but contribution of these proteins in acquisition of stress resistance has only been demonstrated in the case of heat resistance (Flahaut *et al.*, 1997a). The role of these proteins in acquisition of bile tolerance seems inconclusive, because chloramphenicol treatment during exposure to sublethal concentrations of bile acid exhibited no negative effect on the development of bile acid tolerance (Flahaut *et al.*, 1996b). Six proteins were identified as showing commonly upregulated expressions when exposed to at least six different sublethal stress conditions including heat, ethanol, bile salts, SDS, NaCl, H_2O_2, acidic pH and alkaline pH. These proteins were considered as GSPs and thus called Gsp62-Gsp67 (Rince *et al.*, 2000). Of these Gsp66 and Gsp67 were identified as DnaK and GroEL, respectively. No significant homology has been found in Gsp62 and Gsp64 in database searches, while Gsp65 showed homology to organic hydroperoxide resistance proteins (Rince *et al.*, 2000, 2001, 2003). However, the role of these proteins (Gsps62, 64 and 65) in the bile tolerance of this bacterium does not seem to be critical (Rince *et al.*, 2001, 2002, 2003). On the other hand, a knockout of another GSP, Gls24, which was found to be induced in response to glucose starvation, $CdCl_2$ stress and bile salts stress, led to the reduction of bile salts tolerance of this bacterium (Giard *et al.*, 2000). Attempts to screen genes involved in bile tolerance by insertional random mutation led to the identification of ten bile acid-sensitive mutants with varieties of mutations. Some of these are involved in peptidoglycan metabolism, suggesting the importance of cell surface alterations in the bile resistance (Le Breton *et al.*, 2002).

Bile adaptation has also been reported in different LAB, i.e. *Lb. acidophilus* (Kim *et al.* 2001), *Lactococcus lactis* subsp. *lactis*, but not in *Lactococcus lactis* subsp. *cremoris* (Kim *et al.*, 1999). Interestingly, all these bacteria showed higher tolerance to bile acids in their stationary phase than exponential phase (GSR). As in the case of *Ec. faecalis*, *Lb. acidophilus* bile-adapted cells showed cross protectIon to heat, while NaCl-adapted cells became tolerant to both bile and heat (Kim *et al.*, 2001). In bifidobacteria, bile adaptation has been observed in both *Bifidobacterium adolescentis* NCC251 and *Bifidobacterium longum* NCC481 (Schmidt and Zink, 2000).

Physiological studies of the bile acid-resistant mutants

Mutants resistant to bile or bile acids have been isolated and characterized in *Lactococcus*, *Lactobacillus* and *Bifidobacterium*. In this section physiological studies are summarized as the fundamentals of the studies conducted with functional genomics approach.

A cholate-resistant mutant C41–2 was isolated from *Lc. lactis* subsp. *cremoris* MG1363 by the sequential cultures of the bacterial cells in media with increasing concentrations of cholate. A detailed physiological study concluded that the ATP-dependent extrusion of cholate is the basis of cholate resistance in this mutant (Yokota *et al.*, 2000). As we will come back to in the following section, this mutant played an important role in the clarification of MDR transporters as the mechanism of bile acid tolerance in this bacterium (see section '*Lactococcus lactis*'). In lactobacilli, bile resistant derivatives of *Lactobacillus delbrueckii* commonly used in the dairy industry were obtained by exposure to gradually increasing concentrations of ox gall (Guglielmotti *et al.*, 2007). In *Lb. acidophilus* a mutant resistant to both acidic pH (gastric environment) and glyco-DCA (upper intestinal environment) was obtained to improve probiotic relevance of this bacterium during gastrointestinal transit (Chou and Weimer, 1999).

In bifidobacteria, mutants resistant to CA or ox gall were derived from several bifidobacterial strains by progressive transfer after growth in increasing concentration of these compounds (Margolles *et al.*, 2003; Noriega *et al.*, 2004). A series of physiological studies were undertaken to clarify the mechanism of bile tolerance in a representative ox gall-resistant mutant, 4549dOx derived from *B. animalis* subsp. *lactis* IPLA4549. This mutant was found to be resistant to both CA and DCA, as well as ox gall. Furthermore, this mutant (and most of the other ox gall- or cholate-resistant mutants as well) was found to show increased tolerance to low pH and higher ox gall/CA resistance when using unbuffered medium than when using buffered medium, suggesting a link between bile resistance and acid tolerance in bifidobacteria (Noriega *et al.*, 2004). Another interesting change is altered sugar metabolism associated with the ox gall/CA resistance. In the mutant 4549dOx, analysis of the membrane proteins revealed increased expression of fructose-6-phosphate phosphoketolase (F6PPK), a key enzyme of carbohydrate metabolism in bifidobacteria (Sánchez *et al.*, 2004). Fermentation profile of glucose analysed using the resting cell system showed higher acetate/lactate and formate/lactate ratios in the mutant 4549dOx than in the parent IPLA4549, possibly due to the increased F6PPK activity in the mutant. These metabolic shifts are considered to increase theoretical ATP yield of the cell (Sánchez *et al.*, 2004; Ruas-Madiedo *et al.*, 2005). Acquisition of ox gall/CA resistance also induced significant changes in glycoside-hydrolysing enzyme activities. In the mutant 4549dOx, increased activities towards p-nitrophenyl (NP)-*N*-acetyl-β-D-glucosaminide, p-NP-β-D-fucopyranoside and p-NP-α-D-glucopyranoside, and decreased activities towards p-NP-α-D-galactopyranoside and p-NP-β-D-galactopyranoside were observed. Interestingly, p-NP-β-D-glucopyranoside hydrolysing activity, which was not detected in the original IPLA4549 strain, was detected at a remarkable level in the mutant 4549dOx (Noriega *et al.*, 2004). The physiological importance of these changes is not clear but may improve ATP gain through enhanced utilization of otherwise not accessible carbohydrates. In relation to enhanced ATP gain, F_1F_o-ATPase has been implicated to be involved in the bile acid resistance in the mutant 4549dOx (Sánchez, B. *et al.*, 2006). As described previously (see section 'Cell membrane: the main target of bile salts/acid attacks'), deconjugated bile acids are weak acids and their protonated form can pass through the cell membrane without the aid of a transporter. The internalized bile acid then dissociates, releases protons and acidifies the cytoplasmic pH (Kurdi *et al.*, 2003, 2006). A higher ATPase activity would therefore enable the bacteria to better survive bile-challenging conditions by pumping protons out of the cells coupled to the hydrolysis of ATP. This mechanism is also known as the one widely used by various kinds of LAB to tolerate acidic environments (Yokota *et al.*, 1995; Amachi *et al.*, 1998; Kullen and Klaenhammer, 1999). As mentioned, bile-resistant mutants of *B. animalis* tolerated the acid pH better than its parent strain in general (Noriega *et al.*, 2004). In fact, it was shown that ox gall induces the expression of the F_1F_o-ATPase [detected by Western blot analysis using antiserum against β-subunit of F_1-ATPase (AtpD) of *E. coli*] (Table 7.2) and increases the membrane-bound H^+-ATPase activity in both parent strain IPLA4549 and mutant strain 4549dOx. However, in the mutant, the expression and the activity of H^+-ATPase had already reached maximum levels when cells were grown under acidic conditions (pH 5.0), even in the absence of ox gall. This indicated that the F_1F_o-ATPase of the mutant 4549dOx had enhanced acid inducibility and this mutant is intrinsically an acid-resistant mutant. Total ATP content was higher for the mutant in the absence of bile, whereas the presence of bile induced a decrease of intracellular ATP levels, which was much more pronounced for the parent strain. The bile resistant mutant also showed a higher capacity to regulate the intracellular pH than its parent strain, suggesting that the mutant 4549dOx is able to tolerate bile by increasing the intracellular ATP reserve, and by inducing proton pumping by the F_1F_o-ATPase. The higher ATP pool attained in the mutant seems to be related to the shift in the glucose metabolism to increase theoretical ATP yield reported previously in this mutant (Sánchez *et al.*, 2004; Ruas-Madiedo *et al.*, 2005). Note that an acid-resistant mutant of *B. longum* also displayed a higher bile resistance than the parent strains, and a higher induction capacity of the F_1F_o-ATPase enzyme under acidic conditions than its counterpart (Sánchez *et al.*, 2007a). As described, there seems to be a common link between the resistance to acid pH and bile salts in *B. animalis* subsp. *lactis* through F_1F_o-ATPase. However, these results cannot allow us to conclude that bile-resistant phenotypes are always accompanied by acid resistance, because selection of the mutant 4549dOx was conducted by serial transfer of the culture in the unbuffered medium (acidic pH is unavoidable) with increasing concentration of ox gall (Noriega *et al.*, 2004).

Furthermore, it has been shown that acquisition of bile/CA resistance in bifidobacteria induces changes in cell surface properties, such as membrane fluidity and the fatty acid composition (Ruiz *et al.*, 2007), hydrophobicity and adhesion to human intestinal mucus (Gueimonde *et al.*,

2005) and the ability to compete with enteropathogens in human intestinal mucus (Gueimonde *et al.*, 2007). Antibiotic susceptibility changes (Noriega *et al.*, 2005) and increase in BSH activity (Noriega *et al.*, 2006) have also been observed. These studies suggest pleiotropic phenotypes associated with the bile resistance, and indicate the importance of understanding the molecular basis underlying bile resistance.

Functional genomics of bile tolerance

The appearance of several genomes from lactic acid bacteria and bifidobacteria has allowed the scientific community to undertake molecular studies that were unthinkable a decade ago. At the end of 2009, the genome sequences of *Lc. lactis* (subsp. *cremoris* (two strains) and subsp. *lactis*), *Lb. acidophilus, Lactobacillus brevis, Lactobacillus casei* (two strains), *Lb. delbrueckii* subsp. *bulgaricus* (two strains), *Lactobacillus fermentum, Lb. gasseri, Lactobacillus helveticus, Lb. johnsonii, Lactobacillus plantarum, Lb. reuteri* (two strains), *Lactobacillus sakei, Lactobacillus salivarius, B. longum* (subsp. *longum* and subsp. *infantis*), *B. animalis* subsp. *lactis*, and *B. adolescentis*, were available in the NCBI public database. Furthermore, many companies and research institutes are developing their own genome projects, human gut microbiome projects are being carried out (http://nihroadmap.nih.gov/hmp/; http://genome.wustl.edu/hgm/HGM_frontpage.cgi?content_type=overview; www.metahit.eu), and metagenomic analysis of human gut microbiota are being investigated (Gill *et al.*, 2006; Kurokawa *et al.*, 2007; Ley *et al.*, 2008). This will undoubtedly make available an enormous amount of genetic information for gut microbiologists, and will provide an excellent scenario to apply high-throughput methodologies in the emerging field of genomics and proteomics. In this regard, genomic and proteomic research can be extremely useful, as it provides the necessary tools, such as DNA and protein arrays, and massive separation and identification of proteins with chromatographic and electrophoretic techniques coupled to mass spectroscopy, for unravelling the functions of probiotics *in vitro* and *in vivo*. This is opening new avenues to understanding microbe-host, microbe–microbe crosstalks, and the mechanisms that these bacteria use to respond to environmental challenges. In this section, functional genomics of bile tolerance are summarized per LAB/bifidobacteria species.

Lactococcus lactis

Multidrug efflux, whereby a single transporter is capable of recognizing and transporting multiple drugs with no apparent structural similarity, was first described in higher eukaryotes, in which P-glycoprotein was found to provide resistance to anticancer agents. During the early 1990s it became very evident that multidrug efflux systems were also present in microorganisms, playing key roles in the bile resistance and thus pathogenicity of enteric bacteria (Piddock, 2006). From that time, studies on *Lc. lactis* multidrug resistance (MDR) have become a paradigm among Gram-positive bacteria.

Bolhouis and coworkers (Bolhuis *et al.*, 1994) isolated three mutants of *Lc. lactis* subsp. *cremoris* MG1363 for resistance to high concentrations of the cytotoxic drugs, ethidium bromide, daunomycin, and rhodamine 6G, respectively. All these mutants showed cross-resistance to structurally and functionally unrelated drugs, among which were the typical substrates of the mammalian multidrug transporter (P-glycoprotein), such as daunomycin and rhodamine 6G. Later, another mutant, C41–2, showing resistance to CA was derived from MG1363 (see section 'Physiological studies of the bile acid-resistant mutants') (Yokota *et al.*, 2000). Since some molecular studies highlighted the role of MDR transporters in the acquisition of these phenotypes, transcriptome analyses were applied to these four independently isolated mutants to correlate each other through identification of MDR transporter(s) responsible for the resistant phenotype of each mutant. The results demonstrated that all four resistant mutants consistently and specifically over-expressed *lmrC* and *lmrD* genes coding for a heterodimeric ABC-multidrug transporter among the 40 putative MDR transporters encoded in its genome (Lubelski *et al.*, 2004, 2006). In fact, a Δ*lmrCD* mutant was found to be significantly more sensitive to all four drugs, including CA, than its parent, and plasmid-based over-expression of *lmrCD* in the Δ*lmrCD* mutant restores its drug-sensitive phenotype (Lubelski *et al.*, 2006). In the case of CA, the resistance increased several-fold and *LmrCD* efficiently extrudes CA from the cell in an energy-dependent manner, demonstrating that *LmrCD* provides resistance to CA through an extrusion-based mechanism (Zaidi *et al.*, 2008). A survey of the spectrum of bile sensitivity in the ΔlmrCD mutant showed a remarkable increase in the sensitivity to glycoDCA (more than 6-fold) as compared with CA (1.6-fold), versus no change in the sensitivity to CDCA, DCA and LCA, suggesting that glycoDCA is the most important natural substrate for LmrCD (Zaidi *et al.*, 2008). These findings represent the first clear demonstration of the involvement of an ABC-type transporter in bile-acid resistant phenotype in prokaryotes and suggest the involvement of this transporter in cross-resistance mechanism between bile and cytotoxic drugs such as ethidium bromide, daunomycin, and rhodamine 6G.

Sequence analysis of the four drug-resistant mutants showed that the upstream region of *lmrCD* contains mutations in an open reading frame (ORF) named *lmrR*, a putative transcriptional regulator, in a way that a complete LmrR is not produced (Lubelski *et al.*, 2006). Subsequently, LmrR has been shown to bind the promoter region of *lmrCD* and interact with drugs that cause *lmrCD* up-regulation (Agustiandari *et al.*, 2008), indicating that LmrR is a negative regulator of *lmrCD* expression.

Furthermore, in order to search for a bile-acid resistant mechanism other than LmrCD, a stable CA-resistant mutant was derived from the Δ*lmrCD* mutant and transcriptome analysis was conducted to compare gene expression between the two strains growing in the absence of CA (Zaidi *et al.*, 2008). Morphologically the CA-resistant mutant appeared to grow in unusually long strings of cocci clumped together as large aggregate, suggesting alterations in the cell envelope. The microarray analysis showed that the CA-resistant mutant has a different expression of more than 100 genes, including various genes associated with cell envelope biosynthesis (*cdsA*, *murC* and *rgpE*) and sex factor/conjugation (*cluA*), which could explain the different morphology observed for the mutant. Some important changes are summarized in Table 7.2. Some stress response genes such as *groES*, *groEL* and *hslO* (molecular chaperones) and *htrA* (high temperature requirement A; a trypsin-like serine protease with housekeeping function) were upregulated in the mutant. In *Lc. lactis* , HtrA has been characterized as the cell surface protease that degrades abnormal exported proteins and is essential for growth at high temperatures (Poquet *et al.*, 2000; Sriraman and Jayaraman, 2008). In contrast to many housekeeping chaperones/proteases [e.g. GroESL and Clp proteases (see section '*Lactobacillus reuteri*')], HtrA proteins are categorized into ATP-independent chaperone-proteases (Kim and Kim, 2005). Interestingly, no significant alteration in the gene expression of the remaining 40 putative MDR transporters was observed, which reinforces an important role of LmrCD in conferring CA resistance in this bacterium. These results illustrate how *Lc. lactis* can cope with CA stress in the absence of its main detoxifying machinery, LmrCD. An investigation of the presence of the nucleotide sequence for *lmrC* and *lmrD* among 17 independently isolated *Lc. lactis* (subsp. *lactis* and subsp. *cremoris*) strains revealed that these sequences are conserved in almost all these isolates, suggesting an essential role of LmrCD for cell function in *Lc. lactis* inhabiting a natural environment (Flórez *et al.*, 2006).

Lactobacilli

The genus *Lactobacillus* is the largest group among the LAB, including at present more than 110 species and 15 subspecies (Felis and Dellaglio, 2007), although its number increases every year. They are naturally associated with a large variety of nutritive-rich plant- and animal-derived environments. Many lactobacilli can be associated with the human gut, such as those belonging to the acidophilus group (i.e. *Lb. acidophilus, Lactobacillus crispatus, Lb. gasseri* and *Lb. johnsonii*), *Lb. salivarius, Lactobacillus casei* group (*casei, paracasei, rhamnosus* and *zeae*), *Lb. plantarum, Lb. reuteri, Lb. fermentum* and *Lb. brevis* (Song *et al.*, 2000). Among lactobacilli, three species, i.e. *Lb. acidophilus, Lb. reuteri* and *Lb. plantarum* have been the targets of study on the influence of bile on RNA or protein profiles using functional genomics approach. The results are summarized in the following sections for each species.

Lactobacillus reuteri

Microarray analysis of genes involved in bile shock [cells were harvested 15 minutes after 0.5% oxgall addition (~10 mM) to the exponentially growing culture] or bile adaptation (cells were collected 5 hours after the bile shock when the growth resumed) in *Lb. reuteri* ATCC55730 revealed 88 and 84 differentially expressed genes, respectively, with eight up-regulated genes and 12 down-regulated genes being overlapped under both conditions (Whitehead *et al.*, 2008). A subset of nine genes (lr0004–lr1864) listed in Table 7.2 (except lr0597) showing significant overexpression under at least one condition were selected for disruption studies.

Of these, the knockout of a gene specifying putative esterase (lr1516), probably involved in cell wall synthesis, a molecular chaperone ClpL gene (lr1864) but not ClpE gene (lr0004), and a gene for hypothetical protein (lr0085) appeared to significantly increase sensitivity of the mutants to bile shock (exposure to 0.3% oxgall for 30 minutes). Clp chaperones, ATP-dependent chaperone-proteases and members of the Hsp100 family proteins (Schirmer *et al.*, 1996), have been implicated in several stress responses, such as heat and osmotic shock in Gram-positive bacteria (De Dea Lindner *et al.*, 2007; Frees *et al.*, 2007). The results showed that ClpL (considered to work only as Clp ATPase for proper protein folding/disaggregation) was specifically required for *Lb. reuteri* to resist bile shock, while the ClpE (not only working as Clp ATPase but also as the component of Clp protease in association with ClpP) is dispensable under the applied bile-stress conditions. Interestingly, both ClpL gene (lr1864) and a putative esterase gene (lr1516) were induced also by acid pH and appeared to be necessary for increased survival at low pH (Wall *et al.*, 2007). Therefore, both genes display an overlapped function in bile/acid response, suggesting cross resistance to both stresses.

Disruption of the genes specifying two MDR transporters (lr1265 and lr1584), which might work to get rid of bile acids from the cell, showed no effect on the sensitivity to bile shock but appeared to yield less biomass than the parent after growth in the presence of 0.3% bile. These results suggested the importance of cell surface protection (cell wall synthesis and MDR transporters) and a quality control of proteins (ClpL) for bile tolerance in this species. Also, an increased expression of a Dps (the non-specific DNA-binding protein from starved cells), a protein involved in protection of stationary-phase cells from several types of stresses in *E. coli*, including oxidative stress, UV and gamma irradiation, iron and copper toxicity, heat stress, and pH stress (Martinez and Kolter, 1997; Nair and Finkel, 2004), was induced by bile in this bacterium, even though disruption of Dps gene (lr1706) did not significantly affect its sensitivity to either bile shock or bile adaption (Whitehead *et al.*, 2008).

Bile stress response of another strain of *Lb. reuteri* ATCC23272 has also been analysed by proteomic approach using two-dimensional gel electrophoresis (2-DE) (Lee *et al.*, 2008). We have to bear in mind that this system can only access abundantly expressed proteins, which normally include soluble housekeeping proteins (membrane proteins are hardly detected), and thus the results obtained may not represent the whole picture of what is happening under test conditions. Nevertheless, the analysis identified 28 differentially expressed proteins between the cells cultured exponentially in the medium with or without purified bile (0.6 and 1.2 g/l, ~2 mM). Of these, 26 proteins were upregulated, which included those involved in sugar metabolism, transcription-translation, nucleotide metabolism, amino acid biosynthesis, pH homeostasis and stress response, oxidation-reduction reactions, and others (Table 7.2 shows some relevant proteins). Upregulation of glycolytic and pentose phosphate pathways and related enzymes (glucose-6-phosphate isomerase, glucose-6-phosphate dehydrogenase, ribose-5-phosphate isomerase A, L-lactate dehydrogenase, mannitol 2-dehydrogenase) may indicate the importance of enhanced sugar metabolism for energy generation to overcome bile stress by driving many energy-dependent apparatus for maintaining cell homeostasis. An increase of ornithine carbamoyltransferse, a part of the enzyme constituting the arginine-deiminase pathway (Cotter and Hill, 2003), also seems to contribute to additional ATP gain and/or pH homeostasis during bile stress. Increased amounts of RpoD, which is highly conserved in all bacteria and considered to work as the primary sigma factor (Iyer *et al.*, 2004), may indicate the necessity of enhanced protein synthesis under bile stress. RecR and RecN, SOS proteins involved in the reparation of damaged DNA (Webb *et al.*, 1997; Sánchez, H. *et al.*, 2006), were overproduced, which implies occurrence of bile-stress-induced oxidative damage to DNA molecule, as in mammalian cells (see section 'Other possible targets of bile acid attack: macromolecules'). An overproduction of S-adenosylmethionine (SAM) synthetase might reflect possible inhibition of methionine biosynthesis under oxidative stress (Hondorp and Matthews 2004) or a need for SAM for the maintenance of cell homeostasis through involvement of this molecule in rRNA modification, polyamine synthesis and a large number of methylation processes (Rodionov *et al.*, 2004). Upregulation of EzrA (septation ring formation regulator) suggests the importance of cell cycle/division control to cope with bile stress.

Although the results seem to be a mixture of both primary and secondary effects triggered by bile attack, functional genomics illustrated the complex nature of bile-stress response in *Lb. reuteri*, which, in turn, provide us with many clues to identify target molecules and mechanisms of bile tolerance in LAB and bifidobacteria as described in the previous section (see sections 'Cell membrane: the main target of bile salts/acids attack' to 'Other possible targets of bile acid attack: macromolecules').

Lactobacillus acidophilus

Transcriptome study has been undertaken to reveal bile stress response of *Lb. acidophilus* NCFM, an industrially important probiotic strain (Pfeiler *et al.*, 2007). Exponentially growing cells of NCFM strain under anaerobic conditions were exposed to 0.5% (w/v) oxgall (~10 mM) and incubated for 30 minutes under the same conditions. The cells were then collected for total RNA extraction. Comparison of transcriptomes of control cells and bile-stressed cells revealed 289 genes that were differentially expressed with statistical significance. Of these 78 genes were induced, while 168 genes were repressed. Genes involved in many aspects of cell proliferation such as nucleotide metabolism, amino acid metabolism and cell envelope biogenesis were downregulated, reflecting the reduced growth rate observed in the presence of oxgall.

Several of the upregulated genes with physiological relevance are listed in Table 7.2. As seen in *Lc. lactis* and *Lb. reuteri*, several chaperons/proteases (LBA0083, LBA0405, LBA0406, LBA0638, LBA1247, LBA1248) were found to be induced, again suggesting that protein denaturation is one of the deleterious consequences being encountered by bile-treated cells. Remarkable upregulation was detected in the genes for lactose/galactose metabolism (LBA1458, 1459, 1462, 1463, 1467, 1468), which is probably required for energy generation to cope with bile stress utilizing energy-dependent systems such as ATP-dependent chaperons/proteases (e.g. GroESL and Clp protease).

The fact that the culture medium (MRS) does not contain lactose may indicate that the bile worked as a signal to allow the LAB cell to alert its arrival in the small intestine, the environmental niche of this LAB, where lactose derived from milk can be an abundant carbohydrate. It has been demonstrated that in some intestinal bacteria bile is the signal that promotes the secretion of specific proteins essential for adaptation to the gastrointestinal ecosystem (Rivera-Amill *et al.*, 2001). Also interesting was the upregulated expression of mucus binding protein precursor (LBA1460 and LBA1652) and fibrinogen binding proteins (LBA1496), both potentially involved in bacterial adhesion to intestinal cells. These findings also seem to support the above speculation that bile may be a location indicator for *Lb. acidophilus*.

So, how can *Lb. acidophilus* cells sense the presence of bile in the environment? In this regard some important progress has been made, in which the transcriptome analysis identified involvement of a two-component regulatory system (2CRS) in bile tolerance of this LAB. 2CRSs allow bacteria to sense and respond to changes in the environment. They are constituted of a membrane histidine protein kinase (HPK) in charge of getting the signal input, and transferring the phosphoryl groups to a response regulator (RR) responsible for promoting a transcriptional response through its DNA binding domain (Mascher *et al.*, 2006).

The identified 2CRS is included in a 7-kb bile inducible operon encoding, HPK (LBA1430), RR (LBA1431), an oxidoreductase (LBA1427), a transporter (LBA1429), and four hypothetical proteins (LBA1428, LBA1432 and others). Only this 2CRS was bile-inducible out of nine 2CRSs encoded in its genome. Deletion mutants were constructed and the disruptants for HPK gene, RR gene, transporter gene and hypothetical protein gene (LBA1432) displayed a reduced survival in the presence of bile, while those for oxidoreductase gene and hypothetical protein gene (LBA1428) became more resistant to bile than the wild-type. These results indicate that the 2CRS and the transporter play a role in bile tolerance. Furthermore, a transcriptional analysis showed that mutation in the HPK gene had no effect on the induction of the operon, whereas the RR gene mutation showed enhanced induction when the *Lb. acidophilus* was exposed to bile, suggesting that the operon it resides in is negatively controlled by the RR.

Characterization of the transporter-gene (LBA1429)-disruptant revealed growth inhibition specifically by the CA-based bile salts. Although some MDR transporters have been functionally studied in lactobacilli (Sakamoto *et al.*, 2001), a detailed biochemical characterization regarding the influence of MDR efflux pumps on bile resistance has not been performed in lactobacilli. However, the observed effect of the disruption of LBA1429 and the upregulated expression of MDR transporter (LBA1446), as well as the study of deletion mutants of putative MDR efflux pumps in *Lb. reuteri* (lr1265 and lr1584 in 5.2.1.), indirectly suggest a major role of these transporters in bile resistance.

The results of the transcriptome-based study conducted in *Lb. acidophilus* illustrate a new insight into the sensing mechanism of bile molecules through 2CRS and the possible importance of bile salts as the signalling biomolecules for intestinal LAB to adapt into the intestinal environment.

Lactobacillus plantarum

Lb. planarum is a flexible and versatile LAB that can be encountered in many environmental niches including vegetable fermentations, silage fermentations and dairy fermentations. Strain WCFS1, of which the complete genome sequence has been determined (Kleerebezem *et al.*, 2003), is the most intensively studied *Lb. plantarum*. Pharmacokinetic studies showed high persistence of this strain in the human intestine (Vesa *et al.*, 2000), which led to the studies on the bile stress response in WCSF1.

The influence on the cell surface and the cell morphology is one of the common physiological responses upon exposure to bile challenge in Gram-positive bacteria (Leverrier *et al.*, 2003; Taranto *et al.*, 2006; Ruiz *et al.*, 2007; Ruas-Madiedo *et al.*, 2009). Bron and coworkers examined the morphology of the WCFS1 cells grown in MRS liquid medium supplemented with of 0.05% to 0.15% porcine bile by scanning electron microscopy (Bron *et al.*, 2004b). Analysis of cells revealed rough surfaces, clumping together of the cells and the formation of membrane vesicle structures. At higher bile concentrations (0.1–0.15%), some of the cells had a shrunken and empty appearance.

Then, by the application of alr (alanine racemase) complementation-based genome-wide promoter screening technique (Bron *et al.*, 2004a), the ORFs that were inducible in the presence of 0.1% porcine bile in MRS agar plates were identified in strain WCFS1 (Bron *et al.*, 2004b). This approach resulted in the identification of 31 putative genes in the genome of this strain, including six genes identified as unknown. Some relevant genes are listed in Table 7.2. Remarkably, 11 of these ORFs encode proteins related to membrane- or cell wall-associated functions, underlying the importance of cell surface protection in bile resistance. Of the membrane proteins, three possible exporter proteins were found and one of them, specified by lp_3160, was annotated as multidrug transport protein. Included among the cell wall-associated proteins are lysozyme (lp_1158) and two muramidases (lp_3153 and lp_3154), which may indicate that some lytic activities are involved in the expression of bile resistance. Five putative genes identified as bile inducible are those encoding transcriptional regulators. Especially two genes (lp_2484 and lp_3344) encode marR family regulators. In *E. coli* and *Salmonella enterica* serovar Typhimurium, the MarR and

MarA proteins control the transcription of genes for MDR including multidrug efflux proteins. Thus, these regulators might control expression of lp_3160 for a putative multidrug transporter that could export bile for detoxification. Proteins encoded by another five genes identified as bile-inducible are categorized as functioning in redox reactions (four oxidoreductases and a redox protein involved in regulation of disulfide bond formation). Although the exact role of these proteins is not clear, these responses may represent an occurrence of redox imbalance and/or oxidative stress as the result of bile attack.

Microarray analyses of the WCSF1 cells grown on MRS agar supplemented with 0.1% porcine bile have again demonstrated the importance of cell envelope functions and control of redox balance and/or oxidative stress to manage bile toxicity in this bacterium (Bron *et al.*, 2006). As listed in Table 7.2, the bile-induced overexpression of lp_0547 and lp_3687, two genes coding for cell division protein, lp_2019 and lp_2020 both involved in D-alanylation of teichoic acid, were observed as cell envelope-associated alterations. Increased expression of three genes, lp_0254, lp_0255 and lp_0256, involved in methionine/cysteine synthesis, and lp_1253 for glutathione reductase were probably for the management of redox balance and/or oxidative stress. Surprisingly, no putative chaperone was induced in this bacterium under bile stress, which is in contrast to the aforementioned LAB.

Bifidobacteria

The species of the genus *Bifidobacterium* form a coherent phylogenetic group, with more than 30 species and subspecies, which show over 93% similarity to the 16S rRNA sequences (Satokari *et al.*, 2003). In human adults, the most commonly detected bifidobacteria in faecal samples is the *Bifidobacterium catenulatum* group, and then *B. longum*, *B. adolescentis* and *Bifidobacterium bifidum* follow in that order (Matsuki *et al.*, 2002). In breast-fed infants, *Bifidobacterium breve* is the most frequently detected species, and then *Bifidobacterium infantis*, *B. longum* and *B. bifidum* follow in that order (Matsuki *et al.*, 2002). The typical species isolated from functional foods is *B. animalis* subsp. *lactis* (Masco *et al.*, 2005); therefore, strains belonging to these species are the main probiotic targets. Among them, molecular studies on bile stress have been mainly focused on *B. longum* and *B. animalis* subsp. *lactis*. To unravel the molecular mechanisms leading to bile stress response, high-throughput molecular techniques, mainly transcriptomic analysis and proteomic studies, are being used as a first approach to understand the physiological changes involved in this process. However, although significant advances have been achieved very recently (O´Connell Motherway *et al.*, 2008; Yasui *et al.*, 2008), the lack of efficient transformation systems and the paucity of effective molecular tools (e.g. cloning and expression vectors and gene inactivation systems) strongly limited functional studies in *Bifidobacterium*. Despite these bottlenecks, there are a few studies that shed some light on the mechanisms of bifidobacterial response to bile.

Bifidobacterium longum

B. longum is the most intensively studied *Bifidobacterium* in terms of bile stress response. The responses in this bacterium are featured by bile acid extrusion by MDR efflux pumps (Price *et al.*, 2006; Gueimonde *et al.*, 2009), upregulated expression of molecular chaperones (Savijoki *et al.*, 2005) and enhanced carbohydrate catabolism through so-called bifidobacterial shunt (Sánchez *et al.*, 2005). Proteomic approach has been applied as the analytical method in functional genomics of *B. longum* (Sánchez *et al.*, 2005; Savijoki *et al.*, 2005; Ruiz *et al.*, 2009). The relevant proteins are listed in Table 7.1.

In this bacterium two MDR efflux pumps, Ctr and BL0920, have been characterized. The *ctr* gene of *B. longum* NCIMB 702259^{T} codes for a putative sodium-dependent bile acid transporter. When overexpressed in an *E. coli* efflux-negative strain, this protein was responsible for an increase of resistance to glycocholic acid and other antimicrobial compounds (e.g. acriflavine, chloramphenicol, erythromycin and ethidium bromide), and caused the efflux of radioactive cholate (Price *et al.*, 2006). In addition, the product of the gene BL0920, a putative efflux pump of the major facilitator superfamily, was overexpressed more than 30-fold when *B. longum* NCC2705 cells were grown in subinhibitory concentrations of ox bile. The expression of BL0920 in *E. coli* was shown to confer resistance to bile, and transformed cells displayed a higher efflux of fluorescently labelled ursodeoxycholic acid, indicating that active efflux is the mechanism underlying the bile resistance phenotype (Gueimonde *et al.*, 2009). Interestingly, the BL0920 homologue in *B. breve* UCC2003, Bbr_0838, is highly induced in the presence of bile as well, suggesting a similar function (Gueimonde *et al.*, 2009).

There is also a chaperone/protease-mediated bile response in *B. longum*, as in LAB. HtrA, GroEL, and DnaK were found to be present in higher amounts in the cytoplasm of *B. longum* 3A (Savijoki *et al.*, 2005) and DnaK in *B. longum* NCIMB8809 grown in the presence of bile (Sánchez *et al.*, 2005). Remarkably, DnaK has also been implicated in oxgall adaptation in *B. adolescentis* NCC251 (Schmidt and Zink, 2000).

Two important enzymes of the bifidobacterial shunt, Xfp (F6PPK, fructose-6-phosphate phosphoketolase) and Tkt (transketolase), together with proteins involved in the transformation of glyceraldehyde-3-phosphate to lactate via pyruvate (Gap, Pgk, Gpm, Eno, Pyk and Ldh2), were detected at higher levels in *B. longum* NCIMB8809 cells grown in the presence of bile (Sánchez *et al.*, 2005). GalE1 involved in utilization of galactose also appeared to be upregulated (Sánchez *et al.*, 2005). Taken together, these changes suggested enhanced sugar degradation to lactate yielding ATP and reducing equivalents. In fact, glucose consumption rate and the lactic acid/acetic acid production ratio of the resting cells prepared from cells grown in the presence of bile were higher than those of the resting cells prepared from cells grown in the absence of bile (Sánchez *et al.*, 2005). These changes could allow bifidobacterial cells to synthesize more ATP by substrate-level phosphorylation and reducing equivalent. An increased availability of ATP and thus proton-motive force may enhance the above-mentioned cellular machineries (MDR transporters and chaperones) that protect cells from the deleterious action of bile. Increased amounts of reducing equivalent may contribute to relieve possible oxidative stress caused by bile attack.

It is also interesting to note that the recently established cell-envelope proteome of *B. longum* NCIMB8809 cells revealed many theoretical cytoplasmic proteins, including several glycolytic enzymes and ribosomal proteins, probably displaying 'moonlighting' functions (functions different from their known functions in cytoplasm) (Ruiz *et al.*, 2009). Moreover, some of these, Eno and Gap, as well as several ribosomal proteins, were found to be overproduced 2- to 3-fold in the cells grown in the presence of ox bile extract (Ruiz *et al.*, 2009). In relation to this, surface displayed Eno has been shown to be able to capture human plasminogen (proenzyme of the serine protease, plasmin) in *Bifidobacterium lactis* (Candela *et al.*, 2007) and in *Neisseria meningitidis* (Knaust *et al.*, 2007), suggesting its role during the colonization process in the human gut. Although physiological relevance is still not clear, these proteins may be important for the colonization of the gut environment. Additionally, two oligopeptide transport subunits from two different transport systems (OppA and DppD) were detected in higher amounts in membrane protein fraction in response to bile stress (Ruiz *et al.*, 2009). Actually, cells grown in the presence of bile have been demonstrated to remove a tripeptide, Ala-Ala-Ala, from the culture medium at a higher rate than that of the control cells (Ruiz *et al.*, 2009). However, the role of peptide uptake to cope with bile attack is not clearly understood.

Finally, a brief mention should be made of the increased concentration of a response regulator belonging to a 2CRS, BL1000, in *B. longum* NCIMB8809 as a response to bile (Sánchez *et al.*, 2005). 2CRSs have been suggested to be involved in bile sensing in *Lb. acidophilus* (Pfeiler *et al.*, 2007) and other bacteria (Leverrier *et al.*, 2003; Raphael *et al.*, 2005), implying the operation of a similar mechanism of bile sensing in bifidobacteria.

Bifidobacterium animalis subsp. *lactis*

Although not an indigenous member of the human gut microbiota, *B. animalis* subsp. *lactis* is one of the most popular probiotics used commercially in dairy products. Thus, clarification of bile stress response is an important issue to improve the effectiveness of this bacterium as a probiotic from a practical point of view. Proteomic approach has been applied to clarify the bile stress response in *B. animalis* subsp. *lactis*

IPLA4549 and its ox gall-resistant mutant, 4549dOx (Sánchez *et al.,* 2007b). These strains were cultured in the presence or absence of ox gall until exponential phase, and the proteomic analyses were conducted. Twenty-four proteins were considered globally bile responsive, which consisted of sixteen proteins, whose expression varied towards the same direction (up or down) in both strains in response to bile attack and eight proteins whose levels are affected by bile only in the wild type, but in the mutant always up or down regulated. Some of these and other bile responsive proteins, and those not bile responsive but having a higher expression level in the mutant, are listed in Table 7.2. The results also underline those of the physiological studies of these strains described in previous sections ('Known mechanisms to counteract membrane damage by bile salts/acids attack', 'Physiological studies of the bile acid-resistant mutants').

The effect of bile on carbohydrate metabolism in this bacterium has been successfully detected by proteomic approach. The results are summarized in three main stages of alterations: (i) enhanced degradation of complex carbohydrates to feed bifidobacterial shunt to the level of fructose-6-phosphate and D-xyluose 5-phosphate; (ii) increased expression of xylulose-5-phosphate/fructose-6-phosphate phosphoketolase (Xfp), the key enzyme of the bifidobacterial shunt; (iii) a displacement of pyruvate metabolism leading to increased acetate/lactate ratio. Stage (i) includes enhanced hydrolysis of α-glucan (BA10) and α-galactoside (BA15), and enhanced conversion of glucose/galactose to xylulose-5-phosphate/fructose-6-phosphate (BA28, BA50, BA14). Stage (ii) involves three isoforms of Xfp (BA31, BA41 and BA 35). Stage (iii) is due to the enhanced conversion of pyruvate into acetyl-CoA and formate by increased activity of pyruvate formate-lyase (BA32), which may increase acetate formation as well as formate formation. Also overproduction of formyl-CoA transferase (BA29 and BA13) may affect formate metabolism/formation in the presence of oxalate. In fact, previous studies demonstrated higher α-glucosidase activity in the mutant 4549dOx than the wild type (Noriega *et al.,* 2004) and a preferential use of maltose compared with glucose in the mutant (Ruas-Madiedo *et al.,* 2005). Analysis of glucose fermentation by wild type and mutant in the presence or absence of bile showed expected rerouting of the metabolites to different branches, i.e. increased acetate/lactate ratio and formation of formate in the presence of bile in both strains (Sánchez *et al.,* 2007b). These alterations theoretically allow bifidobacteria to synthesize more ATP by substrate-level phosphorylation. These changes are in contrast to those observed in *B. longum* NCIMB8809, in which lactate/acetate ratio increased as a result of the upregulated expression of glycolytic enzymes in response to bile stress.

Among the cell envelope related proteins, two proteins directly related to the cell division process (BA2 and BA9) were detected at higher amounts in the mutant, which is in agreement with one of the notorious consequences of a bile-resistant mutant of bifidobacteria: a decrease in cell size (Margolles *et al.,* 2003). Remarkably, the cell division proteins (BA9) were also induced by bile in *B. longum* (Ruiz *et al.,* 2009) and in *Lb. plantarum* (Bron *et al.,* 2006). Long-chain fatty acid-CoA ligase (BA55) was found to be highly underproduced in both wild type and mutant in the presence of bile and underproduced in the mutant in the absence of bile. This might explain some of the observed alterations in the fatty acid composition of the cytoplasmic membrane both in IPLA4549 and 4549dOx in response to bile (see section 'Physiological studies of the bile acid-resistant mutants'; Ruiz *et al.,* 2007). As mentioned already in the previous sections, gtf01207, encoding a priming glycosyltransferase for EPS production (see section 'Known mechanisms to counteract membrane damage by bile salts/acids attack') (Ruas-Madiedo *et al.,* 2009), and AtpD, H^+-ATPase β-subunit (see section 'Physiological studies of the bile acid-resistant mutants') (Sánchez *et al.,* 2006), were overexpressed.

Six putative chaperones and proteases were found to be overproduced in the presence of bile: ClpB, a trypsin-like serine protease homologue to HtrA, GrpE, GroES, GroEL and DnaK. Some of these were also induced by bile stress in *B. longum* (see section '*Bifidobacterium longum*'). Thus, and despite the lack of molecular studies, these chaperones, together with HtrA, seem to constitute a common chaperone/protease bile-mediated response in bifidobacteria.

B. animalis subsp. *lactis* is also able to trigger mechanisms to cope with the oxidative damage promoted by bile as observed in *Lc. lactis* (see section '*Lactococcus lactis*'), *Lb. reuteri* (see section '*Lactobacillus reuteri*') and *Lb. plantarum* (see section '*Lactobacillus plantarum*'). The concentration of a Dps protein (BA21) (see section '*Lactobacillus reuteri*') is overproduced constitutively in the mutant and is induced in the wild type in response to bile. A thioredoxin-dependent thiol peroxidase (BA22), which may deal with peroxide stress, increased in the wild type by bile stress. Furthermore, complex variations in expression levels of enzymes involved in the last two steps of the methionine biosynthesis, methionine synthase (BA5) and *O*-acetylhomoserine sulfhydrolase (BA40), were observed in response to bile stress, suggesting that methionine metabolism is crucial for the regulation of oxidative stress (Hondorp and Matthews, 2004).

Although differences in experimental conditions have to be taken into consideration (bile: bovine vs. porcine; culture conditions: liquid vs. plate; culture phase: exponential vs. stationary), studies of the bile stress response in *Lc. lactis*, *Lactobacillus* species and *Bifidobacterium* species manifested common mechanisms to manage bile stress among these bacteria: cell surface functions including MDR transporters to protect cells from bile attack, enhanced sugar metabolism for energy generation, chaperones/proteases for quality control of proteins and various proteins involved in the maintenance of redox balance and/or removal of oxidative stress. To summarize, these mechanisms are illustrated collectively in Fig. 7.3.

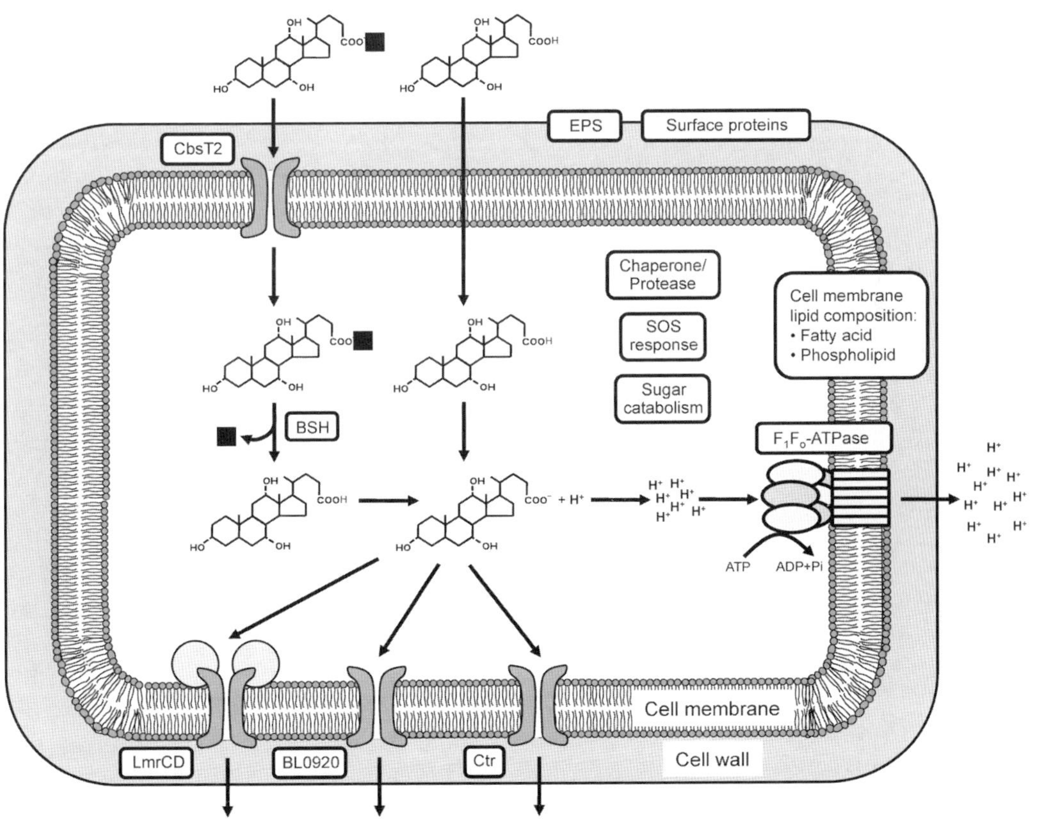

Figure 7.3 The main mechanisms involved in the processes of bile salt tolerance in LAB and bifidobacteria. See text for details. The main mechanisms are displayed within the grey-coloured panels. Bile acids are represented by cholic acid. ■, taurine or glycine; BSH, bile salt hydrolase, for which the role played in bile salt tolerance is still debatable.

Is BSH activity a desirable trait for LAB and bifidobacteria to cope with bile stress?

As mentioned previously (section 'Metabolism of bile salts by gut bacteria') bile salt deconjugation by BSH is carried out by a broad spectrum of intestinal anaerobic bacteria (Jones *et al.*, 2008). As a general rule, all *Bifidobacterium* species seem to be able to hydrolyse bile salts. Lactobacilli, however, associates this activity to the intestinal species. To date, BSH activity has not been detected in bacteria isolated from environments in which bile is absent (e.g. *Lc. lactis* or *Streptococcus thermophilus*) (Tanaka *et al.*, 1999; Begley *et al.*, 2006).

Deconjugation has recently been included by the World Health Organization (WHO) experts as one of the main activities of intestinal microbiota for them to be considered as probiotic microorganisms (FAO/WHO, 2002), mainly because enzymatic deconjugation of bile salts has been related to a reduction of serum cholesterol levels in mammals (Klaver and van der Meer, 1993; Pereira *et al.*, 2003).

Apart from these effects on the host, the physiological role of BSH activity in BSH producers is currently still not clearly understood, but several hypothesis have been proposed (Begley *et al.*, 2006): (i) nutritional role in which glycine or taurine liberated from conjugated bile salts could potentially be used as carbon, nitrogen and energy sources, and/or BSH can hydrolyse phenylacetylated compounds other than conjugated bile salts, thereby serving as a scavenger in the nutritionally poor large intestine, (ii) facilitation of cholesterol incorporation into the cell membrane leading to alteration of membrane characteristics, which offers protection against bile and other stresses (Taranto *et al.*, 1997; Dambekodi and Gilliland, 1998; Taranto *et al.*, 2003), (iii) bile detoxification in which deconjugation yields a free bile acid having a higher pK_a than the corresponding conjugated bile salt, thus less acidification of the internal pH of the BSH producer (De Smet *et al.*, 1995), (iv) enhanced gastrointestinal persistence, in which combined roles of BSH (i), (ii) and (iii) may facilitate survival/persistence of BSH producers in gastrointestinal tract.

Among these effects, by far the most studied is (iii), bile detoxification action of BSHs and its contribution to bile resistance in probiotics. Although some authors have reported no correlation between BSH activity and bile resistance (Moser and Savage, 2001; Ahn *et al.*, 2003; McAuliffe *et al.*, 2005), compiling evidence suggests a role for BSH in bile resistance. A bile-resistant mutant *B. animalis* subsp. *lactis* 4549dOx had a higher content of the enzyme BSH than the wild type (Sánchez *et al.*, 2007b) and displayed a higher hydrolysing activity than the parent strain (Noriega *et al.*, 2006), which suggests a possible relationship between the increased resistance to bile salts and the presence of higher levels of BSH activity, at least in this species. In lactobacilli, some mutants were constructed and BSH activity was then studied using isogenic mutants. *Lb. plantarum* WCFS1 contains four *bsh* genes (*bsh1*, *bsh2*, *bsh3* and *bsh4*). A deletion mutant of one of them, *bsh1*, was significantly more sensitive than the parent strain to glycodeoxycholate (Lambert *et al.*, 2008). Conclusive data from other Gram-positive bacteria also shed light on BSH function: Introduction and expression of a *bsh* gene in a BSH-negative *Listeria innocua* strain increased bile tolerance *in vitro* and enhanced survival in the murine gut *in vivo* (Jones *et al.*, 2008), unequivocally suggesting a role for BSH in bile resistance and persistence in the gut.

In relation to the role in bile detoxification, it is worth noting the substrate specificity of BSH. BSHs predominantly recognize the amino acid moiety, and most BSHs are more active on glycoconjugated than in tauroconjugated bile salts (Christianes *et al.*, 1992; Tanaka *et al.*, 2000; Kim *et al.*, 2004; Begley *et al.*, 2006; Lambert *et al.*, 2008). Since glycoconjugated bile salts are more abundant in human bile, and much more toxic for bacterial cells than the tauroconjugated ones (De Smet *et al.*, 1995; Begley *et al.*, 2005; Noriega *et al.*, 2006; Lambert *et al.*, 2008), this substrate preference may have connotations for the proposed mechanism of bile detoxification.

BSHs have generally been found to be located intracellularly (Begley *et al.*, 2005). Therefore, BSH reaction requires internalization of the conjugated bile salts inside the cells, and also the export of deconjugated product, which is normally more hydrophobic, and thus more toxic, than the substrates. Some studies have dealt with the kinetics of conjugated bile salts uptake. Taranto and coworkers found that taurocholic acid is incorporated inside *Lb. reuteri* cells by an active and saturable transport process (Taranto *et al.*, 2001). The genetic organization of a BSH cluster has been studied in detail in some lactobacilli. Two conjugated bile acid transporters, CbsT1 and CbsT2, were identified in *Lb. johnsonii* 100–100 and *Lb. acidophilus* KS13 (Elkins *et al.*, 2001; Elkins and Savage, 2003). These are gene duplicates encoded in tandem with a BSH (Elkins and Savage, 1998). The three genes probably constitute an operon, and it was demonstrated that CbsT2 is responsible for taurocholic acid uptake via antiporter mechanism with CA (Elkins and Savage, 2003). The possibility of CbsT2 of conferring CA resistance through facilitated diffusion of internal CA has been tested but is not conclusive (Elkins and Savage, 2003). In *B. animalis*, an operon-like structure containing the *bsh* gene and two more open reading frames having a similarity to a decarboxylase and a transporter was also found (Kim and Lee, 2008). On the other hand, it is worthwhile to note that *bsh* genes can also be located independently in the chromosomes of lactobacilli and bifidobacteria (Elkins *et al.*, 2001; Kim *et al.*, 2004; McAuliffe *et al.*, 2005).

In summary, current data do not clearly demonstrate the role(s) of BSH for the microorganism or the human host. However, they point to a key role of BSHs as necessary enzymes in probiotic bacteria for detoxification of bile salts and to favour their survival in the gastrointestinal tract. A better intestinal survival is supposed to bolster the beneficial health-promoting effects of the corresponding bacteria.

Towards a complete understanding of the interaction between LAB and bile

During the last two decades research on *Lc. lactis*, *Lactobacillus* and *Bifidobacterium* has allowed us to establish a reference picture of the enzymatic reactions, and the cell response mechanisms, underlying the interactions between probiotics and bile. The main mechanisms involved in the processes of bile salt tolerance are included in Fig. 7.3. Bile causes a shift in the synthesis of long-chain fatty acids (Taranto *et al.*, 2003; Ruiz *et al.*, 2007), membrane lipids (Yokota, A., Matsubara, H., Kato, S., Morita, N., Sasaki, Y., Ogiya, S., and Fukiya, S., Ninth Symposium on Lactic Acid Bacteria, poster F024, 2008), surface proteins (Ruiz *et al.*, 2009) and exocellular polymers (Lebeer *et al.*, 2007; Ruas-Madiedo *et al.*, 2009). Free bile acids can enter the cytoplasm by passive diffusion (Kurdi, *et al.*, 2000, 2003) and cause membrane damage (Kurdi *et al.*, 2006). Conjugated bile acids, especially taurocholic acid, are internalized into the bacterial cytoplasm through a specific transporter (CbsT2 from *Lb. johnsonii*) (Elkins and Savage, 2003), being cleaved by the BSH rendering taurine or glycine and one deconjugated (free) bile acid moiety (Begley *et al.*, 2006). It seems that the role of BSH played in bile acid resistance needs further clarification. Free bile acids are deprotonated since the intracellular pH is slightly higher than the extracellular pH (Kurdi, *et al.*, 2000, 2003). Ionized bile acids are non-permeable, accumulated according to the alkalinity of the cytoplasm (Kurdi, *et al.*, 2000, 2003), and thus must be excreted by the action of certain transporters, such as the free bile acid transporter LmrCD from *Lc. lactis* (Lubelski *et al.*, 2006; Zaidi *et al.*, 2008), Ctr or BL0920 from *B. longum* (Price *et al.*, 2006; Gueimonde *et al.*, 2009), or facilitated antiport with taurocholic acid through CbsT2 from *Lb. johnsonii* (Elkins and Savage, 2003). Bile acid deprotonation causes cytoplasm acidification (Kurdi *et al.*, 2003, 2006), which is counteracted by proton pumping through F_1F_0-ATPase in *B. animalis* subsp. *lactis* (Sánchez *et al.*, 2006). In addition, the response to bile salts is associated with a modulation of the synthesis of molecular chaperones (Schmidt and Zink, 2000; Sánchez *et al.*, 2005, 2007b; Savijoki *et al.*, 2005; Pfeiler *et al.*, 2007; Whitehead *et al.*, 2008; Zaidi *et al.*, 2008) and SOS proteins (Lee *et al.*, 2008). The amounts of ATP needed for feeding these mechanisms are provided through the enhanced operation of bifidobacterial shunt in bifidobacteria (Sánchez *et al.*, 2005, 2007b), or through the enhanced sugar metabolism in lactobacilli (Pfeiler *et al.*, 2007; Lee *et al.*, 2008).

In summary, bile is considered one of the main toxic challenges that probiotic bacteria have to overcome to stay viable in the human gut, although bile stress is, in many aspects, largely unexplored at molecular level. However detailed physiological analysis, supported by genomic and proteomic data, recently shed some light on the mechanisms of bile tolerance in *Lc. lactis* , lactobacilli and bifidobacteria. Future molecular analysis must take advantage of the enormous amounts of data from human microbiome projects, and of the dozens of genomes of *Lactobacillus* and *Bifidobacterium* species that are currently publically available. Investigating the mechanisms of bile sensing and genetic regulation of bile responses in probiotics is still a pending issue, and bile research might help to generate improved specific strain properties by building robust stress resistance phenotypes that prepare these microorganisms for survival under adverse conditions. Also, bile completely modifies the surface of probiotic bacteria, affecting the synthesis of colonization factors, such as mucin binding proteins or exopolysaccharides, suggesting that a proper understanding of bile–probiotic interactions is critical to understanding the quorum sensing mechanisms of probiotics, and its crosstalk with the host. In turn, bile research will contribute to enable the selection of the most appropriate science-based probiotic strains.

Acknowledgements

The authors are particularly grateful to Wil N. Konings (formerly at University of Groningen, Netherlands), who accepted us to work in his laboratory to study MDR transporters in *Lc. lactis* MG1363, and to Hendrik W. van Veen (University of Cambridge, UK, formerly at University of Groningen) for his valuable guidance at Wil N. Konings' laboratory. Works in the authors' laboratories have been supported by Grants-in-Aid for Scientific Research (B) from the Japan Society for the Promotion of Science (nos. 16380054 and 21380053 to A. Y.) and by grants from the Plan Nacional de I + D + i (AGL2004–06727, AGL2007–61805 and HI2007–0246 to A. M.).

References

Agustiandari, H., Lubelski, J., van den Berg van Saparoea, H.B., Kuipers, O.P., and Driessen, A.J.M. (2008). LmrR is a transcriptional repressor of expression of the multidrug ABC transporter LmrCD in *Lactococcus lactis*. J. Bacteriol. *190*, 759–763.

Ahn, Y.T., Kim, G.B., Lim, K.S., Baek, Y.J., and Kim, H.U. (2003). Deconjugation of bile salts by *Lactobacillus acidophilus* isolates. Int. Dairy J. *13*, 303–311.

Amachi, S., Ishikawa, K., Toyoda, S., Kagawa, Y., Yokota, A., and Tomita, F. (1998). Characterization of a mutant of *Lactococcus lactis* with reduced membrane-bound ATPase activity under acidic conditions. Biosci. Biotechnol. Biochem. *62*, 1574–1580.

Baron, S.F., Franklund, C.V., and Hylemon, P.B. (1991). Cloning, sequencing, and expression of the gene coding for bile acid 7α-hydroxysteroid dehydrogenase from *Eubacterium* sp. strain VPI 12708. J. Bacteriol. *173*, 4558–4569.

Batta, A.K., Salen, G., Arora, R., Shefer, S., Batta, M., and Person, A. (1990). Side chain conjugation prevents bacterial 7-dehydroxylation of bile acids. J. Biol. Chem. *265*,10925–10928.

Begley, M., Gahan, C.G.M., and Hill, C. (2005). The interaction between bacteria and bile. FEMS Microbiol. Rev. *29*, 625–651.

Begley, M., Hill, C., and Gahan, C.G.M. (2006). Bile salt hydrolase activity in probiotics. Appl. Environ. Microbiol. *72*, 1729–1738.

Bennett, M.J., McKnight, S.L., and Coleman, J.P. (2003). Cloning and characterization of the NAD-dependent 7α-hydroxysteroid dehydrogenase from *Bacteroides fragilis*. Curr. Microbiol. *47*, 475–484.

Bernstein, C., Bernstein, H., Payne, C.M., Beard, S.E., and Schneider, J. (1999). Bile salt activation of stress response promoters in *Escherichia coli*. Curr. Microbiol. *39*, 68–72.

Bernstein, H., Bernstein, C., Payne, C.M., Dvorakova, K., and Garewal, H. (2005). Bile acids as carcinogens in human gastrointestinal cancers. Mutat. Res. *589*, 47–65.

Bolhuis, H., Molenaar, D., Poelarends, G., van Veen, H.W., Poolman, B., Driessen, A.J., and Konings, W.N. (1994). Proton motive force-driven and ATP-dependent drug extrusion systems in multidrug-resistant *Lactococcus lactis*. J. Bacteriol. *176*, 6957–6964.

Bron, P.A., Grangette, C., Mercenier, A., de Vos, W.M., and Kleerebezem, M. (2004a). Identification of *Lactobacillus plantarum* genes that are induced in the gastrointestinal tract of mice. J. Bacteriol. *186*, 5721–5729.

Bron, P.A., Marco, M., Hoffer, S.M., Van Mullekom, E., de Vos, W.M., and Kleerebezem, M. (2004b). Genetic characterization of the bile salt response in *Lactobacillus plantarum* and analysis of responsive promoters in vitro and in situ in the gastrointestinal tract. J. Bacteriol. *186*, 7829–7835.

Bron, P.A., Molenaar, D., de Vos, W.M., and Kleerebezem, M. (2006). DNA micro-array-based identification of bile-responsive genes in *Lactobacillus plantarum*. J. Appl. Microbiol. *100*, 728–738.

Calhoun, P., and Hanks, J.B. (1984). Hormonal contributions to biliary secretion. Surg. Gastroenterol. 3, 8–16.

Candela, M., Bergmann, S., Vici, M., Vitali, B., Turroni, S., Eikmanns, B.J., Hammerschmidt, S., and Brigidi,

P. (2007). Binding of human plasminogen to *Bifidobacterium*. J. Bacteriol. *189*, 5929–5936.

Cantafora, A., and Di Biase, A. (1987). Study of biologically relevant physical–chemical properties of bile salts by reverse-phase liquid chromatography. Chromatographia *24*, 277–281.

Carey, M.C. (1984). Bile acids and bile salts: ionization and solubility properties. Hepatology *4*, 66S-71S.

Christiaens, H., Leer, R.J., Pouwels, P.H., and Verstraete, W. (1992). Cloning and expression of a conjugated bile acid hydrolase gene from *Lactobacillus plantarum* by using a direct plate assay. Appl. Environ. Microbiol. *58*, 3792–3798.

Chou, L.S., and Weimer, B. (1999). Isolation and characterization of acid- and bile-tolerant isolates from strains of *Lactobacillus acidophilus*. J. Dairy Sci. *82*, 23–31.

Cotter, P.D., and Hill, C. (2003). Surviving the acid test: responses of Gram-positive bacteria to low pH. Microbiol. Mol. Biol. Rev. *67*, 429–453.

Dambekodi, P.C., and Gilliland, S.E. (1998). Incorporation of cholesterol into the cellular membrane of *Bifidobacterium longum*. J. Dairy Sci. *81*,1818–1824.

De Dea Lindner, J., Canchaya, C., Zhang, Z., Neviani, E., Fitzgerald, G.F., van Sinderen, D., and Ventura, M. (2007). Exploiting *Bifidobacterium* genomes: The molecular basis of stress response. Int. J. Food Microbiol. *120*, 13–24.

Delgado, S., O'Sullivan, E., Fitzgerald, G., and Mayo, B. (2007). Subtractive screening for probiotic properties of *Lactobacillus* species from the human gastrointestinal tract in the search for new probiotics. J. Food Sci. *72*, M310-M315.

Delgado, S., O'Sullivan, E., Fitzgerald, G., and Mayo, B. (2008). *In vitro* evaluation of the probiotic properties of human intestinal *Bifidobacterium* species and selection of new probiotic candidates. J. Appl. Microbiol. *104*, 1119–1127.

Denou, E., Pridmore, R.D., Berger, B., Panoff, J.-M., Arigoni, F., and Brüssow, H. (2008). Identification of genes associated with the long-gut-persistence phenotype of the probiotic *Lactobacillus johnsonii* strain NCC533 using a combination of genomics and transcriptome analysis. J. Bacteriol. *190*, 3161–3168.

De Smet, I., Van Hoorde, L., Vande Woestyne, M., Christiaens, H., and Verstraete, W. (1995). Significance of bile salt hydrolytic activities of lactobacilli. J. Appl. Bacteriol. *79*, 292–301.

Dunne, C., O'Mahony, L., Murphy, L., Thornton, G., Morrissey, D., O'Halloran, S., Feeney, M., Flynn, S., Fitzgerald, G., Daly, C., *et al.* (2001). In vitro selection criteria for probiotic bacteria of human origin: correlation with in vivo findings. Am. J. Clin. Nutr. *73*, 386S-392S.

Ejderhamn, J., Rafter, J.J., and Strandvik, B. (1991). Faecal bile acid excretion in children with inflammatory bowel disease. Gut *32*, 1346–1351.

Elkins, C.A., and Mullis, L.B. (2004). Bile-mediated aminoglycoside sensitivity in *Lactobacillus* species likely results from increased membrane permeability attributable to cholic acid. Appl. Environ. Microbiol. *70*, 7200–7209.

Elkins, C.A., and Savage, D.C. (1998). Identification of genes encoding conjugated bile salt hydrolase and transport in *Lactobacillus johnsonii* 100–100. J. Bacteriol. *180*, 4344–4349.

Elkins, C.A., and Savage, D.C. (2003). CbsT2 from *Lactobacillus johnsonii* 100–100 is a transport protein of the major facilitator superfamily that facilitates bile acid antiport. J. Mol. Microbiol. Biotechnol. *6*, 76–87.

Elkins, C.A., Moser, S.A., and Savage, D.C. (2001). Genes encoding bile salt hydrolases and conjugated bile salt transporters in *Lactobacillus johnsonii* 100–100 and other *Lactobacillus* species. Microbiology *147*, 3403–3412.

FAO/WHO (2002). Report of a joint FAO/WHO working group on drafting guidelines for the evaluation of probiotics in food. London Ontario, Canada, April 30 and May 1, 2002. Available at: http://www.who.int/foodsafety/fs_management/en/probiotic_guidelines.pdf.

Felis, G.E., and Dellaglio, F. (2007). Taxonomy of lactobacilli and bifidobacteria. Curr. Issues Int. Microbiol. *8*, 44–61.

Flahaut, S., Benachour, A., Giard, J.C., Boutibonnes, P., and Auffray, Y. (1996a). Defense against lethal treatments and de novo protein synthesis induced by NaCl in *Enterococcus faecalis* ATCC 19433. Arch. Microbiol. *165*, 317–324.

Flahaut, S., Frere, J., Boutibonnes, P., and Auffray. Y. (1996b). Comparison of the bile salts and sodium dodecyl sulfate stress responses in *Enterococcus faecalis*. Appl. Environ. Microbiol. *62*, 2416–2420.

Flahaut, S., Hartke, A., Giard, J.C., Benachour, A., Boutibonnes, P., and Auffray, Y. (1996c). Relationship between stress response towards bile salts, acid and heat treatment in *Enterococcus faecalis*. FEMS Microbiol. Lett. *138*, 49–54.

Flahaut, S., Frere, J., Boutibonnes, P., and Auffray, Y. (1997a). Relationship between the thermotolerance and the increase of DnaK and GroEL synthesis in *Enterococcus faecalis* ATCC19433. J. Basic Microbiol. *37*, 251–258.

Flahaut, S., Hartke, A., Giard, J.C., and Auffray, Y. (1997b). Alkaline stress response in *Enterococcus faecalis*: Adaptation, cross-protection, and changes in protein synthesis. Appl. Environ. Microbiol. *63*, 812–814.

Flórez, A.B., de los Reyes-Gavilán, C.G., Wind, A., Mayo, B., and Margolles, A. (2006). Ubiquity and diversity of multidrug resistance genes in *Lactococcus lactis* strains isolated between 1936 and 1995. FEMS Microbiol. Lett. *263*, 21–25.

Frees, D., Savijoki, K., Varmanen, P., and Ingmer, H. (2007). Clp ATPases and ClpP proteolytic complexes regulate vital biological processes in low GC, Gram-positive bacteria. Mol. Microbiol. *63*, 1285–1295.

Fuchs M. (2003). Bile acid regulation of hepatic physiology: III. Regulation of bile acid synthesis: past progress and future challenges. Am. J. Physiol. Gastrointest. Liver Physiol. 284, G551–G557.

Fukiya, S., Arata, M., Kawashima, H., Yoshida, D., Kaneko, M., Minamida, K., Watanabe, J., Ogura, Y., Uchida, K., Itoh, K., *et al.* (2009). Conversion of cholic acid and chenodeoxycholic acid into their 7-oxo derivatives by *Bacteroides intestinalis* AM-1 isolated from human feces. FEMS Microbial. Lett. *293*, 263–270.

Giard, J.C., Rince, A., Capiaux, H., Auffray, Y., and Hartke, A. (2000). Inactivation of the stress- and starvation-inducible *gls24* operon has a pleiotrophic effect on cell morphology, stress sensitivity, and gene expression in *Enterococcus faecalis*. J. Bacteriol. *182*, 4512–4520.

Gill, S.R., Pop, M., Deboy, R.T., Eckburg, P.B., Turnbaugh, P.J., Samuel, B.S., Gordon, J.I., Relman, D.A., Fraser-Liggett, C.M., and Nelson, K.E. (2006). Metagenomic analysis of the human distal gut microbiome. Science *312*, 1355–1359.

Gómez Zavaglia, A., Kociubinski, G., Pérez, P., Disalvo, E., and De Antoni, G. (2002). Effect of bile on the lipid composition and surface properties of bifidobacteria. J. Appl. Microbiol. *93*, 794–799.

van de Guchte, M., Serror, P., Chervaux, C., Smokvina, T., Ehrlich, S.D., and Maguin, E. (2002). Stress responses in lactic acid bacteria. Antonie van Leeuwenhoek *82*, 187–216.

Gueimonde, M., Noriega, L., Margolles, A., de los Reyes-Gavilán, C.G., and Salminen, S. (2005). Ability of *Bifidobacterium* strains with acquired resistance to bile to adhere to human intestinal mucus. Int. J. Food Microbiol. *101*, 341–346.

Gueimonde, M., Margolles, A., de los Reyes-Gavilán, C.G., and Salminen, S. (2007). Competitive exclusion of enteropathogens from human intestinal mucus by *Bifidobacterium* strains with acquired resistance to bile – a preliminary study. Int. J. Food Microbiol. *113*, 228–232.

Gueimonde, M., Garrigues, C., van Sinderen, D., de los Reyes-Gavilán, C.G., and Margolles, A. (2009). Bile-inducible efflux transporter from *Bifidobacterium longum* NCC2705, conferring bile resistance. Appl. Environ. Microbiol. *75*, 3153–3160.

Guglielmotti, D., Marcó, M.B., Vinderola C., de Los Reyes-Gavilán, C., Reinheimer, J., and Quiberoni, A. (2007). Spontaneous *Lactobacillus delbrueckii* phage-resistant mutants with acquired bile tolerance. Int. J. Food Microbiol. *119*, 236–242.

Hamilton, J.P., Xie, G., Raufman, J.P., Hogan, S., Griffin,T.L., Packard, C.A., Chatfield,.D.A., Hagey, L.R., Steinbach, J.H., and Hofmann, A.F. (2007). Human cecal bile acids: concentration and spectrum. Am. J. Physiol. Gastrointest. Liver Physiol. *293*, G256–G263.

Heuman, D.M. (1989). Quantitative estimation of the hydrophilic-hydrophobic balance of mixed bile salt solutions. J. Lipid Res. *30*, 719–730.

Hofmann, A.F., and Roda, A. (1984). Physicochemical properties of bile acids and their relationship to biological properties: an overview of the problem. J. Lipid Res. *25*, 1477–1489.

Hondorp, E.R., and Matthews, R.G. (2004). Oxidative stress inactivates cobalamin-independent methionine synthase (MetE) in *Escherichia coli*. PLoS Biol. *2*, 1738–1753.

Hung, D.T., Zhu, J., Sturtevant, D., and Mekalanos, J.J. (2006). Bile acids stimulate biofilm formation in *Vibrio cholerae*. Mol. Microbiol. *59*, 193–201.

Iyer, L.M., Koonin, E.V., and Aravind, L. (2004). Evolution of bacterial RNA polymerase: implications for large-scale bacterial phylogeny, domain accretion, and horizontal gene transfer. Gene *335*, 73–88.

Jones, B.V., Begley, M., Hill, C., Gahan, C.G.M., and Marchesi, J.R. (2008). Functional and comparative metagenomic analysis of bile salt hydrolase activity in the human gut microbiome. Proc. Natl. Acad. Sci. U.S.A. *105*,13580–13585.

Kamp, F., and Hamilton, J. A. (1993). Movement of fatty acids, fatty acid analogues, and bile acids across phospholipid bilayers. Biochemistry *32*, 11074–11086.

Kheadr, E., Dabour, N., Le Lay, C., Lacroix, C., and Fliss, I. (2007). Antibiotic susceptibility profile of bifidobacteria as affected by oxgall, acid, and hydrogen peroxide stress. Antimicrob. Agents Chemother. *51*,169–174.

Kim, D.Y., and Kim, K.K. (2005). Structure and function of HtrA family proteins, the key players in protein quality control. J. Biochem. Mol. Biol. *38*, 266–274.

Kim, G.-B., and Lee, B.H. (2008). Genetic analysis of a bile salt hydrolase in *Bifidobacterium animalis* subsp. *lactis* KL612. J. Appl. Microbiol. *105*, 778–790.

Kim, W.S., Ren, J., and Dunn, N.W. (1999). Differentiation of *Lactococcus lactis* supspecies *lactis* and subspecies *cremoris* strains by their adaptive response to stresses. FEMS Microbiol. Lett. *171*, 57–65.

Kim, W.S., Perl, L., Park, J.H., Tandianus, J.E., and Dunn, N.W. (2001). Assessment of stress response of the probiotic *Lactobacillus acidophilus*. Curr. Microbiol. *43*, 346–350.

Kim, G.-B., Miyamoto, C.M., Meighen, E.A., and Lee, B.H. (2004). Cloning and characterization of the bile salt hydrolase genes (*bsh*) from *Bifidobacterium bifidum* strains. Appl. Environ. Microbiol. *70*, 5603–5612.

Kitahara, M., Takamine, F., Imamura, T., and Benno, Y. (2000). Assignment of *Eubacterium* sp. VPI 12708 and related strains with high bile acid 7α-dehydroxylating activity to *Clostridium scindens* and proposal of *Clostridium hylemonae* sp. nov., isolated from human faeces. Int. J. Syst. Evol. Microbiol. *50*, 971–978.

Kitahara, M., Takamine, F., Imamura, T., and Benno, Y. (2001). *Clostridium hiranonis* sp. nov., a human intestinal bacterium with bile acid 7α –dehydroxylating activity. Int. J. Syst. Evol. Microbiol. *51*, 39–44.

Klaver, F.A.M., and van der Meer, R. (1993). The assumed assimilation of cholesterol by lactobacilli and *Bifidobacterium bifidum* is due to their bile salt-deconjugating activity. Appl. Environ. Microbiol. *59*, 1120–1124.

Kleerebezem, M., Boekhorst, J., van Kranenburg, R., Molenaar, D., Kuipers, O.P., Leer, R., Tarchini, R., Peters, S.A., Sandbrink, H.M., Fiers, M.W.E.J., *et al.* (2003). Complete genome sequence of *Lactobacillus plantarum* WCFS1. Proc. Natl. Acad. Sci. U.S.A. *100*, 1990–1995.

Knaust, A., Weber, M.V.R., Hammerschmidt, S., Bergmann, S., Frosch, M., and Kurzai, O. (2007). Cytosolic proteins contribute to surface plasminogen recruitment of *Neisseria meningitidis*. J. Bacteriol. *189*, 3246–3255.

Kullen, M.J., and Klaenhammer, T.R. (1999). Identification of the pH-inducible, proton-translocating F_1F_0-ATPase (*atpBEFHAGDC*) operon of *Lactobacillus acidophilus* by differential display: gene structure, cloning and characterization. Mol. Microbiol. *33*, 1152–1161.

Kurdi, P., van Veen, H.W., Tanaka, H., Mierau, I., Konings, W.N., Tannock, G.W., Tomita, F., and Yokota, A. (2000). Cholic acid is accumulated spontaneously, driven by membrane ΔpH, in many lactobacilli. J. Bacteriol. *182*, 6525–6528.

Kurdi, P., Tanaka, H., Van Veen, H.W., Asano, K., Tomita, F., and Yokota, A. (2003). Cholic acid accumulation and its diminution by short-chain fatty acids in bifidobacteria. Microbiology *149*, 2031–2037.

Kurdi, P., Kawanishi, K., Mizutani, K., and Yokota, A. (2006). Mechanism of growth inhibition by free bile acids in lactobacilli and bifidobacteria. J. Bacteriol. *188*, 1979–1986.

Kurokawa, K., Itoh, T., Kuwahara, T., Oshima, K., Toh, H., Toyoda, A., Takami, H., Morita, H., Sharma, V.K., Srivastava, T.P., *et al.* (2007). Comparative metagenomics revealed commonly enriched gene sets in human gut microbiomes. DNA Res. *14*, 169–181.

Lambert, J.M., Bongers, R.S., de Vos, W.M., and Kleerebezem, M. (2008). Functional analysis of four bile salt hydrolase and penicillin acylase family members in *Lactobacillus plantarum* WCFS1. Appl. Environ. Microbiol. *74*, 4719–4726.

Le Breton, Y., Maze, A., Hartke, A., Lemarinier, S., Auffray, Y., and Rince, A. (2002). Isolation and characterization of bile salts-sensitive mutants of *Enterococcus faecalis*. Curr. Microbiol. *45*, 434–439.

Lebeer, S., Verhoeven, T.L.A., Perea Vélez, M., Vanderleyden, J., and De Keersmaecker, S.C.J. (2007). Impact of environmental and genetic factors on biofilm formation by the probiotic strain *Lactobacillus rhamnosus* GG. Appl. Environ. Microbiol. *73*, 6768–6775.

Lee, K., Lee, H.G., and Choi, Y.J. (2008). Proteomic analysis of the effect of bile salts on the intestinal and probiotic bacterium *Lactobacillus reuteri*. J. Biotechnol. *137*, 14–19.

Leverrier, P., Dimova, D., Pichereau, V., Auffray, Y., Boyaval, P., and Jan, G. (2003). Susceptibility and adaptive response to bile salts in *Propionibacterium freudenreichii*: physiological and proteomic analysis. Appl. Environ. Microbiol. *69*, 3809–3818.

Ley, R.E., Hamady, M., Lozupone, C., Turnbaugh, P.J., Ramey, R.R., Bircher, J.S., Schlegel, M.L., Tucker, T.A., Schrenzel, M.D., Knight, R., *et al.* (2008). Evolution of mammals and their gut microbes. Science *320*, 1647–1651.

Lubelski, J., Mazurkiewicz, P., van Merkerk, R., Konings, W.N., and Driessen, A.J.M. (2004). *ydaG* and *ydbA* of *Lactococcus lactis* encode a heterodimeric ATP-binding cassette-type multidrug transporter. J. Biol. Chem. *279*, 34449–34455.

Lubelski, J., de Jong, A., van Merkerk, R., Agustiandari, H., Kuipers, O.P., Kok, J., and Driessen, A.J.M. (2006). LmrCD is a major multidrug resistance transporter in *Lactococcus lactis*. Mol. Microbiol. *61*, 771–781.

Macdonald, I.A., Williams, C.N., Mahony, D.E., and Christie, W.M. (1975). NAD- and NADP-dependent 7α-hydroxysteroid dehydrogenases from *Bacteroides fragilis*. Biochim. Biophys. Acta *384*,12–24.

McAuliffe, O., Cano, R.J., and Klaenhammer, T.R. (2005). Genetic analysis of two bile salt hydrolase activities in *Lactobacillus acidophilus* NCFM. Appl. Environ. Microbiol. *71*, 4925–4929.

Margolles, A., García, L., Sánchez, B., Gueimonde, M., and de los Reyes-Gavilán, C.G. (2003). Characterisation of a *Bifidobacterium* strain with acquired resistance to cholate – A preliminary study. Int. J. Food Microbiol. *82*, 191–198.

Martinez, A., and Kolter, R. (1997). Protection of DNA during oxidative stress by the nonspecific DNA-binding protein Dps. J. Bacteriol. *179*, 5188–5194.

Mascher, T., Helmann, J.D., and Unden, G. (2006). Stimulus perception in bacterial signal-transducing histidine kinases. Microbiol. Mol. Biol. Rev. *70*, 910–938.

Masco, L., Huys, G., De Brandt, E., Temmerman, R., and Swings, J. (2005). Culture-dependent and culture-independent qualitative analysis of probiotic products claimed to contain bifidobacteria. Int. J. Food Microbiol. *102*, 221– 230.

Masuda, N., Oda, H., and Tanaka, H. (1983). Purification and characterization of NADP-dependent 7β-hydroxysteroid dehydrogenase from *Peptostreptococcus productus* strain b-52. Biochim. Biophys. Acta *755*, 65–69.

Matsuki, T., Watanabe, K., and Tanaka, R. (2002). Genus- and species-specific PCR primers for the detection and identification of bifidobacteria. In Probiotics and Prebiotics: Where Are We Going? Tannock, G.W., ed. (Caister Academic Press, Wymondham, England), pp. 85–105.

Moser, S.A., and Savage, D.C. (2001). Bile salt hydrolase activity and resistance to toxicity of conjugated bile salts are unrelated properties in lactobacilli. Appl. Environ. Microbiol. *67*, 3476–3480.

Nair, S., and Finkel, S. E. (2004). Dps protects cells against multiple stresses during stationary phase. J. Bacteriol. *186*, 4192–4198.

Noh, D.O., and Gilliland, S.E. (1993). Influence of bile on cellular integrity and β-galactosidase activity of *Lactobacillus acidophilus*. J. Dairy Sci. *76*, 1253–1259.

Noriega, L., Gueimonde, M., Sánchez, B., Margolles, A., and de los Reyes-Gavilán, C.G. (2004). Effect of the adaptation to high bile salts concentrations on glycosidic activity, survival at low PH and cross-resistance to bile salts in *Bifidobacterium*. Int. J. Food Microbiol. *94*, 79–86.

Noriega, L., de los Reyes-Gavilán, C.G., and Margolles, A. (2005). Acquisition of bile salt resistance promotes antibiotic susceptibility changes in *Bifidobacterium*. J. Food Prot. *68*, 1916–1919.

Noriega, L., Cuevas, I., Margolles, A., and de los Reyes-Gavilán, C.G. (2006). Deconjugation and bile salts hydrolase activity by *Bifidobacterium* strains with acquired resistance to bile. Int. Dairy J. *16*, 850–855.

O'Connell-Motherway, M., Fitzgerald, G.F., Neirynck, S., Ryan, S., Steidler, L., and van Sinderen, D. (2008). Characterization of ApuB, an extracellular type II amylopullulanase from *Bifidobacterium breve* UCC2003. Appl. Environ. Microbiol. *74*, 6271–6279.

Pereira, D.I.A., McCartney, A.L., and Gibson, G.R. (2003). An in vitro study of the probiotic potential of a bile-salt-hydrolysing *Lactobacillus fermentum* strain, and determination of its cholesterol-lowering properties. Appl. Environ. Microbiol. *69*, 4743–4752.

Pfeiler, E.A., Azcarate-Peril, M.A., and Klaenhammer, T.R. (2007). Characterization of a novel bile-inducible operon encoding a two-component regulatory system in *Lactobacillus acidophilus*. J. Bacteriol. *189*, 4624–4634.

Pichereau, V., Bourot, S., Flahaut, S., Blanco, C., Auffray, Y., and Bernard, T. (1999). The osmoprotectant glycine betaine inhibits salt-induced cross-tolerance towards lethal treatment in *Enterococcus faecalis*. Microbiology *145*, 427–435.

Piddock, L.J.V. (2006). Multidrug-resistance efflux pumps – not just for resistance. Nat. Rev. Microbiol. *4*, 629–636.

Poquet, I., Saint, V., Seznec, E., Simoes, N., Bolotin, A., and Gruss, A. (2000). HtrA is the unique surface housekeeping protease in *Lactococcus lactis* and is required for natural protein processing. Mol. Microbiol. *35*, 1042–1051.

Price, C.E., Reid, S.J., Driessen, A.J.M., and Abratt, V.R. (2006). The *Bifidobacterium longum* NCIMB 702259^T *ctr* gene codes for a novel cholate transporter. Appl. Environ. Microbiol. *72*, 923–926.

Pumbwe, L., Skilbeck, C. A., Nakano, V., Avila-Campos, M. J., Piazza, R. M. F., and Wexler, H. M. (2007). Bile salts enhance bacterial co-aggregation, bacterial-intestinal epithelial cell adhesion, biofilm formation and antimicrobial resistance of *Bacteroides fragilis*. Microb. Pathog. *43*, 78–87.

Raphael, B.H., Pereira, S., Flom, G.A., Zhang, Q., Ketley, J.M., and Konkel, M.E. (2005). The *Campylobacter jejuni* response regulator, CbrR, modulates sodium deoxycholate resistance and chicken colonization. J. Bacteriol. *187*, 3662–3670.

Ridlon, J.M., Kang, D.J., and Hylemon, P.B. (2006). Bile salt biotransformations by human intestinal bacteria. J. Lipid Res. *47*, 241–259.

Rince, A., Flahaut, S., and Auffray, Y. (2000). Identification of general stress genes in *Enterococcus faecalis*. Int. J. Food Microbiol. *55*, 87–91.

Rince, A., Giard, J.C., Pichereau, V., Flahaut, S., and Auffray, Y. (2001). Identification and characterization of *gsp65*, an organic hydroperoxide resistance (*ohr*) gene encoding a general stress protein in *Enterococcus faecalis*. J. Bacteriol. *183*, 1482–1488.

Rince, A., Uguen, M., Le Breton, Y., Giard, J.C., Flahaut, S., Dufour, A., and Auffray, Y. (2002). The *Enterococcus faecalis* gene encoding the novel general stress protein Gsp62. Microbiology *148*, 703–711.

Rince, A., Breton, Y.L., Verneuil, N., Giard, J.C., Hartke, A., and Auffray, Y. (2003). Physiological and molecular aspects of bile salt response in *Enterococcus faecalis*. Int. J. Food Microbiol. *88*, 207–213.

Rivera-Amill, V., Kim, B.J., Seshu, J., and Konkel, M.E. (2001). Secretion of the virulence-associated *Campylobacter* invasion antigens from *Campylobacter jejuni* requires a stimulatory signal. J. Infect. Dis. *183*, 1607–1616.

Roda, A., Hofmann, A.F., and Mysels, K.J. (1983). The influence of bile salt structure on self-association in aqueous solutions. J. Biol. Chem. *258*, 6362–6370.

Roda, A., Minutello, A., Angellotti, M.A., and Fini, A. (1990). Bile acid structure–activity relationship: evaluation of bile acid lipophilicity using 1-octanol/water partition coefficient and reverse phase HPLC. J. Lipid Res. *31*, 1433–1443.

Rodionov, D.A., Vitreschak, A.G., Mironov, A.A., and Gelfand, M.S. (2004). Comparative genomics of the methionine metabolism in Gram-positive bacteria: a variety of regulatory systems. Nucleic Acids Res. *32*, 3340–3353.

Ruas-Madiedo, P., Hernández-Barranco, A., Margolles, A., and de los Reyes-Gavilán, C.G. (2005). A bile salt-resistant derivative of *Bifidobacterium animalis* has an altered fermentation pattern when grown on glucose and maltose. Appl. Environ. Microbiol. *71*, 6564–6570.

Ruas-Madiedo, P., Gueimonde, M., Arigoni, F., de los Reyes-Gavilán, C.G., and Margolles, A. (2009). Bile affects the synthesis of exopolysaccharides by *Bifidobacterium animalis*. Appl. Environ. Microbiol. *75*, 1204–1207.

Ruiz, L., Sánchez, B., Ruas-Madiedo, P., de Los Reyes-Gavilán, C.G., and Margolles, A. (2007). Cell envelope changes in *Bifidobacterium animalis* subsp. *lactis* as a response to bile. FEMS Microbiol. Lett. *274*, 316–322.

Ruiz, L., Couté, Y., Sánchez, B., de los Reyes-Gavilán, C.G., Sánchez, J.C., and Margolles, A. (2009). The cell-envelope proteome of *Bifidobacterium longum* in an *in vitro* bile environment. Microbiology *155*, 957–967.

Ryan, K.A., Jayaraman, T., Daly, P., Canchaya, C., Curran, S., Fang, F., Quigley, E.M., and O'Toole, P.W. (2008). Isolation of lactobacilli with probiotic properties from the human stomach. Lett. Appl. Microbiol. *47*, 269–274.

Sakamoto, K., Margolles, A., van Veen, H.W., and Konings, W.N. (2001). Hop resistance in the beer spoilage bacterium *Lactobacillus brevis* is mediated by the ATP-binding cassette multidrug transporter HorA. J. Bacteriol. *183*, 5371–5375.

Sánchez, B., Noriega, L., Ruas-Madiedo, P., de los Reyes-Gavilán, C.G., and Margolles, A. (2004). Acquired resistance to bile increases fructose-6-phosphate phosphoketolase activity in *Bifidobacterium*. FEMS Microbiol. Lett. *235*, 35–41.

Sánchez, B., Champomier-Vergès, M.C., Anglade, P., Baraige, F., de Los Reyes-Gavilán, C.G., Margolles, A., and Zagorec, M. (2005). Proteomic analysis of global changes in protein expression during bile salt exposure of *Bifidobacterium longum* NCIMB 8809. J. Bacteriol. *187*, 5799–5808.

Sánchez, H., Kidane, D., Cozar, M.C., Graumann, P.L., and Alonso, J.C. (2006). Recruitment of *Bacillus subtilis* RecN to DNA double-strand breaks in the absence of DNA end processing. J. Bacteriol. *188*, 353–360.

Sánchez, B., de los Reyes-Gavilán, C.G., and Margolles, A. (2006). The F_1F_0-ATPase of *Bifidobacterium animalis* is involved in bile tolerance. Environ. Microbiol. *8*, 1825–1833.

Sánchez, B., Champomier-Vergès, M.C., Collado Mdel, C., Anglade, P., Baraige, F., Sanz, Y., de los Reyes-Gavilán, C.G., Margolles, A., and Zagorec, M. (2007a). Low-pH adaptation and the acid tolerance response of *Bifidobacterium longum* biotype *longum*. Appl. Environ. Microbiol. *73*, 6450–6459.

Sánchez, B., Champomier-Vergès, M.C., Stuer-Lauridsen, B., Ruas-Madiedo, P., Anglade, P., Baraige, F., de los Reyes-Gavilán, C.G., Johansen, E., Zagorec, M., and Margolles, A. (2007b). Adaptation and response of *Bifidobacterium animalis* subsp. *lactis* to bile: a proteomic and physiological approach. Appl. Environ. Microbiol. *73*, 6757–6767.

Sánchez, B., Ruiz, L., de los Reyes-Gavilan, C.G., and Margolles, A. (2008). Proteomics of stress response in *Bifidobacterium*. Front. Biosci. *13*, 6905–6919.

Satokari, R.M., Vaughan, E.E., Smidt, H., Saarela, M., Mättö, J., and de Vos, W.M. (2003). Molecular approaches for the detection and identification of bifidobacteria and lactobacilli in the human gastrointestinal tract. System Appl. Microbiol. *26*, 572–584.

Savijoki, K., Suokko, A., Palva, A., Valmu, L., Kalkkinen, N., and Varmanen, P. (2005). Effect of heat-shock and bile salts on protein synthesis of *Bifidobacterium longum* revealed by [^{35}S]methionine labelling and two-dimensional gel electrophoresis. FEMS Microbiol. Lett. *248*, 207–215.

Schirmer, E.C., Glover, J.R., Singer, M.A., and Lindquist, S. (1996). HSP100/Clp proteins: a common mechanism explains diverse functions. Trends Biochem. Sci. *21*, 289–296.

Schmidt, G., and Zink, R. (2000). Basic features of the stress response in three species of bifidobacteria: *B. longum*, *B. adolescentis*, and *B. breve*. Int. J. Food Microbiol. *55*, 41–45.

Sherrod, J.A., and Hylemon, P.B. (1977). Partial purification and characterization of NAD-dependent 7α-hydroxysteroid dehydrogenase from *Bacteroides thetaiotaomicron*. Biochim. Biophys. Acta *486*, 351–358.

Song, Y.L., Kato, N., Liu, C.X., Matsumiya, Y., Kato, H., and Watanabe, K. (2000). Rapid identification of 11 human intestinal *Lactobacillus* species by multiplex PCR assays using group- and species-specific primers derived from the 16S-23S rRNA intergenic spacer region and its flanking 23S rRNA. FEMS Microbiol. Lett. *187*, 167–173.

Sriraman, K., and Jayaraman, G. (2008). HtrA is essential for efficient secretion of recombinant proteins by *Lactococcus lactis*. Appl. Environ. Microbiol. *74*, 7442–7446.

Takahashi, T., and Morotomi, M. (1994). Absence of cholic acid 7α-dehydroxylase activity in the strains of *Lactobacillus* and *Bifidobacterium*. J. Dairy Sci. *77*, 3275–3286.

Tanaka, H., Doesburg, K., Iwasaki, T., and Mierau, I. (1999). Screening of lactic acid bacteria for bile salt hydrolase activity. J. Dairy Sci. *82*, 2530–2535.

Tanaka, H., Hashiba, H., Kok, J., and Mierau, I. (2000). Bile salt hydrolase of *Bifidobacterium longum* – biochemical and genetic characterization. Appl. Environ. Microbiol. *66*, 2502–2512.

Taranto, M.P., Sesma, F., de Ruiz Holgado, A.P., and de Valdez, G.F. (1997). Bile salt hydrolase plays a key role on cholesterol removal by *Lactobacillus reuteri*. Biotechnol. Lett. *19*, 845–847.

Taranto, M.P., Magni, C., de Mendoza, D., de Valdez, G.F. (2001). Kinetics characterization of taurocholic transport in *Lactobacillus reuteri*. Curr. Microbiol. *42*, 18–20.

Taranto, M.P., Fernandez Murga, M.L., Lorca, G., and de Valdez, G.F. (2003). Bile salts and cholesterol induce changes in the lipid cell membrane of *Lactobacillus reuteri*. J. Appl. Microbiol. *95*, 86–91.

Taranto, M.P., Perez-Martinez, G., and de Valdez, G.F. (2006). Effect of bile acid on the cell membrane functionality of lactic acid bacteria for oral administration. Res. Microbiol. *157*, 720–725.

Thomas, L.A., Veysey, M.J., French., Hylemon, P.B., Murphy, G.M., and Dowling, R.H. (2001). Bile acid metabolism by fresh human colonic contents: a comparison of caecal versus faecal samples. Gut *49*, 835–842.

Treyer, M., Walde, P., and Oberholzer, T. (2002). Permeability enhancement of lipid vesicles to nucleotides by use of sodium cholate: basic studies and application to an enzyme-catalysed reaction occurring inside the vesicles. Langmuir *18*, 1043–1050.

Valdez, G.F., Martos, G., Taranto, M.P., Lorca, G.L., Oliver, G., and Ruiz Holgado, A.P. (1997). Influence of bile on β-galactosidase activity and cell viability of *Lactobacillus reuteri* when subjected to freeze-drying. J. Dairy Sci. *80*, 1955–1958.

Vesa, T., Pochart, P., and Marteau, P. (2000). Pharmacokinetics of *Lactobacillus plantarum* NCIMB 8826, *Lactobacillus fermentum* KLD, and *Lactococcus lactis* MG 1363 in the human gastrointestinal tract. Aliment. Pharmacol. Ther. *14*, 823–828.

Wall, T., Båth, K., Britton, R. A., Jonsson, H., Versalovic, J., and Roos, S. (2007). The early response to acid shock in *Lactobacillus reuteri* involves the ClpL chaperone and a putative cell wall-altering esterase. Appl. Environ. Microbiol. *73*, 3924–3935.

Webb, B.L., Cox, M.M., and Inman, R.B. (1997). Recombinational DNA repair: The RecF and RecR proteins limit the extension of RecA filaments beyond single-strand DNA gaps. Cell *91*, 347–356.

Wells, J.E., Williams, K.B., Whitehead, T.R., Heuman, D.M., and Hylemon, P.B. (2003). Development and application of a polymerase chain reaction assay for the detection and enumeration of bile acid 7α-dehydroxylating bacteria in human feces. Clin. Chim. Acta *331*,127–134.

Whitehead, K., Versalovic, J., Roos, S., and Britton, R.A. (2008). Genomic and genetic characterization of the bile stress response of probiotic *Lactobacillus reuteri* ATCC 55730. Appl. Environ. Microbiol. *74*, 1812–1819.

Yasui, K., Kano, Y., Tanaka, K., Watanabe, K., Shimizu-Kadota, M., Yoshikawa, H., and Suzuki, T. (2008). Improvement of bacterial transformation efficiency using plasmid artificial modification. Nucleic Acids Res. *37*, e3.

Yokota, A., Amachi, S., Ishii, S., and Tomita, F. (1995). Acid sensitivity of a mutant of *Lactococcus lactis* subsp. *lactis* C2 with reduced membrane-bound ATPase activity. Biosci. Biotechnol. Biochem. *59*, 2004–2007.

Yokota, A., Veenstra, M., Kurdi, P., van Veen, H.W., and Konings, W.N. (2000). Cholate resistance in *Lactococcus lactis* is mediated by an ATP-dependent multispecific organic anion transporter. J. Bacteriol. *182*, 5196–5201.

Yoshimoto, T., Higashi, H., Kanatani, A., Lin, X. S., Nagai, H., Oyama, H., Kurazono, K., and Tsuru, D. (1991). Cloning and sequencing of the 7α-hydroxysteroid dehydrogenase gene from *Escherichia coli* HB101 and characterization of the expressed enzyme. J. Bacteriol. *173*, 2173–2179.

Zaidi, A.H., Bakkes, P.J., Lubelski, J., Agustiandari, H., Kuipers. O.P., and Driessen, A.J.M. (2008). The ABC-type multidrug resistance transporter LmrCD is responsible for an extrusion-based mechanism of bile acid resistance in *Lactococcus lactis*. J. Bacteriol. *190*, 7357–7366.

Quality Control of Protein Structure in Lactic Acid Bacteria

8

Shinya Sugimoto and Kenji Sonomoto*

Abstract

In order to maintain cell viability, it is essential to maintain protein homeostasis under normal conditions and during injury or stress. A network of highly conserved molecular chaperones and proteases controls the quality of cellular proteins by repairing and degrading abnormal proteins. The recently completed genome sequencing, combined with the development of advanced molecular techniques, has enabled us to identify these protein quality control systems in various lactic acid bacteria (LAB). Furthermore, mutational and biochemical studies are providing new insights into the physiological significance of protein quality control systems in LAB.

Protein triage by molecular chaperones and ATP-dependent proteases

It was proposed that the polypeptide primary sequence may contain all the information required to attain a native three-dimensional structure (Anfinsen, 1973). However, stress conditions, molecular crowding, and a high concentration of folding intermediates are known to cause errors in the protein folding process, thereby leading to aggregation (Fig. 8.1). The toxic effects of intermediates and of the final forms of the aggregated proteins, and the loss of active proteins are linked to cell death and diseases (Ross and Poirier, 2004). To overcome these problems, organisms have developed specialized protein quality control systems, including molecular chaperones and adenosine triphosphate (ATP)-dependent proteases.

Both molecular chaperones and ATP-dependent proteases recognize and select a variety of misfolded and partially folded proteins that arise from chemical or thermal stresses, intrinsic structural instability, and spontaneous mutations. In general, they bind unfolded or partially folded proteins with great affinity to hydrophobic residues and/or the unstructured backbone regions, which are typically exposed on non-native proteins but are normally concealed when the proteins are in their native state. Classical molecular chaperones modulate protein folding pathways by preventing misfolding and aggregation and promoting proper folding and refolding. ATP-dependent proteases recognize and select their targets that should be degraded at an appropriate time and location. However, their folding mechanism, the regulation of their degradation activities *in vivo*, and the choice of an appropriate pathway are still unclear. These pathways are believed to function stochastically, and the fate of the protein depends on the kinetics of protein–protein interaction with molecular chaperones and the substrate-recognition components of the proteases (Fig. 8.2). The non-covalent interactions of chaperones may not only block intermolecular aggregation by shielding the hydrophobic surfaces of non-native proteins, but also prevent intramolecular misfolding. In the case of incomplete folding, the same or different chaperones, or the regulatory components of proteases can recognize and rebind the non-native proteins, thereby providing another opportunity for proper folding or degradation. If correct folding and rapid degradation are not possible, the protein may interact with other unfolded

*corresponding author

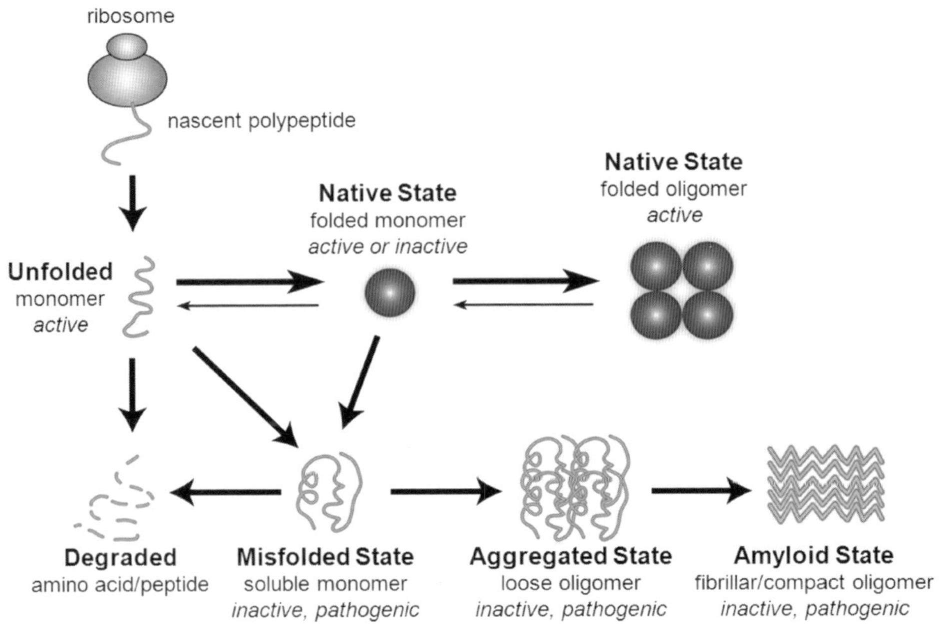

Figure 8.1 Fate of proteins under physiological and stress conditions. Under physiological conditions, newly synthesized unfolded polypeptides may spontaneously fold into active monomers and/or further assemble into discrete oligomers. Some newly synthesized polypeptides and misfolded proteins are known to be degraded by proteases. Under stress conditions, newly synthesized polypeptides may spontaneously misfold into soluble inactive monomers, which may further assemble into loose aggregates or compact aggregates such as amyloids. Active monomers and oligomers may also be denatured and aggregated under these conditions.

or partially folded proteins, leading to the formation of aggregates. Therefore, aggregation seems to be the outcome when the protein quality control systems fail.

Molecular chaperones

Most molecular chaperones are stress-inducible heat shock proteins (Hsps) that can prevent aggregation and facilitate the correct folding of stress-labile proteins. According to the typical molecular mass, Hsps are classified into Hsp100, Hsp90, Hsp70 (Hsp40), Hsp60 (Hsp10), and small Hsps in eukaryotes (co-chaperones); these correspond to ClpB, HtpG, DnaK (DnaJ), GroEL (GroES), and IbpA/IbpB, respectively, in *E. coli*. In addition, the ribosome-associated trigger factor is the first molecular chaperone encountered by nascent polypeptides in prokaryotes (Fig. 8.3).

GroEL and Hsp60 are often called 'chaperonins', since they ubiquitously distributed in nature, and alone among molecular chaperones, they are indispensable for cell survival and growth. GroEL consists of 14 identical subunits that form two heptameric rings stacked back-to-back (Fig. 8.3). GroEL enhances the efficiency of protein-folding reactions by encapsulating the protein substrate in their central cavities. GroEL uses energy derived from ATP binding and hydrolysis to drive a series of conformational changes that enable them to capture, engulf, and then release polypeptide chains. GroEL requires a co-chaperonin GroES which forms a heptameric dome-shaped structure on top of the GroEL rings (Bukau and Horwich, 1998).

DnaK engages in the folding of newly synthesized proteins, the refolding of denatured and aggregated proteins, the protein transport, and the quality control of regulatory proteins. Hsp70 family proteins have a highly conserved

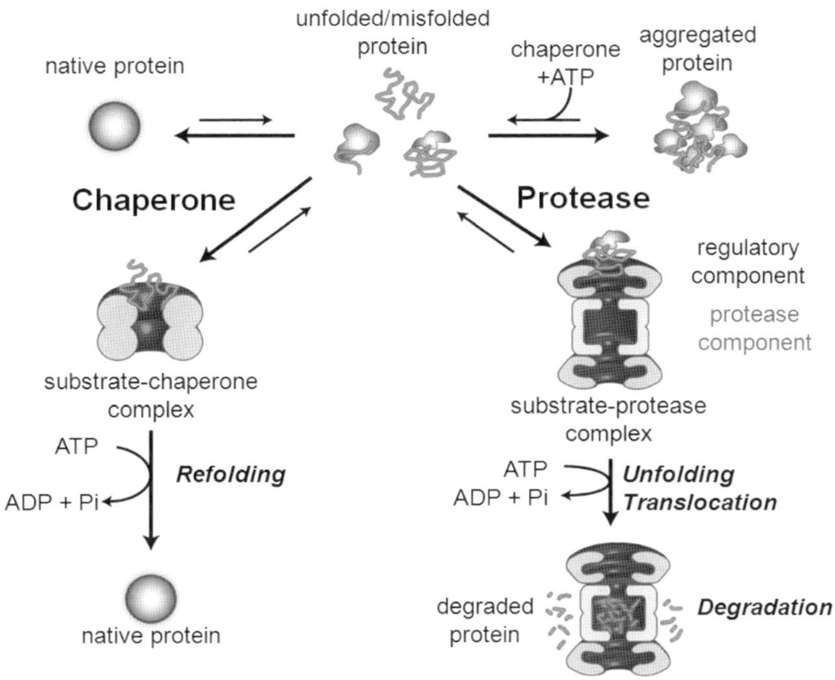

Figure 8.2 Protein triage model for quality control. Substrate proteins are remodelled in the ATP-dependent reactions by any of a variety of molecular chaperones or by the ATP-dependent proteases. The binding of folding intermediates with chaperones may allow them to reach their native conformation. In the case of incomplete protein folding, the substrates released from chaperones may return to the pool to rebind with chaperones or proteases. The binding of proteins with the regulatory components of ATP-dependent proteases may allow them to unfold, translocate to the protease components, and then degrade. Some aggregated proteins may be solubilized into a soluble unfolded state by a specialized chaperone system such as the ClpB-DnaK bichaperone system, and subsequently undergo refolding by the action of chaperones or degradation under the action of proteases.

N-terminal nucleotide-binding domain (NBD, 44 kDa) and a C-terminal substrate-binding domain (SBD, 25 kDa) (Harrison *et al.*, 1997; Pellecchia *et al.*, 2000; Zhu *et al.*, 1996). The ATP-bound state of DnaK shows low substrate affinity causing the rapid binding and release of substrates. In contrast, the ADP-bound state of DnaK shows high substrate affinity and the rates of substrate binding and release are slow. These two states are tightly regulated by the co-chaperones, DnaJ and GrpE. DnaJ stimulates the low intrinsic ATPase activity of DnaK and thus, induces substrate binding to the SBD of DnaK. GrpE functions as a nucleotide-exchange factor to promote ADP-dissociation from the NBD of DnaK and thus, induces substrate release from the SBD of DnaK.

The GroEL (GroEL-GroES) and DnaK (DnaK-DnaJ-GrpE) systems, and the trigger factor assist the *de novo* folding of newly synthesized polypeptides (Houry *et al.*, 1999; Teter *et al.*, 1999). ClpB plays a role in the resolubilization and refolding of the aggregated proteins in cooperation with the DnaK system (Goloubinoff *et al.*, 1999). Unlike other chaperone members, small Hsps show great variation in sequence, size of the single polypeptide, and the oligomer subunit number (Laksanalamai and Robb, 2004). The basic mechanisms of small Hsps include their interaction with non-native proteins to prevent aggregation and to maintain a folding-competent state. This binding results in the formation of small Hsp-substrate complexes, thus enabling the transfer of the substrate to the ATP-dependent

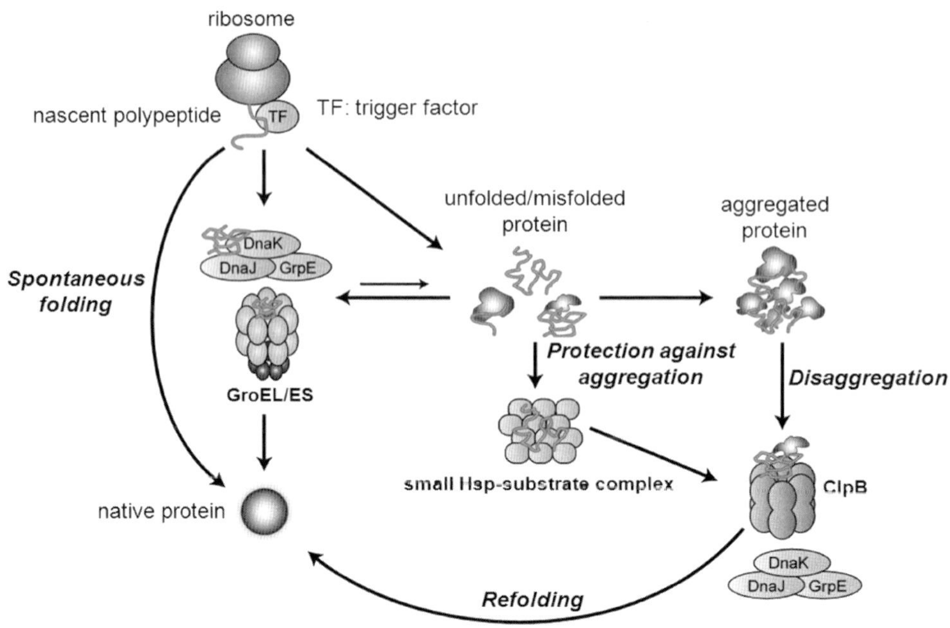

Figure 8.3 Protein folding and disaggregation mediated by classical molecular chaperones. The ribosome-bound trigger factor supports the folding of nascent polypeptides. Some newly synthesized polypeptides can spontaneously fold into a native structure, while others require the support of DnaK and GroEL systems. Aggregate-prone proteins and partially unfolded and/or misfolded proteins can be refolded by DnaK and GroEL systems or captured by small Hsps in a folding-competent state. The protein captured by small Hsps can be transferred to the ClpB-DnaK bichaperone system and then refolded. Aggregated proteins can also be disaggregated and refolded into their native state by the ClpB-DnaK bichaperone system.

chaperones (DnaK and ClpB) for refolding (Fig. 8.3). In *E. coli*, it is known that Hsp33 can protect aggregate-prone proteins under oxidative stress conditions (Winter *et al.*, 2005). In contrast, little is known about the *in vivo* function of HtpG/Hsp90; however, it may facilitate *de novo* protein folding in mildly stressed cells (Thomas and Baneyx, 2000).

Diversity of molecular chaperones in lactic acid bacteria

The trigger factor and the DnaK and GroEL systems have often been identified in lactic acid bacteria (LAB); however, small Hsps, Hsp33, and ClpB are not highly conserved (Table 8.1). Hsp90/HtpG is present in *E. coli* but absent in LAB. It is known that DnaK requires the co-chaperones DnaJ (an ATPase stimulator) and GrpE (a nucleotide exchanger) for its full chaperone activity. It is noteworthy that the number of DnaK and DnaJ homologues varies considerably, although the number of GrpE homologues is constant (only one copy). Genome-wide analyses have revealed six DnaJ homologues in *E. coli*, 22 in *Saccharomyces cerevisiae*, and more than 40 in humans. In general, there are far fewer members in the Hsp70 protein family than in the Hsp40 family (Walsh *et al.*, 2004). Different Hsp40 proteins may interact with distinct Hsp70 proteins. A possible explanation for this evolutionary diversity is that a large number of Hsp40s enable Hsp70s to attain functional versatility. However, only DnaK and DnaJ are found in LAB. Therefore, we hypothesize that diverse Hsp70 functions are not required in LAB and that DnaK alone is capable of satisfying all requirements.

Table 8.1 Conservation of molecular chaperones in LAB[a]

Bacterial species	Trigger factor	Small Hsps	Hsp33	GroEL/ GroES	DnaK/ DnaJ/ GrpE	HtpG	ClpB	ClpL
Bifidobacterium adolescentis	1	0	0	1/1	1/2/1	0	1	0
Bifidobacterium longum	1	1	0	1/1	1/2/1	0	1	0
Enterococcus faecalis	1	0	1	1/1	1/1/1	0	1	1
Lactobacillus acidophilus	1	1	1	1/1	1/1/1	0	0	0
Lactobacillus brevis	1	1	1	1/1	1/1/1	0	1	1
Lactobacillus casei	1	2	1	1/1	1/1/1	0	1	1
Lactobacillus delbrueckii	1	1	1	1/1	1/1/1	0	0	1
Lactobacillus gasseri	1	1	1	1/1	1/1/1	0	0	0
Lactobacillus helveticus	1	1	1	1/1	1/1/1	0	0	0
Lactobacillus johnsonii	1	1	1	1/1	1/1/1	0	0	0
Lactobacillus plantarum	1	3	1	1/1	1/1/1	0	1	1
Lactobacillus reuteri	1	1	1	1/1	1/1/1	0	0	1
Lactobacillus sakei	1	1	1	1/1	1/1/1	0	1	1
Lactobacillus salivarius	1	0	1	1/1	1/1/1	0	1	1
Lactococcus lactis	1	0	1	1/1	1/1/1	0	1	1
Leuconostoc mesenteroides	1	1	1	1/1	1/1/1	0	0	0
Oenococcus oeni	1	1	1	1/1	1/1/1	0	0	2
Pediococcus pentosaceus	1	1	1	1/1	1/1/1	0	1	0
Streptococcus mutans	1	0	1	1/1	1/1/1	0	1	0
Streptococcus pneumoniae	1	0	1	1/1	1/1/1	0	1	1
Streptococcus pyogenes	1	0	1	1/1	1/1/1	0	0	1
Streptococcus thermophilus	1	1	1	1/1	1/1/1	0	0	0
Escherichia coli	1	2	1	1/1	3/6/1	1	1	0

[a]Updated in January 2011.

ATP-dependent proteases

The ATPase components of ATP-dependent proteases belong to ATPase associated with diverse cellular activities (AAA$^+$) family proteins, which is a subfamily of the Walker-type ATPases (or NTPases) (Ogura and Wilkinson, 2004). This family is characterized by a conserved segment of 200–250 amino acids, commonly referred to as an AAA domain, which contains several conserved motifs including those necessary for ATP binding and hydrolysis, the Walker A and Walker B motifs, respectively. One or two AAA domains (termed D1 and D2) follow the N-terminal domain (N-domain) which is considered to be the primary substrate recognition site. One of the functional hallmarks of the regulatory components of the ATP-dependent proteases is their ability to unfold a substrate protein by using the energy generated during ATP hydrolysis. Such unfolding is considered unusual compared with the usual activity of classical molecular chaperones, and the result is the degradation of target proteins rather than their reactivation. During proteolysis, these ATPases play crucial roles in the unfolding of substrate proteins and their delivery to each proteolytic component (Fig. 8.2). Although substrate recognition by prokaryotic ATP-dependent proteases is only partially understood, evidence is accumulating that they interact with specific regions with a hydrophobic patch adjacent to basic residues (Gonciarz-Swiatek *et al.*, 1999; Gonzalez *et al.*, 1998; Smith *et al.*, 1999; Thompson and Maurizi, 1994). In bacteria, protein substrates contain various recognition motifs, called degrons, that target the protein for degradation. These degrons are located at the N-terminus (for example, N-degron), C-terminus (for example, SsrA-tag), or an internal region (Erbse *et al.*, 2006; Flynn *et al.*, 2003).

Clp family

Within caseinolytic protease (Clp) family, ClpX and ClpY possess one AAA domain (single ring architecture), while ClpA and ClpB possess two AAA domains (double ring architecture). Only a subgroup of the Clp AAA^+ ATPases can interact with the protease subunit ClpP (Fig. 8.4). The association of Clp AAA^+ ATPases with ClpP is mediated by a conserved I/V/M-G-F/L tripeptide motif located in the surface of the AAA^+ domain, which faces the subunit ClpP. In *E. coli*, this property is found in ClpA and ClpX but not in ClpB. ClpY (or HslU, an ATPase subunit) docks with ClpQ (or HslV, a protease subunit) to form a ClpYQ (or HslUV) complex. A common feature of Clp family proteins is the formation of ring-shaped oligomers in a nucleotide-dependent or nucleotide-independent manner. The ATPase subunits, ClpA, ClpB, ClpX, and ClpY, form hexamers in the presence of ATP. In contrast, ClpP assembles into a tetradecameric cylinder structure composed of the two back-to-back stacked heptameric rings. In the case of ClpQ, two back-to-back stacked hexameric rings assemble into dodecamer (Frees *et al.*, 2007). The ATPase subunits can be bound on one or both ends of the specific proteolytic cylinder.

Lon and FtsH

In contrast to the Clp family, Lon and FtsH are homohexamers; their subunits carry an ATPase domain and a protease domain (Fig. 8.4). FtsH is anchored to the cytoplasmic membrane, but its ATPase domain and Zn^{2+} metalloprotease domain are located in the cytoplasm. FtsH degrades not only membrane proteins but also

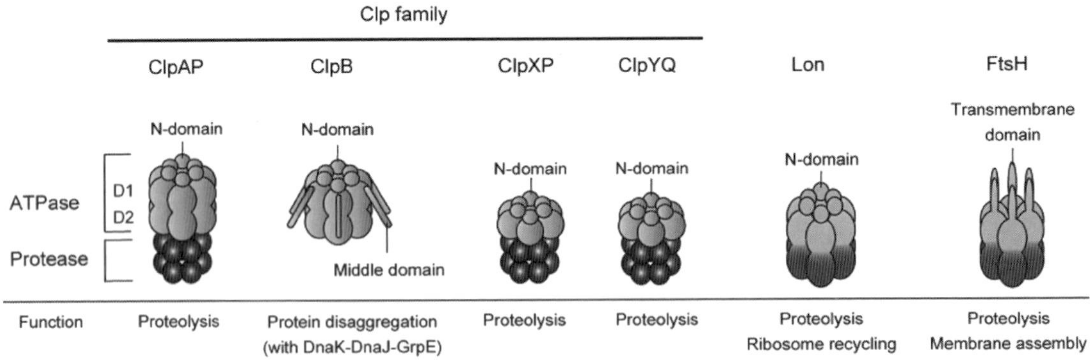

Figure 8.4 Structure of ATP-dependent proteases of *E. coli*. ClpA and ClpB possess two AAA domains (D1 and D2) per subunit and thus form a double ring structure. ClpX and ClpY possess one AAA domain per subunit and thus form a single ring structure. ClpA and ClpX interact with ClpP and form ClpAP and ClpXP proteases, while ClpY interacts with ClpQ and forms a ClpYQ protease. Although ClpB does not interact with ClpP, it cooperates with DnaK system to resolubilize aggregated proteins. The unique middle domain of ClpB is considered to have a crucial role in the disaggregating reaction. In contrast to the Clp family, Lon and FtsH form homohexamers; their subunits carry an ATPase domain and a protease domain. Only FtsH is anchored to the membrane by the N-terminal transmembrane domain. Among them, N-domain is important for substrate recognition and binding.

cytoplasmic proteins. In addition, it is known that some aspects of membrane assembly and protein localization depend on FtsH in a pathway that may be independent of the proteolytic activity. Lon is a cytoplasmic protease, and it degrades both specific substrates and abnormal misfolded proteins. Furthermore, it has been reported that Lon participates in the ribosome recycling (Kuroda *et al.*, 2001).

Diversity of ATP-dependent proteases in LAB

The *ftsH* gene, which is known as the sole essential gene among the ATP-dependent proteases in *E. coli*, is highly conserved in LAB; however, the *lon* gene is not (Table 8.2). Interestingly, a number of Clp family members found in *E. coli* are quite different from those found in LAB. Nonetheless, ClpX and ClpP are well conserved in both LAB and *E. coli*. ClpA is present in Gram-negative bacteria, including *E. coli*, but is absent in Gram-positive bacteria, including LAB. Conversely, Gram-positive bacteria possess ClpC, which is absent in Gram-negative bacteria. The functions of ClpA and ClpC are identical: both of them are ATPase subunits and are associated with their partner ClpP protease. Thus, ClpAP and ClpCP degrade stress-damaged proteins as well as newly synthesized but misfolded proteins. Furthermore, unlike *E. coli*, many LAB possess other proteins from the Clp family, such as ClpE and ClpL. ClpE is believed to function in the proteolytic process because it possesses a ClpP-binding motif. In contrast, ClpL does not possess the ClpP-binding motif and thus, like ClpB, may function in protein repairing rather than proteolysis (Kwon *et al.*, 2003). The ClpYQ (HslUV) protease is not highly conserved in LAB. Thus, in LAB, at least three ATP-dependent proteases (ClpCP, ClpXP, and FtsH) are conserved and may play a central role in proteolysis.

Physiological significance and biochemical properties of protein quality control systems in LAB

Advances in post-genomic research have shown that in LAB, molecular chaperones and ATP-dependent proteases are engaged in diverse stress responses. As reported previously, the expressions of these proteins are induced by various environmental stresses (temperature, pH, oxygen, alcohol, salt, osmotic pressure, organic acid, and antibiotics) (van de Guchte *et al.*, 2002; Sugimoto *et al.*, 2008a). Furthermore, their physiological significance has been examined using two methods: complementation assays with the corresponding mutants of *E. coli* or phenotypic analyses of the knockout or knockdown mutants of LAB.

When the desired mutant is difficult to construct, complementation assays are preferred, as demonstrated with the GroEL and DnaK systems in *Tetragenococcus halophilus* (Tosukhowong *et al.*, 2005; Sugimoto *et al.*, 2008b) and the FtsH system in *Oenococcus oeni* (Bourdineaud *et al.*, 2003). The GroEL system in *T. halophilus* complemented the *E. coli groEL* mutant under heat stress conditions but only when co-expressed with the GroES of *T. halophilus*; this suggests that the GroEL functions together with the GroES in *T. halophilus*. Similarly, FtsH present in *O. oeni* functioned in the corresponding *E. coli* mutant cells. In contrast, even when co-expressed with the deduced co-chaperones DnaJ and GrpE, DnaK in *T. halophilus* did not rescue the thermosensitivity of the *dnaK*-null mutant of *E. coli*. Surprisingly, *in vitro* assays demonstrated that DnaK in *T. halophilus* did not function together with DnaJ and GrpE in ATP hydrolysis and protein-refolding reactions (Sugimoto *et al.* 2008b). Similar results were obtained for *Lactococcus lactis* (S. Matsumoto, unpublished). These results suggest that DnaK does not cooperate with DnaJ and GrpE in LAB; however, further comprehensive studies are required to validate this assumption.

Although many mutational studies have been conducted to reveal the *in vivo* functions of chaperones and proteases in LAB (Table 8.3), none have discussed *groEL*-knockout mutants, probably because of their essential function in LAB as in the case of *E. coli*. Instead, the effect of *groEL-groES* knockdown has been reported in *Streptococcus mutans* (Lemos *et al.*, 2007). When protein levels of GroEL and GroES were reduced to less than 20%, slowed growth, cell aggregation, and impaired biofilm formation were observed.

Table 8.2 Conservation of ATP-dependent proteases in LAB[a]

Bacterial species	Clp family							FtsH	Lon
	ATPase subunit					Protease subunit			
	ClpA	ClpC	ClpE	ClpX	ClpY	ClpP	ClpQ		
Bifidobacterium adolescentis	0	1	0	1	0	2	0	1	1
Bifidobacterium longum	0	1	0	1	0	2	0	1	1
Enterococcus faecalis	0	1	1	1	1	1	1	1	1
Lactobacillus acidophilus	0	1	2	1	1	1	1	1	0
Lactobacillus brevis	0	1	1	1	1	1	1	1	1
Lactobacillus casei	0	1	1	1	1	2	1	1	1
Lactobacillus delbrueckii	0	1	1	1	1	1	1	1	1
Lactobacillus gasseri	0	1	2	1	1	2	1	1	1
Lactobacillus helveticus	0	1	1	1	1	1	1	1	0
Lactobacillus johnsonii	0	1	2	1	1	1	1	1	0
Lactobacillus plantarum	0	2	1	1	1	2	1	1	1
Lactobacillus reuteri	0	1	1	1	0	2	0	1	0
Lactobacillus sakei	0	1	1	1	1	1	1	1	0
Lactobacillus salivarius	0	1	1	1	1	3	1	1	1
Lactococcus lactis	0	1	1	1	0	1	0	1	0
Leuconostoc mesenteroides	0	1	1	1	0	2	0	1	1
Oenococcus oeni	0	1	0	1	0	1	0	1	0
Pediococcus pentosaceus	0	1	0	1	1	2	1	1	0
Streptococcus mutans	0	1	1	1	0	1	0	1	0
Streptococcus pneumoniae	0	1	1	1	0	1	0	1	1
Streptococcus pyogenes	0	1	1	1	0	2	0	1	1
Streptococcus thermophilus	0	1	1	1	0	1	0	1	1
Escherichia coli	1	0	0	1	1	1	1	1	1

[a]Updated in January 2011.

Table 8.3 Effect of mutations in genes related to protein-quality control on the phenotype of LAB

Mutation gene	Species	Phenotype	Reference
dnaK	*Lactococcus lactis*	Heat sensitive	Koch *et al.* (1998)
dnaK-dnaJ[a]	*Streptococcus mutans*	Slow growth, aggregation, impaired biofilm formation, heat sensitive, acid sensitive, susceptible to H_2O_2	Lemos *et al.* (2007)
groEL-groES[b]	*Streptococcus mutans*	Slow growth, aggregation, impaired biofilm formation	Lemos *et al.* (2007)
tig	*Streptococcus mutans*	Acid sensitive, susceptible to H_2O_2, reduced transformation efficiency, reduced biofilm formation in the presence of glucose, enhanced biofilm formation in the presence of sucrose	Wen *et al.* (2005)
clpB	*Lactococcus lactis*	Slightly higher tolerance against puromycin	Ingmer *et al.* (1999)
	Streptococcus mutans	No defect	Kajfasz *et al.* (2009)
clpB-clpL	*Streptococcus mutans*	No defect	Kajfasz *et al.* (2009)
clpL	*Streptococcus pneumoniae*	Heat sensitive	Kwon *et al.* (2003)
	Lactobacillus gasseri	Defect in thermotolerance	Suokko *et al.* (2008)
	Streptococcus mutans	Enhanced biofilm formation in the presence of sucrose, reduced viability, acid sensitive	Kajfasz *et al.* (2009)
clpC	*Lactococcus lactis*	Slightly higher tolerant against puromycin	Ingmer *et al.* (1999)
	Streptococcus pneumoniae	No defect	Robertson *et al.* (2002; Robertson *et al.* (2003)
	Streptococcus mutans	No defect	Lemos and Burne (2002), Kajfasz *et al.* (2009)
clpC-clpE	*Streptococcus mutans*	No defect	Kajfasz *et al.* (2009)

Table 8.3 continued

Mutation gene	Species	Phenotype	Reference
clpE	*Lactococcus lactis*	Puromycin sensitive	Ingmer *et al.* (1999)
	Streptococcus pneumoniae	Heat sensitive	Robertson *et al.* (2003)
	Streptococcus mutans	No defect	Kajfasz *et al.* (2009)
clpX	*Streptococcus mutans*	Slow growth, aggregation, reduced autolysis, acid sensitive, long-term survival, reduced biofilm formation in the presence of glucose, enhanced biofilm formation in the presence of sucrose, decreased infectivity	Kajfasz *et al.* (2009)
clpP	*Lactococcus lactis*	Heat sensitive, puromycin sensitive	Frees *et al.* (2001)
	Streptococcus pneumonia	Poor growth yield, heat sensitive, susceptible to H_2O_2, puromycin sensitive, attenuated virulence	Robertson *et al.* (2002)
	Streptococcus mutans	Slow growth, aggregation, heat sensitive, acid sensitive, streptomycin sensitive, long chain, aggregation, reduced transformation efficiency, long-term survival, reduced biofilm formation in the presence of glucose, enhanced biofilm formation in the presence of sucrose, decreased infectivity	Lemos and Burne (2002) Kajfasz *et al.* (2009)
	Streptococcus pneumoniae	Heat sensitive, attenuated virulence	Kwon *et al.* (2003)
ftsH	*Lactococcus lactis*	Salt sensitive, heat sensitive, cold sensitive	Nilsson *et al.* (1994)
	Lactobacillus plantarum	Slow growth, heat sensitive	Fiocco *et al.* (2009)

[a]Knockdown (<5%).

[b]Knockdown (<20%).

On the other hand, there are many reports about the effect of mutations in the ATP-dependent proteases, especially in the *clp* genes. As summarized in Table 8.3, mutations of the genes involved in protein quality control in LAB show a diverse range of phenotypic changes that affect general stress sensitivity, cell division, and pathogenic virulence in some cases.

In contrast to the *in vivo* functions, insights into the biochemical properties of chaperones and proteases are limited. Recently, Sonomoto and his colleagues reported that in *T. halophilus,* GroEL forms a double toroidal tetradecameric structure that is characteristic of GroEL (Tosukhowong *et al.,* 2005). Furthermore, they also reported that in *T. halophilus,* the structure of ClpB varies depending on stress and the presence of ATP (Sugimoto *et al.,* 2006). ClpB forms ring-shaped hexamers in the presence of ATP, ADP, or its slowly hydrolysable analogue ATPγS. *E. coli* ClpB (Zolkiewski *et al.,* 1999) and *Thermus thermophilus* ClpB (Watanabe *et al.,* 2002) form hexamers in the presence of ATP and ATPγS, while they form 2~5-mers in the presence of ADP. In addition, the oligomeric structures of ClpB homologues also show interspecies differences in the absence of nucleotides (*T. halophilus* ClpB, 1-mers and 2-mers; *T. thermophilus* ClpB, ~5-mers; *E. coli* ClpB, 7-mers). Moreover, when *T. halophilus* ClpB was incubated at a high temperature or in the presence of salt at high concentrations, hexamers dissociated into monomers and dimers (Sugimoto *et al.,* 2006). However, further studies are required to assess whether the structure and function of each chaperone and protease varies between species.

Classification of heat shock regulons

The regulatory systems affecting heat shock genes are well studied in *Bacillus subtilis,* which is frequently used as a model Gram-positive bacterium. *In B. subtilis,* the expressions of approximately 200 genes increase at least 3-fold in response to heat shock (Schumann, 2003). The heat shock genes can be categorized into six classes according to their common regulators (Table 8.4).

Class I genes include the classical chaperones (*dnaK* and *groEL*), which are regulated by a repressor protein HrcA (heat regulation at CIRCE). Their expression involves a sigma A-dependent promoter and a CIRCE (controlling inverted repeat of chaperone expression) element (TTAGCAGTC-N9-GACTGCTAA) besides HrcA. Class II genes include a large number of general stress genes regulated by an alternative sigma factor SigB. However, analyses of the complete LAB genome did not reveal any SigB homologues, suggesting that the stress response systems developed in LAB are different from those in *B. subtilis*. Class III genes, including *clpB, clpC, clpE, clpL,* and *clpP* are controlled by a CtsR (class III stress gene repressor) protein that binds to a specific direct repeat known as the CtsR-box (5′-A/GGTCAAANANA/GGTCAAA-3′). Recently, *ftsH* has been classified under class III (Fiocco *et al.,* 2009). Thus far, class IV contains only the *htpG* gene, which is yet to be identified in LAB. The *htpG* gene is positively controlled by unknown transcriptional activators that bind downstream of the −10 region. Class V genes consist of *htrA* and *htrB*. These genes encode membrane-anchored

Table 8.4 Classes of heat shock regulons in Gram-positive bacteria

Class	Regulator	Regulated genes	Mechanism
I	CIRCE-HrcA	*groEL* operon and *dnaK* operon	Negative
II	σ^B	*gspA*, *csbA*, *katE*, and more	Positive
III	CtsR	*clpB*, *clpC*, *clpE*, *clpL*, *clpP*, *ftsH*, and more	Negative
IV	Unknown	*htpG*	Positive
V	CssRS	*htrA* and *htrB*	Positive
VI	Unknown	*clpX*, *lon*, and more	Unknown

proteases, which are controlled by the CssRS two-component system. This system plays a crucial role in maintaining cell-envelope homeostasis under stress conditions. In LAB, *htrA* genes have been identified, whereas *htrB* and *cssRS* genes have not. Class VI genes include a group of genes whose expression is also responsive to stress, but their induction is not affected by any of the previously mentioned regulators.

References

Anfinsen, C.B. (1973). Principles that govern the folding of protein chains. Science *181*, 223–230.

Bourdineaud, J.P., Nehmé, B., Tesse, S., and Lonvaud-Funel, A. (2003). The *ftsH* gene of the wine bacterium *Oenococcus oeni* is involved in protection against environmental stress. Appl. Environ. Microbiol. *69*, 2512–2520.

Bukau, B., and Horwich, A.L. (1998). The Hsp70 and Hsp60 chaperone machines. Cell *92*, 351–366.

Erbse, A., Schmidt, R., Bornemann, T., Schneider-Mergener, J., Mogk, A., Zahn, R., Dougan, D.A., and Bukau, B. (2006). ClpS is an essential component of the N-end rule pathway in *Escherichia coli*. Nature *439*, 753–756.

Fiocco, D., Collins, M., Muscariello, L., Hols, P., Kleerebezem, M., Msadek, T., and Spano, G. (2009). The *Lactobacillus plantarum ftsH* gene is a novel member of the CtsR stress response regulon. J. Bacteriol. *191*, 1688–1694.

Flynn, J.M., Neher, S.B., Kim, Y.I., Sauer, R.T., and Baker, T.A. (2003). Proteomic discovery of cellular substrates of the ClpXP protease reveals five classes of ClpX-recognition signals. Mol. Cell *11*, 671–683.

Frees, D., Varmanen, P., and Ingmer, H. (2001). Inactivation of a gene that is highly conserved in Gram-positive bacteria stimulates degradation of non-native proteins and concomitantly increases stress tolerance in *Lactococcus lactis*. Mol. Microbiol. *41*, 93–103.

Frees, D., Savijoki, K., Varmanen, P., and Ingmer, H. (2007). Clp ATPases and ClpP proteolytic complexes regulate vital biological processes in low GC, Gram-positive bacteria. Mol. Microbiol. *63*, 1285–1295.

Goloubinoff, P., Mogk, A., Zvi, A.P., Tomoyasu, T., and Bukau, B. (1999). Sequential mechanism of solubilization and refolding of stable protein aggregates by a bichaperone network. Proc. Natl. Acad. Sci. U.S.A. *96*, 13732–13737.

Gonciarz-Swiatek, M., Wawrzynow, A., Um, S.J., Learn, B.A., McMacken, R., Kelley, W.L., Georgopoulos, C., Sliekers, O., and Zylicz, M. (1999). Recognition, targeting, and hydrolysis of the lambda O replication protein by the ClpP/ClpX protease. J. Biol. Chem. *274*, 13999–14005.

Gonzalez, M., Frank, E.G. Levine, A.S., and Woodgate, R. (1998). Lon-mediated proteolysis of the *Escherichia coli* UmuD mutagenesis protein: *in vitro* degradation and identification of residues required for proteolysis. Genes Dev. *12*, 3889–3899.

van de Guchte, M., Serror, P., Chervaux, C., Smokvina, T., Ehrlich, S.D., and Maguin, E. (2002). Stress responses in lactic acid bacteria. Antonie van Leeuwenhoek *82*, 187–216.

Harrison, C.J., Hayer-Hartl, M., Liberto, M.D., Hartl, F.U., and Kuriyan, J. (1997). Crystal structure of the nucleotide exchange factor GrpE bound to the ATPase domain of the molecular chaperone DnaK. Science *276*, 431–435.

Houry, W.A., Frishman, D., Eckerskorn, C., Lottspeich, F., and Hartl, F.U. (1999). Identification of *in vivo* substrates of the chaperonin GroEL. Nature *402*, 147–154.

Ingmer, H., Vogensen, F.K., Hammer, K., and Kilstrup, M. (1999). Disruption and analysis of the *clpB*, *clpC*, and *clpE* genes in *Lactococcus lactis*: ClpE, a new Clp family in Gram-positive bacteria. J. Bacteriol. *181*, 2075–2083.

Kajfasz, J.K., Martinez, A.R., Rivera-Ramos, I., Abranches, J., Koo, H., Quivey, R.G. Jr., and Lemos, J.A. (2009). Role of Clp proteins in expression of virulence properties of *Streptococcus mutans*. J. Bacteriol. *191*, 2060–2068.

Koch, B., Kilstrup, M., Vogensen, F.K., and Hammer, K. (1998). Induced levels of heat shock proteins in a *dnaK* mutant of *Lactococcus lactis*. J. Bacteriol. *180*, 3873–3881.

Kuroda, A., Nomura, K., Ohtomo, R., Kato, J., Ikeda, T., Takiguchi, N., Ohtake, H., and Kornberg, A. (2001). Role of inorganic polyphosphate in promoting ribosomal protein degradation by the Lon protease in *E. coli*. Science *293*, 705–708.

Kwon, H.Y., Kim, S.W., Choi, M.H., Ogunniyi, A.D., Paton, J.C., Park, S.H., Pyo, S.N., and Rhee, D.K. (2003). Effect of heat shock and mutations in ClpL and ClpP on virulence gene expression in *Streptococcus pneumoniae*. Infect. Immun. *71*, 3757–3765.

Laksanalamai, P., and Robb, F.T. (2004). Small heat shock proteins from extremophiles. Extremophiles *8*, 1–11.

Lemos, J.A., and Burne, R.A. (2002). Regulation and physiological significance of ClpC and ClpP in *Streptococcus mutans*. J. Bacteriol. *184*, 6357–6366.

Lemos, J.A., Luzardo, Y., and Burne, R.A. (2007). Physiologic effects of forced down-regulation of *dnaK* and *groEL* expression in *Streptococcus mutans*. J. Bacteriol. *189*, 1582–1588.

Nilsson, D., Lauridsen, A.A., Tomoyasu, T., and Ogura, T. (1994). A *Lactococcus lactis* gene encodes a membrane protein with putative ATPase activity that is homologous to the essential *Escherichia coli ftsH* gene product. Microbiology *140*, 2601–2610.

Ogura, T., and Wilkinson, A.J. (2001). AAA^+ superfamily ATPases: common structure–diverse function. Genes Cells *6*, 575–597.

Pellecchia, M., Montgomery, D.L., Stevens, S.Y., Vander Kooi, C.W., Feng, H.P., Gierasch, L.M., and Zuiderweg, E.R. (2000). Structural insights into substrate binding by the molecular chaperone DnaK. Nat. Struct. Biol. *7*, 298–303.

Robertson, G.T., Ng, W.L., Foley, J., Gilmour, R., and Winkler, M.E. (2002). Global transcriptional analysis of *clpP* mutations of type 2 *Streptococcus pneumoniae* and their effects on physiology and virulence. J. Bacteriol. *184*, 3508–3520.

Robertson, G.T., Ng, W.L., Gilmour, R., and Winkler, M.E. (2003). Essentiality of *clpX*, but not *clpP*, *clpL*, *clpC*, or *clpE*, in *Streptococcus pneumoniae* R6. J. Bacteriol. *185*, 2961–2966.

Ross, C.A., and Poirier, M.A. (2004). Protein aggregation and neurodegenerative disease. Nat. Med. *10*, S10–S17.

Schumann, W. (2003). The *Bacillus subtilis* heat shock stimulon. Cell Stress Chaperones *8*, 207–217.

Smith, C.K., Baker, T.A., and Sauer, R.T. (1999). Lon and Clp family proteases and chaperones share homologous substrate-recognition domains. Proc. Natl. Acad. Sci. U.S.A. *96*, 6678–6682.

Sugimoto, S., Yoshida, H., Mizunoe, Y., Tsuruno, K., Nakayama, J., and Sonomoto, K. (2006). Structural and functional conversion of molecular chaperone ClpB from the Gram-positive halophilic lactic acid bacterium *Tetragenococcus halophilus* mediated by ATP and stress. J. Bacteriol. *188*, 8070–8078.

Sugimoto, S., Abdullah-Al-Mahin, A.A., and Sonomoto, K. (2008a). Molecular chaperones in lactic acid bacteria: physiological consequence and biochemical properties. J. Biosci. Bioeng. *106*, 324–336.

Sugimoto, S., Saruwatari, K., Higashi, C., Tsuruno, K., Matsumoto, S., Nakayama, J., and Sonomoto, K. (2008b). *In vivo* and *in vitro* complementation study comparing the function of DnaK chaperone systems from halophilic lactic acid bacterium *Tetragenococcus halophilus* and *Escherichia coli*. Biosci. Biotechnol. Biochem. *72*, 811–822.

Suokko, A., Poutanen, M., Savijoki, K., Kalkkinen, N., and Varmanen, P. (2008). ClpL is essential for induction of thermotolerance and is potentially part of the HrcA regulon in *Lactobacillus gasseri*. Proteomics *8*, 1029–1041.

Teter, S.A., Houry, W.A., Ang, D., Tradler, T., Rockabrand, D., Fischer, G., Blum, P., Georgopoulos, C., and Hartl, F.U. (1999). Polypeptide flux through bacterial Hsp70: DnaK cooperates with trigger factor in chaperoning nascent chains. Cell *97*, 755–765.

Thomas, J.G., and Baneyx, F. (2000). ClpB and HtpG facilitate *de novo* protein folding in stressed *Escherichia coli* cells. Mol. Microbiol. *36*, 1360–1370.

Thompson, M.W., and Maurizi, M.R. (1994). Activity and specificity of *Escherichia coli* ClpAP protease in cleaving model peptide substrates. J. Biol. Chem. *269*, 18201–18208.

Tosukhowong, A., Nakayama, J., Mizunoe, Y., Sugimoto, S., Fukuda, D., and Sonomoto, K. (2005). Reconstitution and function of *Tetragenococcus halophila* chaperonin 60 tetradecamer. J. Biosci. Bioeng. *99*, 30–37.

Walsh, P., Bursac, D., Law, Y.C., Cyr, D., and Lithgow, T. (2004). The J-protein family: modulating protein assembly, disassembly and translocation. EMBO Rep. *5*, 567–571.

Watanabe, Y.H., Motohashi, K., and Yoshida, M. (2002). Roles of the two ATP binding sites of ClpB from *Thermus thermophilus*. J. Biol. Chem. *277*, 5804–5809.

Wen, Z.T., Suntharaligham, P., Cvitkovitch, D.G., and Burne, R.A. (2005). Trigger factor in *Streptococcus mutans* is involved in stress tolerance, competence development, and biofilm formation. Infect. Immun. *73*, 219–225.

Winter, J., Linke, K., Jatzek, A., and Jakob, U. (2005). Severe oxidative stress causes inactivation of DnaK and activation of the redox-regulated chaperone Hsp33. Mol. Cell *17*, 381–392.

Zhu, X., Zhao, X., Burkholder, W.F., Gragerov, A., Ogata, C.M., Gottesman, M.E., and Hendrickson, W.A. (1996). Structural analysis of substrate binding by the molecular chaperone DnaK. Science *272*, 1606–1614.

Zolkiewski, M., Kessel, M., Ginsburg, A., and Maurizi, M.R. (1999). Nucleotide-dependent oligomerization of ClpB from *Escherichia coli*. Protein Sci. *8*, 1899–1903.

Part III

Bacteriocins from Lactic Acid Bacteria

Classification and Diversity of Bacteriocins

9

Takeshi Zendo and Kenji Sonomoto*

Abstract

Many strains of lactic acid bacteria produce antimicrobial peptides, bacteriocins, which are expected to be used as safe antimicrobial agents as well as food preservatives. Since nisin A was recognized in 1928, diverse bacteriocins have been identified from various species of lactic acid bacteria. Based on their structures and characteristics, bacteriocins were classified mainly into two classes, class I (lantibiotics) and class II, and new classification schemes are being proposed according to their expanding diversity. Class I bacteriocins represented by nisin A contain post-translationally modified amino acids such as lanthionine. Class II bacteriocins are further divided into some subclasses such as class IIa (pediocin-like bacteriocins) and class IIb (two-peptide bacteriocins).

Bacteriocins from lactic acid bacteria

Bacteriocins are antimicrobial peptides or proteins ribosomally synthesized by diverse bacteria. A great number of strains of lactic acid bacteria (LAB) have also been found to produce bacteriocins. It seems that bacteriocins play important roles in bacterial competition and that bacteriocin-producing bacteria have considerable advantages to survive their environments. LAB are often found in foods and unconsciously utilized to manufacture fermented foods. LAB are known to contribute not only improvement of tastes but also preservation of fermented foods by preventing outgrowth of spoilage bacteria. In particular, bacteriocin-producing LAB are thought more advantageous to preserve foods. Actually, the most representative and well-studied bacteriocin, nisin A has been used as a food preservative in more than 50 countries. In addition, new LAB bacteriocins with favourable properties have been explored.

Why LAB bacteriocins?

On the basis of the safety of LAB, their bacteriocins are expected to be antimicrobial agents as well as food preservatives (Jack *et al.*, 1995; Cleveland *et al.*, 2001). However, bacteriocins were often confused with antibiotics. Antibiotics are used for clinical applications and synthesized as secondary metabolites, whereas bacteriocins are used mainly for food preservation and synthesized ribosormally. In addition, although antibiotics generally need micromolar level for their activity, bacteriocins act specifically to target bacteria only at nanomolar level. No side effects and no resistant bacteria have been reported in bacteriocins. One of the reasons is that most of bacteriocins act through pore formation on the target cell membrane at extremely low concentrations. Moreover, since bacteriocins have proteinaceous nature, they can be degraded easily in our bodies and environments. It means that bacteriocins do not remain and disturb environments excessively, although pollution by remaining antibiotics is an emerging problem in environments.

*corresponding author

Conventional classification of LAB bacteriocins

Since nisin A was recognized in 1928 (Rogers 1928), various LAB bacteriocins were discovered and identified, and many researchers have tried to classify bacteriocins. First, Klaenhammer (1993) classified bacteriocins of Gram-positive bacteria including LAB bacteriocins into four classes according to their structures and characteristics. Then, Nes *et al.* (1996, 2000) revised the classification. Their classifications were summarized in Table 9.1.

Klaenhammer's classification included four classes, class I, II, III and IV (1993). Class I bacteriocins, so-called lantibiotics, are heat-stable post-translationally modified peptides that contain multiple rings bridged by lanthionine or 3-methyllanthionine residues (Nagao *et al.*, 2006; Dufour *et al.*, 2007; Asaduzzaman and Sonomoto, 2009). Nisin is the representative in class I bacteriocins. Class II bacteriocins are small heat-stable non-lantibiotic peptides. Class II involves various bacteriocins, and this class was further divided into subclasses, classes IIa, IIb and IIc. Class IIa bacteriocins are *Listeria*-active peptides with a consensus YGNGVXC sequence at their N-termini. The representative class IIa bacteriocin is pediocin PA-1 (AcH), and bacteriocins in this subclass are also called pediocin-like bacteriocins (Ennahar *et al.*, 2000; Drider *et al.*, 2006). Class IIb bacteriocins comprise two peptides, both of which are required for complete antibacterial activity (Garneau *et al.*, 2002; Oppegård *et al.*, 2007). Lactococcin G is the representative class IIb bacteriocin (Nissen-Meyer *et al.* 1992). Class IIc bacteriocins were originally proposed as thiol-activated bacteriocins requiring reduced cysteine residues for activity but further revised as below. Class III bacteriocins are heat-labile proteins, including millericin B that hydrolyses specific peptide bonds in the stem and/or interpeptide bridges in the peptidoglycan layer of susceptible bacteria (Beukes *et al.*, 2000). Class IV bacteriocins are defined as bacteriocins containing sugar or lipid moieties. However, no class IV bacteriocins have been purified, and the classification was revised later.

Nes *et al.* (1996) revised the classification by Klaenhammer. Nes *et al.* defined class IIc as *sec*-dependent secreted bacteriocins, which are

Table 9.1 Classification of bacteriocins from lactic acid bacteria by Klaenhammer (1993) and Nes *et al.* (1996, 2000)

	Classification by		
Class	Klaenhammer (1993)	Nes *et al.* (1996)	Nes and Holo (2000)
I	Lantibiotics, small membrane active-peptides (<5 kDa) containing the unusual amino acids, lanthionine, β-methyllanthionine, and dehydrated residues	Lantibiotics	Lantibiotics
II	Small heat stable non-lanthionine containing membrane-active peptides (<10 kDa)	Small heat stable non-lantibiotics	Nonmodified heat stable bacteriocins
IIa	*Listeria*-active peptides with a consensus sequence in the N-terminal of YGNGVXC	Pediocin-like bacteriocins with strong antilisterial effect	Pediocin-like bacteriocins (listeria active)
IIb	Poration complexes consisting of two peptides	Two-peptide bacteriocins	Two-peptide bacteriocins
IIc[1]	Thiol-activated peptides	*sec*-dependent secreted bacteriocins	Other class II bacteriocins
III	Large heat-labile proteins (>30 kDa)	Large heat-labile proteins	Large heat-labile proteins
IV[1]	Complex bacteriocins, composed of protein plus one or more chemical moieties (lipid, carbohydrate)	–	–

[1]Only the definitions of class IIc and class IV bacteriocins were changed.

secreted by using *sec*-pathway. Then, Nes and Holo (2000) revised the definition of class IIc bacteriocins again, which were defined as the other class II bacteriocins that did not belong to classes IIa and IIb. In addition, Nes *et al.* excluded class IV from the classification, since class IV bacteriocins have not been characterized adequately at the biochemical level.

Newly proposed classification of LAB bacteriocins

The classification scheme suggested by Klaenhammer and revised by Nes *et al.* has been widely accepted. However, with expanding the diversity of LAB bacteriocins, some novel bacteriocins needed new classification scheme.

Class I bacteriocins, so-called lantibiotics, are further grouped into several types or subclasses based on their structures. Jung (1991) separated lantibiotics into type A (elongated flexible peptides) and type B (globular peptides). Since type A included lantibiotics with apparently different structures, some subtypes were suggested (Sahl *et al.*, 1995). However, this classification could not assort some newly discovered lantibiotics. Guder *et al.* (2000) classified lantibiotics into eight groups, and Twomey *et al.* (2002) proposed six groups of LAB lantibiotics. Moreover, Cotter *et al.* (2005a) further divided lantibiotics into 11 groups.

Class II also involved various bacteriocins with unique characteristics such as cyclic structures. van Belkum and Stiles (2000) proposed a new scheme for class II bacteriocins including cyclic bacteriocins. Kemperman *et al.* (2003) suggested a new classification that circular bacteriocins were classified into a new independent class, class V. Cotter *et al.* (2005b) suggested an another classification scheme that circular bacteriocins were categorized in one subclass of class II bacteriocins and that class III bacteriocins were excluded from bacteriocins and moved to a new designation called 'bacteriolysin' (Table 9.2). Nes *et al.* (2007) also suggested a new classification scheme that leaderless bacteriocins and circular bacteriocins were separately classified into subclasses of class II bacteriocins. However, some research groups

Table 9.2 Suggested classification of bacteriocins from lactic acid bacteria by Cotter *et al.* (2005b)

Classification	Remarks	Examples	References
Class I	Lanthionine-containing bacteriocins, lantibiotics. Includes both single- and two-peptide lantibiotics	Nisin A Lacticin 481 Lacticin 3147 (two-peptide)	Gross and Morell (1971) Piard *et al.* (1993) Ryan *et al.* (1999)
Class II	Non-lanthionine-containing bacteriocins. Heterogeneous class of small peptides		
Class IIa	Pediocin-like bacteriocins	Pediocin PA-1/AcH Leucocin A	Henderson *et al.* (1992) Motlagh *et al.* (1992) Hastings *et al.* (1991)
Class IIb	Two-peptide bacteriocins	Lactococcin G Lactacin F	Nissen-Meyer *et al.* (1992) Allison *et al.* (1994)
Class IIc	Cyclic bacteriocins	Enterocin AS-48 Gassericin A Lactocyclicin Q	Samyn *et al.* (1994) Kawai *et al.* (1998) Sawa *et al.* (2009)
Class IId	Non-pediocin single linear bacteriocin	Lactococcin A Lacticin Q	Holo *et al.* (1991) Fujita *et al.* (2007)
Bacteriolysins (former class III)	Non-bacteriocin lytic proteins. Large, heat-labile proteins, often murein hydrolases.	Enterolysin A	Nilsen *et al.* (2003)

suggested that circular bacteriocins where defined as a separate class (Franz *et al.*, 2007; Maqueda *et al.*, 2008).

Thus, classification of bacteriocins, especially in lantibiotics, class III bacteriocins, and circular bacteriocins, is still under discussion. Discovery of new bacteriocins will let us change the classification. In this chapter, we follow the classification by Cotter *et al.* (2005b) temporarily as shown in Table 9.2.

Diversity of bacteriocins in LAB

Bacteriocins are produced by various species of LAB. LAB genera found to produce bacteriocins are as follows; *Lactococcus, Streptococcus, Enterococcus, Pediococcus, Lactobacillus, Carnobacterium, Leuconostoc, Weissella* (Table 9.3). Generally, bacteriocins produced by a species are limited with some exceptions such as pediocin AcH production by *Lactobacillus plantarum* (Ennahar *et al.*, 1996). For example, production of nisin A is limited by strains belonging to *Lactococcus lactis.* As mentioned above, structures of bacteriocins are diverse, and bacteriocins are classified based on their structures. Among the genera of LAB, *Lactococcus, Enterococcus* and *Lactobacillus* produce most of the classes of bacteriocins. Other Gram-positive bacteria of the genera related to LAB, such as *Bacillus, Clostridium, Staphylococcus,*

Table 9.3 Bacteriocin-producing genera of lactic acid bacteria and their bacteriocins

Genus	Representative bacteriocin	Class[1]	Reference
Lactococcus	Nisin A	I	Gross and Morell (1971)
	Lacticin 481	I	Piard *et al.* (1993)
	Lacticin 3147	I	Ryan *et al.* (1999)
	Lactococcin MMFII	IIa	Ferchichi *et al.* (2001)
	Lactococcin G	IIb	Nissen-Meyer *et al.* (1992)
	Lactocyclicin Q	IIc	Sawa *et al.* (2009)
	Lactococcin A	IId	Holo *et al.* (1991)
Streptococcus	Salivaricin A	I	Ross *et al.* (1993)
	Thermophilin 13	IIb	Marciset *et al.* (1997)
	Uberolysin	IIc	Wirawan *et al.* (2007)
	BHT-B	IId	Hyink *et al.* (2005)
Enterococcus	Cytolysin	I	Haas and Gilmore (1999)
	Enterocin A	IIa	Aymerich *et al.* (1996)
	Enterocin 1071	IIb	Franz *et al.* (2002)
	Enterocin AS-48	IIc	Samyn *et al.* (1994)
	Enterocin B	IId	Casaus *et al.* (1997)
Pediococcus	Pediocin PA-1/AcH	IIa	Henderson *et al.* (1992), Motlagh *et al.* (1992)
Lactobacillus	Lactocin S	I	Skaugen *et al.* (1994)
	Curvacin A	IIa	Tichaczek *et al.* (1993)
	Lactacin F	IIb	Allison *et al.* (1994)
	Gassericin A	IIc	Kawai *et al.* (1998)
Carnobacterium	Carnobacteriocin B2	IIa	Quadri *et al.* (1994)
	Carnocyclin A	IIc	Martin-Visscher *et al.* (2008)
	Carnobacteriocin A	IId	Worobo *et al.* (1994)
Leuconostoc	Leucocin A	IIa	Hastings *et al.* (1991)
	Mesentericin 52B	IId	Revol-Junelles *et al.* (1996)
Weissella	Weissellicin 110[2]	IId	Srionnual *et al.* (2007)

[1]Classification is according to Cotter *et al.* (2005b) as shown in Table 9.2.

[2]The complete amino acid sequence has not been determined.

Propionibacterium and *Bifidobacterium*, have also been found to produce bacteriocins satisfying criteria shown above. Since various genera of bacteria have been found to produce bacteriocins, further studies will extend the range of bacteriocin-producing genera and species.

References

Allison, G.E., Fremaux, C., and Klaenhammer, T.R. (1994). Expansion of bacteriocin activity and host range upon complementation of two peptides encoded within the lactacin F operon. J. Bacteriol. *176*, 2235–2241.

Aymerich, T., Holo, H., Håvarstein, L.S., Hugas, M., Garriga, M., and Nes, I.F. (1996). Biochemical and genetic characterization of enterocin A from *Enterococcus faecium*, a new antilisterial bacteriocin in the pediocin family of bacteriocins. Appl. Environ. Microbiol. *62*, 1676–1682.

Asaduzzaman, S.M., and Sonomoto, K. (2009). Lantibiotics: Diverse activities and unique modes of action. J. Biosci. Bioeng. *107*, 475–487.

van Belkum, M.J., and Stiles, M.E. (2000). Nonlantibiotic antibacterial peptides from lactic acid bacteria. Nat. Prod. Rep. *17*, 323–335.

Beukes, M., Bierbaum, G., Sahl, H.-G., and Hastings, J.W. (2000). Purification and partial characterization of a murein hydrolase, millericin B, produced by *Streptococcus milleri* NMSCC 061. Appl. Environ. Microbiol. *66*, 23–28.

Casaus, P., Nilsen, T., Cintas, L.M., Nes, I.F., Hernández, P.E., and Holo, H. (1997). Enterocin B, a new bacteriocin from *Enterococcus faecium* T136 which can act synergistically with enterocin A. Microbiology *143*, 2287–2294.

Cleveland, J., Montville, T.J., Nes, I.F., and Chikindas, M.L. (2001). Bacteriocins: safe, natural antimicrobials for food preservation. Int. J. Food Microbiol. *71*, 1–20.

Cotter, P.D., Hill, C., and Ross, R.P. (2005a). Bacterial lantibiotics: strategies to improve therapeutic potential. Curr. Protein Pept. Sci. *6*, 61–75.

Cotter, P.D., Hill, C., and Ross, R.P. (2005b). Bacteriocins: developing innate immunity for food. Nat. Rev. Microbiol. *3*, 777–788.

Drider, D., Fimland, G., Héchard, Y., McMullen, L.M., and Prévost, H. (2006). The continuing story of class IIa bacteriocins. Microbiol. Mol. Biol. Rev. *70*, 564–582.

Dufour, A., Hindré, T., Haras, D., and Le Pennec, J.P. (2007). The biology of lantibiotics from the lacticin 481 group is coming of age. FEMS Microbiol. Rev. *31*, 134–167.

Ennahar, S., Aoude-Werner, D., Sorokine, O., Van Dorsselaer, A., Bringel, F., Hubert, J.C., and Hasselmann, C. (1996). Production of pediocin AcH by *Lactobacillus plantarum* WHE 92 isolated from cheese. Appl. Environ. Microbiol.. *62*, 4381–4387.

Ennahar, S., Sashihara, T., Sonomoto, K., and Ishizaki, A. (2000). Class IIa bacteriocins: biosynthesis, structure and activity. FEMS Microbiol. Rev. *24*, 85–106.

Ferchichi, M., Frère, J., Mabrouk, K., and Manai, M. (2001). Lactococcin MMFII, a novel class IIa bacteriocin produced by *Lactococcus lactis* MMFII, isolated from a Tunisian dairy product. FEMS Microbiol. Lett. *205*, 49–55.

Franz, C.M., Grube, A., Herrmann, A., Abriouel, H., Stärke, J., Lombardi, A., Tauscher, B., and Holzapfel, W.H. (2002). Biochemical and genetic characterization of the two-peptide bacteriocin enterocin 1071 produced by *Enterococcus faecalis* FAIR-E 309. Appl. Environ. Microbiol. *68*, 2550–2554.

Franz, C.M., van Belkum, M.J., Holzapfel, W.H., Abriouel, H., and Gálvez A. (2007). Diversity of enterococcal bacteriocins and their grouping in a new classification scheme. FEMS Microbiol. Rev. *31*, 293–310.

Fujita, K., Ichimasa, S., Zendo, T., Koga, S., Yoneyama, F., Nakayama, J., and Sonomoto, K. (2007). Structural analysis and characterization of lacticin Q, a novel bacteriocin belonging to a new family of unmodified bacteriocins of Gram-positive bacteria. Appl. Environ. Microbiol. *73*, 2871–2877.

Garneau, S., Martin, N.I., and Vederas, J.C. (2002). Two-peptide bacteriocins produced by lactic acid bacteria. Biochimie *84*, 577–592.

Gross, E., and Morell, J.L. (1971). The structure of nisin. J. Am. Chem. Soc. *93*, 4634–4635.

Guder, A., Wiedemann, I., and Sahl, H.-G. (2000). Posttranslationally modified bacteriocins–the lantibiotics. Biopolymers *55*, 62–73.

Haas, W., and Gilmore, M.S. (1999). Molecular nature of a novel bacterial toxin: the cytolysin of *Enterococcus faecalis*. Med. Microbiol. Immunol. *187*, 183–190.

Hastings, J.W., Sailer, M., Johnson, K., Roy, K.L., Vederas, J.C., and Stiles, M.E. (1991). Characterization of leucocin A-UAL 187 and cloning of the bacteriocin gene from *Leuconostoc gelidum*. J. Bacteriol. *173*, 7491–7500.

Henderson, J.T., Chopko, A.L., and van Wassenaar, P.D. (1992). Purification and primary structure of pediocin PA-1 produced by *Pediococcus acidilactici* PAC-1.0. Arch. Biochem. Biophys. *295*, 5–12.

Holo, H., Nilssen, O., and Nes, I.F. (1991). Lactococcin A, a new bacteriocin from *Lactococcus lactis* subsp. *cremoris*: isolation and characterization of the protein and its gene. J. Bacteriol. *173*, 3879–3887.

Hyink, O., Balakrishnan, M., and Tagg, J.R. (2005). *Streptococcus rattus* strain BHT produces both a class I two-component lantibiotic and a class II bacteriocin. FEMS Microbiol. Lett. *252*, 235–241.

Jack, R.W., Tagg, J.R., and Ray, B. (1995). Bacteriocins of Gram-positive bacteria. Microbiol. Rev. *59*, 171–200.

Jung, G. (1991). Lantibiotics: a survey. In Nisin and novel lantibiotics, Jung, G. and Sahl, H.-G., eds. (Escom, Leiden, Netherlands), pp. 1–34.

Kawai, Y., Saito, T., Kitazawa, H., and Itoh, T. (1998). Gassericin A; an uncommon cyclic bacteriocin produced by *Lactobacillus gasseri* LA39 linked at N- and C-terminal ends. Biosci. Biotechnol. Biochem. *62*, 2438–2440.

Kemperman, R., Kuipers, A., Karsens, H., Nauta, A., Kuipers, O., and Kok, J. (2003). Identification and characterization of two novel clostridial bacteriocins, circularin A and closticin 574. Appl. Environ. Microbiol. *69*, 1589–1597.

Klaenhammer, T.R. (1993). Genetics of bacteriocins produced by lactic acid bacteria. FEMS Microbiol. Rev. *12*, 39–85.

Maqueda, M., Sánchez-Hidalgo, M., Fernández, M., Montalbán-López, M., Valdivia, E., and Martínez-Bueno, M. (2008). Genetic features of circular bacteriocins produced by Gram-positive bacteria. FEMS Microbiol. Rev. *32*, 2–22.

Marciset, O., Jeronimus-Stratingh, M.C., Mollet, B., and Poolman, B. (1997). Thermophilin 13, a nontypical antilisterial poration complex bacteriocin, that functions without a receptor. J. Biol. Chem. *272*, 14277–14284.

Martin-Visscher, L.A., van Belkum, M.J., Garneau-Tsodikova, S., Whittal, R.M., Zheng, J., McMullen, L.M., and Vederas, J.C. (2008). Isolation and characterization of carnocyclin A, a novel circular bacteriocin produced by *Carnobacterium maltaromaticum* UAL307. Appl. Environ. Microbiol. *74*, 4756–4763.

Motlagh, A.M., Bhunia, A.K., Szostek, F., Hansen, T.R., Johnson, M.C., and Ray, B. (1992). Nucleotide and amino acid sequence of *pap*-gene (pediocin AcH production) in *Pediococcus acidilactici* H. Lett. Appl. Microbiol. *15*, 45–48.

Nagao, J., Asaduzzaman, S.M., Aso, Y., Okuda, K., Nakayama, J., and Sonomoto, K. (2006). Lantibiotics: insight and foresight for new paradigm. J. Biosci. Bioeng. *102*, 139–149.

Nes, I.F., and Holo, H. (2000). Class II antimicrobial peptides from lactic acid bacteria. Biopolymers *55*, 50–61.

Nes, I.F., Diep, D.B., Håvarstein, L.S., Brurberg, M.B., Eijsink, V., and Holo H. (1996). Biosynthesis of bacteriocins in lactic acid bacteria. Antonie van Leeuwenhoek *70*, 113–128.

Nes, I.F., Diep, D.B., and Holo, H. (2007). Bacteriocin diversity in *Streptococcus* and *Enterococcus*. J. Bacteriol. *189*, 1189–1198.

Nilsen, T., Nes, I.F., and Holo, H. (2003). Enterolysin A, a cell wall-degrading bacteriocin from *Enterococcus faecalis* LMG 2333. Appl. Environ. Microbiol. *69*, 2975–2984.

Nissen-Meyer, J., Holo, H., Håvarstein, L.S., Sletten, K., and Nes, I.F. (1992). A novel lactococcal bacteriocin whose activity depends on the complementary action of two peptides. J. Bacteriol. *174*, 5686–5692.

Oppegård, C., Rogne, P., Emanuelsen, L., Kristiansen, P.E., Fimland, G., and Nissen-Meyer, J. (2007). The two-peptide class II bacteriocins: structure, production, and mode of action. J. Mol. Microbiol. Biotechnol. *13*, 210–219.

Piard, J.C., Kuipers, O.P., Rollema, H.S., Desmazeaud, M.J., and de Vos, W.M. (1993). Structure, organization, and expression of the *lct* gene for lacticin *481*, a novel lantibiotic produced by *Lactococcus lactis*. J. Biol. Chem. *268*, 16361–16368.

Quadri, L.E., Sailer, M., Roy, K.L., Vederas, J.C., and Stiles, M.E. (1994). Chemical and genetic characterization of bacteriocins produced by *Carnobacterium piscicola* LV17B. J. Biol. Chem. *269*, 12204–12211.

Revol-Junelles, A.M., Mathis, R., Krier, F., Fleury, Y., Delfour, A., and Lefebvre, G. (1996). *Leuconostoc mesenteroides* subsp. *mesenteroides* FR52 synthesizes two distinct bacteriocins. Lett. Appl. Microbiol. *23*, 120–124.

Rogers, L.A. (1928). The inhibiting effect of *Streptococcus lactis* on *Lactobacillus bulgaricus*. J. Bacteriol. *16*, 321–325.

Ross, K.F., Ronson, C.W., and Tagg, J.R. (1993). Isolation and characterization of the lantibiotic salivaricin A and its structural gene *salA* from *Streptococcus salivarius* 20P3. Appl. Environ. Microbiol. *59*, 2014–2021.

Ryan, M.P., Jack, R.W., Josten, M., Sahl, H.-G., Jung, G., Ross, R.P., and Hill, C. (1999). Extensive post-translational modification, including serine to D-alanine conversion, in the two-component lantibiotic, lacticin 3147. J. Biol. Chem. *274*, 37544–37550.

Sahl, H.-G., Jack, R.W., and Bierbaum, G. (1995). Biosynthesis and biological activities of lantibiotics with unique posttranslational modifications. Eur. J. Biochem. *230*, 827–853.

Samyn, B., Martinez-Bueno, M., Devreese, B., Maqueda, M., Gálvez, A., Valdivia, E., Coyette, J., and Van Beeumen, J. (1994). The cyclic structure of the enterococcal peptide antibiotic AS-48. FEBS Lett. *352*, 87–90.

Sawa, N., Zendo, T., Kiyofuji, J., Fujita, K., Himeno, K., Nakayama, J., and Sonomoto, K. (2009). Identification and characterization of lactocyclicin Q, a novel cyclic bacteriocin produced by *Lactococcus* sp. strain QU 12. Appl. Environ. Microbiol. *75*, 1552–1558.

Skaugen, M., Nissen-Meyer, J., Jung, G., Stevanovic, S., Sletten, K., Inger, C., Abildgaard, M., and Nes, I.F. (1994). *In vivo* conversion of L-serine to D-alanine in a ribosomally synthesized polypeptide. J. Biol. Chem. *269*, 27183–27185.

Srionnual, S., Yanagida, F., Lin, L.H., Hsiao, K.N., and Chen, Y.S. (2007). Weissellicin 110, a newly discovered bacteriocin from *Weissella cibaria* 110, isolated from plaa-som, a fermented fish product from Thailand. Appl. Environ. Microbiol. *73*, 2247–2250.

Tichaczek, P.S., Vogel, R.F., and Hammes, W.P. (1993). Cloning and sequencing of *curA* encoding curvacin A, the bacteriocin produced by *Lactobacillus curvatus* LTH1174. Arch. Microbiol. *160*, 279–283.

Twomey, D., Ross, R.P., Ryan, M., Meaney, B., and Hill, C. (2002). Lantibiotics produced by lactic acid bacteria: structure, function and applications. Antonie van Leeuwenhoek *82*, 165–185.

Wirawan, R.E., Swanson, K.M., Kleffmann, T., Jack, R.W., and Tagg, J.R. (2007). Uberolysin: a novel cyclic bacteriocin produced by *Streptococcus uberis*. Microbiology *153*, 1619–1630.

Worobo, R.W., Henkel, T., Sailer, M., Roy, K.L., Vederas, J.C., and Stiles, M.E. (1994). Characteristics and genetic determinant of a hydrophobic peptide bacteriocin, carnobacteriocin A, produced by *Carnobacterium piscicola* LV17A. Microbiology *140*, 517–526.

Lactococcal Bacteriocins

10

Fuminori Yoneyama, Takeshi Zendo and Kenji Sonomoto*

Abstract

Lactococcus is one of the most important genera of lactic acid bacteria (LAB), because of the widespread use in dairy fermentation foods. Since nisin produced by *Lactococcus lactis* was discovered in 1928 (Rogers and Whittier, 1928), many bacteriocin-producing *Lactococcus* strains have been reported and studied so far (Cotter *et al.,* 2005a). Here, recent studies on lactococcal bacteriocins, mainly nisin, are summarized in their diversities and structures, biosynthesis, antimicrobial mechanisms, and applications.

Diversity and structure of lactococcal bacteriocins

Bacteriocins are classified based on structures and biochemical features (as described in the last section), and lactococcal bacteriocins divided into all classes are found as listed in Table 10.1. Nisin variants, lacticin 481, and lacticin 3147 belong to the class I also called lantibiotics (Chatterjee *et al.,* 2005). Nisin shows antimicrobial activities against broad range Gram-positive bacteria. The structures of nisin A, nisin Z, and nisin Q were determined in 1971 (Gross and Morell), 1991 (Mulders *et al.*), and 2008 (Fukao *et al.*), respectively (Fig. 10.1A). The three nisins comprise 34 amino acids including post-translationally modified (unusual) amino acids, and their positions are identical among the three variants. The three nisins have linear shapes and contains five thioether linkages. N-terminal two and C-teminal three bridges are linked by flexible hinge region. The three nisins show very similar antimicrobial features and biochemical properties (Yoneyama *et al.,* 2008). Nisin F is a new variant reported recently, and the structure gene shows high homology to those of other variants (de Kwaadsteniet *et al.,* 2008).

Lacticin 481 (Piard *et al.,* 1993), which comprises 27 amino acids, also belongs to lantibiotic group, but its structure differs from nisin variants (Fig. 10.1A). Lacticin 481 consists of N-terminal flexible and C-terminal globular regions. Lacticin 3147 (Martin *et al.,* 2004) is a so-called 'two-peptide lantibiotic', and consists of two peptides (A1 and A2), which are required for full activity, as in the case of class IIb bacteriocin described below. Interestingly, D-alanines are introduced into lacticin 3147 in addition to unusual amino acids.

The structures of other lactococcal bacteriocins divided into classes IIa-d are shown in Fig. 10.1B. Lactococcin MMFII is a class IIa bacteriocin produced by *L. lactis* MMFII isolated from a Tunisian dairy product (Ferchichi *et al.,* 2001). As other class IIa bacteriocins, lactococcin MMFII possesses anti-*Listeria* activity, N-terminal YGNGV consensus motif, and heat stability. The antimicrobial spectrum is also similar to those of other class IIa bacteriocins. Lactococcin G and lactococcin Q belong to class IIb bacteriocins, and the structures of two bacteriocins are very similar (Nissen-Meyer *et al.,* 1992; Zendo *et al.,* 2006). The both lactococcin G and lactococcin Q show specific antimicrobial activity against *L. lactis* strains. Interestingly, the original two–peptide combination of lactococcin G is exchangeable for the corresponding peptide of lactococcin Q, hybrid combinations being also active (Zendo *et*

*corresponding author

Table 10.1 Classifications, features, and modes of action of lactococcal bacteriocins

Bacteriocin	Class	Spectrum	General feature and mode of action
Nisin (variants; A, Z, Q)	I	Wide	The well-characterized bacteriocin. Lipid II-mediated actions (pore formation: inhibition of peptidoglycan biosynthesis: Septum segregation)
Lacticin 481	I	Middle	Non-nisin type lantibiotic
Lacticin 3147	I	Wide	Two-peptide lantibiotic. Lipid II-mediated actions (Ion leakage via small pore: Inhibition of peptidoglycan biosynthesis)
Lactococcin MMFII	IIa	Specific	Bactericidal activity
Lactococcin G	IIb	Specific	Synergistic activity via α and β peptides. Cationic ion leakage. α-helical structure
Lactococcin Q	IIb	Specific	A two-peptide bacteriocin that is similar to lactococcin G. α-helical structure
Lactocyclicin Q	IIc	Wide	Cyclic bacteriocin with a linkage between the N- and C-termini
Lactococcin A	IId	Specific	Ion leakage with high cell selectivity
Lacticin Q	IId	Wide	No signal peptide. High molecular stability. Pore formation in the absence of specific receptor
Lacticin Z	IId	Wide	A bacteriocin that is similar to lacticin Q
Lactococcin 972	IId	Specific	*sec*-mediated secretion. Bacteriostatic activity. Inhibition of peptidoglycan biosynthesis via interaction to lipid II

al., 2006). Lactocyclicin Q is a recently reported class IIc bacteriocin produced by *Lactococcus* sp. QU 12, which has a cyclic structure with a linkage between the N- and C-termini by a peptide bond (Sawa *et al.*, 2009). Lactocyclicin Q kills widely Gram-positive bacteria. Lactococcin A is a class IId bacteriocins that are unmodified peptides not to be categorized as classes IIa–c (Holo *et al.*, 1991). Lactococcin A shows a specific activity only against lactococci, and the most sensitive *L. lactis* indicator strain is inhibited by 7-pM peptide treatment. Lacticin Q is a recently reported class IId bacteriocin produced by *L. lactis* QU 5 (Fujita *et al.*, 2007). As nisin, lacticin Q has a highly antimicrobial activity in the nanomolar range against widely Gram-positive bacteria. In addition, lacticin Q shows high-level heat and pH stabilities including alkaline range in which nisin is inactivated. Lacticin Z is a homologue of lacticin Q, and their biochemical features are similar (Iwatani *et al.*, 2007). Lactococcin 972 is a class IId bacteriocin consisting of 66 amino acid residues (Martinez *et al.*, 1999). As mentioned below, lipid II-mediated bacteriostatic action has been reported recently.

The tertiary structures of nisin (van den Hooven *et al.*, 1996a), lacticin 3147 (Martin *et al.*, 2004), and lactococcin G (Rogne *et al.*, 2008) were determined by NMR studies. Especially for nisin, the receptor binding structural manner (see antimicrobial mechanism of lactococcal bacteriocins) was well characterized in addition to the solution structure (Hsu *et al.*, 2002, 2004). Nisin Q and lacticin 481 have never been determined the tertiary structures, but their positions of the thioether linkages and the dehydrated residues, and the covalent structures were determined by NMR studies (Fukao *et al.*, 2008; van den Hooven *et al.*, 1996b). About the other unmodified lactococcal bacteriocins, the nucleotide sequences of their structure genes elucidating the covalent structures are revealed except lactococcin MMFII. Although the lack of the information about structure gene, the structure of lactococcin MMFII was determined, because, the molecular weight calculated from amino acid sequence determined by Edman degradation agreed with the analysed molecular weight by MALDI-TOF MS (matrix-associated laser desorption/ionization time-of-flight mass spectrometry). The structure of lactocyclicin Q including N- and C-terminal linkage was elucidated by satisfactory experimental data (Sawa *et al.*, 2009). The amino acids sequences of the fragments digested by chemical reagents were linked

A

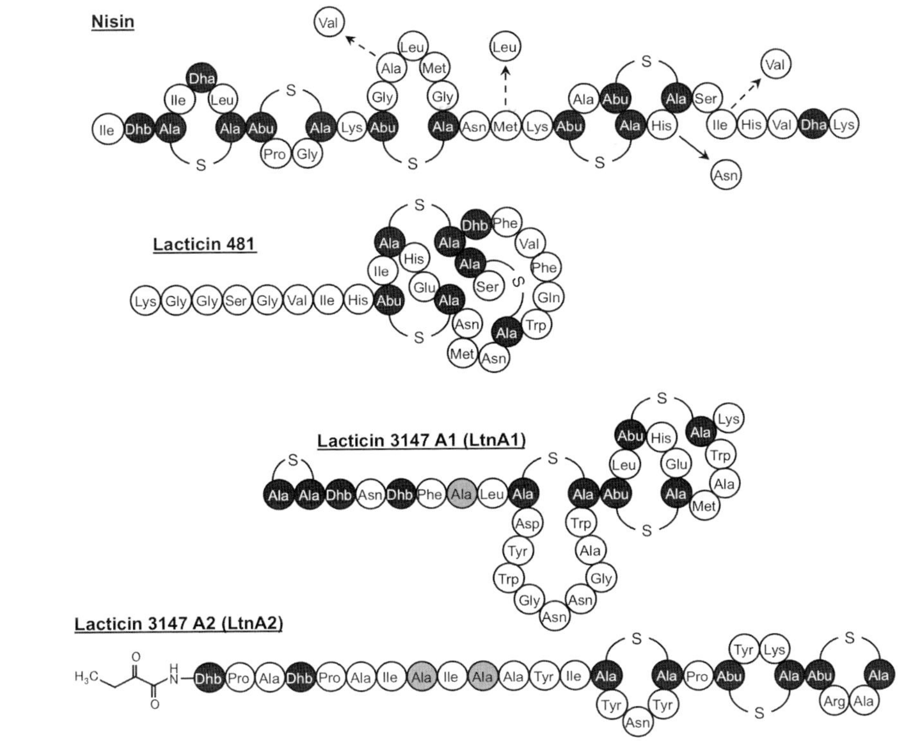

B

Bacteriocin	Signal peptide	Mature region
Lactococcin MMFII	-	TSYGNGVHCNKSKCWIDVSELETYKAGTVSNPKDILW
Lactococcin Gα	MKELSEKELRECVGG	GTWDDIGQGIGRVAYWVGKAMGNMSDVNQASRINRKKKH
Lactococcin Gβ	MKNNNNFFKGMEIIEDQELVSITGG	KKWGWLAWVDPAYEFIKGFGKGAIKEGNKDKWKNI
Lactococcin Qα	MKELSEKELRECVGG	SIWGDIGQGVGKAAYWVGKAMGNMSDVNQASRINRKKKH
Lactococcin Qβ	MKNNNNNFFKDMEIIEDQELVSITGG	KKWGWLAWVEPAGEFLKGFGKGAIKEGNKDKWKNI
Lactocyclicin Q	MK	LIDHLGAPRWAVDTILGAIAVGNLASWVLALVPGPGWAVKAGLATAAAIVKHQGKAAAAAW
Lactococcin A	MKNQLNFNIVSDEELSEANGG	KLTFIQSTAAGDLYYNTNTHKYVYQQTQNAFGAAANTIVNGWMGGAAGGFGLHH
Lacticin Q	-	MAGFLKVVQLLAKYGSKAVQWAWANKGKILDWLNAGQAIDWVVSKIKQILGIK
Lacticin Z	-	MAGFLKVVQILAKYGSKAVQWAWANKGKILDWINAGQAIDWVVEKIKQILGIK
Lactococcin 972	MKTKSLVLALSAVTLFSAGGIVAQA	EGTWQHGYGVSSAYSNYHHGSKTHSATVVNNNTGRQGKDTQRAGVWAKATVGRNLTEKASFYYNFW

Figure 10.1 (A) Structure of lactococcal lantibiotics nisin A, lacticin 481, and lacticin 3147 (LtnA1 and LtnA2). The black- and grey-filled amino acids indicate modified residues and D-alanines, respectively. In the nisin A structure, amino acid that is pointed by solid arrow is the substituted residue in both nisins Z and Q, and those by dotted arrows are in nisin Q. Dha, 2,3-dehydroalanine; Dhb, 2,3-dehydrobutyrine; Ala-S-Ala, lanthionine; Abu-S-Ala, 3-methyllanthionine; and S, the sulfur atom of the thioether linkage. (B) Amino acid sequences of lactococcal bacteriocins (non-lantibiotic type). The dotted line indicates the cleavage site of signal peptide. Signal peptide of lactococcin MMFII is undetermined. No signal peptides of lacticin Q and lacticin Z exist.

with each other and all the amino acid sequences were chemically determined. In addition, the chemically defined structure did not conflict with other data such as the nucleotide sequence of the structure gene and the mass spectrum. Leastwise, the structures of lactococcal bacteriocins are well-determined in despite of the diversity.

Abundant variety of structures, covering all classes, has been discovered from *Lactococcus* strains as mentioned above. It is assumed that horizontal gene transfer is closely related to expansion of lactococcal as well as other bacteriocin production. For example, the genes of nisin and lacticin 481 are located on transposon elements

(Immonen *et al.*, 1995; Dufour *et al.*, 2000). Lacticin 3147, lactococcin A, and lactococcin 972 are encoded on plasmids (Holo *et al.*, 1991; Martinez *et al.*, 1999; McAuliffe *et al.*, 2000a). Further screening will discover structurally novel lactococcal bacteriocins. In addition, peptides structurally similar to lactococcal bacteriocins are often found even from strains belonging to the other bacterial genera.

Biosynthesis, immunity and regulation of lactococcal bacteriocins

Nisin

For bacteriocin production, several genes are required, and the bacteriocin-related genes generally form operon and/or cluster. Biosynthesis procedures of nisin are illustrated in Fig. 10.2. Similar biosynthesis mechanisms have been characterized among three nisin variants, A, Z, and Q (Yoneyama *et al.*, 2008). Therefore that of nisin A is shown as a representative of the variants (Siezen *et al.*, 1996). A gene cluster containing 11 *orfs* (*nisA, nisB, nisT, nisC, nisI, nisP, nisR, nisK, nisF, nisE,* and *nisG*) are required for the production of mature nisin molecules. Functions of their translated products are precursor (NisA), modification (NisB, NisC), secretion (NisT), processing (NisP), regulation (NisR, NisK), and immunity (NisF, NisE, NisG, and NisI). First, precursor NisA is ribosomally synthesized as 57 amino acid residues consisting of N-terminal signal region (23 residues) and C-terminal pro-peptide region (34 residues). NisB takes on the function of dehydration of serine and threonine residues in the precursor NisA, and NisC couples these dehydrated residues to cysteines. Subsequently, modified NisA is transferred to the outside of the membrane by the ATP-binding cassette (ABC) transporter NisT. Finally, NisP, a serine protease, cleaves modified NisA to produce mature nisin and signal peptide. NisR and NisK belong to the two-component regulatory system consisting of response regulator and histidine kinase, which receives nisin as the signal stimulus (auto-inducing peptide) and induces the gene expressions of the nisin gene cluster. Based on its high sensitivity to nisin, this system is applied to an expression vector for LAB as the nisin-controlled gene expression (NICE) system (Mierau and Kleerebezem, 2005). NisF, NisE, and NisG (subunits of the ABC transporter), and NisI (a lipoprotein) serve as a protection system against nisin itself, and these kinds of mechanisms are termed immunity. It was suggested that ABC transporter (export) and adsorption (NisI) systems cooperatively function (Stein *et al.*, 2003). Recent advanced studies revealed that the N-terminal signal peptide of NisA performs a significant role in the recognition by modification enzymes (Kuipers *et al.*, 2004; Kluskens *et al.*, 2005).

Lacticin 481

In the case of lacticin 481, the biosynthesis manner is slightly different. The dehydration and cyclization reactions are catalysed by the single bifunctional enzyme, LctM, and membrane-transport and cleavage of signal peptide are also performed by a single bifunctional membrane protein, LctT (Xie *et al.*, 2004; Chatterjee *et al.*, 2005). Recently, the *in vitro* reconstitution of the biosynthesis of lacticin 481 was reported, and the approaches allow further investigation of lantibiotic modification procedures and engineering of lantibiotic analogues (Xie *et al.*, 2004). No regulatory genes are identified in the gene cluster of lacticin 481. It was previously reported that transcriptionally regulated lacticin 481 production was induced by the acidification due to lactic acid and by low pH medium (Hindre *et al.*, 2004). Immunity of lacticin 481 is mediated by the ABC-transporter system consisting of LctF, LctE, and LctG.

Lacticin 3147

In the lacticin 3147 biosynthesis, there are some interesting features that have been well characterized. Since lacticin 3147 is a two-peptide lantibiotic, two precursors from individual structural genes, LtnA1 and LtnA2, are ribosomally synthesized. LtnA1 pre-peptide is dehydrated and cyclized by the specific modification enzyme LtnM1, and LtnA2 is done by LtnM2 (McAuliffe *et al.*, 2000a). As described above, mature lacticin 3147 peptides contain D-alanines. For D-alanine formation, specific L-serines in the precursors are dehydrated and converted to

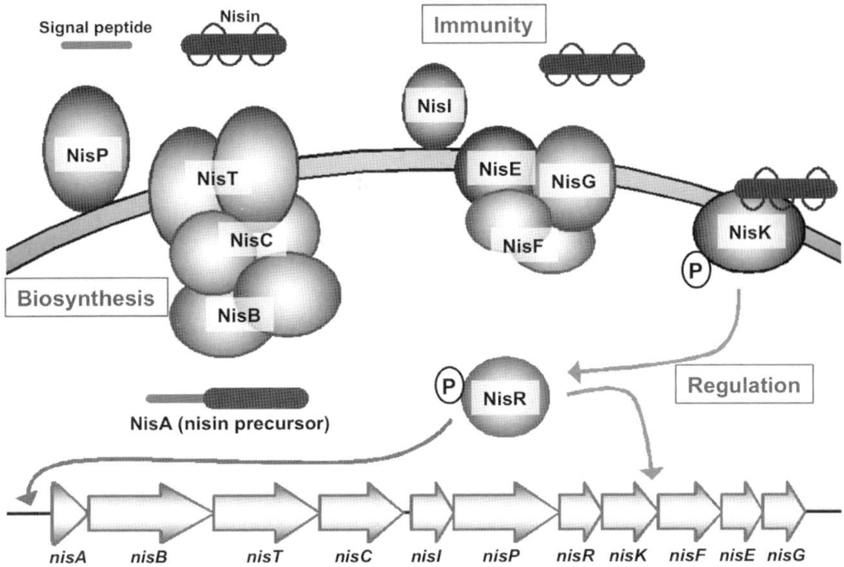

Figure 10.2 The model of nisin biosynthesis, immunity, and regulation. The precursor NisA is ribosomally synthesized, and then modified by NisB and NisC. The modified precursor is transported by NisT, and the signal peptide is cleaved by NisP, resulting in formation of active nisin molecule. NisI and ABC-transporter subunits consisting of NisF, NisE, and NisG act as self-protection system so-called immunity. Nisin also acts as an auto-inducing peptide of the gene regulation system via phosphorylation relay by NisR and NisK. The regulation system activates the genes expression of the nisin gene cluster.

dehydroalanines by LtnM1 or LtnM2. Subsequently, a dehydrogenase LtnJ forms D-alanines from the dehydroalanines (Ryan *et al.*, 1999; Cotter *et al.*, 2005b). In the regulation system of lacticin 3147, an operon consisting of precursor, modification, and transport genes is constitutively expressed. On the other hand, another operon containing immunity and regulator genes is downregulated by the transcriptional repressor LtnR (McAuliffe *et al.*, 2001). It is presumed that when LtnR sufficiently downregulated immunity genes, the immunity proteins already accumulated within the cell. Only LtnI, a putative membrane-associated protein (116 amino acids), provides the immunity phenotype against lacticin 3147, although homologues of immunity-associated ABC transporter subunits, LtnF and LtnE, are encoded on the cluster (McAuliffe *et al.*, 2000b).

Lantibiotic engineering

Lantibiotics have a high possibility for the application in medical and food industries owing to their properties. The lactococcal lantibiotic precursors consist of an N-terminal signal peptide followed by a C-terminal propeptide moiety that undergoes several post-translational modification events. Recently, it has been reported that the post-translational modifications are introduced to various peptides in addition to original precursor by lantibiotic modification enzymes and signal peptides both *in vivo* and *in vitro*. Thus, enzymes involved in lantibiotic biosynthesis may have high potentiality for bioactive peptide engineering by introduction of unusual amino acids into desired peptides, which might envisage the founding a universal approach to advance for structural design of novel peptides, termed lantibiotic engineering (Chatterjee *et al.*, 2005).

Other bacteriocins

Other lactococcal bacteriocins have never been well-studied on their biosynthesis. Even so, the gene clusters of lactococcin A and lactococcin 972 were sequenced, and their putative biosynthesis mechanisms were proposed. Four genes are identified in the gene cluster of lactococcin A (Holo *et al.*, 1991). Immediately downstream of precursor *lcnA, lciA* providing immunity phenotype against lactococcin A is encoded. In addition, *lcnC* and *lcnD,* that are essential genes for processing and transport of lactococcin A, are located on immediately upstream of *lcnA*. In the gene cluster of lactococcin 972, two genes encoding precursor and immunity protein were found (Martinez *et al.,* 1999). Lactococcin 972 is known as *sec*-dependent bacteriocin production. Lactococcin 972 uses *sec* pathway, which is a common protein export system in bacteria, for processing and transport. A *sec*-dependent processing signal sequence is found in the N-terminal of the lactococcin 972 precursor.

Antimicrobial mechanisms of lactococcal bacteriocins

All of reported lactococcal bacteriocins show bactericidal or bacteriostatic action only against Gram-positive bacteria (without synergism between bacteriocin itself and other factors), and the actions are summarized in Table 10.1. Many studies on the antimicrobial mechanism of lactococcal bacteriocins have been reported so far. Most of them act at cytoplasmic membrane resulting in membrane permeabilization and leakage of cytoplasmic compounds.

Nisin

Numerous studies about nisin were carried out on the antimicrobial mechanism. Nisin shows strong antimicrobial activity against specific Gram-positive bacteria in nanomolar concentration ranges, and the membrane-permeabilizing activity has also been observed under similar level of peptide concentrations. However, it was previously wondered that higher micromolar concentration of nisin was required for membrane permeabilization against artificial phospholipid bilayers, liposomes (Breukink *et al.,* 1997). Recently, the reason of difference in nisin activity between biological and artificial membranes is discovered (Breukink *et al.,* 1999). Nisin needs initial receptors, so-called docking molecules. A peptidoglycan precursor, lipid II, is the docking molecule of nisin for ion and ATP efflux via pore formation. The proposed nisin pore-forming model is shown in Fig. 10.3A. The nisin–lipid II interaction characterized by NMR suggests that N-terminal of nisin binds to pyrophosphate part of lipid II, and C-terminal of nisin penetrates membrane owing to flexibility of central part of hinge region (Hsu *et al.,* 2004). However, biological roles of hinge and C-teminal of nisin have never been evidenced (Hsu *et al.,* 2002). The pore diameter of nisin is around 2.5 nm, which can pass through not only small

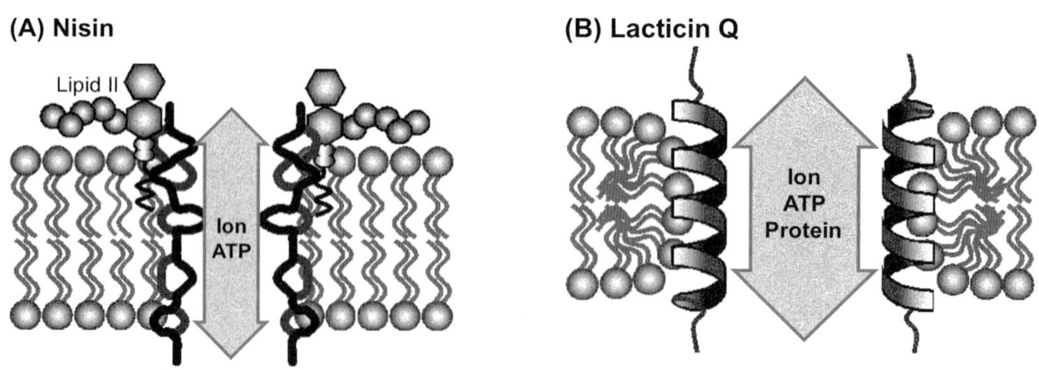

Figure 10.3 The pore-forming models of antimicrobial mechanisms of nisin (A) and lacticin Q (B). (A) The N-terminal of nisin binds to a peptidoglycan precursor, lipid II. The C-terminal of nisin penetrates cytoplasmic membrane resulting in pore formation. The diameter of the pore is around 2.5 nm, which induces ion and ATP leakage. Nisin-lipid II interaction also causes inhibition of peptidoglycan biosynthesis. (B) Lacticin Q induces ion and ATP leakages via pore formation in the absence of specific receptor. Lacticin Q forms α-helical structure on the membrane. The diameter of the pore is 4.6–6.6 nm, which also induces protein leakage.

ions but also ATP (Wiedemann *et al.*, 2004). In addition, recent study using fluorescence-labelled lipid II suggested that eight nisin and four lipid II molecules form a pore (Hasper *et al.*, 2004).

Furthermore, other antimicrobial mechanisms of nisin have been reported. It was found that nisin mutants showing antimicrobial activities did not exert pore-forming activity. The discrepancy led the second antimicrobial mechanism of nisin, inhibition of peptidoglycan biosynthesis (Wiedemann *et al.*, 2001). Since lipid II is a peptidoglycan precursor, nisin-lipid II binding inhibits peptidoglycan formation, and the activity is also observed *in vitro*. Most recently, the third antimicrobial mechanism of nisin, segregation of lipid II, was discovered. Abundant lipid II molecules are required in cell division, and most lipid II is located on the septum. However, nisin changes and disperses the localization of lipid II (Hasper *et al.*, 2006). Bacterial cells treated by nisin thus causes significant problem on cell wall synthesis, and are led to cell death.

Lacticin 3147

Antimicrobial mechanism of lacticin 481 is poorly understood. On the other hand, lipid II-mediated action of lacticin 3147 has been well characterized. A two-peptide lantibiotic, lacticin 3147 consisting of LtnA1 and LtnA2, shows synergistic activity described above, and the optimum peptides molar ratio is 1:1 (Morgan *et al.*, 2005). Interestingly, the specific lanthionine bridging pattern of LtnA1 resembles a lantibiotic mersacidin produced by *Bacillus* spp., which kills Gram-positive bacteria owing to binding to lipid II leading inhibition of peptidoglycan biosynthesis. Thus, participation of lipid II to antimicrobial mechanism of lacticin 3147 has been suspected and evidenced. Lacticin 3147 can bind to lipid II containing liposomes when the both peptides are added, and lipid II is also essential molecule for pore formation and inhibition of peptidoglycan biosynthesis (Wiedemann *et al.*, 2006). Pore diameter of lacticin 3147 is around 0.6 nm that induces ion leakage, but not ATP leakage. Furthermore, it is suggested that LtnA1 firstly binds to lipid II, and the resulting interaction activates formation of higher affinity complex among LtnA1, LtnA2, and lipid II that lead pore formation.

Lactococcin G, lactococcin Q and lactococcin A

Although antimicrobial mechanism of lactococcin G, lactococcin Q, lactococcin A, and lactococcin 972 are characterized, that of lactococcin MMFII and lactocyclycin Q are still unknown. As described above, amino acid sequences of lactociccin G and lactococcin Q, structures of two-peptide bacteriocins, are very similar. Both of mechanism may be similar, since the combinations of the peptides are exchangeable (Zendo *et al.*, 2006). Equivalent amount of both peptides of lactococcin G are needed for antimicrobial effect as lacticin 3147. Lactococcin G induces monovalent cations leakage, leading disruption of membrane potential, such as Na^+, K^+, Li^+, Cs^+, and Rb^+, but not for H^+, divalent cations, and anions (Moll *et al.*, 1996). Both lactococcin G and lactococcin Q form amphiphilic α-helix, and the three-dimensional structures of lactococcin G peptides in micelles were determined by NMR. Recently, it was reported that lactococcin G possesses GXXXG motifs in α-helix structures of both peptides, and structure–activity relationship study reveals the importance of the regions (Oppegard *et al.*, 2008). A membrane penetration model due to a parallel transmembrane helix–helix bundle is proposed. Existence of docking molecules of lactococcin G is gathered from the high cell selectivity. Action manner of lactococcin A such as membrane permeation is similar to that of lactococcin G (van Belkum *et al.*, 1991). Lactococcin A does not induce membrane permeation against liposomes made from extracted lipid of sensitive cells. As lactococcin G, specific receptor of lactococcin A is discussed.

Lacticin Q

Antimicrobial mechanism of lacticin Q, is compared with that of nisin, which is recently reported, because the antimicrobial features of lacticin Q are similar to those of nisin (Yoneyama *et al.*, 2009a). Both bacteriocins show antimicrobial activity in the nanomolar range, and single peptides are required for the full activities. In addition, both peptides induce ATP leakage from target cells owing to large size pore formation, although other many bacteriocins form small pores that induce ion leakage. However, lipid II was not required for lacticin Q activity. In addition, liposomal study indicated that lacticin Q induced high membrane permeabilization in the absence of specific receptors (Fig. 10.3B). Structural feature of lacticin Q is also different from that of nisin. Lacticin Q is identified to contain two amphiphilic α-helixes, and the similar helical structures in the presence of and absence of liposomes are analysed. On the other hand, nisin does not form typical secondary structure. The antimicrobial features of lacticin Q are more similar to antimicrobial peptides from multicellular eukaryotes than to lactococcal bacteriocins including nisin (Zasloff, 2002). However, many antimicrobial peptides from multicellular eukaryotes show antimicrobial activities in the micromolar range. Further research has just done that lacticin Q causes the leakage of small proteins. The proposed pore formation model of lacticin Q is as follows: (i) quick binding to outer membrane leaflets; (ii) the formation of at least 4.6-nm pores, causing protein leakage with lipid flip-flop; and (iii) the migration of lacticin Q molecules from the outer to the inner membrane leaflets. Consequently, they termed the novel pore model in the antimicrobial mechanism of lacticin Q a 'huge toroidal pore' (Yoneyama *et al.*, 2009b). Furthermore, negatively charged lipids accelerated the binding of lacticin Q on the membranes and the resulting pore formation. However, the acceleration was not an essential factor in the killing activity of lacticin Q, since pore-forming activities against electrically neutral and negatively charged liposomes occurred similarly (Yoneyama *et al.*, 2010).

Lactococcin 972

A non-lantibiotic, lactococcin 972, specifically interacts with lipid II to lead inhibition of septum biosynthesis despite of structural difference from lantibiotics that uses lipid II for the activity (Martinez *et al.*, 2008). Lactococcin 972 inhibits enzymes related to peptidoglycan biosynthesis that requires lipid II as substrate, although lactococcin 972 does not induce membrane permeabilization. Interestingly, lactococcin 972–lipid II interaction did not affect the nisin pore-forming activity, suggests that lipid II-binding site of lactococcin 972 is different from that of nisin. Lactociccin 972 is the first non-lantibiotic bacteriocin that binds to lipid II.

Application of lactococcal bacterioicins

Recently, food industry has dramatically developed in concert with technological advances to comply with consumer habits. On the other hand, diversification and globalization of food require more safe and functional food preservatives that do not affect original food taste, and one of the most hopeful candidates are LAB bacteriocins owing to the inhibitory effects for foodborne pathogenic and spoilage bacteria. Especially, lactococcal bacteriocins have higher potential, since *Lactococcus* strains have been widely used in diary fermentation foods leading the evidence of their safety. Lactococcal bacteriocins are considered as safety compounds for human, because of no report about toxicity. In addition, nisin and *Lactococcus lactis* are known as generally recognized as safe (GRAS). Consequently, many studies have been performed so far to actualize biopreservative using lactococcal bacteriocins (Galvez *et al.*, 2008).

About nisin and lacticin 3147, considerable number of experimental work has been performed on application of bacteriocins in foods. Strong antimicrobial effects and high stability of the bacteriocins are desirable features for food preservatives. Nisin can be purchased for food additive as in the commercial form 'Nisaplin'. Both bacteriocins are effective in many foods system to control target cells. Many studies are carried out for meat, milk, and dairy products. In addition, recent works indicate potency of the bacteriocins as preservative of fish and vegetable foods. Combination of bacteriocins and other factors, such as pulsed electric fields, high hydrostatic pressure, and organic acids, enhances the potency of bacteriocins. The combinations make it possible to control Gram-negative bacteria such as *Escherichia* and *Pseudomonas*, which can not be inhibited by simple treatment of bacteriocins.

As described above, nisin possesses higher advantages for microbial control. However, deactivation of nisin under neutral-alkaline pH conditions occasionally disturbs widespread use. Furthermore, to avoid emergence of bacteriocin-resistant bacteria, it is preferable that nisin and other bacteriocins cover disadvantages each other. Lacticin Q is one of the most promising bacteriocin for combination of nisin owing to the high stability at alkaline pH. In addition, the antimicrobial mechanism of lacticin Q differs from that of nisin as described above. The different antimicrobial mechanism and spectra indicate that lacticin Q affects nisin-resistant bacteria. As just described, more fundamental and application studies are needed to make biopreservatives using bacteriocins come true in the food industry.

References

van Belkum, M.J., Kok, J., Venema, G., Holo, H., Nes, I.F., Konings, W.N., and Abee, T. (1991). The bacteriocin lactococcin A specifically increases permeability of lactococcal cytoplasmic membranes in a voltage-independent, protein-mediated manner. J. Bacteriol. *173,* 7934–7941.

Breukink, E., van Kraaij, C., Demel, R.A., Siezen, R.J., Kuipers, O.P., and de Kruijff, B. (1997). The C-terminal region of nisin is responsible for the initial interaction of nisin with the target membrane. Biochemistry *36,* 6968–6976.

Breukink, E., Wiedemann, I., van Kraaij, C., Kuipers, O.P., Sahl, H.-G., and de Kruijff, B. (1999). Use of the cell wall precursor lipid II by a pore-forming peptide antibiotic. Science *286,* 2361–2364.

Chatterjee, C., Paul, M., Xie, L., and van der Donk, W.A. (2005). Biosynthesis and mode of action of lantibiotics. Chem. Rev. *105,* 633–684.

Cotter, P.D., Hill, C., and Ross, R.P. (2005a). Bacteriocins: developing innate immunity for food. Nat. Rev. Microbiol. 3 777–788.

Cotter, P.D., O'Connor, P.M., Draper, L.A., Lawton, E.M., Deegan, L.H., Hill, C., and Ross, R.P. (2005b). Posttranslational conversion of L-serines to D-alanines is vital for optimal production and activity of the lantibiotic lacticin 3147. Proc. Natl. Acad. Sci. U.S.A. *102,* 18584–18589.

Dufour, A., Rince, A., Uguen, P., and Le Pennec, J.P. (2000). IS1675, a novel lactococcal insertion element, forms a transposon-like structure including the lacticin 481 lantibiotic operon. J. Bacteriol. *182,* 5600–5605.

Ferchichi, M., Frere, J., Mabrouk, K., and Manai, M. (2001). Lactococcin MMFII, a novel class IIa bacteriocin produced by *Lactococcus lactis* MMFII, isolated from a Tunisian dairy product. FEMS Microbiol. Lett. *205,* 49–55.

Fujita, K., Ichimasa, S., Zendo, T., Koga, S., Yoneyama, F., Nakayama, J., and Sonomoto, K. (2007). Structural analysis and characterization of lacticin Q, a novel bacteriocin belonging to a new family of unmodified bacteriocins of Gram-positive bacteria. Appl. Environ. Microbiol. *73,* 2871–2877.

Fukao, M., Obita, T., Yoneyama, F., Kohda, D., Zendo, T., Nakayama, J., and Sonomoto, K. (2008). Complete covalent structure of nisin Q, new natural nisin variant, containing post-translationally modified amino acids. Biosci. Biotechnol. Biochem. *72,* 1750–1755.

Galvez, A., Lopez, R.L., Abriouel, H., Valdivia, E., and Omar, N.B. (2008). Application of bacteriocins in the control of foodborne pathogenic and spoilage bacteria. Crit. Rev. Biotechnol. *28,* 125–152.

Gross, E., and Morell, J.L. (1971). The structure of nisin. J. Am. Chem. Soc. *93,* 4634–4635.

Hasper, H.E., de Kruijff, B., and Breukink, E. (2004). Assembly and stability of nisin-lipid II pores. Biochemistry *43,* 11567–11575.

Hasper, H.E., Kramer, N.E., Smith, J.L., Hillman, J.D., Zachariah, C., Kuipers, O.P., de Kruijff, B., and Breukink, E. (2006). An alternative bactericidal mechanism of action for lantibiotic peptides that target lipid II. Science *313,* 1636–1637.

Hindre, T., Le Pennec, J.P., Haras, D., and Dufour, A. (2004). Regulation of lantibiotic lacticin 481 production at the transcriptional level by acid pH. FEMS Microbiol. Lett. *231,* 291–298.

Holo, H., Nilssen, O., and Nes, I.F. (1991). Lactococcin A, a new bacteriocin from *Lactococcus lactis* subsp. *cremoris*: isolation and characterization of the protein and its gene. J. Bacteriol. *173,* 3879–3887.

van den Hooven, H.W., Doeland, C.C., van de Kamp, M., Konings, R.N., Hilbers, C.W., and van de Ven, F.J. (1996a). Three-dimensional structure of the lantibiotic nisin in the presence of membrane-mimetic micelles of dodecylphosphocholine and of sodium dodecylsulfate. Eur. J. Biochem. *235,* 382–393.

van den Hooven, H.W., Lagerwerf, F.M., Heerma, W., Haverkamp, J., Piard, J.C., Hilbers, C.W., Siezen, R.J., Kuipers, O.P., and Rollema, H.S. (1996b). The structure of the lantibiotic lacticin 481 produced by *Lactococcus lactis*: location of the thioether bridges. FEBS Lett *391,* 317–322.

Hsu, S.T., Breukink, E., de Kruijff, B., Kaptein, R., Bonvin, A.M., and van Nuland, N.A. (2002). Mapping the targeted membrane pore formation mechanism by solution NMR: the nisin Z and lipid II interaction in SDS micelles. Biochemistry *41,* 7670–7676.

Hsu, S.T., Breukink, E., Tischenko, E., Lutters, M.A., de Kruijff, B., Kaptein, R., Bonvin, A.M., and van Nuland, N.A. (2004). The nisin-lipid II complex reveals a pyrophosphate cage that provides a blueprint for novel antibiotics. Nat. Struct. Mol. Biol. *11,* 963–967.

Immonen, T., Ye, S., Ra, R., Qiao, M., Paulin, L., and Saris, P.E. (1995). The codon usage of the nisZ operon in *Lactococcus lactis* N8 suggests a non-lactococcal origin of the conjugative nisin-sucrose transposon. DNA Seq. 5 203–218.

Iwatani, S., Zendo, T., Yoneyama, F., Nakayama, J., and Sonomoto, K. (2007). Characterization and structure analysis of a novel bacteriocin, lacticin Z, produced by *Lactococcus lactis* QU 14. Biosci. Biotechnol. Biochem. *71,* 1984–1992.

Kluskens, L.D., Kuipers, A., Rink, R., de Boef, E., Fekken, S., Driessen, A.J., Kuipers, O.P., and Moll, G.N. (2005). Post-translational modification of therapeutic peptides by NisB, the dehydratase of the lantibiotic nisin. Biochemistry *44,* 12827–12834.

Kuipers, A., de Boef, E., Rink, R., Fekken, S., Kluskens, L.D., Driessen, A.J., Leenhouts, K., Kuipers, O.P., and Moll, G.N. (2004). NisT, the transporter of the lantibiotic nisin, can transport fully modified, dehydrated, and unmodified prenisin and fusions of the leader peptide with non-lantibiotic peptides. J. Biol. Chem. *279,* 22176–22182.

de Kwaadsteniet, M., Ten Doeschate, K., and Dicks, L.M. (2008). Characterization of the structural gene encoding nisin F, a new lantibiotic produced by a *Lactococcus lactis* subsp. *lactis* isolate from freshwater catfish (Clarias gariepinus). Appl. Environ. Microbiol. *74,* 547–549.

Martin, N.I., Sprules, T., Carpenter, M.R., Cotter, P.D., Hill, C., Ross, R.P., and Vederas, J.C. (2004). Structural characterization of lacticin 3147, a two-peptide lantibiotic with synergistic activity. Biochemistry *43,* 3049–3056.

Martinez, B., Bottiger, T., Schneider, T., Rodriguez, A., Sahl, H.-G., and Wiedemann, I. (2008). Specific interaction of the unmodified bacteriocin Lactococcin 972 with the cell wall precursor lipid II. Appl. Environ. Microbiol. *74,* 4666–4670.

Martinez, B., Fernandez, M., Suarez, J.E., and Rodriguez, A. (1999). Synthesis of lactococcin 972, a bacteriocin produced by *Lactococcus lactis* IPLA 972, depends on the expression of a plasmid-encoded bicistronic operon. Microbiology *145,* 3155–3161.

McAuliffe, O., Hill, C., and Ross, R.P. (2000a). Each peptide of the two-component lantibiotic lacticin 3147 requires a separate modification enzyme for activity. Microbiology *146,* 2147–2154.

McAuliffe, O., Hill, C., and Ross, R.P. (2000b). Identification and overexpression of ltnl, a novel gene which confers immunity to the two-component lantibiotic lacticin 3147. Microbiology *146,* 129–138.

McAuliffe, O., O'Keeffe, T., Hill, C., and Ross, R.P. (2001). Regulation of immunity to the two-component lantibiotic, lacticin 3147, by the transcriptional repressor LtnR. Mol. Microbiol. *39,* 982–993.

Mierau, I., and Kleerebezem, M. (2005). 10 years of the nisin-controlled gene expression system (NICE) in *Lactococcus lactis.* Appl. Microbiol. Biotechnol. *68,* 705–717.

Moll, G., Ubbink-Kok, T., Hildeng-Hauge, H., Nissen-Meyer, J., Nes, I.F., Konings, W.N., and Driessen, A.J. (1996). Lactococcin G is a potassium ion-conducting, two-component bacteriocin. J. Bacteriol. *178,* 600–605.

Morgan, S.M., O'Connor P., M., Cotter, P.D., Ross, R.P., and Hill, C. (2005). Sequential actions of the two component peptides of the lantibiotic lacticin 3147 explain its antimicrobial activity at nanomolar concentrations. Antimicrob. Agents Chemother. *49,* 2606–2611.

Mulders, J.W., Boerrigter, I.J., Rollema, H.S., Siezen, R.J., and de Vos, W.M. (1991). Identification and characterization of the lantibiotic nisin Z, a natural nisin variant. Eur. J. Biochem. *201,* 581–584.

Nissen-Meyer, J., Holo, H., Havarstein, L.S., Sletten, K., and Nes, I.F. (1992). A novel lactococcal bacteriocin whose activity depends on the complementary action of two peptides. J. Bacteriol. *174,* 5686–5692.

Oppegard, C., Schmidt, J., Kristiansen, P.E., and Nissen-Meyer, J. (2008). Mutational analysis of putative helix–helix interacting GxxxG-motifs and tryptophan residues in the two-peptide bacteriocin lactococcin G. Biochemistry *47,* 5242–5249.

Piard, J.C., Kuipers, O.P., Rollema, H.S., Desmazeaud, M.J., and de Vos, W.M. (1993). Structure, organization, and expression of the lct gene for lacticin 481, a novel lantibiotic produced by *Lactococcus lactis.* J. Biol. Chem. *268,* 16361–16368.

Rogers, L.A., and Whittier, E.O. (1928). Limiting factors in the lactic fermentation. J. Bacteriol. *16,* 211–229.

Rogne, P., Fimland, G., Nissen-Meyer, J., and Kristiansen, P.E. (2008). Three-dimensional structure of the two peptides that constitute the two-peptide bacteriocin lactococcin G. Biochim. Biophys. Acta 1784, 543–554.

Ryan, M.P., Jack, R.W., Josten, M., Sahl, H.-G., Jung, G., Ross, R.P., and Hill, C. (1999). Extensive post-translational modification, including serine to D-alanine conversion, in the two-component lantibiotic, lacticin 3147. J. Biol. Chem. *274,* 37544–37550.

Sawa, N., Zendo, T., Kiyofuji, J., Fujita, K., Himeno, K., Nakayama, J., and Sonomoto, K. (2009). Identification and characterization of lactocyclicin Q, a novel cyclic bacteriocin produced by Lactococcus sp. strain QU 12. Appl. Environ. Microbiol. *75,* 1552–1558.

Siezen, R.J., Kuipers, O.P., and de Vos, W.M. (1996). Comparison of lantibiotic gene clusters and encoded proteins. Antonie van Leeuwenhoek *69,* 171–184.

Stein, T., Heinzmann, S., Solovieva, I., and Entian, K.D. (2003). Function of *Lactococcus lactis* nisin immunity genes nisI and nisFEG after coordinated expression in the surrogate host Bacillus subtilis. J. Biol. Chem. *278,* 89–94.

Wiedemann, I., Breukink, E., van Kraaij, C., Kuipers, O.P., Bierbaum, G., de Kruijff, B., and Sahl, H.-G. (2001). Specific binding of nisin to the peptidoglycan precursor lipid II combines pore formation and inhibition of cell wall biosynthesis for potent antibiotic activity. J. Biol. Chem. *276,* 1772–1779.

Wiedemann, I., Benz, R., and Sahl, H.-G. (2004). Lipid II-mediated pore formation by the peptide antibiotic nisin: a black lipid membrane study. J. Bacteriol. *186,* 3259–3261.

Wiedemann, I., Bottiger, T., Bonelli, R.R., Wiese, A., Hagge, S.O., Gutsmann, T., Seydel, U., Deegan, L., Hill, C., Ross, P., *et al.* (2006). The mode of action of the lantibiotic lacticin 3147--a complex mechanism involving specific interaction of two peptides and the cell wall precursor lipid II. Mol. Microbiol. *61,* 285–296.

Xie, L., Miller, L.M., Chatterjee, C., Averin, O., Kelleher, N.L., and van der Donk, W.A. (2004). Lacticin 481: *in vitro* reconstitution of lantibiotic synthetase activity. Science *303,* 679–681.

Yoneyama, F., Fukao, M., Zendo, T., Nakayama, J., and Sonomoto, K. (2008). Biosynthetic characterization and biochemical features of the third natural nisin variant, nisin Q, produced by *Lactococcus lactis* 61–14. J. Appl. Microbiol. *105,* 1982–1990.

Yoneyama, F., Imura, Y., Ichimasa, S., Fujita, K., Zendo, T., Nakayama, J., Matsuzaki, K., and Sonomoto, K. (2009a). Lacticin Q, a lactococcal bacteriocin, causes high-level membrane permeability in the absence of specific receptors. Appl. Environ. Microbiol. *75*, 538–541.

Yoneyama, F., Imura, Ohno, K., Zendo, T., Nakayama, J., Matsuzaki, K., and Sonomoto, K. (2009b). Peptide-lipid huge toroidal pore, a new antimicrobial mechanism mediated by a lactococcal bacteriocin, lacticin Q. Antimicrob. Agents Chemother. *53*, 3211–3217.

Yoneyama, F., Shioya, K., Zendo, T., Nakayama, J., and Sonomoto, K. (2010). Effect of negatively charged lipid on membrane-lacticin Q interaction and resulting pore formation. Biosci. Biotechnol. Biochem. *74*, 218–221.

Zasloff, M. (2002). Antimicrobial peptides of multicellular organisms. Nature *415*, 389–395.

Zendo, T., Koga, S., Shigeri, Y., Nakayama, J., and Sonomoto, K. (2006). Lactococcin Q, a novel two-peptide bacteriocin produced by *Lactococcus lactis* QU 4. Appl. Environ. Microbiol. *72*, 3383–3389.

Lactobacilli Bacteriocins

11

Yasushi Kawai and Tadao Saito*

Abstract

Lactobacilli and lactococci are a major lactic acid bacteria (LAB) and are well known probiotics. The lactobacilli bacteriocins belong to class I–III, as mentioned in Chapter 9, and act as good biopresevatives. In particular, lactobacilli is commonly detected in the intestines of human and animals in addition to soil, water, plants and food, and the bacteriocins and producing strains are expected to be used in various fields. In this chapter, sakacins, the lactacin F family, Abp-118, plantaricins and gassericin A are examined as several recent representatives, and the antibacterial spectra, mode of action, the chemical structures, genes for biosynthesis and immunity, and the application for food are described.

Introduction

Bacteriocins produced by lactobacilli from the lactic acid bacteria (LAB) group have been reported since about 1990 in the same time period when similar bacteriocins were reported in other bacterial species. Almost all of the bacteriocins from lactobacilli belong to class II bacteriocins except for a class III helveticin J produced by *Lactobacillus helveticus* 481 (37 kDa, 333 amino acid residues, using an unknown antibacterial mechanism; Joerger and Klaenhammer, 1990). Interestingly, there have been few reports of class I lantibiotics from lactobacilli. In this chapter, the chemical structure, genetics, mode of action, immunity, and topics concerning lactobacilli bacteriocins are described from the recent decade. Other lactobacilli bacteriocins are referred to from many published papers and reviews.

Sakacin A (curvacin A), sakacin P and sakacin Q (class IIa)

Sakacin A is an anti-listerial class IIa bacteriocin produced by *Lb. sake* Lb706. The 41 amino acid residues of sakacin A were chemically sequenced and its molecular weight was found to be 4308.7 using MS analysis (Holck *et al.*, 1992). Cloning and sequencing of the gene cluster (8668-bp) containing the structural gene for sakacin A (*sapA*) and a mutation deletion analyses shows two operons for sakacin A biosynthesis: *sapA* (the structural gene) and *saiA* (a putative immunity protein) encoding 59 amino acid residues and 90 amino acid residues, respectively; and the second has a three component regulatory gene system, *sap-Ph* (a 23-amino-acid cationic pheromone peptide), *sapK* (a putative histidine kinase) and *sapR* (a response regulator), and the ATP-binding cassette transporter genes, *sapTE* (Axelsson and Holck, 1995). Northern (RNA) analysis suggests the phosphorylated SapR may be the transcriptional activator found in a 35-bp sequence upstream of the putative promoters for *sapA* and *sapK*. *Lb. sake* strain Lb706 produces sakacin A at 25 or 30°C where production is reduced or absent at 33.5–35°C. However, at 33–34°C, the production is recovered by adding isolated Sap-Ph to the growth medium. No sakacin bacteriocin activity was detected in a Sap-Ph knock-out Lb706 mutant strain suggesting sakacin A production is

*corresponding author

dependent on Sap-Ph and temperature (Diep *et al.*, 2000).

Curvacin A is a class IIa membrane-permeabilizing bacteriocin from *Lb. curvatus* LTH1174 isolated from fermented sausage that inhibits the growth of food pathogens such as *Listeria monocytogenes* and *Enterococcus faecalis* (Nieto Lozano *et al.*, 1992). The structural gene for curvacin A, *curA* (encoding 59 amino acids), is on a 60 kb plasmid of *Lb. curvatus* LTH1174 and was sequenced. Curvacin A (41 amino acid residues) is identical to sakacin A. Curvacin A shows a similar response to temperature as Sap-Ph from *Lb. sake* Lb706 indicating other related genes for curvacin A biosynthesis may be similar to the Lb706 strain (Tichaczek *et al.*, 1993). The three-dimensional structure of curvacin A in lipid micelles (DPC: dodecylphosphocholine) was studied using nuclear magnetic resonance (NMR) spectroscopy and shows curvacin A has an S-shaped β-sheet-like domain (residues 1–16) stabilized by a disulfide bridge and a few hydrogen bonds at the cationic and hydrophilic N-terminus, and a double hinged-helix structure with two hinges (residues 16–18 and residues 26–28) in front of the α-helices giving a hydrophilic heximer helix (residues 19–24) and an amphiphilic/hydrophobic 11-mer helix (residues 29–39) having the putative role of piercing targeted cell membranes (Haugen *et al.*, 2005) (Fig. 11.1). This arrangement suggests curvacin A is different from the other pediocin-like class IIa bacteriocins and belongs in a subgroup of the pediocin-like bacteriocins, i.e. enterocin P.

Mutational analysis of curvacin A from 55 variants obtained using site-directed mutagenesis and four different indicator cells suggest its orientation in the target cell membranes is as follows: the hydrophilic short helix (residues 19–24) of the bacteriocin inserts in the interface region of the target cell membrane where Ala22 is close to the hydrophobic core of the membrane (Haugen *et al.*, 2008). The hinge region (residues 16–18) forms a turn where Gly28 is located near the border between the interface and the hydrophobic regions. Then, the hydrophobic C-terminal helix (residues 29–41) is inserted into the target cell membrane leading to cell death. A helix–helix-interacting GxxxGxxG motif (G33, G37, and G40) in the helix (residues 29–41) is described to be important for exerting curvacin A activity and was studied using amino acid replacement test for each GxxxGxxG residue with mutational analysis changes to other amino acid residues.

Cell growth of *Lb. curvatus* LTH 1174 and the activity of curvacin A produced in the culture were dependant on temperature and pH similar to *Lb. sake* Lb706 and sakacin A. The optimum temperature for cell growth is 34°C and the optimum temperature for curvacin A production is 20–27°C. This indicates *Lb. curvatus* LTH 1174 could be used as a candidate for bacteriocin-production in starters for the fermentation process of European dry sausage at temperatures around 25°C (Messens *et al.*, 2003). Verluyten *et al.* investigated the cell growth and the bacteriocin production under various concentrations of sodium chloride and complex nutrient sources in modified MRS medium. Sodium chloride (2%, w/v) clearly decreased bacteriocin production although the growth of *Lb. curvatus* LTH 1174 was unaffected (Verluyten *et al.*, 2004a). Higher nutrient concentrations improved growth having

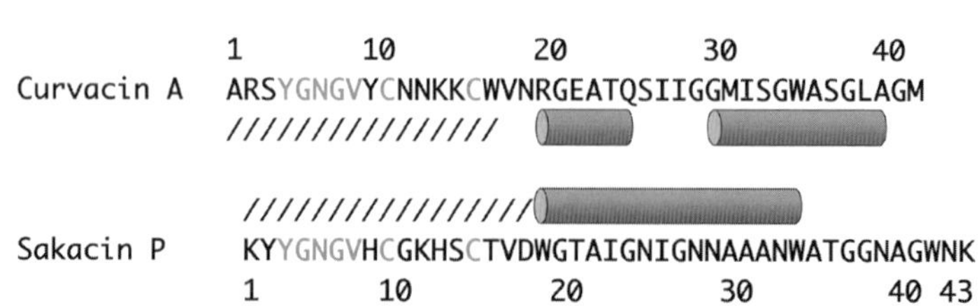

Figure 11.1 The amino acid sequence of curvacin A and sakacin P. The column bars and ///// marks show α-helical regions and S-shaped β-sheet-like domains determined using NMR analysis, respectively (Uteng *et al.*, 2003; Haugen *et al.*, 2005).

a higher attainable biomass concentration and a higher bacteriocin activity in the supernatant. Addition of each separate component in the complex nutrient source showed bacterial media peptone was a stimulator for growth and bacteriocin production (Verluyten *et al.,* 2004b). For the effects of different spices used in production of fermented sausages on curvacin A production, nutmeg decreased bacteriocin production along with cell growth; whereas garlic enhanced bacteriocin production. This suggests using spices and bacteriocinogenic starter cultures is effective in fermented sausage manufacture (Verluyten *et al.,* 2004c).

Sakacin P is an anti-listerial class IIa bacteriocin produced by *Lb. sake* LTH 673 isolated from fermented dry sausages. Cloning and sequencing of the structural gene for sakacin P (*sppA*) shows sakacin P (43 amino acid residues) may be synthesized as a precursor peptide of 61 amino acid residues (Tichaczek *et al.,* 1994). The cloning and sequencing of the chromosomal DNA fragment (7.6 kbp) of strain LTH 673 shows the sakacin P gene cluster has six ORFs where their function was deduced by computer analysis: *sppK,* a histidine kinase; *sppR,* response regulator; *sppA,* sakacin P precursor; *spiA,* immunity protein; and *sppTE,* the transport proteins (Hühne *et al.,* 1996).

Lb. sakei Lb674, a sakacin P producer, has another bacteriocin (sakacin Q) gene (*sppQ*) containing a putative immunity gene (*spiQ*), a gene of unknown function (*orf4*) and an immunity gene with a frame-shift mutation (*orf5*) in the inducible promoter downstream of *spp* gene clusters (Mathiesen *et al.,* 2005). The precursor peptide (67 amino acid residues) may be cleaved at the Gly–Gly consensus motif and the mature sakacin Q (49 amino acid residues) is synthesized. The antimicrobial spectrum of sakacin Q is different from sakacin P. Cocolin and Rantsiou (2007) performed PCR detection of the genes for sakacin P and sakacin Q (*sppA* and *sppQ,* respectively) and found them on chromosomes of other strains of *Lb. curvatus,* L442 and LTH1174, that have plasmids and are curvacin A and anti-listerial bacteriocin producers, respectively. Although the *sppQ* from L442 and LTH1174 was short at about 15 nucleotides in comparison to the *sppQ* from *Lb. sakei*, there was no difference in the activity of these sakacin Qs against several sensitive strains; and SppQ was expressed in both plasmid-cured and wild-type LTH1174 strains. Conversely, sakacin P was expressed in only *Lb. curvatus* L442; however, the expression of the *sppA* gene was detected after curing the megaplasmid from strain LTH1174.

The three-dimensional structure of sakacin P in DPC micelles and in trifluoroethanol (TFE) was determined using NMR spectroscopy. The cationic N-terminal region (residues 1–17) has an S-shaped conformation followed by a central amphiphilic α-helix (residues 18–33) and a C-terminal tail (residues 34–43) lacking secondary structural motifs. However, the C-terminal tails fold back onto the central α-helix in DPC to form hairpin-like structures in the C-terminal halves (Uteng *et al.,* 2003) (Fig. 11.1). Class IIa bacteriocins are known to share high sequence similarity in these structural relationships. Measurement of the CD spectra at different temperatures and 3D structure in several test solutions (i.e. TFE) shows the class IIa bacteriocins, pediocin PA-1 and curvacin A (sakacin A) in addition to sakacin P, have a central α-helix region (especially in the C-terminal hydrophobic residues) that is important for target recognition and activity; and have a more variable β-sheet and coiled N-terminus (Kaur *et al.,* 2004). Pediocin PA-1 produced by *Pediococcus acidilactici* strains is well studied and is a representative of the class IIa bacteriocins having a second disulfide bond at the C-terminus in addition to a highly conserved first N-terminal disulfide bond (see chapter 12). Although pediocin PA-1 maintains its overall structure at a wide range of temperatures, partial disruption of the helical section caused at higher temperatures in curvacin A and sakacin P lacking the second C-terminal disulfide bond that are 30–50 times less antimicrobially potent at 37°C than at 25°C. This activity loss may be due to the structural changes in the helical region at different temperatures.

Bacteriocin producers have protein(s) or peptide(s) to protect themselves from their own bacteriocins. A representative is the ABC transporter and its accessory proteins to pump out the bacteriocins produced to protect the cell from invasion of bacteriocins from outside of the producing cells (Ramírez-Rodrigo and Maqueda, 2003; Cheigh and Pyun, 2005). Many bacteriocin producers also make specific extra immunity proteins (peptides) that interestingly are more diverse in primary structures when comparing the bacteriocin and immunity proteins in the class IIa bacteriocins. They range in size from 81 to 115 amino acid residues without the deduced trans-membrane segments and may be located at the inner surface of the cell membrane and act intracellularly (Fimland *et al.*, 2002; Martin-Visscher *et al.*, 2008). NMR and crystal structure analysis show some class IIa immunity proteins have four α-helices and show similar 3D structure despite different primary structures. In 2007, Diep *et al.* reported the function and mechanism of an immunity protein (LciA) for the class II bacteriocin, lactococcin A, from *Lactococcus lactis*. LciA. It prevented the action of the lactococcin A by forming a complex with the receptor proteins (the mannose phosphotransferase system) and the bacteriocin. Further studies concerning the mechanism for other immunity proteins from class IIa bacteriocins is required.

Lactacin F family bacteriocins (class IIb)

Many strains of the *Lb. acidophilus* group LAB such as *Lb. gasseri* and *Lb. johnsonii* naturally inhabit human and animal intestinal tracts (Reuter, 2001) and produce a lot of available bacteriocin. They are expected to be used as potential probiotics (Itoh *et al.*, 1995; Kawai *et al.*, 1997; Bogovic-Matijasić *et al.*, 1998). Lactacin F produced by *Lb. johnsonii* VPI11088 (NCC533) is a representative of the class IIb lactobacilli bacteriocins that have two peptides, LafA and LafX (so-called lactacin F-family bacteriocins) (Fig. 11.2). Purification of the mature active LafA, cloning the structural genes (*lafA* and *lafX*) and molecular analysis of *laf* operon (*lafAXI*) are reported (Muriana and Klaenhammer, 1991a,b; Fremaux *et al.*, 1993). The mechanism of activity for lactacin F suggests pore formation in the target cytoplasmic membrane in the absence of a proton motive force causes a loss of cellular K+, depolarization of the cell membrane, and hydrolysis of internal ATP (Abee *et al.*, 1994). A study of the complementary LafA and LafX using expression analysis shows VPI11088 variants lack the *laf* operon (Allison *et al.*, 1994); and heterologous expression of lactacin F in

```
                        1       10        20        30        40        50     57
LafA                    RNNWQTNVGGAVGSAMIGATVGGTICGPACAVAGAHYLPILWTGVTAATGGFGKIRK
                        ::::  : ::  :    :   :   :::::   :::: :: : ::::::::::::::
GatA, LF221 OrfB3       RNNWAANIGGVGGATVAGWALGNAVCGPACGFVGAHYVPIAWAGVTAATGGFGKIRK
                        ::::: :: :       : ::    ::: :::::::     :  :    : ::::::
AmyL alpha              RNNWAGNTIGIVSSAATGAALGSAICGPGCGFVGAHWGAVGWTAVASFSGAFGKIRK

                        1       10        20        30        40        50
AmyL beta, lactobin A   NRWTNAY-SAALGCAVPGVKYGKKLGGVWGAVIGGVGGAAVCGLAGYVRKG
                        :::     ::: :    : :  :  :  ::: : ::::::  :  ::
LafX                    NRWGDTVLSAASGAGT-GIKACKSFGP-WGMAICGVGGAAIGGYFGYTHN
                        : ::  :  :: ::   :   :  ::: :::  :: :::::::: ::  :
GatX, LF221B            NKWGNAVIGAATGATR-GVSWCRGFGP-WGMTACGLGGAAIGGYLGYKSN
                        1       10        20        30        40      48

                        1       10        20        30        40        50 53
LF221A                  NNVNWGSVAGSCGKGAVMEIYFGNPILGCANGAATSLVLQTASGIYKNYQKKR
```

Figure 11.2 Sequence homology of the primary structures for the lactacin F family of bacteriocins: lactacin F (LafA, LafX), gassericin T (GatA, GatX), acidocin LF221 (LF221 OrfB3, LF221B and LF221A) and amylovorin L (AmyLα, AmyLβ = lactobin A) (Fremaux *et al.*, 1993; Kawai *et al.*, 2000; Majhenic *et al.*, 2004; Moreno *et al.*, 2008). The underline shows an α-helix trans-membrane region using the SOSUI program (http://bp.nuap.nagoya-u.ac.jp/sosui/).

Carnobacterium piscicola LV17, a carnobacteriocin producer, suggests the processing and export of both bacteriocins may be a consensus mechanism (Allison *et al.,* 1995). LafI, an immunity protein, was expressed in lactacin F-sensitive strains such as *Lb. acidophilus* ATCC 4356, *Lb. acidophilus* NCFM/N2, *Lb. fermentum* NCDO1750, *Lb. gasseri* ATCC 33323, and *Lb. johnsonii* ATCC 33200; and shows an immunity function against lactacin F. This may be a food-grade genetic marker for use with *Lactobacillus* species (Allison and Klaenhammer, 1996). For lactacin F, no reports are found for application in foods after 1996. However, some lactacin F-family bacteriocins (Fig. 11.2) and the producers described below have been studied for use as biopreservatives and probiotics.

Gassericin T produced by *Lb. gasseri* SBT 2055 from human faeces is closely related to the lactacin F-family and has bactericidal activity against the food-borne pathogens: *List. monocytogenes, Bacillus cereus* and *Staphylococcus aureus* (Kawai *et al.,* 1997). The GatA peptide of gassericin T (mature GatA/X) was purified and the molecular size was estimated to be ca. 5400 using SDS-PAGE analysis (Kawai *et al.,* 2000). N-terminal complete chemical sequencing shows mature GatA (57 amino acid residues) has a disulfide bond in the molecule and no modified amino acid residues. Cloning and sequencing of a 2.9-kbp HindIII–HindIII fragment from chromosomal DNA for gassericin T shows six ORFs including the structural genes of gassericin T (*gatAX*); and their function was deduced using computer analysis. OFR1 (*gatT*) is a putative ABC-transporter ATP-binding protein with six trans-membrane regions; OFR2 (*gatC*) has an unknown function (possibly a putative accessory protein for GatT); OFR3 (*gatZ*) may not be expressed; and ORF6 (*gatI*) is a putative immunity protein differing from LafI in primary structure. ORF4 (*gatA*) and ORF5 (*gatX*) encode each bacteriocin that are composed of 75 and 65 amino acid residues including the double-glycine motif at the cleavage site, respectively. Mature GatA has 63% sequence similarity to the mature LafA peptide of lactacin F (57 amino acid residues); and the sequences around the processing site, alignment of glycine residues and C-terminal area are conserved. The putative mature GatX (48 amino acid residues) has 60% similarity to mature LafX (48 amino acid residues) than to amylovorin β of amylovorin L (50 amino acid residues, described below); although the length of the leader peptide (17 amino acids) for GatX is different from LafX (14 amino acids). Mature GatX and anti-bacterial activity were not detected in the culture supernatant suggesting mature GatX has no activity with the indicator strain, *Lb. delbrueckii* subsp. *bulgaricus* JCM1002^{T} (Kawai *et al.,* 2006). Another gassericin T producer, *Lb. gasseri* LA158 isolated from the faeces of a human infant, secretes a quantity of gassericin T in MRS broth but minimally in milk-based media such as reconstituted skim milk and cheese whey. Arakawa *et al.* shows the inhibition of gassericin T production was caused by divalent cations dependent on concentration. This was reversed by adding trisodium citrate dihydrate, a food-grade chelator of divalent cations. Addition of Tween 80 increased gassericin T production in cheese-whey-based medium supplemented with proteose peptone and trisodium citrate dihydrate. These methods may promote the effective use of gassericin T for food preservation and in animal formulations (Arakwara *et al.* 2009).

Lb. gasseri LF221 isolated from the faeces of a human child produces two putative heat-stable class IIb bacteriocins named acidocin LF221A and LF221B that inhibit the growth of some LAB strains and some pathogenic and food-spoilage species such as *B. cereus, Clostridium* sp., *List. innocua, Staph. aureus, Streptococcus* sp. (Bogovic-Matijasić *et al.,* 1998). Cloning and sequencing of the chromosomal DNA fragments of LF221 strain for LF221A (858 bp) and B (1772 bp) show three ORFs and five ORFs, respectively (Majhenic *et al.,* 2004) where the function of the genes was deduced using computer analysis: LF221A, part of *orfA1,* encodes a putative complementary peptide for LF221A; *orfA2* encodes the structural gene for LF221A and is composed of 69 amino acid residues with a 16-amino-acid leader peptide; and *orfA3* encodes a putative immunity protein. Homology analysis of LF221A showed similarities to one or both mature peptides for other class IIb bacteriocins, i.e. BrcB (52% identity) of brochocin C (BrcAB; McCormick *et al.,* 1998). Acidocin LF221B is identical to mature GatX of gassericin T, and a part of OrfB1 and OrfB2, 3,

4, 5 were similar to GatCZAXI of gassericin T, respectively. Using *in vivo* studies with *Lb. gasseri* LF221 as a probiotic strain, the LF221 strain was detected in the faeces of mice and piglets fed with LF221; and the total count of lactobacilli in faecal samples was similar to that of the normal animals. LF221 strain was also detected at least 6 weeks after inoculation into semi-hard cheese (Rogelj *et al.*, 2002). The viable cells of *Lb. gasseri* LF221 and K7 strains (another LF221 producer isolated from a human; Canzek *et al.*, 2003) were observed in the faeces and the mucosa of piglets fed with each strain. Both strains were found in the faeces for 6 days after ceasing inoculation indicating both strains adhered to intestinal mucosa and were alive in the intestines (Matijasic *et al.*, 2006).

Amylovorin L (amylovorin α/β) is a class IIb bacteriocin produced by *Lb. amylovorus* DCE 471. Amylovorin β was purified using a rapid and simple three-step protocol including ammonium sulfate precipitation, chloroform/methanol extraction/precipitation and reversed-phase HPLC. The molecular mass of amylovorin was determined to be 4876.9 using electrospray mass analyses (Callewaert *et al.*, 1999). The N-terminal amino acid sequence had significant homology with LafX (GatX and LF221B) of lactacin F (gasssericins T and LF221); and was identical to the lactobin A produced by *Lb. amylovorus* LMG P-13139 isolated from maize steep liquor (Contreras *et al.*, 1997; De Vuyst *et al.*, 2004). The DNA region for biosynthesis of amylovorin L was sequenced and 12 predicted ORFs of 7577 bp (*amyL*) containing the structural genes (*amyLαβ*) were found. AmyLα shows 47% and 46% similarity to GatA of gassericin T and LafA of lactacin F, respectively; and AmyLβ shows 62% similarity to LafX. Both of the mature amylovorin L peptides may be cleaved at the typical bacteriocin-processing glycine–glycine consensus motifs. No genes for the three-component regulatory system were deduced where amylovorin L was produced constitutively and stably because the regulatory system is absent (Moreno *et al.*, 2008). Amylovorin L production increases 2-fold if the culture has acidification control. A study of the nutritional requirements for bacteriocin production using fed-batch fermentations shows amylovorin L production and adsorption on the producer cells were inhibited using high concentrations of a complex nitrogen source (Callewaert and Vuyst, 2000). The cell growth and bacteriocin production by *Lb. amylovorus* DCE 471, a sourdough starter culture, depend on the concentration of sodium chloride where 5–10 NaCl g/l in the type II sourdough fermentations enhances the growth and bacteriocin production (Neysens *et al.*, 2003). Investigations concerning the effects of both oxygen and carbon dioxide under controlled conditions (37°C and pH 5.4) shows mild aeration of type II sourdoughs may promote bacterial growth and amylovorin L production (Neysens and De Vuyst, 2005). Assays of sugars relevant to sourdough fermentation on the kinetics of *Lb. amylovorus* DCE 471 indicated a diversity in the energy source and sugar combinations stimulates the production of amylovorin L and cell growth (Leroy *et al.*, 2006).

ABP-118 (class IIb)

ABP-118 is a broad-spectrum bacteriocin from *Lb. salivarius* subsp. *salivarius* UCC118 isolated from the ileal-caecal region of the human gastrointestinal tract; and is the first example of a bacteriocin from a human probiotic bacterium (Flynn *et al.*, 2002). ABP-118 was purified to homogeneity using chromatography from MRS broth culture-supernatant and found to be a Class IIb bacteriocin composed of Abp118α exhibiting the antimicrobial activity and Abp118β that enhanced the antimicrobial activity. A DNA region for ABP-118 biosynthesis was identified in a 10,769-bp region of the chromosomal DNA. Analysis of this region shows nine ORFs related to ABP-118 production containing the structural genes (*abp118α* and *abp118β*) were deduced from a total of 18 ORFs. The gene conferring strain UCC118 immunity against the action of ABP-118, *abpIM*, encodes for a peptide of 55 amino acid residues; and does not show similarity to known bacteriocin-related genes. Usually an immunity gene is identified downstream of the structural gene(s) and sometimes the primary structure has low similarity to the bacteriocin and from other bacterial species. Located downstream of *abpIM*, five ORFs involved in bacteriocin regulation and transport were identified. AbpIP,

AbpK, and AbpR, a three-component regulatory system, are small pheromone peptides for the regulation of ABP-118 production and include a histidine kinase and the response regulator. AbpT containing the ABC transporter nucleotide-binding domain is homologous to the putative ATP-binding protein for ABC-transporter found in the plantaricin A system described below (Diep *et al.*, 1996; Risøen *et al.*, 1998); and AbpD, a putative accessory transport protein, shows similarity (40%) to the sakacin A system (Axelsson and Holck, 1995).

Probiotics enhance the health in their hosts, humans and domestic animals; however, the mechanisms of the probiotic strains are little characterized and described. The total 2.13 Mbp genome of *Lb. salivarius* UCC118 was sequenced including the chromosome (1.83 Mbp) and three plasmids (242, 44, and 20.4 kbp). The gene for ABP-118 bacteriocin was located on the megaplasmid, pMP118 (242 kbp), having a bile salt hydrolase and a presumptive conjugation locus. This indicates the megaplasmid, pMP118, is related to the probiotic properties of the UCC118 strain (Claesson *et al.*, 2006). Corr *et al.* (2007) demonstrated *Lb. salivarius* UCC118 produces ABP-118 *in vivo* and significantly protects mice against infection with the food borne pathogen, *List. monocytogenes*. A mutant of *Lb. salivarius* UCC118, a non-bacteriocin producer (delta *abp118*), did not protect mice against infection with two indicator strains of *List. monocytogenes*, EGDe and LO28. The original UCC118 strain had no effects against mice infected with a mutant strain of *List. monocytogenes* containing and expressing *abpIM*, the ABP-118 immunity protein. This data indicates bacteriocin production is the primary mediator of protection against this organism and its antimicrobial effect is a direct result of antagonism between *Lb. salivarius* and the pathogen. This is the first complete report showing a bacteriocinogenic probiotic bacterium and its bacteriocin enhances the health of the host.

Plantaricin

There are reports of over 25 kinds of plantaricins produced by *Lb. plantarum*; and interestingly class I, class IIa and IIb bacteriocins have been detected in *Lb. plantarum* as well as *Lc. lactis* (Table 11.1). Here below, we describe the plantaricins studied as well as major LAB bacteriocins.

Lb. plantarum C11 produces six plantaricins containing bacteriocin-like peptides [plantaricin A (PlnA), PlnE, PlnF, PlnJ, PlnK, and PlnN]. Each plantaricin was purified, characterized using amino acid sequencing and mass spectrometry; and the genes for biosynthesis were cloned and sequenced. PlnA, a small cationic bacteriocin, also works as a strain-specific pheromone that induces bacteriocin production. PlnEF and PlnJK, class IIb bacteriocins, have increased activities at least 1000 times using a 1:1 combination in comparison to the activity of each peptide alone; although neither PlnE nor PlnF exerts complementary effects with PlnJ or PlnK. No bacteriocin activity is detected for PlnN (Diep *et al.*, 1996; Anderssen *et al.*, 1998).

Plantaricin A was purified and the partial amino acid and DNA sequencing were performed (Nissen-Meyer *et al.*, 1993). The structural gene for plantaricin A (*plnA*: encoding a 48-amino-acid precursor peptide) was cloned and sequenced; and *plnB* (a histidine kinase) and *plnCD* (two response regulators), a putative two-component signal transduction system were detected downstream of *plnA* (Diep *et al.*, 1994). Plantaricin A was found to be an inducer of bacteriocin production in *Lb. plantarum* C11 where bacteriocin production in a mutant of non-producing *Lb. plantarum* C11 was induced using 1 ng/ml plantaricin A (Diep *et al.*, 1995). Northern (RNA) and sequencing analyses demonstrate plantaricin A induced the transcription of four plantaricin operons in addition to *plnABCD* as follows: *plnEFI* and *plnJKLR* are operons containing two class IIb bacteriocins (PlnEF and PlnJK, respectively) with double-glycine-type leaders; and *plnI* and *plnL* encode putative bacteriocin immunity proteins. *plnMNOP* encodes plantaricin N (PlnN) and its putative immunity protein (PlnMP); and *plnGHSTUV* contains an ABC transporter (PlnG) and its accessory protein (PlnH) is involved in processing and export of the LAB bacteriocins (Diep *et al.*, 1996). The response regulators, PlnC and PlnD, bind specifically to the direct repeats in the promoter of the *plnABCD* operon and to similar sequences in the promoter regions of the

Table 11.1 Bacteriocins produced by *Lactobacillus plantarum*

Plantaricin	Strain	Source	Remarks	Year
SIK-83	SIK-83	Fermented carrots	Mode of action	1988
T	LPCO10	Green olive fermentation		1993
S	LPCO10	Green olive fermentation	Described in this chapter	1995
A	C11		Described in this chapter	1994
EF	C11		Class IIb, described in this chapter	1996
JK	C11		Class IIb, described in this chapter	1996
C	LL441	Cabrales cheese	Class I, described in this chapter	1994
SA6	SA6	Fermented sausage	Optimal pH 4–7	1995
UG1	UG1	Dry sausage	Stable pH range 4.5–7.0	1996
1.25	TMW1.25	Raw fermented sausage	Class IIb, homology to ebrevicin 27	1996
D	BFE 905	Waldorf salad	Stable pH range 2–10	1998
LP84	NCIM 2084		Bactericidal and lytic effect	1998
423	423	Sorghum beer	Class IIa, described in this chapter	1998
NA		Vegetable	Stable pH range 2–10	1998
ST31	ST31	Sourdough	MW 2755 ± 0.3	1999
W	LMG 2379	Pinot Noir wine in Oregon	Two-latibiotic, in this chapter	2001
35d	134	Italian sausage	Bactericidal, ca. 4.5 kDa	2001
C19	C19		Class IIa, MW 3845.3	2001
NC8	NC8		Class IIb, described in this chapter	2003
TF711	TF711	Raw Tenerife goats' cheese	Ca. 2.5 kDa, stable pH range 1~9	2005
149	NRIC 149	Pineapple	Pln149a by SPPS-Fmoc chemistry	2007
J23	J23	Grape must sample, Spain	PlnEF and PlnJ	2007
C7	C7	Kimchi	3,8 kDa, class II?	2007
AA135	AA135			2008
J51	J51	Rioja red wine	Putative two-peptide and PlnAEF	2008
LR/14	LR/14		Class I?, in this chapter	2008
LP31	LP31	Dry-fermented sausage	Described in this chapter	2009
MG	KLDS1.0391	Chinese traditional fermented cream	*Salmonella*, pore formation	2010
154	LTF154	Fermented sausage	Ca. 3.0 kDa	1994
N	KKY12	Nigerian fermented foods		1995

two bacteriocin operons (*plnEFI* and *plnJKLR*) (Risøen *et al.*, 1998). However, PlnC and PlnD have opposite functions: transcription and bacteriocin production were activated by PlnC and were repressed by PlnD (Diep *et al.*, 2001).

The positively charged plantaricin A targets the negatively charged membrane of sensitive cells, dissipates the membrane potential, and causes a trans-membrane pH gradient. The antimicrobial activity is mediated through the formation of the strongly amphiphilic α-helix structure; and the pheromone activity is mediated through the amphiphilic helix and a receptor protein PlnB (a histidine kinase) (Hauge *et al.*, 1998). The three-dimensional solution structure of plantaricin A (26 amino acid residues) using NMR in DPC micelles shows the N-terminal end (pentamer fragment) as the pheromone peptide was unstructured; and the amphiphilic α-helix (residues 12–21 or 23) exerting the membrane-permeabilizing antimicrobial activity was found in the C-terminal portion (17-mer fragment) indicating the α-helical structure has the initial interaction with membrane lipids; and then the α-helix is positioned in the membrane interface to display antimicrobial activity and pheromone activity (Kristiansen *et al.*, 2005). With PlnA, the interaction with the membrane is dependent on the membrane lipid composition where in killed human leukaemic T cells (both necrosis and apoptosis) formed supra-molecular protein-lipid amyloid-like fibres on the membranes (Zhao *et al.*, 2006). The membrane-permeabilizing effects of PlnA on cultured cancerous and normal rat anterior pituitary cells using patch-clamp techniques and microfluorometry found the primary cultures of normal rat anterior pituitary cells were insensitive to the peptide; and PlnA differentiates between plasma membranes and membrane leaflets (Sand *et al.*, 2007).

The class IIb bacterioicins, plantaricin EF and JK, form pores in the membranes of target cells and dissipate the trans-membrane electrical potential and pH gradient; however, the mode of action is different between the plantaricin EF pores conducting small monovalent cations and the plantaricin JK pores conducting specific anions (Moll *et al.*, 1999). Circular dichroism indicates each peptide for plantaricin EFJK adopted a partly α-helix structure in the presence of trifluoroethanol, micelles of dodecylphosphocholine, and negatively charged dioleoylphosphoglycerol liposomes. A negative charge in the phospholipids on liposomes is important for a structure–inducing interaction between plantaricins and the membranes (Hauge *et al.*, 1999). Three-dimensional solution structure using NMR in DPC micelles shows PlnE (33 amino acid residues) has an N-terminal amphiphilic α-helix (residues 10–21) and a C-terminal amphiphilic α-helix-like structure (residues 25–31); and PlnF (34 amino acid residues) has a polar and large amphiphilic α-helix (residues 7–32) (Fimland *et al.*, 2008). Using the two putative GxxxG motifs (residues 5–9 and 20–24) in PlnE and one putative motif (residues 30–34) in PlnF, plantaricin EF may work on the target cell membrane using the helix–helix interactions with the GxxxG motifs (Senes *et al.*, 2000); however, the GxxxG motif of PlnF is located outside of the α-helix differing from lactococcin G (Oppegård *et al.*, 2008) (Fig. 11.3).

Plantaricin C is a heat stable bacteriocin produced by *Lb. plantarum* LL441 isolated from a dairy origin (González *et al.*, 1994); and is a pore-forming bacteriocin that functions in a voltage-independent manner; and does not require a specific protein receptor in the target membrane (Gonzalez *et al.*, 1996). Turner *et al.* (1999) report plantaricin C (27 amino acid residues) is the first lantibiotic bacteriocin (class I) from

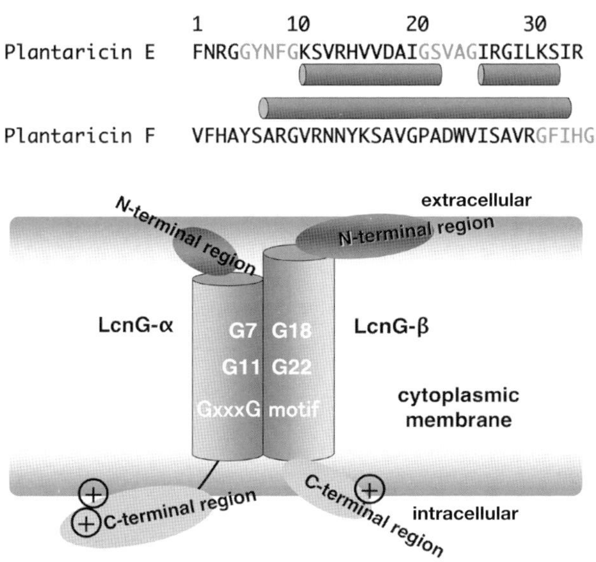

Figure 11.3 The amino acid sequence of class IIb plantaricin EF. The column bars show the α-helix regions determined using NMR analysis (Fimland *et al.*, 2008). The figure shows the membrane-permeabilizing model for lactococcin G (LcnG-α and LcnG-β) (first class IIb bacteriocin produced by *Lactococcus lactis* LMG 2081) interacting with GxxxG-motifs of G7xxxG11 in LcnG-α and G18xxxG22 in LcnG-β (Oppegård *et al.*, 2008).

Lb. plantarum; and has one dehydroalanine, one lanthionine (residues 7–27) and three β-methyllanthionine residues (residues 12–15, 13–18 and 23–26). Three-dimensional solution structure using NMR shows the six NH_2-terminal residues are highly positive charged with four lysine residues and the structure at positions 7 and 27 are linked by lanthionine bridges that form a globular conformation containing an amphipathic α-helix in the central region with a charge distribution. Further analysis for the mode of action of pore formation with plantaricin C using the cell wall precursor lipid II, liposomes and indicator cells, suggests plantaricin C may prevent lipid II synthesis and the FemX reaction (the addition of the first Gly to the pentapeptide side chain of lipid II) by the formation of a PlnC-lipid I and/or PlnC-lipid II complex that differs from the other pore forming lantibiotic, nisin (Wiedemann *et al.*, 2006).

Another putative lantibiotic bacteriocin produced by *Lb. plantarum* is plantaricin W from strain LMG2379 that has two peptides, Plwα (29 amino acid residues) and Plwβ (32 amino acid residues). They have synergistic activity when in a 1:1 ratio. Cloning and sequencing of the structural gene for plantaricin W (*plwα* and *plwβ*) shows plantaricin W is similar to lacticin 3147, a representative of the two-peptide lantibiotics (Ryan *et al.*, 1999) where the serine, threonine and cysteine residues are involved in intramolecular thioether bond formation and are conserved at the C-terminal ends of both Plwα and Plwβ (Holo *et al.*, 2001).

A plasmid-encoded class IIa bacteriocin, plantaricin 423, produced by *Lb. plantarum* strain 423 isolated from sorghum beer is pH- and heat-stable and is inhibitory to several food spoilage bacteria and food-borne pathogens including *B. cereus*, *Clostridium sporogenes*, *Ent. faecalis*, *Lactobacillus* spp., *Listeria* spp. *Pediococcus* spp., and *Staphylococcus* spp. (van Reenen *et al.*, 1998). The complete DNA sequence of the plasmid, pPLA4, containing the plantaricin 423-encoding region shows *plaA* encoding a 56-amino acid pre-peptide with a 19-amino acid N-terminal leader peptide (mature PlaA is 37-amino acid residues) and *plaB* encoding an immunity protein having 109 amino acid residues that share a high sequence similarity to the pediocin PA-1 operon (Marugg *et al.*, 1992). However, no cross-reactivity was detected between the immunity proteins of plantaricin 423 and pediocin PA-1. *plaC* and *plaD* are identical to *pedC* and *pedD* of the pediocin PA-1 operon, respectively (van Reenen *et al.*, 2003; van Reenen *et al.*, 2006). Using fluorescent *in situ* hybridization (FISH), *Lb. plantarum* 423 adhered strongly to the ileum and posterior colon in a gastro-intestinal model where plantaricin 423 produced in the ileum section inhibits the growth of *Ent. faecalis* indicator strains (Maré *et al.*, 2006). The mode of action was bactericidal and may destabilize the cell membrane where the adsorption of plantaricin 423 onto target cells was from 17% to 67%; and some indicator strains treated with Triton X-100, Triton X-114 and chloroform increased the adsorption of plantaricin 423 (Todorov *et al.*, 2006).

Plantaricin S from *Lb. plantarum* LPCO10, a class IIb bacteriocin (PlsA and PlsB), has full bacteriocin activity and was purified to homogeneity (Jiménez-Díaz *et al.*, 1995). By cloning and sequencing of a 4.5-kb region of chromosomal DNA, the genes for the biosynthesis of plantaricin S were found: two structural genes of plantaricin S (*plsA* and *plsB*) contain the N-terminal leader sequence with the double-glycine motif, ORF2 (a putative immunity protein), several genes responsible for the regulation of bacteriocin production such as ORF8 (a putative histidine protein kinase) and ORF14 (a putative response regulator gene) (Stephens *et al.*, 1998). Among 68 strains of *Lb. plantarum* isolated from different olive processing plants in southern Spain, all 15 strains producing antibacterial substances were shown to be plantaricin S producers using PCR detection of the *plsA* and *plsB* genes; and the N-terminal amino acid sequencing after purification suggests plantaricin S is the major bacteriocin produced by *Lb. plantanum* in the traditional olive fermentation process (Maldonado *et al.*, 2002). To use the bacteriocins for food application from *Lb. plantarum* such as plantaricin S, optimization of the bacteriocin production is important. Leal-Sánchez *et al.* (2002) determined the optimum conditions for plantaricin S production as follows: temperature, 22–27°C; salt concentration, 2.3–2.5%; LPCO10 inoculum volume, $10^{7.3}$–$10^{7.4}$ CFU/ml; fixed 2% glucose concentration; and with no aeration. This is about 3.2×10^4 times higher in plantaricin

S production compared with the conventional cultivation method for olives.

Plantaricin NC8 from *Lb. plantarum* NC8, a class IIb bacteriocin, has PLNC8α and PLNC8β and was purified and chemically sequenced. Mass spectrometry analysis show molecular weights of 3587 for PLNC8α and 4000 for PLNC8β. The PLNC8 operon was found using cloning and sequenced. Two structural genes of plantaricin NC8, *plNC8A* and *plNC8B*, encode bacteriocin precursors of 47 and 55 amino acids residues where the mature PLNC8α and PLNC8β have 29 and 34 amino acids residues after the release N-terminal leader sequence at the double-glycine motif, respectively; and *orfC* encodes a putative immunity protein (Maldonado *et al.*, 2003). Sequencing an 8.1-kb DNA region downstream of the PLNC8 operon and computer analysis show four operons for bacteriocin production with high homology to the plantaricin cluster in *Lb. plantarum* C11 that contain a three-component regulatory operon (a quorum-sensing mechanism) as an induction factor (PLNC8IF), a novel histidine kinase, and the response regulator identical to PlnD (from the C11 strain). Determining homologous and heterologous expression of PLNC8IF shows constitutive PLNC8 production in culture supernatants that promote the bacteriocin production from *Lb. plantarum* NC8 (Maldonado *et al.*, 2004a). PLNC8 production from the NC8 strain was induced by both alive and dead cells from the Gram-positive strains tested, i.e. *Bacillus, Enterococcus, Lactobacillus, Lactococcus, Leuconostoc, Listeria, Pediococcus, Staphylococcus,* and *Streptococcus*; however no induction occurred with their culture-supernatants. Conversely, culture supernatants promote the production of PLNC8 auto-inducing activity in *Lb. plantarum* NC8 in which the molecules were diffusible, heat resistant, and proteinaceous but were not bacteriocins. The specific Gram-positive bacteria mentioned above may contribute in maintaining PLNC8 production (Maldonado *et al.*, 2004b).

Many LAB bacteriocins are controlled using a quorum-sensing mechanism, the so-called three-component regulatory system indicating bacteriocin production differs under various circumstances and conditions. Bacteriocin production in two plantaricin producers, *Lb. plantarum* NC8 and *Lb. plantarum* WCFS1 have two different QS-regulating systems where the QS-knockout mutants were affected on solid media rather than in liquid media (Maldonado-Barragán *et al.*, 2009). This indicates the selection of QS-regulated bacteriocinogenic LAB and the bacteriocins they produce have to be under optimum conditions; and may be important considerations for use in biopreservation on solid substrates such as vegetable fermentations or liquids, e.g. milk.

Recently, several new plantaricins were reported. *L. plantarum* LR/14 produces a putative class I, plantaricin LR14 (α and β peptides, molecular mass 3012.46 and 5605.74, respectively). The purified peptides were highly heat stable and active in acidic pH conditions. Each N-terminal amino acid was blocked and the partial sequencing after CNBr digestion shows plantaricin LR14 is a novel bacteriocin (Tiwari and Srivastava, 2008). The antimicrobial spectrum of a bacteriocin from *Lb. plantarum* strain LP 31 isolated from Argentinian dry-fermented sausage is diversely active against *Pseudomonas* sp., *Staph. aureus, B. cereus* and *List. monocytogenes*. The molecular weight is 1558.85 using MALDI-TOF mass spectrometry; and the bacteriocin may be 14 amino acid residues using amino acid composition (Müller *et al.*, 2009). This small unique heat-stable bacteriocin requires further study and has high potential as a biopreservative.

Gassericin A (class IIc)

About ten circular bacteriocins linked at the N- and C-terminal amino acids (class IIc) are reported. There are two major groups using their primary structures, biochemical characteristics, and genetic arrangement. One group is in the family of enterocin AS-48 (Samyn *et al.*, 1994, described in Chapter 12), the first circular bacteriocins including circularin A (Kemperman *et al.*, 2003), uberolysin (Wirawan *et al.*, 2007), carnocyclin A (Martin-Visscher *et al.*, 2008) and lactocyclicin Q (Sawa *et al.*, 2009). The other group is the gassericin A family (Kawai *et al.*, 1998a,b) of a second bacteriocin found in 1998, acidocin B (Leer *et al.*, 1995), reutericin 6 [a primary structure 100% identical to gassericin A

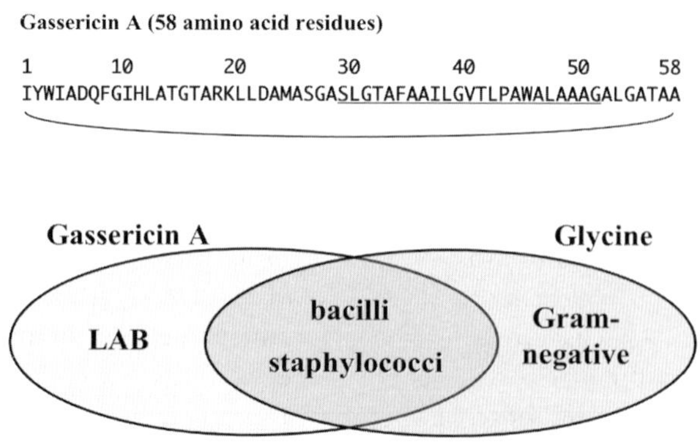

Figure 11.4 The primary structure of gassericin A (58 amino acid residues, Kawai *et al.*, 1998a,b). The underline shows an α helix transmembrane region using the SOSUI program (http://bp.nuap.nagoya-u.ac.jp/sosui/). Schematic diagram for complete growth inhibition of custard cream spoilage bacteria strains using a combination of gassericin A and glycine (Arakawa *et al.* 2009b).

(Kawai *et al.*, 2001; 2004)] and butyrivibriocin AR10 (Kalmokoff *et al.*, 2003).

Gassericin A is a hydrophobic and heat and pH-stable bacteriocin produced by *Lb. gasseri* LA39 isolated from the faeces of a human infant (Toba *et al.*, 1991) and has bactericidal activity against the food-borne pathogens: *List. monocytogenes*, *B. cereus* and *Staph. aureus* (Itoh *et al.*, 1995). Gassericin A was purified, characterized and its structural gene (*gaaA*) was cloned and sequenced (Kawai *et al.*, 1994; 1998a). Determination of the complete primary chemical structure of gassericin A shows this bacteriocin (MW 5653) belongs to class IIc having 58 amino acid residues linked at the N and C termini (Kawai *et al.*, 1998b) (Fig. 11.4). Heterologous expression in *E. coli* JM109 shows the specific activity of the expressed gassericin A having N- and C-termini decreased to 1/173 of the native circular gassericin A (Kawai *et al.*, 2003). To study the mode of action using an indicator strain *Lb. delbrueckii* subsp. *bulgaricus* JCM1002^T and a liposome of L-α-phosphatidylcholine, gasssericin A acts in the cytoplasmic membrane of a target cell and causes the death of the cell using an efflux of potassium ions not ATP (MW 507) (Kawai *et al.*, 2004).

The DNA region for production, secretion and immunity of gassericin A was sequenced. Seven ORFs (*gaaBCADITE*) at 3.5 kbp (*gaa*) contain the structural gene (*gaaA*). Homologous expression of GaaI, a small positively charged hydrophobic peptide of 53 amino acids contains a putative trans-membrane segment; and using *Lb. gasseri* JCM1131^T we found *gaaI* encodes an immunity peptide for gassericin A (Kawai *et al.*, 2009). Almost all immunity peptides in the class IIc bacteriocins (from 49 to 56 amino acid residues) such as GaaI, AS-48D1 for enterocin AS-48 (Martínez-Bueno *et al.*, 1998), and CirE for circularin A (Kemperman *et al.*, 2003) usually have one or two putative trans-membrane segments (TMSs) located in the cell membranes of the producing strains. The regions outside the TMSs contain positively charged amino acid residues such as lysines and arginines exposed to the outer and/or inner surface of the cytoplasmic membrane; and may electrostatically prevent the attachment of the positively charged bacteriocins. *gaaT* (a putative ATP-binding protein as an ABC-transporter) and *gaaE* (a putative membrane protein with six trans-membrane regions) may encode for the export of gassericin A out of the producer cells and/or may work on the auxiliary immunity function reported for several bacteriocin genes. *gaaBCD* may encode for membrane-associated proteins; however, there is no speculation as to their functions because there is no similarity in the protein databases. Potential promoter sequences were predicted in the *gaa* region, especially in *gaaB* and *gaaC*, indicating *gaaB* and/or *gaaC* may be involved in the complete expression of gassericin A.

Seven *gaa* genes are similar to some genes of other putative circular bacteriocins such as the acidocin B from *Lb. acidophilus* M46 (Leer *et al.*, 1995) and the butyrivibriocin AR10 from the ruminal bacterium, *Butyrivibrio fibrisolvens* AR10 (Kalmokoff *et al.*, 2003). Approximately 2 kb of the 2200-bp sequence of the acidocin B gene locus (*acd*) lack the genes corresponding to *gaaTE* and is 98.4% identical to the *gaa* locus. However, the gene corresponding to GaaI, the small immunity peptide, was lost by a deduced frame-shift mutation. Although the genetic arrangement is different in the two circular bacteriocin genes, the five consecutive ORFs, *gaaADITE*, have ca. 50% similarity to genes for butyrivibriocin AR10 production, excluding BviE, a putative immunity protein (Kalmokoff *et al.*, 2003) and BviD, a putative transporter element or the immunity peptides that show low similarity to GaaI and GaaD, respectively. No similarities for ORF1 (response regulator-like protein) and ORF2 (histidine kinase-like protein) in the butyrivibriocin AR10 locus (*bvi*) that are putative two-component regulator genes were found in the *gaa* gene; while genes like *gaaB* and *gaaC* in both *gaa* and *acd* loci were not present in the *bvi* sequence. The gassericin A family of bacteriocins were isolated from various bacterial species from several countries suggesting the bacteriocin genes may be associated with transferable genetic elements.

Natural, safe and low-cost media are required to cultivate bacteriocin producers to supply bacteriocins as biopreservatives in the food industry and in animal formulations. However, few strains of probiotic *Lb. gasseri* are cultivated in natural media such as milk-based media and therefore gassericins have limited use in these applications. Arakawa *et al.* (2008a) developed a method to cultivate *Lb. gasseri* LA39 in reconstituted skim milk and cheese whey using proteose peptone as a nutrient supplement, and they prepared the bacteriocin. This result indicates gassericin A could be used as a safe biopreservative for foods and a natural drugs therapy for bovine mastitis. The excess cheese whey by-product from the cheese making industry can be used as an inexpensive media substrate. Arakawa *et al.* report gassericin A has broader antibacterial spectra than nisin A against 12 Gram-positive strains among isolated custard cream spoilage bacteria using an *in vitro* assay for growth inhibition of all custard cream spoilage strains containing Gram-negative isolates (Arakawa *et al.*, 2008b). With the addition of glycine, bacteriostatic against many strains except lactic acid bacteria (Arakawa *et al.*, 2009b), a practical preservative using gassericin A and glycine in custard creams (*in vivo*) was developed. Effects were shown against four isolated strains, *B. cereus*, *Lc. lactis* subsp. *lactis*, *Achromobacter denitrificans*, and *Pseudomonas fluorescens* (Fig. 11.4). This work may be helpful for future use of bacteriocins in the food industry.

References

Abee, T., Klaenhammer, T.R., and Letellier, L. (1994). Kinetic studies of the action of lactacin F, a bacteriocin produced by *Lactobacillus johnsonii* that forms poration complexes in the cytoplasmic membrane. Appl. Environ. Microbiol. *60*, 1006–1013.

Allison, G.E., and Klaenhammer, T.R. (1996). Functional analysis of the gene encoding immunity to lactacin F, lafI, and its use as a *Lactobacillus*-specific, food-grade genetic marker. Appl. Environ. Microbiol. *62*, 4450–4460.

Allison, G.E., Fremaux, C., and Klaenhammer, T.R. (1994). Expansion of bacteriocin activity and host range upon complementation of two peptides encoded within the lactacin F operon. J. Bacteriol. *176*, 2235–2241.

Allison, G.E., Worobo, R.W., Stiles, M.E., and Klaenhammer, T.R. (1995). Heterologous expression of the lactacin F peptides by *Carnobacterium piscicola* LV17. Appl. Environ. Microbiol. *61*, 1371–1377.

Anderssen, E.L., Diep, D.B., Nes, I.F., Eijsink, V.G., and Nissen-Meyer, J. (1998). Antagonistic activity of *Lactobacillus plantarum* C11: two new two-peptide bacteriocins, plantaricins EF and JK, and the induction factor plantaricin A. Appl. Environ. Microbiol. *64*, 2269–2272.

Arakawa, K., Kawai, Y., Fujitani, K., Nishimura, J., Kitazawa, H., Komine, K., Kai, K., and Saito, T. (2008a). Bacteriocin production of probiotic *Lactobacillus gasseri* LA39 isolated from human feces in milk-based media. Ani. Sci. J. *79*, 634–640.

Arakawa, K., Kawai, Y., Iioka, H., Tanioka, M., Nishimura, J., Kitazawa, H., Tsurumi, K., and Saito, T. (2008b). Microbial community analysis of food-spoilage bacteria in commercial custard creams using culture-dependent and -independent methods. J. Dairy Sci. *91*, 2938–2946.

Arakawa, K., Kawai, Y., Nishimura, J., Kitazawa, H., and Saito, T. (2009a). Negative effect of divalent metal cations on production of gassericin T, a bacteriocin produced by *Lactobacillus gasseri*, in milk-based media. Int. Dairy J. *19*, 612–616.

Arakawa, K., Kawai, Y., Iioka, H., Tanioka, M., Nishimura, J., Kitazawa, H., Tsurumi, K., and Saito, T. (2009b). Effects of gassericins A and T, bacteriocins produced by *Lactobacillus gasseri*, with glycine on custard cream preservation. J. Dairy Sci. *92*, 2365–2372.

Axelsson, L., and Holck, A. (1995). The genes involved in production of and immunity to sakacin A, a bacteriocin from *Lactobacillus sake* Lb706. J. Bacteriol. *177*, 2125–2137.

Bogovic-Matijasić, B., Rogelj, I., Nes, I.F., and Holo, H. (1998). Isolation and characterization of two bacteriocins of *Lactobacillus acidophilus* LF221. Appl. Microbiol. Biotechnol. *49*, 606–612.

Callewaert, R., Holo, H., Devreese, B., Van Beeumen, J., Nes, I., and De Vuyst, L. (1999). Characterization and production of amylovorin L471, a bacteriocin purified from *Lactobacillus amylovorus* DCE 471 by a novel three-step method. Microbiology *145*, 2559–2568.

Callewaert, R., and De Vuyst, L. (2000). Bacteriocin production with *Lactobacillus amylovorus* DCE 471 is improved and stabilized by fed-batch fermentation. Appl. Environ. Microbiol. *66*, 606–613.

Canzek Majhenic, A., Bogovic Matijasić, B., and Rogelj, I. (2003). Chromosomal location of the genetic determinants for bacteriocins produced by *Lactobacillus gasseri* K7. J. Dairy Res. *70*, 199–203.

Cheigh, C.I., and Pyun, Y.R. (2005). Nisin biosynthesis and its properties. Biotechnol. Lett. *27*, 1641–1648.

Claesson, M.J., Li, Y., Leahy, S., Canchaya, C., van Pijkeren, J.P., Cerdeño-Tárraga, A.M., Parkhill, J., Flynn, S., O'Sullivan, G.C., Collins, J.K., *et al.* (2006). Multireplicon genome architecture of *Lactobacillus salivarius*. Proc. Natl. Acad. Sci. U.S.A. *103*, 6718–6723.

Cocolin, L., and Rantsiou, K. (2007). Sequencing and expression analysis of sakacin genes in *Lactobacillus curvatus* strains. Appl. Microbiol. Biotechnol. *76*, 1403–1411.

Contreras, B.G., De Vuyst, L., Devreese, B., Busanyova, K., Raymaeckers, J., Bosman, F., Sablon, E., and Vandamme, E.J. (1997). Isolation, purification, and amino acid sequence of lactobin A, one of the two bacteriocins produced by *Lactobacillus amylovorus* LMG P-13139. Appl. Environ. Microbiol. *63*, 13–20.

Corr, S.C., Li, Y., Riedel, C.U., O'Toole, P.W., Hill, C., and Gahan, C.G. (2007). Bacteriocin production as a mechanism for the antiinfective activity of *Lactobacillus salivarius* UCC118. Proc. Natl. Acad. Sci. U.S.A. *104*, 7617–7621.

Díaz, M., Valdivia, E., Martínez-Bueno, M., Fernández, M., Soler-González, A.S., Ramírez-Rodrigo, H., and Maqueda, M. (2003). Characterization of a new operon, *as-48EFGH*, from the *as-48* gene cluster involved in immunity to enterocin AS-48. Appl. Environ. Microbiol. *69*, 1229–1236.

Diep, D.B., Håvarstein, L.S., Nissen-Meyer, J., and Nes, I.F. (1994). The gene encoding plantaricin A, a bacteriocin from *Lactobacillus plantarum* C11, is located on the same transcription unit as an agr-like regulatory system. Appl. Environ. Microbiol. *60*, 160–166.

Diep, D.B., Håvarstein, L.S., and Nes, I.F. (1995). A bacteriocin-like peptide induces bacteriocin synthesis in *Lactobacillus plantarum* C11. Mol. Microbiol. *18*, 631–639.

Diep, D.B., Håvarstein, L.S., and Nes, I.F. (1996). Characterization of the locus responsible for the bacteriocin production in *Lactobacillus plantarum* C11. J. Bacteriol. *178*, 4472–4483.

Diep, D.B., Axelsson, L., Grefsli, C., and Nes, I.F. (2000). The synthesis of the bacteriocin sakacin A is a temperature-sensitive process regulated by a pheromone peptide through a three-component regulatory system. Microbiology *146*, 2155–2160.

Diep, D.B., Johnsborg, O., Risøen, P.A., and Nes, I.F. (2001). Evidence for dual functionality of the operon plnABCD in the regulation of bacteriocin production in *Lactobacillus plantarum*. Mol. Microbiol. *41*, 633–644.

Diep, D.B., Skaugen, M., Salehian, Z., Holo, H., and Nes, I.F. (2007). Common mechanisms of target cell recognition and immunity for class II bacteriocins. Proc. Natl. Acad. Sci. U.S.A. *104*, 2384–2389.

Fimland, G., Eijsink, V.G.H., and Nissen-Meyer, J. (2002). Comparative studies of immunity proteins of pediocin-like bacteriocins. Microbiology *148*, 3661–3670.

Fimland, N., Rogne, P., Fimland, G., Nissen-Meyer, J., and Kristiansen, P.E. (2008). Three-dimensional structure of the two peptides that constitute the two-peptide bacteriocin plantaricin EF. Biochim. Biophys. Acta. *1784*, 1711–1719.

Flynn, S., van Sinderen, D., Thornton, G.M., Holo, H., Nes, I.F., and Collins, J.K. (2002). Characterization of the genetic locus responsible for the production of ABP-118, a novel bacteriocin produced by the probiotic bacterium *Lactobacillus salivarius* subsp. *salivarius* UCC118. Microbiology *148*, 973–984.

Fremaux, C., Ahn, C., and Klaenhammer, T.R. (1993). Molecular analysis of the lactacin F operon. Appl. Environ. Microbiol. *59*, 3906–3915.

González, B., Arca, P., Mayo, B., and Suárez, J.E. (1994). Detection, purification, and partial characterization of plantaricin C, a bacteriocin produced by a *Lactobacillus plantarum* strain of dairy origin. Appl. Environ. Microbiol. *60*, 2158–2163.

González, B., Glaasker, E., Kunji, E., Driessen, A., Suarez, J.E., and Konings, W.N. (1996). Bactericidal mode of action of plantaricin C. Appl. Environ. Microbiol. *62*, 2701–2709.

Hauge, H.H., Mantzilas, D., Moll, G.N., Konings, W.N., Driessen, A.J., Eijsink, V.G., and Nissen-Meyer, J. (1998). Plantaricin A is an amphiphilic alpha-helical bacteriocin-like pheromone which exerts antimicrobial and pheromone activities through different mechanisms. Biochemistry *37*, 16026–16032.

Hauge, H.H., Mantzilas, D., Eijsink, V.G., and Nissen-Meyer, J. (1999). Membrane-mimicking entities induce structuring of the two-peptide bacteriocins plantaricin E/F and plantaricin J/K. J. Bacteriol. *181*, 740–747.

Haugen, H.S., Fimland, G., Nissen-Meyer, J., and Kristiansen, P.E. (2005). Three-dimensional structure in lipid micelles of the pediocin-like antimicrobial peptide curvacin A. Biochemistry *44*, 16149–16157.

Haugen, H.S., Kristiansen, P.E., Fimland, G., and Nissen-Meyer, J. (2008). Mutational analysis of the class IIa bacteriocin curvacin A and its orientation in target cell membranes. Appl. Environ. Microbiol. *74*, 6766–6773.

Holck, A., Axelsson, L., Birkeland, S.E., Aukrust, T., and Blom, H. (1992). Purification and amino acid sequence of sakacin A, a bacteriocin from *Lactobacillus sake* Lb706. J. Gen. Microbiol. *138*, 2715–2720.

Holo, H., Jeknic,.Z., Daeschel, M., Stevanovic, S., and Nes, I.F. (2001). Plantaricin W from *Lactobacillus plantarum* belongs to a new family of two-peptide lantibiotics. Microbiology *147*, 643–651.

Hühne, K., Axelsson, L., Holck, A., and Kröckel, L. (1996). Analysis of the sakacin P gene cluster from *Lactobacillus sake* Lb674 and its expression in sakacin-negative Lb. sake strains. Microbiology *142*, 1437–1448.

Itoh, T., Fujimoto, Y., Kawai, Y., Toba, T., and Saito, T. (1995). Inhibition of food-borne pathogenic bacteria by bacteriocins from *Lactobacillus gasseri*. Lett. Appl. Microbiol. *21*, 137–141.

Jiménez-Díaz, R., Ruiz-Barba, J.L., Cathcart, D.P., Holo, H., Nes, I.F., Sletten, K.H., and Warner, P.J. (1995). Purification and partial amino acid sequence of plantaricin S, a bacteriocin produced by *Lactobacillus plantarum* LPCO10, the activity of which depends on the complementary action of two peptides. Appl. Environ. Microbiol. *61*, 4459–4463.

Joerger, M.C., and Klaenhammer, T.R. (1990). Cloning, expression, and nucleotide sequence of the *Lactobacillus helveticus* 481 gene encoding the bacteriocin helveticin J. J. Bacteriol. *172*, 6339–6347.

Kalmokoff, M.L., Cyr, T.D., Hefford, M.A., Whitford, M.F., and Teather, R.M. (2003). Butyrivibriocin AR10, a new cyclic bacteriocin produced by the ruminal anaerobe *Butyrivibrio fibrisolvens* AR10: characterization of the gene and peptide. Can. J. Microbiol. *49*, 763–773.

Kaur, K., Andrew, L.C., Wishart, D.S., and Vederas, J.C. (2004). Dynamic relationships among type IIa bacteriocins: temperature effects on antimicrobial activity and on structure of the C-terminal amphipathic alpha helix as a receptor-binding region. Biochemistry *43*, 9009–9020.

Kawai, Y., Saito, T., Toba, T., Samant, S.K., and Itoh, T. (1994). Isolation and characterization of a highly hydrophobic new bacteriocin (gassericin A) from *Lactobacillus gasseri* LA39. Biosci. Biotechnol. Biochem. *58*, 1218–1221.

Kawai, Y., Saito, T., Uemura, J., and Itoh, T. (1997). Rapid detection method for bacteriocin and distribution of bacteriocin-producing strains in *Lactobacillus acidophilus* group lactic acid bacteria isolated from human feces. Biosci. Biotechnol. Biochem. *61*, 179–182.

Kawai, Y., Saito, T., Suzuki, M., and Itoh, T. (1998a). Sequence analysis by cloning of the structural gene of gassericin A, a hydrophobic bacteriocin produced by *Lactobacillus gasseri* LA39. Biosci. Biotechnol. Biochem. *62*, 887–892.

Kawai, Y., Saito, T., Kitazawa, H., and Itoh, T. (1998b). Gassericin A; an uncommon cyclic bacteriocin produced by *Lactobacillus gasseri* LA39 linked at N- and C-terminal ends. Biosci. Biotechnol. Biochem. *62*, 2438–2440.

Kawai, Y., Saitoh, B., Takahashi, O., Kitazawa, H., Saito, T., Nakajima, H., and Itoh T. (2000). Primary amino acid and DNA sequences of gassericin T, a lactacin F-family bacteriocin produced by *Lactobacillus gasseri* SBT2055. Biosci. Biotechnol. Biochem. *64*, 2201–2208.

Kawai, Y., Ishii, Y., Uemura, K., Kitazawa, H., Saito, T., and Itoh, T. (2001). *Lactobacillus reuteri* LA6 and *Lactobacillus gasseri* LA39 isolated from feces of the same human infant produce identical cyclic bacteriocin. Food Microbiol. *18*, 407–415.

Kawai, Y., Arakawa, K., Itoh, A., Saitoh, B., Ishiii, Y., Nishimura, J., Kitazawa, H., Itoh, T., and Saito, T. (2003). Heterologous expression of gassericin A, a bacteriocin produced by *Lactobacillus gasseri* LA39. Ani. Sci. J., *74*, 45–51.

Kawai, Y., Ishii, Y., Arakawa, K., Uemura, K., Saitoh, B., Nishimura, J., Kitazawa, H., Yamazaki, Y., Tateno, Y., Itoh, T., *et al.* (2004). Structural and functional differences in two cyclic bacteriocins with the same sequences produced by lactobacilli. Appl. Environ. Microbiol. *70*, 2906–2911.

Kawai, Y., Takahashi, R., Arakawa, K., and Saito, T. (2006). Structure and function of gassericin A and T produced by probiotic lactic acid bacteria. Jap. J. Lactic Acid Bacteria *17*, 32–39.

Kawai, Y., Kusnadi, J., Kemperman, R., Kok, J., Ito, Y., Endo, M., Arakawa, K., Uchida, H., Nishimura, J., Kitazawa, H., and Saito, T. (2009). DNA sequencing and homologous expression of a small peptide conferring immunity to gassericin A, a circular bacteriocin produced by *Lactobacillus gasseri* LA39. Appl. Environ. Microbiol. *75*, 1324–1330.

Kemperman, R., Jonker, M., Nauta, A., Kuipers, O.P., and Kok, J. (2003). Functional analysis of the gene cluster involved in production of the bacteriocin circularin A by *Clostridium beijerinckii* ATCC 25752. Appl. Environ. Microbiol. *69*, 5839–5848.

Kemperman, R., Kuipers, A., Karsens, H., Nauta, A., Kuipers, O., and Kok, J. (2003). Identification and characterization of two novel clostridial bacteriocins, circularin A and closticin 574. Appl. Environ. Microbiol. *69*, 1589–1597.

Kristiansen, P.E., Fimland, G., Mantzilas, D., and Nissen-Meyer, J. (2005). Structure and mode of action of the membrane-permeabilizing antimicrobial peptide pheromone plantaricin A. J. Biol. Chem. *280*, 22945–22950.

Leal-Sánchez, M.V., Jiménez-Díaz, R., Maldonado-Barragán, A., Garrido-Fernández, A., and Ruiz-Barba, J.L. (2002). Optimization of bacteriocin production by batch fermentation of *Lactobacillus plantarum* LPCO10. Appl. Environ. Microbiol. *68*, 4465–4471.

Leer, R.J., van der Vossen, J.M., van Giezen, M., van Noort, J.M., and Pouwels, P.H. (1995). Genetic analysis of acidocin B, a novel bacteriocin produced by *Lactobacillus acidophilus*. Microbiology *141*, 1629–1635.

Leroy, F., De Winter, T., Adriany, T., Neysens, P., and De Vuyst, L. (2006). Sugars relevant for sourdough fermentation stimulate growth of and bacteriocin production by *Lactobacillus amylovorus* DCE 471. Int. J. Food Microbiol. *112*, 102–111.

McCormick, J.K., Poon, A., Sailer, M., Gao, Y., Roy, K.L., McMullen, L.M., Vederas, J.C., Stiles, M.E., and Van Belkum, M.J. (1998). Genetic characterization and heterologous expression of brochocin-C, an antibotulinal, two-peptide bacteriocin produced by *Brochothrix campestris* ATCC43754. Appl. Environ. Microbiol. *64*, 4757–4766.

Majhenic, A.C., Venema, K., Allison, G.E., Matijasić, B.B., Rogelj, I., and Klaenhammer, T.R. (2004). DNA analysis of the genes encoding acidocin LF221 A and acidocin LF221 B, two bacteriocins produced by *Lactobacillus gasseri* LF221. Appl. Microbiol. Biotechnol. *63*, 705–714.

Maldonado, A., Ruiz-Barba, J.L., Floriano, B., and Jiménez-Díaz, R. (2002). The locus responsible for production of plantaricin S, a class IIb bacteriocin produced by *Lactobacillus plantarum* LPCO10, is widely distributed among wild-type *Lact. plantarum* strains isolated from olive fermentations. Int. J. Food Microbiol. *77*, 117–124.

Maldonado, A., Ruiz-Barba, J.L., and Jiménez-Díaz, R. (2003). Purification and genetic characterization of plantaricin NC8, a novel coculture-inducible two-peptide bacteriocin from *Lactobacillus plantarum* NC8. Appl. Environ. Microbiol. *69*, 383–389.

Maldonado, A., Ruiz-Barba, J.L., and Jiménez-Díaz, R. (2004a). Production of plantaricin NC8 by *Lactobacillus plantarum* NC8 is induced in the presence of different types of Gram-positive bacteria. Arch. Microbiol. *181*, 8–16.

Maldonado, A., Jiménez-Díaz, R., and Ruiz-Barba, J.L. (2004b). Induction of plantaricin production in *Lactobacillus plantarum* NC8 after coculture with specific Gram-positive bacteria is mediated by an autoinduction mechanism. J. Bacteriol. *186*, 1556–1564.

Maldonado-Barragán, A., Ruiz-Barba, J.L., and Jiménez-Díaz, R. (2009). Knockout of three-component regulatory systems reveals that the apparently constitutive plantaricin-production phenotype shown by *Lactobacillus plantarum* on solid medium is regulated via quorum sensing. Int. J. Food Microbiol. *130*, 35–42.

Maré, L., Wolfaardt, G.M., and Dicks, L.M. (2006). Adhesion of *Lactobacillus plantarum* 423 and *Lactobacillus salivarius* 241 to the intestinal tract of piglets, as recorded with fluorescent in situ hybridization (FISH), and production of plantaricin 423 by cells colonized to the ileum. J. Appl. Microbiol. *100*, 838–845.

Martin-Visscher, L.A., van Belkum, M.J., Garneau-Tsodikova, S., Whittal, R.M., Zheng, J., McMullen, L.M., and Vederas, J.C. (2008). Isolation and characterization of carnocyclin a, a novel circular bacteriocin produced by *Carnobacterium maltaromaticum* UAL307. Appl. Environ. Microbiol. *74*, 4756–4763.

Martin-Visscher, L.A., Sprules, T., Gursky, L.J., and Vederas, J.C. (2008). Nuclear magnetic resonance solution structure of PisI, a group B immunity protein that provides protection against the type IIa bacteriocin piscicolin 126, PisA. Biochemistry *47*, 6427–6436.

Martínez-Bueno, M., Valdivia, E., Gálvez, A., Coyette, J., and Maqueda, M. (1998). Analysis of the gene cluster involved in production and immunity of the peptide antibiotic AS-48 in *Enterococcus faecalis*. Mol. Microbiol. *27*, 347–358.

Marugg, J.D., Gonzalez, C.F., Kunka, B.S., Ledeboer, A.M., Pucci, M.J., Toonen, M.Y., Walker, S.A., Zoetmulder, L.C., and Vandenbergh, P.A. (1992). Cloning, expression, and nucleotide sequence of genes involved in production of pediocin PA-1, and bacteriocin from *Pediococcus acidilactici* PAC1.0. Appl. Environ. Microbiol. *58*, 2360–2367.

Mathiesen, G., Huehne, K., Kroeckel, L., Axelsson, L., and Eijsink, V.G. (2005). Characterization of a new bacteriocin operon in sakacin P-producing *Lactobacillus sakei* showing strong translational coupling between the bacteriocin and immunity genes. Appl. Environ. Microbiol. *71*, 3565–2574.

Matijasic, B.B., Stojkovic, S., and Rogelj, I. (2006). Survival and *in vivo* adhesion of human isolates *Lactobacillus gasseri* LF221 and K7 in weaned piglets and their effects on coliforms, clostridia and lactobacilli viable counts in faeces and mucosa. J. Dairy Res. 73, 417–422.

Messens, W., Verluyten, J., Leroy, F., and De Vuyst, L. (2003). Modelling growth and bacteriocin production by *Lactobacillus curvatus* LTH 1174 in response to temperature and pH values used for European sausage fermentation processes. Int. J. Food. Microbiol. *81*, 41–52.

Moll, G.N., van den Akker, E., Hauge, H.H., Nissen-Meyer, J., Nes, I.F., Konings, W.N., and Driessen, A.J. (1999). Complementary and overlapping selectivity of the two-peptide bacteriocins plantaricin EF and JK. J. Bacteriol. *181*, 4848–4852.

Moreno, M.R., Baert, B., Denayer, S., Cornelis, P., and De Vuyst, L. (2008). Characterization of the amylovorin locus of *Lactobacillus amylovorus* DCE 471, producer of a bacteriocin active against *Pseudomonas aeruginosa*, in combination with colistin and pyocins. FEMS Microbiol. Lett. *286*, 199–206.

Muriana, P.M., and Klaenhammer, T.R. (1991). Purification and partial characterization of lactacin F, a bacteriocin produced by *Lactobacillus acidophilus* 11088. Appl. Environ. Microbiol. *57*, 114–121.

Muriana, P.M., and Klaenhammer, T.R. (1991b). Cloning, phenotypic expression, and DNA sequence of the gene for lactacin F, an antimicrobial peptide produced by *Lactobacillus* spp. J. Bacteriol. *173*, 1779–1788.

Müller, D.M., Carrasco, M.S., Tonarelli, G.G., and Simonetta, A.C. (2009). Characterization and purification of a new bacteriocin with a broad inhibitory spectrum produced by *Lactobacillus plantarum* lp 31 strain isolated from dry-fermented sausage. J. Appl. Microbiol. *106*, 2031–2040.

Neysens, P., and De Vuyst, L. (2005). Carbon dioxide stimulates the production of amylovorin L by *Lactobacillus amylovorus* DCE 471, while enhanced aeration causes biphasic kinetics of growth and bacteriocin production. Int. J. Food Microbiol. *105*, 191–202.

Neysens, P., Messens, W., and De Vuyst, L. (2003). Effect of sodium chloride on growth and bacteriocin production by *Lactobacillus amylovorus* DCE 471. Int. J. Food Microbiol. *88*, 29–39.

Nieto Lozano, J.C., Meyer, J.N., Sletten, K., Peláz, C., and Nes, I.F. (1992). Purification and amino acid sequence of a bacteriocin produced by *Pediococcus acidilactici*. J. Gen. Microbiol. *138*, 1985–1990.

Nissen-Meyer, J., Larsen, A.G., Sletten, K., Daeschel, M., and Nes, I.F. (1993). Purification and characterization of plantaricin A, a *Lactobacillus plantarum* bacteriocin whose activity depends on the action of two peptides. J. Gen. Microbiol. *139*, 1973–1978.

Oppegård, C., Schmidt, J., Kristiansen, P.E., and Nissen-Meyer, J. (2008). Mutational analysis of putative helix–helix interacting GxxxG-motifs and tryptophan residues in the two-peptide bacteriocin lactococcin G. Biochemistry *47*, 5242–5249.

van Reenen, C.A., Dicks, L.M., and Chikindas, M.L. (1998). Isolation, purification and partial characterization of plantaricin 423, a bacteriocin produced by *Lactobacillus plantarum*. J. Appl. Microbiol. *84*, 1131–1137.

van Reenen, C.A., Chikindas, M.L., Van Zyl, W.H., and Dicks, L.M. (2003). Characterization andheterologous expression of a class IIa bacteriocin, plantaricin 423 from *Lactobacillus plantarum* 423, in *Saccharomyces cerevisiae*. Int. J. Food Microbiol. *81*, 29–40.

van Reenen, C.A., Van Zyl, W.H., and Dicks, L.M. (2006). Expression of the immunity protein of plantaricin 423, produced by *Lactobacillus plantarum* 423, and analysis of the plasmid encoding the bacteriocin. Appl. Environ. Microbiol. *72*, 7644–7651.

Reuter, G. (2001). The *Lactobacillus* and *Bifidobacterium* microflora of the human intestine: composition and succession. Curr. Issues Intest. Microbiol. *2*, 43–53.

Risøen, P.A., Håvarstein, L.S., Diep, D.B., and Nes, I.F. (1998). Identification of the DNA-binding sites for two response regulators involved in control of bacteriocin synthesis in *Lactobacillus plantarum* C11. Mol. Gen. Genet. *259*, 224–232.

Rogelj, I., Bogovic Matijasić, B., Canzek Majhenic, A., and Stojković, S. (2002). The survival and persistence of *Lactobacillus acidophilus* LF221 in different ecosystems. Int. J. Food Microbiol. *76*, 83–91.

Ryan, M.P., Jack, R.W., Josten, M., Sahl, H.G., Jung, G., Ross, R.P., and Hill, C. (1999). Extensive post-translational modification, including serine to D-alanine conversion, in the two-component lantibiotic, lacticin 3147. J. Biol. Chem. *274*, 37544–37550.

Samyn, B., Martínez-Bueno, M., Devreese, B., Maqueda, M., Gálvez, A., Valdivia, E., Coyette, J., and Van Beeumen, J. (1994). The cyclic structure of the enterococal peptide antibiotic AS-48. FEBS Lett. *352*, 87–90.

Sand, S.L., Haug, T.M., Nissen-Meyer, J., and Sand, O. (2007). The bacterial peptide pheromone plantaricin A permeabilizes cancerous, but not normal, rat pituitary cells and differentiates between the outer and inner membrane leaflet. J. Membr. Biol. *216*, 61–71.

Sawa, N., Zendo, T., Kiyofuji, J., Fujita, K., Himeno, K., Nakayama, J., and Sonomoto, K. (2009). Identification and characterization of lactocyclicin Q, a novel cyclic bacteriocin produced by *Lactococcus* sp. strain QU 12. Appl. Environ. Microbiol. *75*, 1552–1558.

Senes, A., Gerstein, M., and Engelman, D.M. (2000). Statistical analysis of amino acid patterns in transmembrane helices: the GxxxG motif occurs frequently and in association with beta-branched residues at neighboring positions. J. Mol. Biol. *296*, 921–936.

Stephens, S.K., Floriano, B., Cathcart, D.P., Bayley, S.A., Witt, V.F., Jiménez-Díaz, R., Warner, P.J., and Ruiz-Barba, J.L. (1998). Molecular analysis of the locus responsible for production of plantaricin S, a two-peptide bacteriocin produced by *Lactobacillus plantarum* LPCO10. Appl. Environ. Microbiol. *64*, 1871–1877.

Tichaczek, P.S., Vogel, R.F., and Hammes, W.P. (1993). Cloning and sequencing of curA encoding curvacin A, the bacteriocin produced by *Lactobacillus curvatus* LTH1174. Arch. Microbiol. *160*, 279–283.

Tichaczek, P.S., Vogel, R.F., and Hammes, W.P. (1994). Cloning and sequencing of sakP encoding sakacin P, the bacteriocin produced by *Lactobacillus sake* LTH 673. Microbiology *140*, 361–367.

Tiwari, S.K., and Srivastava, S. (2008). Purification and characterization of plantaricin LR14: a novel bacteriocin produced by *Lactobacillus plantarum* LR/14. Appl. Microbiol. Biotechnol. *79*, 759–767.

Toba, T., Yoshioka, E., and Itoh, T. (1991). Potential of *Lactobacillus gasseri* isolated from infant faeces to produce bacteriocin. Lett. Appl. Microbiol. *12*, 228–231.

Todorov, S.D., and Dicks, L.M. (2006). Parameters affecting the adsorption of plantaricin 423, a bacteriocin produced by *Lactobacillus plantarum* 423 isolated from sorghum beer. Biotechnol. J. *1*, 405–409.

Turner, D.L., Brennan, L., Meyer, H.E., Lohaus, C., Siethoff, C., Costa, H.S., Gonzalez, B., Santos, H., and Suárez, J.E. (1999). Solution structure of plantaricin C, a novel lantibiotic. Eur. J. Biochem. *264*, 833–839.

Uteng, M., Hauge, H.H., Markwick, P.R., Fimland, G., Mantzilas, D., and Nissen-Meyer, J. (2003). Muhle-Goll C. Three-dimensional structure in lipid micelles of the pediocin-like antimicrobial peptide sakacin P and a sakacin P variant that is structurally stabilized by an inserted C-terminal disulfide bridge. Biochemistry *42*, 11417–11426.

Verluyten, J., Messens, W., and de Vuyst, L. (2004a). Sodium chloride reduces production of curvacin A, a bacteriocin produced by *Lactobacillus curvatus* strain LTH 1174, originating from fermented sausage. Appl. Environ. Microbiol. *70*, 2271–2278.

Verluyten, J., Leroy, F., and de Vuyst, L. (2004b). Influence of complex nutrient source on growth of and curvacin a production by sausage isolate *Lactobacillus curvatus* LTH 1174. Appl. Environ. Microbiol. *70*, 5081–5088.

Verluyten, J., Leroy, F., and de Vuyst, L. (2004c). Effects of different spices used in production of fermented sausages on growth of and curvacin A production by *Lactobacillus curvatus* LTH 1174. Appl. Environ. Microbiol. *70*, 4807–4813.

de Vuyst, L., Avonts, L., Neysens, P., Hoste, B., Vancanneyt, M., Swings, J., and Callewaert, R. (2004). The lactobin A and amylovorin L471 encoding genes are identical, and their distribution seems to be restricted to the species *Lactobacillus amylovorus* that is of interest for cereal fermentations. Int. J. Food Microbiol. *90*, 93–106.

Wiedemann, I., Böttiger, T., Bonelli, R.R., Schneider, T., Sahl, H.-G., and Martínez, B. (2006). Lipid II-based antimicrobial activity of the lantibiotic plantaricin C. Appl. Environ. Microbiol. *72*, 2809–2814.

Wirawan, R.E., Swanson, K.M., Kleffmann, T., Jack R.W., and Tagg, J.R. (2007). Uberolysin: a novel cyclic bacteriocin produced by *Streptococcus uberis*. Microbiology *153*, 1619–30.

Zhao, H., Sood, R., Jutila, A., Bose, S., Fimland, G., Nissen-Meyer, J., and Kinnunen, P.K. (2006). Interaction of the antimicrobial peptide pheromone Plantaricin A with model membranes: implications for a novel mechanism of action. Biochim. Biophys. Acta *1758*, 1461–1474.

Other Bacteriocins

12

Takeshi Zendo, Kenji Sonomoto, Yasushi Kawai and Tadao Saito*

Abstract

Many lactococcal and lactobacilli strains produce various available bacteriocins such as nisin, the representative bacteriocin in lactic acid bacteria (LAB), lacticin 3147, and plantaricins as mentioned in Chapter 10 and 11. Recently, pediocin PA-1/AcH (class IIa, well studied) from some *Pediococcus* strains and enterocin AS-48 (the first class IIc) from *Enterococcus faecalis* have been most likely candidates to be used for biopreservatives as the second bacteriocin next to nisin. In this chapter, the research advances of both bacteriocins for the antibacterial spectra, mode of action, and food application in addition to the chemical structures and genes for biosynthesis and immunity, are described.

Pediocin-like bacteriocins

Pediocin PA-1/AcH is one of the representative bacteriocins of lactic acid bacteria (LAB). Pediocin PA-1/AcH and their homologues are characterized in conserved sequence of YGNGVXC in their N-termini and categorized in class IIa of LAB bacteriocins, also called pediocin-like or pediocin-type bacteriocins (Klaenhammer, 1993; Cotter *et al.*, 2005). Since class IIa bacteriocins have strong activity against *Listeria monocytogenes*, which causes serious food poisoning, they are expected to be used as food preservatives. Until now, various class IIa bacteriocins have been discovered and studied on their properties and applications (Ennahar *et al.*, 2000; Drider *et al.*, 2006).

The most representative, pediocin PA-1 and pediocin AcH were studied separately by different groups, and their same structure was determined in the same year, 1992 (Henderson *et al.*, 1992; Motlagh *et al.*, 1992). We call this bacteriocin pediocin PA-1/AcH in this section. Generally, a certain bacteriocin is produced by strains belonging to a specific species. However, pediocin PA-1/AcH production has been observed not only in the original species, *Pediococcus acidilactici*, but also *P. pentosaceus* and *P. parvulus* (Bennik *et al.*, 1997; Miller *et al.*, 2005). In addition, *Lactobacillus plantarum* was also found to produce pediocin PA-1/AcH (Ennahar *et al.*, 1996). This is the very rare case that a certain bacteriocin is produced by strains belonging to different genera and species.

Class IIa bacteriocins is widely produced by LAB and related species. Until now, the production has been confirmed in LAB genera of *Pediococcus, Enterococcus, Lactobacillus, Carnobacterium, Leuconostoc,* and *Lactococcus,* and in non-LAB genera of *Bifidobacterium, Bacillus,* and *Listeria* (Table 12.1). On the other hand, class IIa bacteriocins generally show high activity against *Listeria, Enterococcus, Lactobacillus, Leuconostoc,* and some other Gram-positive bacteria (Ennahar *et al.*, 2000; Drider *et al.*, 2006). Whereas the antibacterial spectrum of pediocin PA-1/AcH is narrower than nisin, pediocin PA-1/AcH shows higher activity against *L. monocytogenes* than nisin, which accelerates the trials of its application as food preservatives.

The sizes of class IIa bacteriocins range from 36 to 48 in amino acids, and their molecular masses are generally in the range between 4000 and 5000. Their primary structures and alignment are shown in Fig. 12.1. They consist of highly conserved

*corresponding author

Table 12.1 Pediocin-like bacteriocin-producing species and their bacteriocins

Species	Bacteriocin	Mass (Da) (amino acids)	Reference
Pediococcus acidilactici	Pediocin PA-1/AcH	4624 (44)	Henderson *et al.* (1992), Motlagh *et al.* (1992
Pediococcus pentosaceus	Pediocin PA-1/AcH	4624 (44)	Miller *et al.* (2005)
Pediococcus parvulus	Pediocin PA-1/AcH	4624 (44)	Bennik *et al.* (1997)
Enterococcus faecium	Enterocin A	4829 (47)	Aymerich *et al.* (1996)
Enterococcus faecalis	Enterocin SE-K4	5356 (48)	Eguchi *et al.* (2001)
Enterococcus mundtii	Mundticin	4287 (43)	Bennik *et al.* (1998)
Lactobacillus curvatus	Curvacin A	4306 (41)	Tichaczek *et al.* (1993)
Lactobacillus sakei	Sakacin P	4434 (43)	Tichaczek *et al.* (1994)
Lactobacillus plantarum	Pediocin PA-1/AcH	4624 (44)	Ennahar *et al.* (1996)
Lactobacillus plantarum	Plantaricin 423	3931 (37)	Van Reenen *et al.* (2003)
Carnobacterium piscicola	Carnobacteriocin B2	4967 (48)	Quadri *et al.* (1994)
Carnobacterium divergens	Divercin V41	4509 (43)	Métivier *et al.* (1998)
Leuconostoc gelidum	Leucocin A	3390 (37)	Hastings *et al.* (1991)
Leuconostoc mesenteroides	Mesentericin Y105	3868 (37)	Fremaux *et al.* (1995)
Lactococcus lactis	Lactococcin MMFII	4142 (37)	Ferchichi *et al.* (2001)
Bacillus coagulans	Coagulin	4612 (44)	Le Marrec *et al.* (2000)
Bifidobacterium bifidum	Bifidocin B	3800 (36)	Yildirim *et al.* (1999)
Listeria innocua	Listeriocin 743A	4484 (43)	Kalmokoff *et al.* (2001)

```
Enterocin A           TTHSGKYYGNGVYCTKNKCTVDWAKATTCIAGMSIGGFLGGAIPGKC
Divercin V41              TKYYGNGVYCNSKKCWVDWGQASGCIGQTVVGGWLGGAIPGKC
Pediocin PA-1/AcH          KYYGNGVTCGKHSCSVDWGKATTCIINNGAMAWATGGHQGNHKC
Coagulin                   KYYGNGVTCGKHSCSVDWGKATTCIINNGAMAWATGGHQGTHKC
Bifidocin B                KYYGNGVTCGLHDCRVDRGKATCGIINNGGMWGDIG
Mundticin                  KYYGNGVSCNKKGCSVDWGKAIGIIGNNSAANLATGGAAGWKS
Listeriocin 743A           KSYGNGVQCNKKKCWVDWGSAISTIGNNSAANWATGGAAGWKS
Sakacin P                  KYYGNGVHCGKHSCTVDWGTAIGNIGNNAAANWATGGNAGWNK
Leucocin A                 KYYGNGVHCTKSGCSVNWGEAFSAGVHRLANGGNGFW
Mesentericin Y105          KYYGNGVHCTKSGCSVNWGEAASAGIHRLANGGNGFW
Plantaricin 423            KYYGNGVTCGKHSCSVNWGQAFSCSVSHLANFGHGKC
Lactococcin MMFII          TSYGNGVHCNKSKCWIDVSELETYKAGTVSNPKDILW
Enterocin SE-K4           ATYYGNGVYCNKQKCWVDWSRARSEIIDRGVKAYVNGFTKVLGGIGGR
Carnobacteriocin B2        VNYGNGVSCSKTKCSVNWGQAFQERYTAGINSFVSGVASGAGSIGRRP
Curvacin A                ARSYGNGVYCNNKKCWVNRGEATQSIIGGMISGWASGLAGM

Consensus                    YGNGV C   C
```

Figure 12.1 Sequence alignment of pediocin-like bacteriocins. N-terminal conserved amino acid residues are shown in bold letters.

N-terminal halves and variable C-terminal halves. In their N-terminal parts, the consensus sequence, YGNGVxCxxxC, is observed. Cystein residues in the consensus sequence form a disulfide bridge. Some of class IIa bacteriocins including pediocin PA-1/AcH possess another disulfide bridge in the C-terminal part. Based on the variation of the C-terminal parts, some researchers suggested further grouping on the class IIa bacteriocins (Fimland *et al.*, 2005). These disulfide bridges are important to keep their structures and activity. Class IIa bacteriocins with a C-terminal additional disulfide bridge often show higher activity than those with a single disulfide bridge (Eijsink *et al.*, 1998). The three-dimensional (3D) structures of some class IIa bacteriocins have been elucidated by nuclear magnetic resonance (NMR), which contain an N-terminal β-sheet domain and a C-terminal hairpin domain consisting of a central amphiphilic α-helix and a C-terminal tail (Fimland *et al.*, 2005; Drider *et al.*, 2006).

Biosynthesis of pediocin PA-1/AcH involves four genes, *pedA* (prepediocin), *pedB* (immunity), *pedC* (accessory protein), and *pedD* (ABC transporter) (Marrugg *et al.*, 1992). Membrane proteins, PedD and PedC, cleave an 18-amino-acid N-terminal leader peptide and secret mature pediocin PA-1/AcH. PedB confers immunity that protects producer cells against pediocin PA-1/AcH produced by themselves. Biosynthesis of other class IIa bacteriocins also involves similar genes and proteins (Ennahar *et al.*, 2000). For some other class IIa bacteriocins, additional regulatory genes are also involved, which control bacteriocin production under the three component regulatory system (Ennahar *et al.*, 2000; Eijsink *et al.*, 2002; Drider *et al.*, 2006).

Pediocin PA-1/AcH exerts the activity by leakage of ions, amino acids and other low-molecular-weight compounds from sensitive cells (Bhunia *et al.*, 1991; Chikindas *et al.*, 1993). Recently, the IIC subunit of mannose permease (EII_t^{Man}) of sensitive strains was found to be a receptor for class IIa bacteriocins (Ramnath *et al.*, 2004). Furthermore, the immunity proteins were proved to form complex with mannose permease subunits (Diep *et al.*, 2007).

A foodborne pathogen, *L. monocytogenes* can grow even under refrigerating temperatures, which makes it difficult to control *L. monocytogenes*. Since their high activity against *L. monocytogenes*, pediocin PA-1/AcH and class IIa bacteriocins are expected as food preservatives. Their producing strains are also expected as antimicrobial starter cultures that can produce bacteriocins *in situ*. For example, a pediocin PA-1/AcH-producing strain, *L. plantarum* WHE92, is widely utilized as a starter culture (Rodríguez *et al.*, 2002). Some examples of their applications are listed in Table 12.2. Although it is effective even when used alone, some studies proved that pediocin PA-1/AcH is more effective in combination with some other factors (Cleveland *et al.*, 2001). The pediocin PA-1/AcH-containing fermentate, ALTA™

Table 12.2 Applications of pediocin PA-1/AcH and pediocin-like bacteriocins

Bacteriocin	Application and target organisms	Reference
Pediocin PA-1/AcH	A pediocin-producing *Lactobacillus plantarum* strain inhibited *Listeria monocytogenes* in cheese surface	Loessner *et al.* (2003)
Pediocin PA-1/AcH	A pediocin-producing *Pediococcus pentosaceus* inhibited *Salmonella anatum* synergistically with nitrite and garlic during sausage fermentation	Swetwiwathana *et al.* (2007)
Pediocin PA-1/AcH	High pressure and high temperature synergistically reduced *L. monocytogenes*, *Staphylococcus aureus*, *Escherichia coli* O157:H7, and some other bacteria	Kalchayanand *et al.* (1998)
Muntdicin	A mundticin-producing *E. mundtii* inhibited *L. monocytogenes* in refrigerated, modified atmosphere stored mungbean sprouts.	Bennik *et al.* (1999)
Divercin V41	A divercin-producing *Carnobacterium divergens* and crude divercin prevented *L. monocytogenes* in vacuum-packed cold smoke salmon	Duffes *et al.* (1999)
Sakacin P	Sakacin P was produced *in situ*, and sakacin P and a sakacin-producing strain inhibited *L. monocytogenes* in chicken cold cuts	Katla *et al.* (2002)
Leucocin A	A leucocin-producing *Leuconostoc gelidum* delayed spoilage by sulfide-producing *L. sake* in chill-stored, vacuum-packaged beef	Leisner *et al.* (1996)
Enterocin A	A food grade lactic acid bacteria, *Lactococcus lactis*, heterologously producing enterocin A inhibited *L. monocytogenes* in cottage cheese	Liu *et al.* (2008)

2341, has already been commercially exploited (Rodríguez *et al.*, 2002). This commercial food gradient is reported to extend the shelf-life of a variety of foods, especially ready-to-eat meat products.

Enterocin AS-48

Enterocin AS-48 produced by *Enterococcus faecalis* subsp. *liquefaciens* strain S-48 is the first class IIc circular bacteriocin described (Gálvez *et al.*, 1986; Samyn *et al.*, 1994); has a broad-spectrum bactericidal activity against Gram-positive food-spoilage and pathogenic bacteria such as *Bacillus* subsp., *Clostridium* subsp., *Brochothrix thermosphacta*, *Staphylococcus aureus* and *Listeria monocytogenes* (Gálvez *et al.*, 1989a; Mendoza *et al.*, 1999; Abriouel *et al.*, 2002); and also against some Gram-negative bacterial species such as *Salmonella choleraesuis* and *Escherichia coli* (Gálvez *et al.*, 1989b; Abriouel *et al.*, 1998). The growth of enterotoxigenic *Staph. aureus* CECT 976 and Gram-negative bacteria such as *Sal. enterica* CECT915 and *E. coli* O157 was inhibited using a combination of heat treatment, organic acids, nitrites, chelating agents, phenolic compounds, high-pressure treatment, or high-intensity pulsed-electric field treatment (Ananou *et al.*, 2004; Rodriguez *et al.*, 2005; Ananou *et al.*, 2005; Grande et. al., 2007a; Martínez Viedma *et al.*, 2008). Enterocin AS-48 is produced by several other *Enterococcus* strains such as *Ent. faecalis* INIA4 (Enterocin 4), strain EFS2 (enterococcin EFS2), an unnamed bacteria (bacteriocin 21) and *Ent. faecium* 7C5 (Joosten *et al.*, 1996; Maisnier-Patin *et al.*, 1996; Tomita *et al.*, 1997; Folli *et al.*, 2003); except for AS-48RJ from *Ent. faecium* RJ16, an AS-48 variant replaces the Glu residue at position 20 of AS-48 to Val making it inactive (Abriouel *et al.*, 2005).

The DNA sequence of the AS-48 structural gene was determined (Martínez-Bueno *et al.*, 1994) and the circular structure was determined using amino acid sequencing after protease digestion of fragments and mass-spectrometry (Samyn *et al.*, 1994). This was the first report of a class IIc circular bacteriocin. The structural gene, *as-48A*, encodes a 105-amino-acid precursor peptide having a 35-amino-acid signal peptide and a 70-amino-acid mature peptide (MW 7149). The *as-48* gene cluster is located on a conjugative, pheromone-response plasmid, pMB2 (68 kb), in *Ent. faecalis* S-48 (Martínez-Bueno *et al.*, 1990); and AS-48RJ from *Ent. faecium* 7C5 is only located on the chromosome (Abriouel *et al.*, 2005). The complete expression of AS-48 and its immunity are performed using the *as-48A, B, C, C1, D, D1, E, F, G* and *H* genes (Martínez-Bueno *et al.*, 1998; Díaz *et al.*, 2003). The putative function of each gene for AS-48 was deduced using their bioinformatics e.g. hydrophobicity and trans-membrane-spanning domains (TMSs). AS-48B is a hydrophobic membrane protein involved in transport or pore formation. Deletion analysis shows *as-48B* plays an essential role for AS-48 biogenesis; whereas *as-48C* is related to AS-48 production and resistance (Fernández *et al.*, 2008). Using computer analysis, AS-48D has the functional domain, ATP-binding cassette (ABC) transporter, and the six TMSs of AS-48C1; this suggests AS-48D works with the ABC transporter (Martínez-Bueno *et al.*, 1998). Studies using mutants, AS-48C1D showing low levels of immunity was a transporter for exporting newly synthesized AS-48 that indicates AS-48EFGH is another multi ABC transporter related to immunity functioning against its own extra-cellular bacteriocin (Díaz *et al.*, 2003) as reported for NisFEG in nisin (Stein *et al.*, 2003). No genes for a quorum-sensing mechanism were found in the *as-48* region. Using transcriptional analysis with transformants introduced in the *as-48* gene cluster, two mRNA transcripts (T1: *as-48ABC* and T2: *as-48C1DD1EFGH*) were detected. For transcript T1 (TABC), the mRNA product (TA) was very stable to degradation by endonucleases but TBC may be easily digested. The T2 transcript contains the T3 transcript of the last four genes (TEFGH) controlled by an internal promoter (Díaz *et al.*, 2003).

The interaction and the pore formation between positively charged bacteriocins and the negatively charged cytoplasmic membranes have been well studied as described above (Chapters 10 and 11). Enterocin AS-48 after interacting with the cell membrane of susceptible bacteria causes ion efflux in a voltage-independent manner with a collapse of the membrane potential that leads to cell death. The three-dimensional solution structure using 1H NMR shows AS-48 (70-amino acid residues) has five α-helical regions: each domain, 9–21 (α1), 25–34 (α2), 37–45 (α3), 51–62 (α4) and 64–5 (α5), may be involved in the structural stability and structure–function relationships (Langdon *et al.*, 1998; González *et al.*, 2000; Cobos *et al.*, 2001). The positive charged cluster in the side chains in helix α4 residues and in the turn linking helix α4 to helix α5 are candidates to perform antibacterial activity by promoting pore formation in the target cell membranes. X-ray diffraction analysis found AS-48 formed an oligomeric structure that transited from the water-soluble DF-I to the membrane-bound DF-II at the membrane surface where it exerts its antibacterial activity in the membrane (Sánchez-Barrena *et al.*, 2003). A 21 amino acid residue AS-48 fragment (a putative high positively charged membrane-interacting region) had no bacteriocin activity and formed a native-like helical hairpin conformation similar to the intact native bacteriocin. This suggests the other hydrophobic region(s) of AS-48 were needed to present the full activity for AS-48 causing membrane disruption (Jiménez *et al.*, 2005). The helix contents of four AS-48 variants were examined where each glutamic residue was replaced with alanine similar to the native molecule when examining CD spectra; three of the four variants showed weaker activity indicating the negative charge of Glu residues played an essential role in AS-48 target-cell specificity and bactericidal activity (Sánchez-Hidalgo *et al.*, 2008). Three AS-48 fragments, AS10/11 (70 amino acid residues) carrying a single nick, AS43–27 (55 residues) and AS42,43–10 (38 residues) were obtained using limited proteolysis of AS-48 (Fig. 12.2). The far-UV CD spectra shows AS10/11, AS43–27 and AS-48 have an α-helical conformation; and AS42,43–10 has a lower helical content suggesting an unfolded conformation. Although the three peptides of AS-48 fragments were active against the sensitive strains, the stability to thermal denaturation during freezing and thawing was at least 300 times less than that of the native AS-48. These findings indicate circularization to stabilize the native structure is not required for the bactericidal activity; and a minimal AS-48 domain retaining the activity exist (Montalbán-López *et al.*, 2008) and are very important to study many seamless peptides in addition to the class IIc bacteriocins. There are over 100 naturally occurring circular proteins and peptides found in bacteria, plants and mammals; and not in human. Usually the circular structure may be involved to stability against pH, heating and proteolytic degradation. However, little is known about the mechanisms of secretion and circularization or the involved enzymes and their auxiliary proteins (Craik, 2006). The class IIc bacteriocins such as AS-48 and gassericin A are translated by only one linear precursor peptide after transcription; and

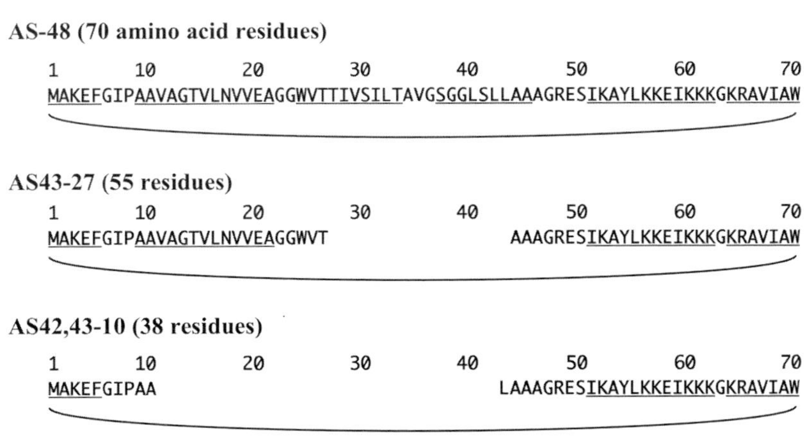

Figure 12.2 The primary structure of enterocin AS-48 (70 amino acid residues) and two fragments of AS43–27 (55 residues) and AS42,43–10 (38 residues) obtained after limited proteolysis (Montalbán-López *et al.*, 2008). The underlines show α-helical regions determined using 1H NMR analysis (Langdon *et al.*, 1998; González *et al.*, 2000) except for the putative α-helical regions of AS43–27 and AS42,43–10.

should be good candidates for analysis of these seamless peptides.

Enterocin AS-48 acting against many Gram-positive and some Gram-negative bacteria is a candidate as a natural food preservative and antimicrobial agent. There have been many reports concerning the effects and available application of AS-48 as a potential biopreservative as described below: the efficient control of the psychrotrophic and toxigenic strain *B. cereus* LWL1 in milk and in a nonfat hard cow's cheese by AS-48 producer strain *Ent. faecalis* A-48–32. (Muñoz *et al.*, 2005); the control of *Alicyclobacillus acidoterrestris*, a spoilage-causing bacterium of fruit juices, using AS-48 in natural orange and apple juices (Grande *et al.*, 2005a); the control of *List. monocytogenes* CECT 4032 in sausage using semi-purified AS-48 and adding the enterocin AS-48 producer strains *Ent. faecalis* A-48–32 and *Ent. faecium* S-32–81 (Ananou *et al.*, 2005); study of the stability of enterocin AS-48 in fruit (orange, apple, grapefruit, pear, peach, pineapple, and kiwi) and vegetable (cabbage, cauliflower, lettuce, green beans, celery, and avocado) juices (Grande *et al.*, 2005b); the effect of immersion solutions containing AS-48 alone, or plus potassium permanganate or in combination with acetic acid, citric acid, sodium propionate, and potassium sorbate on survival and proliferation of *List. monocytogenes* CECT 4032 inoculated on fresh alfalfa sprouts, soybean sprouts, and green asparagus (Molinos *et al.*, 2005); the inhibition of enterotoxin production and inactivation of endospores in toxicogenic psychrotrophic *B. cereus* LWL1 in boiled rice and in a commercial infant rice-based gruel dissolved in whole milk (Grande *et al.*, 2006a); tests with heat treatment (80–95°C for 5 minutes) against *B. coagulans* CECT 12 and 561 in three vegetable canned foods such as tomato paste (pH 4.64), syrup from canned peaches (pH 3.97), and juice from canned pineapple (pH 3.65) (Lucas *et al.*, 2006); the activity of the bacteriocin in two commercial apple ciders against ropy-forming *B. licheniformis* LMG 19409 from cider (Grande *et al.*, 2006b); the antimicrobial activity alone and in combination with phenolic compounds (hydrocinnamic acid and carvacrol) against *Staph. aureus* in vegetable sauces such as napoletana, pesto, carbonara and green sauce (Grande *et al.*, 2007a); the combination effects with phenolic compounds (carvacrol, eugenol, geraniol, and hydrocinnamic acid) for biopreservation of ready-to-eat vegetable foods such as natural vegetable cream, asparagus cream, traditional soup, homemade-style traditional soup, vegetable soup, and vichyssoise against aerobic mesophilic endospore-forming bacteria such as *B. cereus* LWL1, *B. macroides*, *Paenibacillus polymyxa*, and *Pb. amylolyticus* (Grande *et al.*, 2007b); the growth inhibition of exopolysaccharide producing LAB strains of *Lb. dioliovorans*, *Pediococcus parvulus* and 3-hydroxypropionaldehyde (3-HPA)-producing *Lb. collinoides* in a fresh-made apple juice, a commercial Basque apple cider, and a commercial Asturian apple cider (Martínez-Viedma *et al.*, 2008); the synergistic antimicrobial effect with polyphosphoric and decontamination of soybean sprouts against Gram-negative bacteria such as *Sal. enterica*, *E. coli* O157:H7, *Shigella* spp., *Enterobacter aerogenes*, *Yersinia enterocolitica*, *Aeromonas hydrophila* and *Pseudomonas fluorescens* (Cobo Molinos *et al.*, 2008a); the combination effects with other antimicrobials such as sodium hypochlorite, peracetic acid or hexadecylpyridinium chloride for decontamination of *B. cereus* and *B. weihenstephanensis* on raw vegetables such as alfalfa, soybean sprouts and green asparagus (Cobo Molinos *et al.*, 2008b); the effect of AS-48 and in combination with carvacrol and *n*-propyl *p*-hydroxybenzoate against *List. monocytogenes* CECT 4032 in fruits and fruit juice such as strawberries, raspberries, blackberries, melon, watermelon, pear, and kiwi (Molinos *et al.*, 2008); the inhibition of two *Geobacillus stearothermophilus* strains and its endospores in canned foods (maize and peas) stored at 45°C for 30 days, and in coconut milk (Viedma *et al.*, 2009a); and the effect as a preservative in acidic sport and energy drinks adjusted to pH 5.0 against *List. monocytogenes*, *Staph. aureus*, *B. cereus*, and *B. licheniformis* (Viedma *et al.*, 2009b). For the control of *Staph. aureus* in milk, the combined effect of AS-48 and/or nisin with the use of high-intensity pulsed-electric field (HIPEF) was reported; and the addition of AS-48 or AS-48 and nisin was more effective before than after HIPEF treatment (Sobrino-Lopez *et al.*, 2009). These reports indicate AS-48 should be the second bacteriocin next to nisin in addition to pediocin-like bacteriocins (class IIa) to be used as a biopreservative.

References

Abriouel, H., Valdivia, E., Gálvez, A., and Maqueda, M. (1998). Response of *Salmonella choleraesuis* LT2 spheroplasts and permeabilized cells to the bacteriocin AS-48. Appl. Environ. Microbiol. *64*, 4623–4626.

Abriouel, H., Maqueda, M., Gálvez, A., Martínez-Bueno, M., and Valdivia, E. (2002). Inhibition of bacterial growth, enterotoxin production, and spore outgrowth in strains of *Bacillus cereus* by bacteriocin AS-48. Appl. Environ. Microbiol. *68*, 1473–1477.

Abriouel, H., Lucas, R., Ben Omar, N., Valdivia, E., Maqueda, M., Martínez-Cañamero, M., and Gálvez, A. (2005). Enterocin AS-48RJ: a variant of enterocin AS-48 chromosomally encoded by *Enterococcus faecium* RJ16 isolated from food. Syst. Appl. Microbiol. *28*, 383–397.

Ananou, S., Valdivia, E., Martínez-Bueno, M., Gálvez, A., and Maqueda, M. (2004). Effect of combined physico-chemical preservatives on enterocin AS-48 activity against the enterotoxigenic *Staphylococcus aureus* CECT 976 strain. J. Appl. Microbiol. *97*, 48–56.

Ananou, S., Garriga, M., Hugas, M., Maqueda, M., Martínez-Bueno, M., Gálvez, A., and Valdivia, E. (2005a). Control of *Listeria monocytogenes* in model sausages by enterocin AS-48. Int. J. Food Microbiol. *103*, 179–190.

Ananou, S., Gálvez, A., Martínez-Bueno, M., Maqueda, M., and Valdivia, E. (2005b). Synergistic effect on enterocin AS-48 in combination with outer membrane permeabilizing treatments against *Escherichia coli* O157: H7. J. Appl. Microbiol. *99*, 1364–1372.

Aymerich, T., Holo, H., Håvarstein, L.S., Hugas, M., Garriga, M., and Nes, I.F. (1996). Biochemical and genetic characterization of enterocin A from *Enterococcus faecium*, a new antilisterial bacteriocin in the pediocin family of bacteriocins. Appl. Environ. Microbiol. *62*, 1676–1682.

Bennik, M., Smid, E.J., and Gorris, L. (1997). Vegetable-associated *Pediococcus parvulus* produces pediocin PA-1. Appl. Environ. Microbiol. *63*, 2074–2076.

Bennik, M.H., Vanloo, B., Brasseur, R., Gorris, L.G., and Smid, E.J. (1998). A novel bacteriocin with a YGNGV motif from vegetable-associated *Enterococcus mundtii*: full characterization and interaction with target organisms. Biochim. Biophys. Acta. *1373*, 47–58.

Bennik, M.H., van Overbeek, W., Smid, E.J., and Gorris, L.G. (1999). Biopreservation in modified atmosphere stored mungbean sprouts: the use of vegetable-associated bacteriocinogenic lactic acid bacteria to control the growth of *Listeria monocytogenes*. Lett. Appl. Microbiol. *28*, 226–232.

Bhunia, A.K., Johnson, M.C., Ray, B., and Kalchayanand, N. (1991). Mode of action of pediocin AcH from *Pediococcus acidilactici* H on sensitive bacterial strains. J. Appl. Bacteriol. *70*, 25–33.

Chikindas, M.L., Garcia-Garcera, M.J., Driessen, A.J., Ledeboer, A.M., Nissen-Meyer, J., Nes, I.F., Abee, T., Konings, W.N., and Venema, G. (1993). Pediocin PA-1, a bacteriocin from *Pediococcus acidilactici* PAC1.0, forms hydrophilic pores in the cytoplasmic membrane of target cells. Appl. Environ. Microbiol. *59*, 3577–3584.

Cleveland, J., Montville, T.J., Nes, I.F., and Chikindas, M.L. (2001). Bacteriocins: safe, natural antimicrobials for food preservation. Int. J. Food Microbiol. *71*, 1–20.

Cobos, E., Filimonov, V.V., Gálvez, A., Maqueda, M., Valdivia, E., Martínez, J.C., and Mateo, P.L. (2001). AS-48: a circular protein with an extremely stable globular structure. FEBS Lett. *505*, 379–382.

Cobo Molinos, A., Abriouel, H., López, R.L., Valdivia, E., Omar, N.B., and Gálvez, A. (2008a). Combined physico-chemical treatments based on enterocin AS-48 for inactivation of Gram-negative bacteria in soybean sprouts. Food Chem. Toxicol. *46*, 2912–2921.

Cobo Molinos, A., Abriouel, H., Lucas López, R., Ben Omar, N., Valdivia, E., and Gálvez, A. (2008b). Inhibition of *Bacillus cereus* and *Bacillus weihenstephanensis* in raw vegetables by application of washing solutions containing enterocin AS-48 alone and in combination with other antimicrobials. Food Microbiol. *25*, 762–770.

Cotter, P.D., Hill, C., and Ross, R.P. (2005). Bacteriocins: developing innate immunity for food. Nat. Rev. Microbiol. *3*, 777–788.

Craik, D.J. (2006). Seamless proteins tie up their loose ends. Science *311*, 1563–1564.

Díaz, M., Valdivia, E., Martínez-Bueno, M., Fernández, M., Soler-González, A.S., Ramírez-Rodrigo, H., and Maqueda, M. (2003). Characterization of a new operon, *as-48EFGH*, from the *as-48* gene cluster involved in immunity to enterocin AS-48. Appl. Environ. Microbiol. *69*, 1229–1236.

Diep, D.B., Skaugen, M., Salehian, Z., Holo, H., and Nes, I.F. (2007). Common mechanisms of target cell recognition and immunity for class II bacteriocins. Proc. Natl. Acad. Sci. U.S.A. *104*, 2384–2389.

Drider, D., Fimland, G., Héchard, Y., McMullen, L.M., and Prévost, H. (2006). The continuing story of class IIa bacteriocins. Microbiol. Mol. Biol. Rev. *70*, 564–582.

Duffes, F., Corre, C., Leroi, F., Dousset, X., and Boyaval, P. (1999). Inhibition of *Listeria monocytogenes* by *in situ* produced and semipurified bacteriocins of *Carnobacterium* spp. on vacuum-packed, refrigerated cold-smoked salmon. J. Food Prot. *62*, 1394–1403.

Eguchi, T., Kaminaka, K., Shima, J., Kawamoto, S., Mori, K., Choi, S.H., Doi, K., Ohmomo, S., and Ogata, S. (2001). Isolation and characterization of enterocin SE-K4 produced by thermophilic enterococci, *Enterococcus faecalis* K-4. Biosci. Biotechnol. Biochem. *65*, 247–253.

Eijsink, V.G., Skeie, M., Middelhoven, P.H., Brurberg, M.B., and Nes, I.F. (1998). Comparative studies of class IIa bacteriocins of lactic acid bacteria. Appl. Environ. Microbiol. *64*, 3275–3281.

Eijsink, V.G., Axelsson, L., Diep, D.B., Håvarstein, L.S., Holo, H., and Nes, I.F. (2002). Production of class II bacteriocins by lactic acid bacteria; an example of biological warfare and communication. Antonie Van Leeuwenhoek *81*, 639–654.

Ennahar, S., Aoude-Werner, D., Sorokine, O., Van Dorsselaer, A., Bringel, F., Hubert, J.C., and Hasselmann, C. (1996). Production of pediocin AcH by *Lactobacillus plantarum* WHE 92 isolated from cheese. Appl. Environ. Microbiol. *62*, 4381–4387.

Ennahar, S., Sashihara, T., Sonomoto, K., and Ishizaki, A. (2000). Class IIa bacteriocins: biosynthesis, structure and activity. FEMS Microbiol. Rev. *24*, 85–106.

Ferchichi, M., Frère, J., Mabrouk, K., and Manai, M. (2001). Lactococcin MMFII, a novel class IIa bacteriocin produced by *Lactococcus lactis* MMFII, isolated from a Tunisian dairy product. FEMS Microbiol. Lett. *205*, 49–55.

Fernández, M., Sánchez-Hidalgo, M., García-Quintáns, N., Martínez-Bueno, M., Valdivia, E., López, P., and Maqueda, M. (2008). Processing of as-48ABC RNA in AS-48 enterocin production by *Enterococcus faecalis*. J. Bacteriol. *190*, 240–250.

Fimland, G., Johnsen, L., Dalhus, B., and Nissen-Meyer, J. (2005). Pediocin-like antimicrobial peptides (class IIa bacteriocins) and their immunity proteins: biosynthesis, structure, and mode of action. J. Peptide Sci. *11*, 688–696.

Folli, C., Ramazzina, I., Arcidiaco, P., Stoppini, M., and Berni, R. (2003). Purification of bacteriocin AS-48 from an *Enterococcus faecium* strain and analysis of the gene cluster involved in its production. FEMS Microbiol. Lett. *221*, 143–149.

Fremaux, C., Héchard, Y., and Cenatiempo, Y. (1995). Mesentericin Y105 gene clusters in *Leuconostoc mesenteroides* Y105. Microbiology. *141*, 1637–1645.

Hastings, J.W., Sailer, M., Johnson, K., Roy, K.L., Vederas, J.C., and Stiles, M.E. (1991). Characterization of leucocin A-UAL 187 and cloning of the bacteriocin gene from *Leuconostoc gelidum*. J. Bacteriol. *173*, 7491–7500.

Gálvez, A., Maqueda, M., Valdivia, E., Quesada, A., and Montoya, E. (1986). Characterization and partial purification of a broad spectrum antibiotic AS-48 produced by *Streptococcus faecalis*. Can. J. Microbiol. *32*, 765–771.

Gálvez, A., Maqueda, M., Martínez-Bueno, M., and Valdivia, E. (1989a). Bactericidal and bacteriolytic action of peptide antibiotic AS-48 against Gram-positive and Gram-negative bacteria and other organisms. Res. Microbiol. *140*, 57–68.

Gálvez, A., Valdivia, E., Martínez, M., and Maqueda, M. (1989b). Bactericidal action of peptide antibiotic AS-48 against *Escherichia coli* K-12. Can. J. Microbiol. *35*, 318–321.

González, C., Langdon, G.M., Bruix, M., Gálvez, A., Valdivia, E., Maqueda, M. and Rico, M. (2000). Bacteriocin AS-48, a cyclic polypeptide structurally and functionally related to mammalian NK-lysin. Proc. Natl. Acad. Sci. U.S.A. *97*, 11221–11226.

Grande, M.J., Lucas, R., Abriouel, H., Omar, N.B., Maqueda, M., Martínez-Bueno, M., Martínez-Cañamero, M., Valdivia, E., and Gálvez, A. (2005a). Control of *Alicyclobacillus acidoterrestris* in fruit juices by enterocin AS-48. Int. J. Food Microbiol. *104*, 289–297.

Grande, M.J., Lucas, R., Valdivia, E., Abriouel, H., Maqueda, M., Omar, N.B., Martínez-Cañamero, M., and Gálvezi, A. (2005b). Stability of enterocin AS-48 in fruit and vegetable juices. J. Food Prot. *68*, 2085–2094.

Grande, M.J., Lucas, R., Abriouel, H., Valdivia, E., Omar, N.B., Maqueda, M., Martínez-Bueno, M., Martínez-Cañamero, M., and Gálvez, A. (2006a). Inhibition of toxicogenic *Bacillus cereus* in rice-based foods by enterocin AS-48. Int. J. Food Microbiol. *106*, 185–194.

Grande, M.J., Lucas, R., Abriouel, H., Valdivia, E., Ben Omar, N., Maqueda, M., Martínez-Cañamero, M., and Gálvez, A. (2006b). Inhibition of *Bacillus licheniformis* LMG 19409 from ropy cider by enterocin AS-48. J. Appl. Microbiol. *101*, 422–428.

Grande, M.J., López, R.L., Abriouel, H., Valdivia, E., Ben Omar, N., Maqueda, M., Martínez-Cañamero, M., and Gálvez, A. (2007a). Treatment of vegetable sauces with enterocin AS-48 alone or in combination with phenolic compounds to inhibit proliferation of *Staphylococcus aureus*. J. Food Prot. *70*, 405–411.

Grande, M.J., Abriouel, H., Lucas López, R., Valdivia, E., Ben Omar, N., Martínez-Cañamero, M., and Gálvez, A. (2007b).Efficacy of enterocin AS-48 against bacilli in ready-to-eat vegetable soups and purees. J. Food Prot. *70*, 2339–2345.

Henderson, J.T., Chopko, A.L., and van Wassenaar, P.D. (1992). Purification and primary structure of pediocin PA-1 produced by *Pediococcus acidilactici* PAC-1.0. Arch. Biochem. Biophys. *295*, 5–12.

Jiménez, M.A., Barrachi-Saccilott, A.C., Valdivia, E., Maqueda, M., and Rico, M. (2005). Design, NMR characterization and activity of a 21-residue peptide fragment of bacteriocin AS-48 containing its putative membrane interacting region. J. Pept. Sci. *11*, 29–36.

Joosten, H.M., Nunez, M., Devreese, B., Van Beeumen, J., and Marugg, J.D. (1996). Purification and characterization of enterocin 4, a bacteriocin produced by *Enterococcus faecalis* INIA 4. Appl. Environ. Microbiol. *62*, 4220–4223.

Kalchayanand, N., Sikes, A., Dunne, C.P., and Ray, B. (1998). Interaction of hydrostatic pressure, time and temperature of pressurization and pediocin AcH on inactivation of foodborne bacteria. J. Food Prot. *61*, 425–431.

Kalmokoff, M.L., Banerjee, S.K., Cyr, T., Hefford, M.A., and Gleeson, T. (2001). Identification of a new plasmid-encoded sec-dependent bacteriocin produced by *Listeria innocua* 743. Appl. Environ. Microbiol. *67*, 4041–4047.

Katla, T., Møretrø, T., Sveen, I., Aasen, I.M., Axelsson, L., Rørvik, L.M., and Naterstad, K. (2002). Inhibition of *Listeria monocytogenes* in chicken cold cuts by addition of sakacin P and sakacin P-producing *Lactobacillus sakei*. J. Appl. Microbiol. 93, 191–196.

Klaenhammer, T.R. (1993). Genetics of bacteriocins produced by lactic acid bacteria. FEMS Microbiol. Rev. *12*, 39–85.

Langdon, G.M., Bruix, M., Gálvez, A., Valdivia, E., Maqueda, M., and Rico, M. (1998). Sequence-specific 1H assignment and secondary structure of the bacteriocin AS-48 cyclic peptide. J. Biomol. NMR *1*, 173–175.

Leisner, J.J., Greer, G.G., and Stiles, M.E. (1996). Control of beef spoilage by a sulfide-producing *Lactobacillus sake* strain with bacteriocinogenic *Leuconostoc gelidum* UAL187 during anaerobic storage at 2 degrees C. Appl. Environ. Microbiol. *62*, 2610–2614.

Le Marrec, C., Hyronimus, B., Bressollier, P., Verneuil, B., and Urdaci, M.C. (2000). Biochemical and genetic characterization of coagulin, a new antilisterial bacteriocin in the pediocin family of bacteriocins, produced by *Bacillus coagulans* I(4). Appl. Environ. Microbiol. *66*, 5213–5220.

Liu, L., O'Conner, P., Cotter, P.D., Hill, C., and Ross, R.P. (2008). Controlling *Listeria monocytogenes* in cottage cheese through heterologous production of enterocin A by *Lactococcus lactis*. J. Appl. Microbiol. *104*, 1059–1066.

Loessner, M., Guenther, S., Steffan, S., and Scherer, S. (2003). A pediocin-producing *Lactobacillus plantarum* strain inhibits *Listeria monocytogenes* in a multispecies cheese surface microbial ripening consortium. Appl. Environ. Microbiol. *69*, 1854–1857.

Lucas, R., Grande, M.A., Abriouel, H., Maqueda, M., Ben Omar, N., Valdivia, E., Martínez-Cañamero, M., and Gálvez, A. (2006). Application of the broad-spectrum bacteriocin enterocin AS-48 to inhibit *Bacillus coagulans* in canned fruit and vegetable foods. Food Chem. Toxicol. *44*, 1774–1781.

Maisnier-Patin, S., Forni, E., and Richard, J. (1996). Purification, partial characterization and mode of action of enterocin EFS2, an antilisterial bacteriocin produced by a strain of *Enterococcus faecalis* isolated from a cheese. Int. J. Food Microbiol. *30*, 255–270.

Martínez-Bueno, M., Gálvez, A., Valdivia, E., and Maqueda, M. (1990). A transferable plasmid associated with AS-48 production in *Enterococcus faecalis*. J. Bacteriol. *172*, 2817–2818.

Martínez–Bueno, M., Maqueda, M., Gálvez, A., Samyn, B., Van Beeumen, J., Coyette, J., and Valdivia, E. (1994). Determination of the gene sequence and molecular structure of the enterococcal peptide antibiotic AS-48. J. Bacteriol. *176*, 6334–6339.

Martínez-Bueno, M., Valdivia, E., Gálvez, A., Coyette, J., and Maqueda, M. (1998). Analysis of the gene cluster involved in production and immunity of the peptide antibiotic AS-48 in *Enterococcus faecalis*. Mol. Microbiol. *27*, 347–358.

Martínez-Viedma, P., Abriouel, H., Omar, N.B., Valdivia, E., López, R.L., and Gálvez, A. (2008). Inactivation of exopolysaccharide and 3-hydroxypropionaldehyde-producing lactic acid bacteria in apple juice and apple cider by enterocin AS-48. Food Chem. Toxicol. *46*, 1143–1151.

Martínez Viedma, P., Sobrino López, A., Ben Omar, N., Abriouel, H., Lucas López, R., Valdivia, E., Martín Belloso, O., and Gálvez, A. (2008). Enhanced bactericidal effect of enterocin AS-48 in combination with high-intensity pulsed-electric field treatment against *Salmonella enterica* in apple juice. Int. J. Food. Microbiol. *128*, 244–249.

Marugg, J.D., González, C.F., Kunka, B.S., Ledeboer, A.M., Pucci, M.J., Toonen, M.Y., Walker, S.A., Zoetmulder, L.C.M., and Vandenbergh, P.A. (1992). Cloning, expression, and nucleotide sequence of genes involved in production of pediocin PA-1, a bacteriocin from *Pediococcus acidilactici* PAC1.0. Appl. Environ. Microbiol. *58*, 2360–2367.

Mendoza, F., Maqueda, M., Gálvez, A., Martínez-Bueno, M., and Valdivia, E. (1999). Antilisterial activity of peptide AS-48 and study of changes induced in the cell envelope properties of an AS-48-adapted strain of *Listeria monocytogenes*. Appl. Environ. Microbiol. *65*, 618–625.

Métivier, A., Pilet, M.F., Dousset, X., Sorokine, O., Anglade, P., Zagorec, M., Piard, J.C., Marion, D., Cenatiempo, Y., and Fremaux, C. (1998). Divercin V41, a new bacteriocin with two disulfide bonds produced by *Carnobacterium divergens* V41: primary structure and genomic organization. Microbiology *144*, 2837–2844.

Miller, K.W., Ray, P., Steinmetz, T., Hanekamp, T., and Ray, B. (2005). Gene organization and sequences of pediocin AcH/PA-1 production operons in *Pediococcus* and *Lactobacillus* plasmids. Lett. Appl. Microbiol. *40*, 56–62.

Molinos, A.C., Abriouel, H., Ben Omar, N., Valdiviam, E., López, R.L., Maqueda, M., Cañamero, M.M., and Gálvez, A. (2005). Effect of immersion solutions containing enterocin AS-48 on *Listeria monocytogenes* in vegetable foods. Appl. Environ. Microbiol. *71*, 7781–7787.

Molinos, A.C., Abriouel, H., Ben Omar, N., Lucas, R., Valdivia, E., and Gálvez, A. (2008). Inactivation of *Listeria monocytogenes* in raw fruits by enterocin AS-48. J. Food Prot. *71*, 2460–2467.

Montalbán-López, M., Spolaore, B., Pinato, O., Martínez-Bueno, M., Valdivia, E., Maqueda, M., and Fontana, A. (2008). Characterization of linear forms of the circular enterocin AS-48 obtained by limited proteolysis. FEBS Lett. *582*, 3237–3242.

Motlagh, A.M., Bhunia, A.K., Szostek, F., Hansen, T.R., Johnson, M.C., and Ray, B. (1992). Nucleotide and amino acid sequence of *pap*-gene (pediocin AcH production) in *Pediococcus acidilactici* H. Lett. Appl. Microbiol. *15*, 45–48.

Muñoz, A., Maqueda, M., Gálvez, A., Martínez-Bueno, M., Rodríguez, A., and Valdivia, E. (2004). Biocontrol of psychrotrophic enterotoxigenic *Bacillus cereus* in a nonfat hard cheese by an enterococcal strain-producing enterocin AS-48. J. Food Prot. *67*, 1517–1521.

Quadri, L.E., Sailer, M., Roy, K.L., Vederas, J.C., and Stiles, M.E. (1994). Chemical and genetic characterization of bacteriocins produced by *Carnobacterium piscicola* LV17B. J. Biol. Chem. *269*, 12204–12211.

Ramnath, M., Arous, S., Gravesen, A., Hastings, J.W., and Héchard, Y. (2004). Expression of *mptC* of *Listeria monocytogenes* induces sensitivity to class IIa bacteriocins in *Lactococcus lactis*. Microbiology *150*, 2663–2668.

van Reenen, C.A., Chikindas, M.L., Van Zyl W.H., and Dicks, L.M. (2003). Characterization and heterologous expression of a class IIa bacteriocin, plantaricin 423 from *Lactobacillus plantarum* 423, in *Saccharomyces cerevisiae*. Int. J. Food Microbiol. *81*, 29–40.

Rodríguez, J.M., Martínez, M.I., and Kok, J. (2002). Pediocin PA-1, a wide-spectrum bacteriocin from lactic acid bacteria. Crit. Rev. Food Sci. Nutr. *42*, 91–121.

Rodriguez, E., Arques, J.L., Nuñez, M., Gaya, P., and Medina, M. (2005). Combined effect of high-pressure treatments and bacteriocin-producing lactic acid bacteria on inactivation of *Escherichia coli* O157:H7 in raw-milk cheese. Appl. Environ. Microbiol. *71*, 3399–3404.

Samyn, B., Martínez-Bueno, M., Devreese, B., Maqueda, M., Gálvez, A., Valdivia, E., Coyette, J., and Van Beeumen, J. (1994). The cyclic structure of the enterococal peptide antibiotic AS-48. FEBS Lett. *352*, 87–90.

Sánchez-Barrena, M., Martínez-Ripoll, G., Gálvez, A., Valdivia, E., Maqueda, M., Cruz, V., and Albert, A. (2003). Structure of bacteriocin AS-48: from soluble state to membrane bound state. J. Mol. Biol. *334*, 541–549.

Sánchez-Hidalgo, M., Martínez-Bueno, M., Fernández-Escamilla, A.M., Valdivia, E., Serrano, L., and Maqueda, M. (2008). Effect of replacing glutamic residues upon the biological activity and stability of the circular enterocin AS-48. J. Antimicrob. Chemother. *61*, 1256–1265.

Sobrino-Lopez, A., Viedma-Martínez, P., Abriouel, H., Valdivia, E., Gálvez, A., and Martin-Belloso, O. (2009). The effect of adding antimicrobial peptides to milk inoculated with Staphylococcus aureus and processed by high-intensity pulsed-electric field. J. Dairy Sci. *92*, 2514–2523.

Stein, T., Heinzmann, S., Solovieva, I., and Entian, K.D. (2003). Function of *Lactococcus lactis* nisin immunity genes *nisI* and *nisFEG* after coordinated expression in the surrogate host *Bacillus subtilis*. J. Biol. Chem. *278*, 89–94.

Swetwiwathana, A., Lotong, N., Nakayama, J., and Sonomoto, K. (2007). Maturation of Nahm – a Thai fermented meat product. Effect of pediocin PA-1 producer (*Pediococcus pentosaceus* TISTR 536) as starter culture, nitrite and garlic on *Salmonella anatum* during Nham fermentation. Fleischwirtschaft Int. *22*, 46–49.

Tichaczek, P.S., Vogel, R.F., and Hammes, W.P. (1993). Cloning and sequencing of *curA* encoding curvacin A, the bacteriocin produced by *Lactobacillus curvatus* LTH1174. Arch. Microbiol. *160*, 279–283.

Tichaczek, P.S., Vogel, R.F., and Hammes, W.P. (1994). Cloning and sequencing of *sakP* encoding sakacin P, the bacteriocin produced by *Lactobacillus sake* LTH 673. Microbiology. *140*, 361–367.

Tomita, H., Fujimoto, S., Tanimoto, K., and Ike, Y. (1997). Cloning and genetic and sequence analyses of the bacteriocin 21 determinant encoded on the *Enterococcus faecalis* pheromone-responsive conjugative plasmid pPD1. J. Bacteriol. *179*, 7843–7855.

Viedma, P.M., Abriouel, H., Ben Omar, N., López, R.L., Valdivia, E., and Gálvez, A. (2009a). Inactivation of *Geobacillus stearothermophilus* in canned food and coconut milk samples by addition of enterocin AS-48. Food Microbiol. *26*, 289–293.

Viedma, P.M., Abriouel, H., Ben Omar, N., López, R.L., Valdivia, E., and Gálvez, A. (2009b). Antibacterial protection by enterocin AS-48 in sport and energy drinks with less acidic pH values. J. Food Prot. *72*, 881–884.

Yildirim, Z., Winters, D.K., and Johnson, M.G. (1999). Purification, amino acid sequence and mode of action of bifidocin B produced by *Bifidobacterium bifidum* NCFB 1454. J. Appl. Microbiol. *86*, 45–54.

Bacteriocins: Remarks and Future Studies

13

Yasushi Kawai and Tadao Saito*

Abstract

Bacteriocins have been reported in a large number of bacteria and are used by these organisms in the struggle for survival and bacterial communication. The bacteriocins produced by lactic acid bacteria (LAB) have potential applications to prevent the growth of harmful bacteria in humans. The bacteriocin-producing LAB strains isolated from foods and human origins are expected to be effective probiotic candidates. In this chapter, the facts, topics and problems to overcome for LAB bacteriocins and the producing strains: heterologous expression studies, nisin resistant strains and the appearance of resistant strains, applications for foods as biopreservatives and bovine mastitis as a clinical cure, positively effects in human hosts and their microbiota, are described and discussed.

Bacteriocins

Because of the inhibitory spectrum, safety, heat resistance and pH stability, the class I and II bacteriocins produced by lactic acid bacteria (LAB) have potential applications as biopreservatives to prevent the growth of food-borne pathogens and food spoilage bacteria in food products. The bacteriocins must have high bactericidal activity against the target bacteria and cooperate with other antibiotic agents. And the bacteriocin-producing LAB strains isolated from foods especially from human origins are expected to be good probiotic candidates that may help in preventing the growth of harmful bacteria in food and the human intestine. However, there are several obstacles to overcome before we may effectively use LAB bacteriocins.

To use bacteriocins in food production as a secondary bacteriocin to nisin, a low-cost and easily mass-produced preparation of bacteriocin is needed. Using expression studies of several bacteriocins, the recovery amounts in the supernatants and the specific activity drastically decrease compared with native bacteriocins from the original strains suggesting the expressed bacteriocins may not form the native structure. This especially occurs in the expression of specific modified bacteriocins such as AS-48 and gassericin A where it is difficult to retain the full set of genes for biosynthesis. This is especially shown in heterologous expression of enterocin AS-48 that was not successful in generally recognized as safe (GRAS) LAB strains of *Lactococcus* and *Lactobacillus* except for *Enterococcus* (Fernández *et al.*, 2007); and the activity of the non-circular gassericin A (MW 5671) expressed in *Escherichia coli* decreased to 1/173 compared with the native bacteriocin (MW 5653) despite the same migration using SDS-PAGE (MW ca. 3800) with the compact globular form (Kawai *et al.*, 2003). Although selection and trials for suitable host, promoters and plasmid vectors are useful for heterologous expression, determining the appropriate cultivation conditions and food grade media using native bacteriocin producers may be more realistic in overcoming production problems.

Listeria monocytogenes grows at 4°C and sometimes causes systemic inflammatory response syndrome in infants, children, old people, pregnant women, and immuno-compromised patients. *List. monocytogenes* is a serious food spoilage bacteria especially for the meat processing industry.

*corresponding author

The control of this bacterium using bacteriocins such as nisin, lacticin 3147 and pediocin-like class IIa bacteriocins requires development. Nisin has a wide bactericidal activity and is the most representative bacteriocin used in over 50 countries for food biopreservation; but nisin is not suitable for all situations. For instance, nisin exerts activity at only pH 3.0–7.0 and is insoluble at alkaline pHs (Liu *et al.,* 1990). Ye *et al.* report sodium lactate, not nisin, was most effective when using chitosan-coated plastic films on ham steaks and cold smoked salmon to control *List. monocytogenes* with five GRAS antimicrobials (nisin, sodium lactate, sodium diacetate, potassium sorbate, and sodium benzoate) that enhanced the anti-listerial effectiveness (Ye *et al.,* 2008a,b). This indicates the need to assay the antimicrobial effect in each food material using bacteriocins.

Another major problem is the appearance of nisin (bacteriocin) resistant strains. Since Davies *et al.* (1994) reported the amount of nisin adsorbed to target cells may reflect the nisin sensitivity of the strain; and indicated the nisin resistance may be a reduction of suitable adsorption sites, there are reports of many nisin resistant strains, and the putative mechanisms. Sensitive strains have acquired nisin resistance by successive mild exposure of nisin through the expression of resistant proteins on cell wall and cytoplasmic membrane biosynthesis (Maisnier-Patin *et al.,* 1996; Davies *et al.,* 1996; Dielbandhoesing *et al.,* 1998; Gravesen *et al.,* 2001; Mantovani *et al.,* 2001; Cao *et al.,* 2004); and the composition change in membrane fatty acids and phospholipids as a physiological adaptation (Mazzotta *et al.,* 1997; Verheul *et al.,* 1997; Mazzotta *et al.,* 1999). For the nisin resistance protein (NSR), Jarvis *et al.* (1967, 1971) reported several nisin-resistant strains of *Bacillus* spp. had nisin-inactivating enzymes and one of them from *B. cereus* was identified to be a dehydroalanine reductase. The gene of another, NSRs (*nsr*), encoded a 35 kDa protein located on a 60 kb plasmid of the nisin non-producer of *Lc. lactis* subsp. *diacetylactis* DRC3 (McKay *et al.,* 1984; Froseth *et al.,* 1988, 1991); and were identified to an ORF of 957 bases located on a 56-kb plasmid of *Lc. lactis* M189 (Liu *et al.,* 1997). Further *in vitro* analysis for the former 35-kDa NSR found the NSR inactivated nisin by removing the six C-terminal amino acid residues where the cleaved nisin (residues 1–28) showed reduction in the pore-forming potency in the membrane and reduced affinity to the indicator cell. *In vivo* analysis indicated the NSR is a nisin-degrading protease located on the cell membrane in non-nisin-producing *L. lactis* (Sun *et al.,* 2009). Therefore, the appearance of bacteriocin resistant strains may be unavoidable where the careful use of nisin should be performed.

Nisin forms a pore in the cytoplasmic membrane through lipid II as the nisin receptor; and exerts antibacterial activity. However, resistance by Gram-positive bacteria to nisin is not determined by the amount of lipid II on the indicator cells (Kramer *et al.,* 2004). Whole-genome DNA micro-array analysis of *Lc. lactis* IL1403 and its high-resistance mutant indicated the factors exhibiting the nisin resistance mechanisms are complicated and four major mechanisms are suggested: preventing nisin to reach the cytoplasmic membrane; promoting the binding of nisin to the cell wall but not the cell membrane; preventing the insertion of nisin into the membrane; and transporting nisin out of the membrane (Kramer *et al.,* 2006). Nisin sensitivity to an *Ent. faecalis* deletion mutant lacking the *dltA* gene for D-alanine lipoteichoic acid was increased by the absence of D-alanine in the lipoteichoic acid similar to test mutant strains of several Gram-positive bacteria suggesting D-alanine esters of lipoteichoic acid holds a net positive charge on the bacterial cell surface to prevent the attack of cationic antimicrobial peptides such as nisin (Fabretti *et al.,* 2006; Kovács *et al.,* 2006). In *Lc. lactis,* an increase of D-alanine esters and galactose in lipoteichoic acid causes a positive charged cell wall; and a decrease in nisin binding to lipid II; the authors concluded to be a major defence mechanism against nisin (Kramer *et al.,* 2008). No similar mechanism with the immunity protein of NisI that enhances the resistance by co-operation with NisFEG as an ABC transporter (Takala *et al.,* 2006) has been found in these resistant strains.

One of the major fields for application of nisin and lacticin 3147 show several reports to cure bovine mastitis. Nisin showed similar effects as a clinical cure with the antibiotic, gentamicin; and no nisin-resistant strains were found in 17 wild

Staphylococcus aureus isolates in spite of 82.5% and 35.3% detection rates of resistance to penicillin and to gentamicin, respectively (Cao *et al.*, 2007). An application trial of nisin to the nipple and mammary areola of human women with clinical signs of staphylococcal mastitis decreased the viable cell counts of staphylococci in comparison with a control group indicating nisin is an effective antibiotic for the treatment of staphylococcal mastitis (Fernández *et al.*, 2008). There are over 100 different mastitis pathogens causing mammalian mastitis where control using bacteriocins requires further study.

Almost all of the bacteriocins discussed can be used as biopreservatives because bacteriocins from probiotics are less foreign than those from bacterial strains in foods. Although few bacteria inhabit the human stomach under low pH conditions, *Helicobacter pylori* lives in gastric mucus and is the major aetiologic agent in chronic gastritis, peptic ulcer and gastric cancer (Malaty *et al.*, 2003). Recently, five lactobacilli species of *Lb. gasseri*, *Lb. fermentum*, *Lb. vaginalis*, *Lb. reuteri* and *Lb. salivarius* have been isolated from human gastric biopsies and showed a wide resistance against acid and bile (Ryan *et al.*, 2008). However, the popular bacterial species used as bacteriocin producers are limited to *Lb. gasseri* and *Lb. salivarius* are among five species already described above. And it is unclear whether bacteriocin producers and the produced bacteriocins give good effects in human hosts and positively affect their microbiota. Few results for dynamics of bacteriocins in animal intestine are reported as the first *in vivo* study that bacteriocin (ABP-118 form *Lb. salivarius* UCC118) can protect mice against infection of food-borne pathogen *List. monocytogenes* (Corr *et al.*, 2007) was introduced in Chapter 11. There have been many obstacles to determine probiotics and determining the bacteriocins enhance human health in the host and maintain appropriate microbiota. Recently, several bacteriocins produced by food spoilage bacteria and bacterial strains isolated from untapped sources except for foods have been found, which may have beneficial probiotic and/or biotechnology properties. Although it will take several developments to be applied to foods, the approach that food spoilage bacteria should be inhibited by the bacteriocins from the same genus and related bacteria is a reasonable approach and available; and the new development to evaluate the safe use of the bacteriocins are easily and at low cost.

References

Cao, M., and Helmann, J.D. (2004). The Bacillus subtilis extracytoplasmic-function sigmaX factor regulates modification of the cell envelope and resistance to cationic antimicrobial peptides. J. Bacteriol. *186*, 1136–1146.

Cao, L.T., Wu, J.Q., Xie, F., Hu, S.H., and Mo, Y. (2007). Efficacy of nisin in treatment of clinical mastitis in lactating dairy cows. J. Dairy Sci. *90*, 3980–3985.

Davies, E.A., and Adams, M.R. (1994). Resistance of *Listeria monocytogenes* to the bacteriocin nisin. Int. J. Food Microbiol. *21*, 341–347.

Davies, E.A., Falahee, M.B., and Adams, M.R. (1996). Involvement of the cell envelope of *Listeria monocytogenes* in the acquisition of nisin resistance. J. Appl. Bacteriol. *81*, 139–146.

Dielbandhoesing, S.K., Zhang, H., Caro, L.H., van der Vaart, J.M., Klis, F.M., Verrips, C.T., and Brul, S. (1998). Specific cell wall proteins confer resistance to nisin upon yeast cells. Appl. Environ. Microbiol. *64*, 4047–4052.

Fabretti, F., Theilacker, C., Baldassarri, L., Kaczynski, Z., Kropec, A., Holst, O., and Huebner, J. (2006). Alanine esters of enterococcal lipoteichoic acid play a role in biofilm formation and resistance to antimicrobial peptides. Infect. Immun. *74*, 4164–4171.

Froseth, B.R., Herman, R..E., and McKay, L.L. (1988). Cloning of nisin resistance determinant and replication origin on 7.6-kilobase EcoRI fragment of pNP40 from *Streptococcus lactis* subsp. *diacetylactis* DRC3. Appl. Environ. Microbiol. *54*, 2136–2139.

Fernández, M., Martínez-Bueno, M., Martín, M.C., Valdivia, E., and Maqueda, M. (2007). Heterologous expression of enterocin AS-48 in several strains of lactic acid bacteria. J. Appl. Microbiol. *102*, 1350–1361.

Fernández, L., Delgado, S., Herrero, H., Maldonado, A., and Rodríguez. J.M. (2008). The bacteriocin nisin, an effective agent for the treatment of staphylococcal mastitis during lactation. J. Hum. Lact. *24*, 311–316.

Froseth, B.R., and McKay, L.L. (1991). Molecular characterization of the nisin resistance region of *Lactococcus lactis* subsp. *lactis* biovar *diacetylactis* DRC3. Appl. Environ. Microbiol. *57*, 804–811.

Gravesen, A., Sorensen, K., Aarestrup, F.M., and Knochel, S. (2001). Spontaneous nisin-resistant *Listeria monocytogenes* mutants with increased expression of a putative penicillin-binding protein and their sensitivity to various antibiotics. Microb. Drug Resist. *7*, 127–135.

Jarvis, B. (1967). Resistance to nisin and production of nisin-inactivating enzymes by several *Bacillus* species. J. Gen. Microbiol. *47*, 33–48.

Jarvis, B., and J. Farr. (1971). Partial purification, specificity and mechanism of action of the nisin-inactivating enzyme from *Bacillus cereus*. Biochim. Biophys. Acta. *227*, 232–240.

Kovács, M., Halfmann, A., Fedtke, I., Heintz, M., Peschel, A., Vollmer, W., Hakenbeck, R., and Brückner, R. (2006). A functional dlt operon, encoding proteins required for incorporation of d-alanine in teichoic acids in Gram-positive bacteria, confers resistance to cationic antimicrobial peptides in *Streptococcus pneumoniae*. J. Bacteriol. *188*, 5797–5805.

Kramer, N.E., Smid, E.J., Kok, J., de Kruijff, B., Kuipers, O.P., and Breukink, E. (2004). Resistance of Gram-positive bacteria to nisin is not determined by lipid II levels. FEMS Microbiol. Lett. *239*, 157–161.

Kramer, N.E., van Hijum, S.A., Knol, J., Kok, J., and Kuipers, O.P. (2006). Transcriptome analysis reveals mechanisms by which *Lactococcus lactis* acquires nisin resistance. Antimicrob. Agents Chemother. *50*, 1753–1761.

Kramer, N.E., Hasper, H.E., van den Bogaard, P.T., Morath, S., de Kruijff, B., Hartung, T., Smid, E.J., Breukink, E., Kok, J., and Kuipers O.P. (2008). Increased D-alanylation of lipoteichoic acid and a thickened septum are main determinants in the nisin resistance mechanism of *Lactococcus lactis*. Microbiology *154*, 1755–1762.

Liu, W., and Hansen, J.N. (1990). Some chemical and physical properties of nisin, a small-protein antibiotic produced by *Lactococcus lactis*. Appl. Environ. Microbiol. *56*, 2551–2558.

Liu, C.Q., Harvey, M.L., and Dunn, N.W. (1997). Cloning of a gene encoding nisin resistance from *Lactococcus lactis* subsp. *lactis* M189 which is transcribed from an extended 210 promoter. J. Gen. Appl. Microbiol. *43*, 67–73.

Maisnier-Patin, S., and Richard, J. (1996). Cell wall changes in nisin-resistant variants of *Listeria innocua* grown in the presence of high nisin concentrations. FEMS Microbiol. Lett. *140*, 29–35.

Malaty, H.M. and Nyrén, O. (2003). Epidemiology of *Helicobacter pylori* infection. Helicobacter *8*, 8–12.

Mantovani, H.C., and Russell, J.B. (2001). Nisin resistance of *Streptococcus bovis*. Appl. Environ. Microbiol. *67*, 808–813.

Mazzotta, A.S., and Montville, T.J. (1997). Nisin induces changes in membrane fatty acid composition of *Listeria monocytogenes* nisin-resistant strains at 10 degrees C and 30 degrees C. J. Appl. Microbiol. *82*, 32–38.

Mazzotta, A.S., and Montville, T.J. (1999). Characterization of fatty acid composition, spore germination, and thermal resistance in a nisin-resistant mutant of *Clostridium botulinum* 169B and in the wild-type strain. Appl. Environ. Microbiol. *65*, 659–664.

McKay, L.L., and Baldwin, K.A. (1984). Conjugative 40-megadalton plasmid in *Streptococcus lactis* subsp. diacetylactis DRC3 is associated with resistance to nisin and bacteriophage. Appl. Environ. Microbiol., *47*, 68–74.

Ryan, K.A., Jayaraman, T., Daly, P., Canchaya, C., Curran, S., Fang, F., Quigley, E.M., and O'Toole, P.W. (2008). Isolation of lactobacilli with probiotic properties from the human stomach. Lett. Appl. Microbiol. *47*, 269–274.

Sun, Z., Zhong, J., Liang, X., Liu, J., Chen, X., and Huan, L. (2009). Novel mechanism for nisin resistance via proteolytic degradation of nisin by the nisin resistance protein NSR. Antimicrob. Agents Chemother. *53*, 1964–1973.

Takala, T.M., and Saris, P.E. (2006). C terminus of NisI provides specificity to nisin. Microbiology *152*, 3543–3549.

Verheul, A., Russell, N.J., Van'T Hof, R., Rombouts, F.M., and Abee, T. (1997). Modifications of membrane phospholipid composition in nisin-resistant *Listeria monocytogenes* Scott A. Appl. Environ. Microbiol. *63*, 3451–3457.

Ye, M., Neetoo, H., and Chen, H. (2008a). Control of *Listeria monocytogenes* on ham steaks by antimicrobials incorporated into chitosan-coated plastic films. Food Microbiol. 25, 260–268.

Ye, M., Neetoo, H., and Chen, H. (2008b). Effectiveness of chitosan-coated plastic films incorporating antimicrobials in inhibition of *Listeria monocytogenes* on cold-smoked salmon. Int.J Food Microbiol. *127*, 235–240.

Part IV

Lactic Acid for Bioplastics

Production of Optically Pure Lactic Acid for Bioplastics

14

Amira M. Hamdan and Kenji Sonomoto*

Abstract

Lactic acid has long been used for the fermentation and preservation of foodstuffs for human consumption. Lactic acid has two optical isomers: L(+)-lactic acid and D(−)-lactic acid. Lactic acid is classified as GRAS (generally recognized as safe) for use as a food additive by the US FDA (Food and Drug Administration), but D(−)-lactic acid is known to be harmful to human metabolism at times, and it can result in acidosis and decalcification. The optical purity of lactic acid is crucial to the physical properties of poly(lactic acid) (PLA), and an optically pure L(+)- or D(−)-lactic acid, rather than racemic DL-lactic acid, can be polymerized to a highly crystalline PLA that is suitable for commercial uses as a bioplastic. Therefore, the biotechnological production of lactic acid has received a significant amount of interest recently, since it offers a solution to the environmental pollution caused by the petrochemical industry and an alternative to petrochemical resources, which are limited.

Lactic acid

Lactic acid, $CH_3CHOHCOOH$, was first discovered in sour milk by the Swedish chemist Scheele in 1780, who initially considered it to be a component of milk (Reddy *et al.*, 2008). In 1789, Lavoisier named this milk component 'acide lactique', which is now the terminology for lactic acid. Lactic acid is present in many foods, either naturally or as a product of *in situ* microbial fermentation (as seen in Sauerkraut, yoghurt, buttermilk, sourdough breads, and many other fermented foods). It is also a major metabolic intermediate in most living organisms from anaerobic prokaryotes to human beings. Lactic acid is produced in mammalian muscles during glycogenolysis and is involved in Kreb's cycle through the intermediates pyruvic acid and acetyl-CoA (Fambri *et al.*, 1997).

Owing to its many potential applications, lactic acid has received a significant amount of attention. Lactic acid mainly has food and food-related applications, and in the USA, approximately 85% of the demand for lactic acid comes from the food industry. The rest (~15%) of the demand is from non-food industries. The current uses and applications of lactic acid can be divided into four main categories: food, cosmetic, pharmaceutical, and chemical applications. The potential applications of lactic acid are illustrated in Fig.14.1 (Wee *et al.*, 2006).

Polymers of lactic acid (PLA) are biodegradable thermoplastics. Recently, PLA has been hailed as an eco-friendly alternative to petroleum-based plastic containers. These polymers are transparent, which is important for applicability as packaging material, and they have good shelf life because they degrade slowly by hydrolysis. Moreover, they have potential uses in a wide variety of consumer products, such as paper coatings, films, moulded articles, foamed articles, and fibres (Lunt, 1998; Drumright *et al.*, 2000; Södergård and Stolt, 2002; Auras *et al.*, 2004).

Currently, the worldwide consumption of lactic acid is estimated to be 130,000–150,000 (metric) tonnes per year, and the commercial prices of food-grade lactic acid range between 1.38 US$/kg (for lactic acid of 50% purity) and

*corresponding author

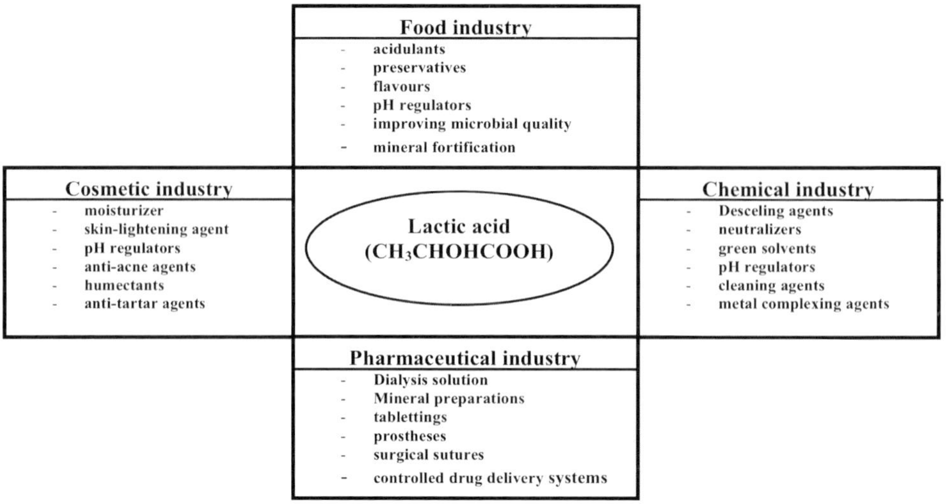

Figure 14.1 Diagram depicting the commercial uses and applications of lactic acid and its salts (Wee *et al.*, 2006).

1.54 US\$/kg (for lactic acid of 88% purity). Technical-grade lactic acid with 88% purity costs as much as 1.59 US\$/kg (Mirasol, 1999; Wee *et al.*, 2006). Lactic acid consumption in the chemical industry, which includes the manufacture of PLA polymers and new green solvents such as ethyl lactate, is expected to expand by 19% each year (Jarvis, 2003).

Poly(L-lactic acid) and poly(D-lactic acid) polymers

Poly(L-lactic acid) (PLLA) is considered as the next-generation plastic material and is now being manufactured on a commercial scale. However, the melting temperature (T_m) of PLLA, which is below 180°C, is slightly lower than that of the current petroleum-based polymers, which makes its wider application more difficult. Recently, stereocomplex-type poly(lactic acid) (sc-PLA) has emerged as a novel heat-resistant polymeric material, because it is characterized by a high *Tm* as well as retention of sustainability. Stereocomplex (sc) formation was first found in a mixture of PLLA and its enantiomer, poly(D-lactic acid) (PDLA) (Ikada *et al.*, 1987). Until now, several approaches have been proposed to improve sc formation in a polymer blend of PLLA and PDLA, for example, by modifying the miscibility of PLLA and PDLA or preventing the concomitant homochiral crystallization (Duan *et al.*, 2006; Fukushima *et al.*, 2007). As mentioned in the report by Tsuji (2005), stereocomplexation gives rise to PLA-based materials with higher mechanical performance, thermal resistance, and hydrolysis resistance; using this process, various types of biodegradable materials such as bottles for hot fill applications, microwaveable food containers, injection moulded parts for cars and electronics, and textile fibres that withstand ironing, hydrogels and drug delivery systems particles can be produced.

Technological production of optically pure lactic acid

Lactic acid can be manufactured by (a) chemical synthesis and (b) carbohydrate fermentation (Fig. 14.2).

Of the 150,000 tonnes of lactic acid produced worldwide every year about 90% is produced by bacterial fermentation and the rest is produced synthetically by the hydrolysis of lactonitrile. Since lactic acid exists as two optical isomers, i.e. in the D- and L-forms, fermentative production of lactic acid is advantageous, because an optically pure product can be obtained by choosing a strain of lactic acid bacteria (LAB) that produces only one of the isomers; however, synthetic production always results in a racemic mixture of lactic

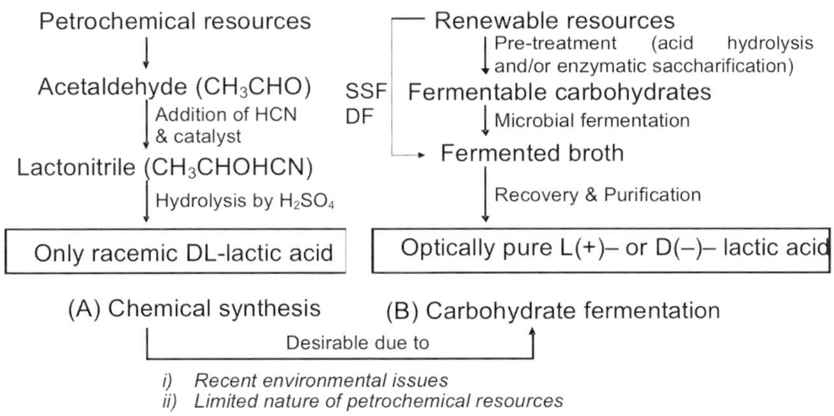

Figure 14.2 Overview of the two methods used for the manufacturing of lactic acid: chemical synthesis (A) and carbohydrate fermentation (B). SSF and DF represent simultaneous saccharification and fermentation and direct fermentation, respectively. (Modified from Wee *et al.*, 2006.)

acid. The optical purity of lactic acid is crucial for the physical properties of PLA, and an optically pure L(+)- or D(−)-lactic acid, rather than racemic DL-lactic acid, can be polymerized to a highly crystalline PLA that is suitable for commercial uses (Lunt, 1998; Södergård and Stolt, 2002). Therefore, the biotechnological production of lactic acid has received a significant amount of attention recently, since it offers a solution to the environmental pollution caused by the petrochemical industry and an alternative to petrochemical resources, which are limited.

Production of PLA

The polymer is formed either by (1) direct condensation of lactic acid or (2) via the cyclic intermediate dimer (lactide), through a ring opening process (Fig. 14.3) (Drumright *et al.*, 2000).

Polycondensation of lactic acid

This process was originally used by Carothers who discovered PLA in 1932. The approach involves the removal of water by condensation and the use of solvent under high vacuum and temperature (Carothers *et al.*, 1932). With this route, only low to intermediate molecular weight polymers can be produced, mainly owing to the difficulties of removing water and impurities. Other disadvantages of this route are the relatively large reactor required, the need for evaporation, recovery of the solvent and increased colour and racemization.

Ring-opening polymerization

This method is a better way to produce a high molecular weight polymer. The process is based on removing water under milder conditions, without the use of costly and environmentally unfriendly solvents, to produce a cyclic intermediate dimer, referred to as lactide. This monomer is readily purified under vacuum distillation. By controlling the purity of the dimer, it is possible to produce a wide range of molecular weights. Next, the pre-polymer is converted to a mixture of lactide stereoisomers using tin catalysis to enhance the rate and selectivity of the intramolecular cyclization reaction. The molten lactide mixture is then purified by vacuum distillation. Finally, PLA high polymer is produced. After the polymerization is complete, any remaining monomer is removed under vacuum and recycled to the beginning of the process.

PLA: life cycle analysis

In recent years, products, processes, and technology have been judged for their overall sustainability. Those that use large quantities of energy, deplete natural resources, pollute the environment, or emit greenhouse gases are viewed as less sustainable. An excellent method to quantify the sustainability of a product or process is to do a life cycle analysis (LCA). LCA looks at all factors from raw materials, transportation, product

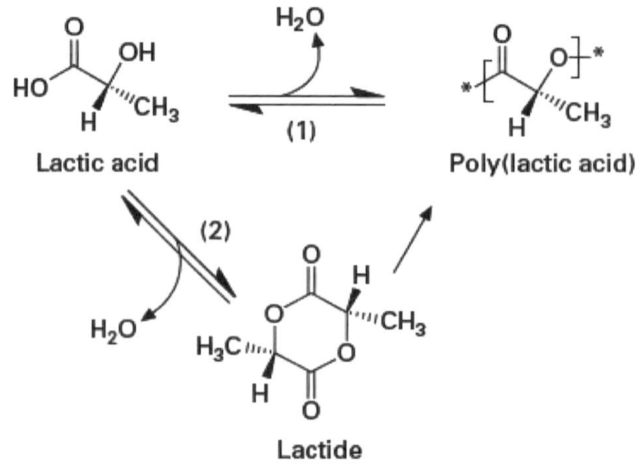

Figure 14.3 Polymerization routes to poly(lactic acid) (Drumright *et al.*, 2000).

manufacture, end use, to disposal. The energy, air and water pollution, as well as the hazards of the chemicals are assessed and judged against alternative products or solutions. Each step from cradle to grave is quantified and the total impact on the environment in each of the areas is thus known. LCA calculations indicate that PLA has a greenhouse gas emission rate of about 1600 kg CO_2/metric ton, while polypropylene, polystyrene, and nylon have greenhouse gas values of 1850, 2740, and 7150 kg CO_2/metric ton, respectively (Henton *et al.*, 2005). These PLA value, because CO_2 generated during PLA biodegradation is balanced by an equal amount taken from the atmosphere during the growth of plant feedstocks. In contrast, petrochemical-based polymers contribute to volatile organic compound (VOC) emissions and CO_2 generation when incinerated.

Lactic acid-producing microorganisms

Microorganisms that can produce lactic acid can be divided into two groups: bacteria and fungi (Litchfield, 1996; Wee *et al.*, 2006). Even though filamentous fungi such as *Rhizopus* utilize glucose aerobically to produce lactic acid, most investigations of lactic acid production were carried out with LAB (Zhou *et al.*, 1999; Tay and Yang, 2002). *Rhizopus* species such as *R. oryzae* and *R. arrhizus* have amylolytic enzyme activity, which enables them to convert starch directly to L(+)-lactic acid (Oda *et al.*, 2002). Fungal fermentation has some advantages: *R. oryzae* requires only a simple medium and produces L(+)-lactic acid, but it also requires vigorous aeration because it is an obligate aerobe (Tay and Yang, 2002). The low production rate, i.e. below 3 g/(l/h), that is obtained with fungal fermentation is probably due to the low reaction rate caused by mass transfer limitation (Park *et al.*, 1998). The lower product yield from fungal fermentation is attributed partially to the formation of by-products such as fumaric acid and ethanol (Tay and Yang, 2002). Although there have been persistent attempts to produce lactic acid through fungal fermentation, LAB have been commonly used for the production of lactic acid owing to the aforementioned disadvantages of fungal fermentation (Wee *et al.*, 2006).

Lactic acid bacteria

LAB belong to a group of Gram-positive anaerobic bacteria that produce lactic acid as their main fermentation product, which is released into the culture medium. LAB can be classified into two groups: homofermentative and heterofermentative. Only the homofermentative LAB, which are capable of converting glucose exclusively into lactic acid, are available for the commercial production of lactic acid (Hofvendahl and Hahn-Hägerdal, 2000; Yun *et al.*, 2003). LAB have an enormous capacity for the degradation of different carbohydrate substrates from surplus agricultural crops and industrial wastes resulting from the production of various value-added products (Vishnu *et al.*, 2006). LAB are divided into three groups based on fermentation patterns (Hofvendahl and Hahn-Hägerdal, 2000; Reddy *et al.*, 2008):

- Homofermentative: produce more than 85% lactic acid from glucose. They ferment 1 mol of glucose to 2 mol of lactic acid, generating a net yield of 2 mol of ATP per molecule of glucose metabolized. Lactic acid is the major product of this fermentation (Fig. 14.4).
- Heterofermentative: produce only 50% lactic acid. These ferment 1 mol of glucose to 1 mol of lactic acid, 1 mol of ethanol, and 1 mol of CO_2. One mole of ATP is generated per mole of glucose, resulting in less growth per mole of glucose metabolized (Fig. 14.4).
- Less well known heterofermentative species which produce DL-lactic acid, acetic acid and carbon dioxide.

Raw materials for biotechnological production of lactic acid

In order to ensure that the biotechnological production of lactic acid is feasible, cheap raw materials are a necessity; this is because polymer producers and other industrial users usually require large quantities of lactic acid at a relatively low cost. The raw materials used for lactic acid production should have the following characteristics: low cost, low levels of contaminants, rapid production rate, high yield, little or no by-product formation, ability to get fermented with little or no pre-treatment and year-round availability. When refined materials are used for production, the cost of product purification is significantly reduced. However, this is still economically unfavourable because refined carbohydrates are so expensive that they eventually result in higher production costs (Hofvendahl and Hahn-Hägerdal, 2000). Over the years, a large number of carbohydrates and nitrogenous materials have been studied for the fermentative production of lactic acid. They have been investigated so as to determine materials with the following characteristics: fast

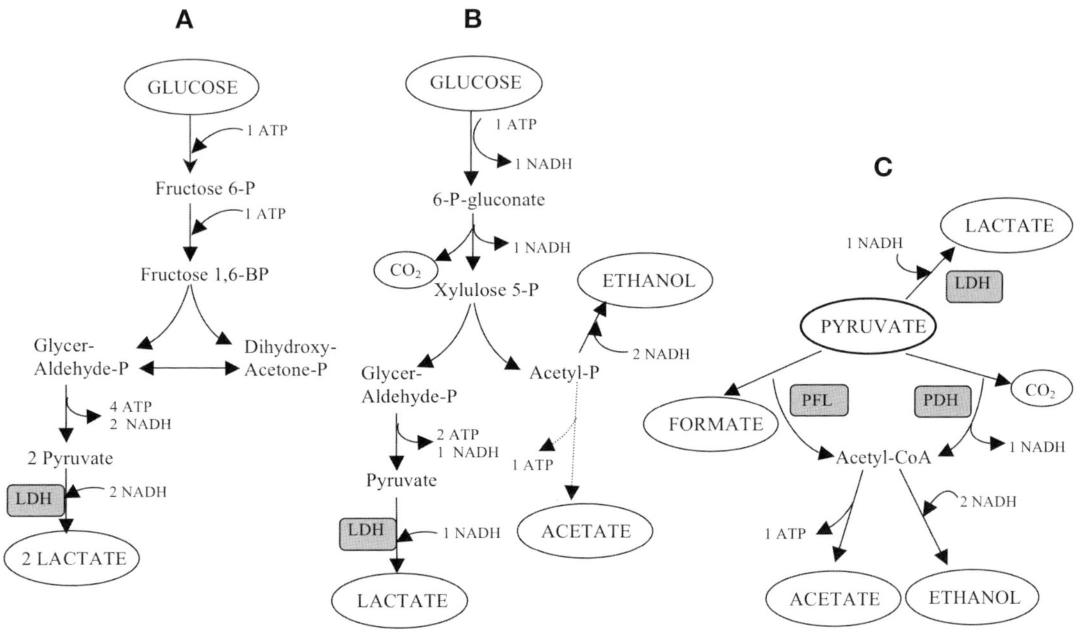

Figure 14.4 Catabolic pathways in lactic acid bacteria. Homofermentation (A), heterofermentation (B) and mixed acid fermentation (C). P, phosphate; BP, bisphosphate; LDH, lactate dehydrogenase; PFL, pyruvate formate lyase; PDH, pyruvate dehydrogenase (Hofvendahl and Hahn-Hägerdal, 2000).

fermentation rate, high lactic acid yields, negligible by-product formation, less pre-treatment, easy downstream processing, low cost, optimum biomass production, ease of availability, etc. The choice of the raw material to be used depends on the microorganisms being studied and also on the product desired. The microorganisms selected for recent investigations of the biotechnological production of lactic acid are listed in Table 14.1.

Fermentation approaches to lactic acid production

Batch, fed-batch, repeated batch, and continuous fermentations are the most frequently used methods for lactic acid production. Higher lactic acid concentrations may be obtained in batch and fed-batch cultures than in continuous cultures, whereas higher productivity may be achieved with the use of continuous cultures (Hofvendahl and Hahn-Hägerdal, 2000). Another advantage of continuous culture over batch culture is that with continuous culture, it is possible to continue the process for a longer period of time. The cell recycle system, together with the repeated batch and continuous processes, enables the achievement of higher cell concentration and productivity (Kwon *et al.*, 2001; Oh *et al.*, 2003).

In addition, immobilization of cells has been one of the means for high cell retention in the bioreactor (Senthuran *et al.*, 1999). Several materials, such as Ca-alginate gels, poly(ethyleneimine), and plastic composite support, have been used for the immobilization of LAB in order to produce lactic acid (Goksungur and Guvenc, 1999; Senthuran *et al.*, 1999; Cotton *et al.*, 2001).

Lactic acid production processes traditionally suffer from end-product inhibition. Undissociated lactic acid molecules pass through the bacterial membrane and dissociate inside the cell. The inhibition mechanism of lactic acid is probably related to the solubility of the undissociated lactic acid within the cytoplasmic membrane and the insolubility of the dissociated lactate, which causes acidification of the cytoplasm and failure of proton motive forces. It eventually influences the transmembrane pH gradient and decreases the amount of energy available for cell growth (Goncalves *et al.*, 1997). Therefore, to alleviate the inhibitory effect of lactic acid during the fermentation, it must be selectively removed *in situ* from the fermentation broth. Recently, various techniques, such as solvent extraction and electrodialysis, have been used to remove the lactic acid simultaneously as it is formed (Hano *et al.*, 1993; Nomura *et al.*, 1998; Iyer and Lee, 1999; Kim and Moon, 2001; Min-Tian *et al.*, 2005).

Efforts towards manufacturing lactic acid and lactic acid-based polymers

In a world that is becoming increasingly sensitive to the need to protect our environment, the ability to manufacture products from sustainable resources and which are fully compostable at the end of their useful life, is an exciting and attractive proposition. Fossil fuel-derived plastics are non-renewable, often threaten public health, have devastating impacts on marine life, and increase reliance on imported fossil fuel-based feedstocks. Therefore, virgin fossil fuel-based plastics must inevitably be entirely replaced with sustainable, renewable materials-either recycled, biobased, or both. Overall recycling of fossil fuel-based plastics is carried out to a very low extent. In contrast, PLA is derived from 100% renewable sources such as maize, and the polymer is biodegradable, compostable, and non-toxic to the human body and the environment. Therefore, the development of bioplastics holds great promise in that many of these sustainability problems may be mitigated, offering renewability, biodegradation, and a way out of the use of harmful additives.

A number of companies have made efforts towards manufacturing lactic acid-based polymers and their products. L-lactic acid polymers exist in three different forms-solids that can be used for filling the gaps in bones, solids with enough tensile strength to produce sutures (stitching material), and the glue form that is mainly applied in joining membranes or thin skins in humans. Other important properties of PLA are its high durability against UV radiation, easily dyeable and resistance to stain in standard tests with coffee, cola, lipstick and mustard. Cargill Dow LLC, awarded the 2002 Presidential Green Chemistry Award for their process to produce NatureWorks PLA, have

Table 14.1 Reports about recent investigations on the biotechnological production of optically pure lactic acid from cheap raw materials

Lactic acid isomer	Optical purity (%)	Strain	Lactic acid (g/l)	Lactic acid yield (g/g)	Productivity (g/l/h)	Fermentation mode	Operational stability (h)	Raw material	Carbon source	Reference
L-	98.6	*Enterococcus faecium* No. 78	11.7	0.93	3.04	Continuous via cell-recycle and bleeding (DF)	84	Sago	Starch	Shibata *et al.* (2007)
L-	–	*Lactobacillus amylophilus* GV6	–	0.92	–	Batch (DF)	–	Flour	Starch	Altaf *et al.* (2005)
L-	96	*Lactobacillus casei* and *Enterococcus casseliflavus*	95	0.63	–	Batch, cocultivation	192	–	Xylose and Glucose	Taniguchi *et al.* (2004)
L-	–	*Lactococcus lactis* IO-1	33.3 20.6	0.66 0.54	–	Batch Continuous	50 –	– –	Xylose Xylose	Tanaka *et al.* (2002) Tanaka *et al.* (2002)
L-	–	*Enterococcus faecalis* RKY1	–	0.93	5.36	Batch (SSF)	–	Barley, wheat, corn	Starch	Oh *et al.* (2005)
L-	–	*Lactococcus lactis* IO-1	25 30	0.47 0.66	– –	Batch Batch	38 45	– –	Xylose Xylose and glucose	Ishizaki *et al.* (1992) Ishizaki *et al.* (1992)
D-	–	*Leuconostoc lactis* SHO-47	2.3	–	–	Batch (DF)	12	–	Xylooligo-saccharides	Ohara *et al.* (2006)
D-	–	*Leuconostoc lactis* SHO-54	2.2	–	–	Batch (DF)	12	–	Xylooligo-saccharides	Ohara *et al.* (2006)
D-	95	*Lactobacillus delbrueckii* subsp. *delbrueckii*	–	0.78	–	Batch (SSF)	36	Defatted rice	Starch	Tanaka *et al.* (2006)
D-	97.5	*Lactobacillus delbrueckii*	–	0.7	–	Batch (SSF)	96	Rice	Starch	Fukushima *et al.* (2004)
D-	–	*Lactobacillus coryniformis* subsp. *torquens*	–	0.89	0.5	Batch (SSF)	48	Filter paper	Cellulose	Yanez *et al.* (2003)

DF, direct fermentation; SSF, simultaneous saccharification and fermentation.

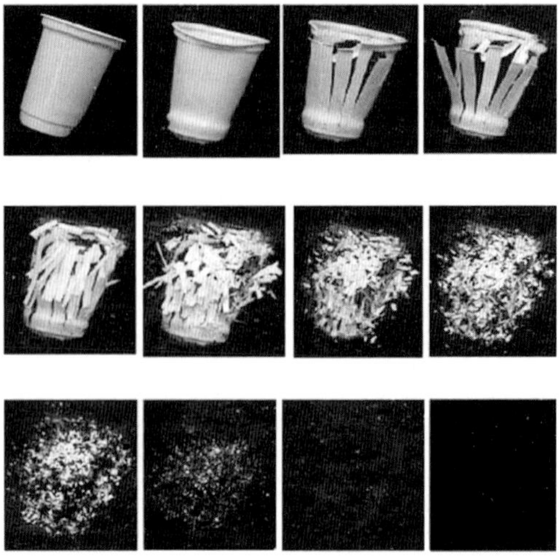

Figure 14.5 Photos showing plastics made out of Cargill Dow's corn-resin pellets, which break down in an industrial compost facility in about 45 days (Katsnelson, 2005).

the largest polylactid-producing company with an annual capacity of 140,000 tonnes located in Blair, USA (Narayanan *et al.*, 2004; Henton *et al.*, 2005). Fig. 14.5 shows the breakdown of plastic cup pellets made of Cargill Dow's maize resin (Katsnelson, 2005).

Biodegradation of PLA polymers

Under conditions of high temperature and high humidity, PLA degrades quickly and disintegrate within weeks to months. The primary mechanism of degradation is hydrolysis, followed by bacterial attack on the fragmented residues. The environmental degradation of PLA occurs by a two-step process (Lunt, 1998). During the initial phases of degradation, the high molecular weight polyester chains hydrolyse to lower molecular weight oligomers. The rate of hydrolysis is accelerated by acids or bases and is dependent on moisture content and temperature. Article dimensions, crystallinity, and blends will affect the rate of degradation. Next, microorganisms in the environment continue the degradation process by converting these lower molecular weight components to carbon dioxide, water, and humus, with complete degradation occurring within only few weeks under typical compost conditions. PLA products rapidly degrade in both aerobic and anaerobic composting conditions (Drumright *et al.*, 2000). In addition, PLA articles can be readily hydrolysed with water to form lactic acid, which is then purified and polymerized to remake prime polymer, which offers a preferred method for PLA recycling (Henton *et al.*, 2005).

References

Altaf, M., Naveena, B., and Reddy, G. (2005). Screening of inexpensive nitrogen sources for production of L(+) lactic acid from starch by amylolytic *Lactobacillus amylophilus* GV6 in single step fermentation. Food Technol. Biotechnol. *43*, 235–239.

Auras, R., Harte, B., and Selke, S. (2004). An overview of polylactides as packaging materials. Macromol. Biosci. *4*, 835–864.

Carothers, W., Dorough, G., and van Natta, F. (1932). Studies of polymerization and ring formation. X. The reversible polymerization of six-membered cyclic esters. J. Am. Chem. Soc. *54*, 761–772.

Cotton, J., Pometto III, A., and Gvozdenovic-Jeremic, J. (2001). Continuous lactic acid fermentation using plastic composite support biofilm reactor. Appl. Microbiol. Biotechnol. *57*, 626 – 630.

Drumright, R., Gruber, P., and Henton, D. (2000). Polylactic acid technology. Adv. Mater. *12*, 1841–1846.

Duan, Y., Liu, J., Sato, H., Zhang, J., Tsuji, H., Ozaki, Y., and Yan, S. (2006). Molecular weight dependence of the Poly(L-lactide)/Poly(D-lactide) stereocomplex at the air–water interface. Biomacromolecules *7*, 2728–2735.

Fambri, L., Pegoretti, A., Fenner, R., Incardona, S., and Migliaresi, C. (1997). Biodegradable fibers of poly(L-lactic acid) produced by melt spinning. Polymer *38*, 79–85.

Fukushima, K., Sogo, K., Miura, S., and Kimura, Y. (2004). Production of D-lactic acid by bacterial fermentation of rice starch. Macromol. Biosci. *4*, 1021–1027.

Fukushima, K., Chang, Y.-H., and Kimura, Y. (2007). Enhanced stereocomplex formation of Poly(L-lactic acid) and Poly(L-lactic acid) in the presence of stereoblock Poly(lactic acid). Macromol. Biosci. *7*, 829–835.

Goksungur, Y., and Guvenc, U. (1999). Production of lactic acid from beet molasses by calcium alginate immobilized *Lactobacillus delbrueckii* IFO 3202. J. Chem. Technol. Biotechnol. *74*, 131–136.

Goncalves, L., Ramos, A., Almeida, J., Xavier, A., and Carrondo, M. (1997). Elucidation of the mechanism of lactic acid growth inhibition and production in batch cultures of *Lactobacillus rhamnosus*. Appl. Microbiol. Biotechnol. *48*, 346–350.

Hano, T., Matsumoto, M., Uenoyama, S., Ohtake, T., Kawano, Y., and Miura, S. (1993). Separation of lactic acid from fermented broth by solvent extraction. Bioseparation *3*, 321–326.

Henton, D., Gruber, P., Lunt, J., and Randall, J. (2005). Polylactic acid technology. In Natural Fibers, Biopolymers, and Biocomposites, Mohanty, A.K., Misra, M., and Drzal, L.T., eds. (Taylor & Francis Group, Boca Raton, Florida), pp. 527–578.

Hofvendahl, K., and Hahn-Hägerdal, B. (2000). Factors affecting the fermentative lactic acid production from renewable resources. Enzyme Microb. Technol. *26*, 87–107.

Ikada, Y., Jamshidi, K., Tsuji, H., and Hyon, S. (1987). Stereocomplex formation between enantiomeric poly(lactides). Macromolecules *20*, 904–906.

Ishizaki, A., Ueda, T., Tanaka, K., and Stanbury, P. (1992). L-lactate production from xylose employing *Lactococcus lactis* IO-1. Biotechnol. Lett. *14*, 599–604.

Iyer, P., and Lee, Y. (1999). Simultaneous saccharification and extractive fermentation of lignocellulosic materials into lactic acid in a two-zone fermentor-extractor system. Appl. Biochem. Biotechnol. *78*, 409–419.

Jarvis, L. (2003). Prospects for lactic acid are healthy as demand for all end uses grows. Chem. Market Rep. *263*, 12.

Katsnelson, A. (2005). Dow pulls out of bioplastics due to slow sector maturation. Nature Biotechnol. *23*, 638.

Kim, Y., and Moon, S. (2001). Lactic acid recovery from fermentation broth using one stage electrodialysis. J. Chem. Technol. Biotechnol. *76*, 169–178.

Kwon, S., Yoo, I., Lee, W., Chang, H., and Chang, Y. (2001). High-rate continuous production of lactic acid by *Lactobacillus rhamnosus* in a two-stage membrane cell-recycle bioreactor. Biotechnol. Bioeng. *73*, 25–34.

Litchfield, J. (1996). Microbiological production of lactic acid. Adv. Appl. Microbiol. *42*, 45–95.

Lunt, J. (1998). Large-scale production, properties and commercial applications of polylactic acid polymers. Polymer Degrad. Stabil. *59*, 145–152.

Min-Tian, G., Koide, M., Gotou, R., Takanashi, H., Hirata, M., and Hano, T. (2005). Development of a continuous fermentation system for production of lactic acid by *Lactobacillus rhamnosus*. Process Biochem. *40*, 1033–1036.

Mirasol, F. (1999). Lactic acid prices falter as competition toughen. Chem. Market Rep. *255*, 16.

Narayanan, N., Roychoudhury, P., and Srivastava, A. (2004). L(+)lactic acid fermentation and its product polymerization. Electronic J. Biotechnol. *7*, 167–179.

Nomura, Y., Iwahara, M., Hallsworth, J., Tanaka, T., and Ishizaki, A. (1998). High-speed conversion of xylose to L-lactate by electrodialysis bioprocess. J. Biotechnol. *60*, 131–135.

Oda, Y., Saito, K., Yamauchi, H., and Mori, M. (2002). Lactic acid fermentation of potato pulp by the fungus *Rhizopus oryzae*. Curr. Microbiol. *45*, 1–4.

Oh, H., Wee, Y., Yun, J.-S., and Ryu, H. (2003). Lactic acid production through cell-recycle repeated-batch bioreactor. Appl. Biochem. Biotechnol. *107*, 603–613.

Oh, H., Wee, Y., Yun, J.-S., Han, S., Jung, S., and Ryu, H. (2005). Lactic acid production from agricultural resources as cheap raw materials. Bioresource Technol. *96*, 1492–1498.

Ohara, H., Owaki, M., and Sonomoto, K. (2006). Xylooligosaccharide fermentation with *Leuconostoc lactis*. J. Biosci. Bioeng. *101*, 415–420.

Park, E., Kosakai, M., and Okabe, M. (1998). Efficient production of L(+)-lactic acid using mycelial cotton-like flocs of *Rhizopus oryzae* in an air-lift bioreactor. Biotechnol. Progr. *14*, 699–704.

Reddy, G., Altaf, M., Naveena, B., Venkateshwar, M., and Vijay Kumar, E. (2008). Amylolytic bacterial lactic acid fermentation – A review. Biotechnol. Adv. *26*, 22–34.

Senthuran, A., Senthuran, V., Hatti-Kaul, R., and Mattiasson, B. (1999). Lactic acid production by immobilized *Lactobacillus casei* in recycle batch reactor: A step towards optimization. J. Biotechnol. *73*, 61–70.

Shibata, K., Flores, D., Kobayashi, G., and Sonomoto, K. (2007). Direct L-lactic acid fermentation with sago starch by a novel amylolytic lactic acid bacterium, *Enterococcus faecium*. Enzyme Microb. Technol. *41*, 149–155.

Södergård, A., and Stolt, M. (2002). Properties of lactic acid based polymers and their correlation with composition. Prog. Polym. Sci. *27*, 1123–1163.

Tanaka, K., Komiyama, A., Sonomoto, K., Ishizaki, A., Hall, S., and Stanbury, P. (2002). Two different pathways for D-xylose metabolism and the effects of xylose concentration on the yield coefficient of L-lactate in mixed-acid fermentation by the lactic acid bacterium *Lactococcus lactis* IO-1. Appl. Microbiol. Biotechnol. *60*, 160–167.

Tanaka, T., Hoshima, M., Tanabe, S., Sakai, K., Ohtsubo, S., and Taniguchi, M. (2006). Production of D-lactic acid from defatted rice bran by simultaneous saccharification and fermentation. Bioresource Technol. *97*, 211–217.

Taniguchi, M., Tokunaga, T., Horiuchi, K., Hoshino, K., Sakai, K., and Tanaka, T. (2004). Production of L-lactic acid from a mixture of xylose and glucose by co-cultivation of lactic acid bacteria. Appl. Microbiol. Biotechnol. *66*, 160–165.

Tay, A., and Yang, S. (2002). Production of L(+)-lactic acid from glucose and starch by immobilized cells of *Rhizopus oryzae* in a rotating fibrous bed bioreactor. Biotechnol. Bioeng. *80*, 1–12.

Tsuji, H. (2005). Poly(lactide) stereocomplexes: formation, structure, properties, degradation, and applications. Macromol. Biosci. *5*, 567–597.

Vishnu, C., Naveena, B., Altaf, M., Venkateshwar, M., and Reddy, G. (2006). Amylopullulanase – A novel enzyme of *L. amylophilus* GV6 in direct fermentation of starch to L(+) lactic acid. Enzyme Microb. Technol. *38*, 545–550.

Wee, Y.-J., Kim, J.-N., and Ryu, H.-W. (2006). Biotechnological production of lactic acid and its recent applications. Food Technol. Biotechnol. *44*, 163–172.

Yanez, R., Moldes, A., Alonso, J., and Parajo, J. (2003). Production of D(-)-lactic acid from cellulose by simultaneous saccharification and fermentation using *Lactobacillus coryniformis* subsp. *torquens*. Biotechnol. Lett. *25*, 1161–1164.

Yun, J.-S., Wee, Y.-J., and Ryu, H.-W. (2003). Production of optically pure L(+)-lactic acid from various carbohydrates by batch fermentation of *Enterococcus faecalis* RKY1. Enzyme Microb.Technol. 33, 416–423.

Zhou, Y., Domínguez, J., Cao, N., Du, J., and Tsao, G. (1999). Optimization of L-lactic acid production from glucose by *Rhizopus oryzae* ATCC 52311. Appl. Biochem. Biotechnol. *78*, 401–407.

Part V

Health Benefit of Lactobacilli and Bifidobacteria

Antihypertensive Metabolites from Lactic Acid Bacteria

15

Naoyuki Yamamoto

Abstract

Many studies have suggested milk fermented with lactic acid bacteria has beneficial effects on the health status of animals and humans. This chapter reviews the potential of antihypertensive peptides in milk fermented with lactic acid bacteria. Most of the antihypertensive effects of these peptides can be explained by their inhibition of angiotensin converting enzyme. Well-characterized antihypertensive peptides released by *Lactobacillus helveticus* fermented milk are mainly reviewed in this chapter. Studies on Val-Pro-Pro (VPP) and Ile-Pro-Pro (IPP), released in *L. helveticus* fermented milk, and their use in animal and clinical studies are discussed in more detail. In addition, processing of VPP and IPP by proteolytic enzymes of *L. helveticus* is discussed. A long peptide containing VPP and IPP sequences generated from β-casein by an extracellular proteinase was thought to be processed intracellularly by the following enzymes: an endopeptidase was expected for C-terminal processing, and an aminopeptidase and X-prolyl dipeptidyl aminopeptidase for N-terminal processing. Genome analyses of these proteolytic enzymes in *L. helveticus* enabled comparisons to corresponding genome sequences reported in other lactic acid bacteria; from this, it was suggested that the proteolytic system to process VPP and IPP is specific to *L. helveticus*. The regulatory system observed in the production of VPP and IPP during *L. helveticus* fermentation is also discussed.

Introduction

Many studies have suggested lactic acid bacteria and cultured dairy products have beneficial effects on the nutrition and health status of animals and humans. Microbe fermentation involves the proteolytic processing of proteins to release peptides for use as a nitrogen source. Lactic acid bacteria are suitable microbes for milk fermentation because they have a proteolytic system, which decomposes casein and contains lactose digestive enzymes. Peptides released from casein by an extracellular proteinase in lactic acid bacteria have been reported (Smid *et al.*, 1991; Pritchard *et al.*, 1993; Tan *et al* 1993; Yamamoto *et al.*, 1993, 1994a). Some of the peptides which accumulate in the fermented media during the fermentation process are good sources of physiologically functional peptide (bioactive peptides). Generally, to prepare bioactive peptides, food ingredients such as milk (Maruyama *et al.*, 1987a,b; Kohmura *et al.*, 1989), egg (Miguel *et al.*, 2004), wheat (Matsui *et al.*, 1999), maize (Miyoshi *et al.*, 1991) or fish (Fujita *et al.* 1999; Kawasaki *et al.* 2000) are hydrolysed by commercially available proteases. Milk proteins, such as bovine casein and whey proteins, are often used in the preparation of bioactive peptides because they are inexpensive, safe and readily available. Various bioactive peptides, such as opioid peptides (Teschemacher *et al.*, 1997), immunostimulating peptides (Migliore-Samour, 1989), mineral soluble peptides (FitzGerald, 1998), antihypertensive peptides (Meisel *et al.*, 1999; Yamamoto *et al.*, 2003) and antimicrobial peptides (Bellamy *et al.*, 1992) have been isolated from enzymatic hydrolysates of raw food materials and fermented food products.

In this chapter, antihypertensive effects detected in milk fermented with lactic acid bacteria are briefly reviewed before a more detailed discussion on antihypertensive peptides isolated from *Lactobacillus helveticus*: in particular, processing of the antihypertensive peptides specifically detected in the fermented milk. The regulatory pathways involved in Val-Pro-Pro (VPP) and Ile-Pro-Pro (IPP) processing in *L. helveticus* are also discussed.

Antihypertensive peptides in fermented milk

Hypertension is a major risk factor in cardiovascular diseases, such as heart disease and stroke. In order to reduce the incidence of disease, pharmacological substances can be used to decrease high blood pressure to within the normal range. Angiotensin I-converting enzyme (kininase II; EC 3.4.15.1) (ACE) is predominantly expressed as a membrane-bound ectoenzyme in vascular endothelial cells and several other cell types, including absorptive epithelia, neuroepithelia and male germinal cells (Miguel *et al.*, 2004; Matsui *et al*, 1999). ACE is a dipeptidyl carboxypeptidase, which catalyses the production of a vasoconstrictor, angiotensin II, and inactivates a vasodilator, bradykinin (Miyoshi *et al.*, 1991; Fujita *et al.*, 1999). The first inhibitors to ACE were reported as naturally occurring peptides isolated from snake venom (Kawasaki *et al.*, 2000; Cladwell *et al.*, 1976). Since then, antihypertensive peptides have been extensively studied and reviewed (Yamamoto *et al.*, 2003; Takano 2002).

An antihypertensive effect was discovered in various kinds of fermented milk, however, the effect was greatest in *L. helveticus* fermented milk (Yamamoto *et al*, 1994b) (Table 15.1). In our study, an antihypertensive effect related to ACE

Table 15.1 Antihypertensive effects in spontaneously hypertensive rats and ACE inhibitory activities of various fermented milk

Strain	Peptide conc. (%)	Proteinase act. (U/ml)	ACEI act. (U/ml)[1]	Change in SBP (mmHg)[2]
Non-fermented milk (lactobacilli)	0.00	–	0	−5.0 ± 7.3
L. helveticus CP790	0.19	230	58	−27.4 ± 13.3**
L. helveticus CP611	0.25	367	70	−20.0 ± 9.6**
L. helveticus CP615	0.18	420	51	−23.0 ± 13.4**
L. helveticus JCM1006	0.15	182	26	−15.2 ± 9.3*
L. helveticus JCM1120	0.10	112	34	− 6.5 ± 10.8
L. helveticus JCM1004	0.21	186	48	−29.3 ± 13.6**
L. delbrueckii subsp. *bulgaricus* CP973	0.19	105	22	−0.8 ± 8.2
L. delbrueckii subsp. *bulgaricus* JCM1002	0.11	124	28	−4.5 ± 4.0
L. casei CP680	0.01	35	3	−0.2 ± 6.6
L. casei JCM1134	0.00	28	9	−7.0 ± 11.2
L. casei JCM1136	0.09	25	18	−9.6 ± 7.2
L. acidophilus JCM1132	0.00	28	8	−8.7 ± 7.8
L. delbrueckii subsp. *lactis*JCM1105 (streptococci)	0.08	18	16	−3.3 ± 3.5
S. thermophilus CP1007 (lactococci)	0.02	25	3	−2.4 ± 8.1
L. lactis subsp. *lactis* CP684	0.00	35	4	−7.3 ± 10.5
L. lactis subsp. *cremoris* CP312	0.02	18	4	−5.8 ± 13.9

[1]ACEI activity, peptides that show 50% inhibition of ACE activity was defined as one unit.

[2]SBP, systolic blood pressure.

Significant differences from the control, $^{**}P<0.01$, $^{*}P<0.05$.

inhibitory peptides was found in sour milk produced by *L. helveticus* (Nakamura *et al.*, 1995a,b). Two ACE inhibitory peptides were purified from sour milk and identified as VPP and IPP. The ACE inhibitory activities of VPP and IPP were high, with IC_{50} values of 9 μM and 5 μM, respectively, compared with other reported ACE inhibitory peptides (Table 15.2). The amino acid sequences of VPP and IPP were found in the primary structure of bovine β-casein (84–86) (74–76) and κ-casein (108–110), respectively. These peptides were produced during fermentation (Nakamura *et al.*, 1995a), but were not found in the hydrolysate of casein after digestion with an extracellular proteinase of *L. helveticus* (Yamamoto *et al.*, 1993). Spontaneously hypertensive rats (SHR) are a useful animal model to evaluate the antihypertensive activity of ACE inhibitory peptide. The systolic blood pressure of SHR reaches over 230 mmHg while the systolic blood pressure of a normotensive rat reaches approximately 150 mmHg at the age of 25 weeks. Some of the orally administered ACE inhibitory peptides have demonstrated strong antihypertensive effects in SHR (Table 15.2). Peptides with the highest ACE inhibitory activities were relatively short peptides with proline residue(s) at the C-terminus. Oral administration of *L. helveticus* fermented milk containing VPP and IPP to SHR, with a single dose of 5 ml/kg body weight, significantly decreased systolic blood pressure between 4 and 8 hours after administration (Nakamura *et al.*, 1995b). The antihypertensive effects of chemically synthesized VPP and IPP were also observed between 2 and 8 hours after administration and the effects were dose-dependent. Furthermore, a dose-dependent antihypertensive effect of these two chemically synthesized peptides was also observed from 0.1 to 10 mg/kg of body weight (Nakamura *et al.*, 1995b).

Table 15.2 Antihypertensive peptides derived from caseins by proteolytic action

Peptode	Source	Preparation	IC50 (μm)[1]	Dose (mg/kg)	SBP (mmHg)
Enzymatic hydrolysate					
FFVAPFPEVFGK	αs1-casein	Trypsin	77	100	−13.0
AVPYPQR	β-Casein	Trypsin	15	100	−10.0
TTMPLW	αs1-casein	Trypsin	16	100	−13.6
LKPNM	Aldolase	Thermolysin	2.4	60	−23
LKP	Aldolase	Chicken muscle	0.32	60	−18
IPA	β-Lactogloblin	Proteinase K	141	8	−31
VYPFPG	β-Casein	Proteinase K	221	8	−22
GKP	β-Microglobulin	Proteinase K	352	8	−26
FP	β-Casein, albumin	Proteinase K	315	8	−27
YKVPQL	αs1-casein	Proteinase	22	1	−12.5
Fermented products					
RF	Sake lees	Brewing	ND	100	−17
VW	Sake lees	Brewing	1.4	100	−10
YW	Sake lees	Brewing	10.5	100	−28
VY	Sake	Brewing	7.1	100	−31
IYPRY	Sake	Brewing	4.1	100	−19
VPP	β-Casein	Fermentation	9	1.6	−20
IPP	β- and κ-casein	Fermentation	5	1	−15.1
YP	αs1, β- and κ-casein	Fermentation	720	1	−27.4

[1]IC_{50}, peptide concentration that shows 50% inhibition of ACE activity.

ND, not described; SBP, systolic blood pressure.

Clinical effects of the fermented milk

In the first Japanese study using fermented milk, hypertensive subjects were randomly assigned to two groups: one group ingested 95 ml of fermented milk, containing 3.4 mg of VPP and IPP, daily for 8 weeks; the other group ingested the same amount of artificially acidified milk as a placebo for 8 weeks (Hata *et al.,* 1996). In the fermented milk group, systolic blood pressure decreased significantly between 4 and 8 weeks after the beginning of ingestion, but not in the placebo group (Hata *et al.,* 1996). In addition, many clinical tests have been performed using Japanese subjects with different blood pressure levels, the studies confirmed mild and prolonged effects in hypertensive subjects following oral administration of bioactive milk (Hata *et al.,* 1996; Kajimoto *et al.,* 2001a,b; Nakamura *et al.,* 2004). In a pilot study conducted in Finland, an antihypertensive effect was also observed in a group ingesting *L. helveticus* fermented milk containing VPP and IPP (Seppo *et al.,* 2002, 2003). In addition, a recent study indicated significant beneficial effects in hypertensive patients ingesting *L. helveticus* fermented milk over a long period of time (21 weeks): a significant decrease in systolic blood pressure (6.7 ± 3.0 mmHg) by comparative study with a placebo group (Seppo *et al.,* 2002).

Casein hydrolysis by extracellular proteinase

As previously described, many kinds of proteolytic enzymes have been reported in lactic acid bacteria, and have been extensively reviewed (Pritchard *et al.,* 1993; Kunji *et al.,* 1996; Christensen *et al.,* 1999). The components of the proteolytic systems of lactic acid bacteria are divided into three groups, (i) the extracellular proteinase that catalyses casein breakdown to peptides, (ii) peptidases that hydrolyse peptides to amino acids and (iii) a peptide transport system. A number of proteinases have been reported in lactococci, which are mainly used in cheese making. The extracellular proteinase activity is linked to cell growth and may be essential in utilization of milk protein for growth. The unprocessed proteinase of *Lactococcus lactis* consists of approximately 1950 amino acid residues, and the mature proteinase is a serine type enzyme with a molecular mass between 180 and 190 kDa (Kok *et al,* 1988). The gene encoding a proteinase, *prtP*, has been cloned and sequenced from several *L. lactis* strains (Kok *et al.,* 1988; Kiwaki *et al.,* 1989; de Vos *et al,* 1989). These proteinases are processed to active enzymes by removal of N-terminal polypeptides from the pre-proteinase, with the help of a maturation protein, PrtM.

The first step in casein decomposition is typically caused by an extracellular proteinase, and further digestion to amino acids is catalysed by many kinds of intracellular peptidases (Pritchard *et al.,* 1993; Kunji *et al.,* 1996). Among the lactic acid bacteria, *L. helveticus* has the highest extracellular proteinase activity and the ability to release the highest amount of peptides in fermented milk (Yamamoto *et al.,* 1994b). *L. helveticus* strains were classified into two types based on differences in the extracellular proteinase (Yamamoto *et al.,* 1998). One type has an enzyme with a molecular weight of 170 kDa, with homology to the lactococcus enzyme, and the other type has an enzyme with a molecular weight of 45 kDa (Yamamoto *et al.,* 1993, 1995). Polyclonal antibodies raised against the mature 170 kDa proteinase reacted not only with the 170 kDa enzyme but also a 53 kDa protein. The 53 kDa protein is thought to be a degradation product from the 170 kDa active enzyme. A gene encoding a proteinase with homology to the *Lactococcus* proteinase was cloned and sequenced from *L. helveticus* CNRZ32 (Pederson *et al.,* 1999). A gene encoding the PrtM was also cloned from *L. helveticus* CNRZ32. In addition, a gene encoding a small type of proteinase with a molecular weight of 45 kDa was cloned from *L. helveticus* CM4 strain (Yamamoto *et al.,* 1999). As shown in Fig. 15.1, the 46-kDa pre-proteinase (later designated PrtY; see Table 15.3) was activated to the 45-kDa active proteinase by release of seven amino acids from the N-terminus, with the help of a maturation protein (PrtM, not yet identified in the CM4 genome) (Yamamoto *et al.,* 1995).

The existence of two types of proteinases with different substrate specificity towards casein was suggested in the *L. helveticus* strains (Ono *et al.,* 1997). However, there were no clear relationships

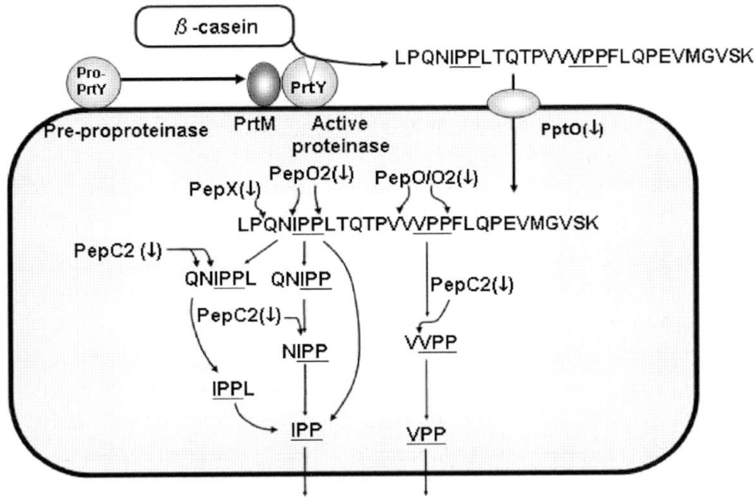

Figure 15.1 Postulated proteolytic system for Val-Pro-Pro and Ile-Pro-Pro processing and its down-regulation by amino acids in *Lactobacillus helveticus* CM4. An extracellular proteinase (PrtY), activated by a maturation protein (PrtM), plays an important role in decomposition of milk casein. Enzymes for endo-type processing (PepO and PepO2) and for N-terminal processing (PepX and PepC2) and an oligopeptide transporter (PrtO) were down regulated by addition of amino acids to the DNA microarray analysis (↓). Peptides containing Val-Pro-Pro and Ile-Pro-Pro sequences in the fermented milk were detected by LC-MS/MS analysis.

between proteinase specificity and the antihypertensive effects, and the ACE inhibitory activities of the fermented milk, which depended on the strain of *L. helveticus* used. The extracellular proteinase activity of each *L. helveticus* strain almost correlated with ACE inhibitory activity in the fermented milk (Table 15.1). These results strongly suggested that proteolysis of casein by the extracellular proteinase was the most important parameter in the processing of active components (Yamamoto *et al.*, 1994a). This is also supported by the fact that *L. helveticus* CM4, which was selected for its strong proteinase activity compared with other *L. helveticus* strains, has the ability to release ACE inhibitory peptides in fermented milk (Yamamoto *et al.*, 1997b). CM4 showed the highest production of antihypertensive peptides among the tested *L. helveticus* strains (Yamamoto *et al.*, 1997b). The proposed importance of the proteinase was also supported by the fact that a proteinase negative mutant was not able to generate antihypertensive peptides in fermented milk, whereas the wild-type strain did generate antihypertensive peptides (Yamamoto *et al.*, 1994a). Strain CM4, which had the highest antihypertensive peptide production in fermented milk, has potential for use as a functional food product.

As shown in Fig. 15.1, a long β-casein peptide with a 28 amino acid residue, including VPP and IPP sequences, was generated by the proteolytic action of an extracellular proteinase (Yamamoto *et al.*, 1993). The long peptide could then be hydrolysed to a 20 and an 8 amino acid peptide by the same proteinase. Proteinase activity is easily repressed by accumulated peptides produced by the proteinase in the fermented milk. Enzyme activity is also inactivated by a decrease in pH during fermentation. Thus, the majority of casein degradation by the extracellular proteinase must occur at the beginning of fermentation.

Intracellular processing of the antihypertensive peptides

Pathways for the intracellular processing of the antihypertensive peptides in *L. helveticus* CM4 are illustrated in Fig. 15.1. The long peptide released by the extracellular proteinase from β-casein is incorporated into the cell by an oligopeptide transporter (pptO), and processed intracellularly to VPP and IPP by peptidases. However, the details

Table 15.3 Genes encoding proteolytic enzymes reported in *Lactobacillus helveticus strains*

			Molecular weight (kDa)		
Proteolytic enzyme	Substrate	Gene	CM4	CNRZ32	53/7
Proteinase	No specificity	*prtY*	47.0	–	–
		prtH	–	204.0	–
Aminopeptidase	Xaa-|-Xbb-Xcc-	*pepC1*	51.4	51.4	51.4
		pepC2	53.0	–	–
		pepN	95.8	97.0	95.9
Aminopeptidase P	Xaa-|-Pro-	*pepP*	41.4	–	–
Aminopeptidase A	Glu-|-Xaa-	*pepA*	40.1	–	–
Endopeptidase	No specificity	*pepE*	50.0	52.1	–
		pepE2	50.3	50.0	–
		pepF	68.1	68.0	–
		pepO	73.6	71.2	–
		pepO2	73.8	71.4	–
		pepO3	73.1	72.5	–
XPDAP	Xaa-Pro-|-Xbb-Xcc-	*pepX*	90.5	90.5	–
Dipeptidase	Xaa-|-Xbb	*pepDA*	53.4	53.5	53.5
		pepD	54.0	–	–
		pepD	48.0	–	–
		pepD	94.9	–	–
		pepD	53.5	–	–
		pepV	51.5	51.5	–
Tripeptidase	Xaa-|-Xbb-Xcc	*pepT*	47.1	–	150.0
Prolidase	Xaa-|-Pro	*pepQ*	41.2	–	–
Prolinase	Pro-|-Xaa	*pepPN*	35.0	35.0	35.0
Proline iminopeptidase	Pro-|-Xaa-Xbb-	*pepI*	33.9	–	33.8

–, not reported.

of processing the long peptide to VPP and IPP are unclear. A carboxyl peptidase is not able to release the C-terminal amino acid (Ueno *et al.*, 2004) if a Pro-Pro-X sequence exists at the end. Recently, a key enzyme that catalyses carboxyl terminal processing to produce VPP and IPP was detected and purified from the CM4 strain (Ueno *et al.*, 2004). Amino terminal sequence analysis of the purified enzyme (Ueno *et al.*, 2004) and deduced amino acid sequence of the gene (Yamamoto *et al.*, 2004) showed it had homology to an endopeptidase (pepO) from *L. helveticus* CNRZ32. The enzyme can catalyse C-terminal processing of Val-Pro-Pro-Phe-Leu and Ile-Pro-Pro-Leu-Thr to VPP and IPP, respectively.

Unfortunately, the amino terminal processing reactions catalysed by peptidases to release VPP and IPP are still unclear. Peptidases of *L. helveticus* and their genes were recently well reviewed (Christensen *et al.*, 1999). Based on the previous reported characteristics of many peptidases from *L. helveticus*, processing of the N-terminal sequence of VPP and IPP has been deduced and is summarized in Fig. 15.1. Generally, amino peptidase shows broad specificity towards amino acids at the N-terminus, however, amino terminal processing seems to terminate if a proline residue is present at the N-terminal end of the peptide. X-prolyl dipeptidyl aminopeptidase (XPDAP), however, is able to release a di-peptide, Xaa-Pro, from the N-terminus. Alternatively, aminopeptidase may stop the catalysis at a Xaa-Pro-Pro- sequence if it is present. Therefore, N-terminal processing to release VPP and IPP may be catalysed by specific aminopeptidases and XPDAP. For more detailed understanding of the processing of these peptides in *L. helveticus*, the productivities of these peptides in transformant strains expressing each postulated peptidase gene or disruption of the corresponding peptidase genes from the wild type strain should be studied.

Comparison to other lactic acid bacteria

Recently, we completed the whole genome sequence of *L. helveticus* CM4, and the results revealed the presence of 2171 open reading frames in 2,028,493 bp of whole DNA sequence (T. Shinoda *et al.*, manuscript in preparation). To understand the processing of antihypertensive peptides from casein by the proteolytic system in *L. helveticus* CM4, genes for all proteinases and peptidases detected in the whole genome sequence were identified and are summarized in Table 15.3. As shown in Table 15.3, one type of proteinase gene previously cloned from strain CP790, and named *prtY*, is present in the CM4 genome sequence. However, a large proteinase gene, *prtH*, corresponding to a 200 kDa proteinase, and its maturation gene (*prtM*), reported by Pederson *et al.* (1999), were not detected in the CM4 genome. This observation supports our previous results that there are two types of extracellular proteinases, and *L. helveticus* can be classified by its proteinase type, as shown in Table 15.3. Twenty-two kinds of genes corresponding to peptidases were detected in the whole genome. Previously reported peptidase genes from *L. helveticus* CNRZ32 and *L. helveticus* 53/7 are also summarized in Table 15.3. At least eight new peptidase genes, *pepC2, pepP, pepA and pepQ* genes and four kinds of *pepD* genes, were identified in the genome of *L. helveticus* CM4. Six kinds of endopeptidase genes were found in the CM4 genome sequence, however, purification of a key enzyme involved in C-terminal processing, pepO, suggested it was the most important enzyme among the six detected endopeptidase genes. For N-terminal processing, only one gene for XPDAP (*pepX*) and five kinds of aminopeptidase genes were detected. Proteinase (*prtY* and *prtH*), endopeptidase (*pepE, pepE2, pepF, pepO, pepO2* and *pepO3*), XPDAP (*pepX*) and aminopeptidase (*pepC1, pepC2* and *pepN*) genes were selected for comparison to genes from other lactic acid bacteria. The comparative analysis of these genes to other species revealed that the CM4 peptidases were more homologous to those in *Lactobacillus acidophilus* NCFM strain (Altermann *et al.*, 2005) (Table 15.4). In addition, homologies to CM4 peptidases were detected in peptidases from *Lactobacillus gasseri* ATCC33323 (DOE Joint Genome Institute, 2002) and *Lactobacillus johnsonii* NC533 (Pridmore *et al.*, 2004). These observations suggest that the *L. acidophilus* group has similar peptidases, endopeptidase, XPDAP, and aminopeptidase to *L. helveticus* CM4 and might have the ability to process VPP and IPP if the initial decomposition of casein by an extracellular proteinase is accelerated in the fermentation process.

Repression of VPP and IPP processing

For growth of lactic acid bacteria in milk, the proteolytic system is activated in the milk medium because of a limited amount of amino acids. However, during fermentation in the milk medium, the proteolytic system of lactic acid bacteria is repressed by accumulated peptides in the fermented milk. The amount of VPP and IPP in the *L. helveticus* fermented milk was also repressed if amino acids were added to the fermented milk (N. Yamaguchi *et al.*, manuscript in preparation). Microarray analysis of the whole *L. helveticus* CM4 genome suggested extracellular proteinase, endopeptidases, XPDAP, some aminopeptidases and some kinds of peptide transporters were suppressed by addition of amino acids and this was linked to VPP and IPP processing (Fig. 15.1). Regulatory systems (codY system) that repress the proteolytic enzymes and transporters in the presence of amino acids were reported in *Lactococcus lactis* (den Hengst *et al.*, 2005). However, no codY-like protein in the *Lactobacillus* genome has been reported. Thus, a novel regulatory system must be involved in the regulation of VPP and IPP processing in *L. helveticus*.

Other fermented milk

γ-Aminobutyric acid (GABA) is an amine produced by decarboxylation of glutamate and has a blood pressure lowering effect on humans

Table 15.4 Proteolytic enzymes in various lactic acid bacteria with homology to enzymes in *Lactobacillus helveticus* CM4, which are expected to have roles in processing Val-Pro-Pro and Ile-Pro-Pro

	Proteinase		Aminopeptidase			XPDAP	Endopeptidase					
	prtY	prtH	pepN	pepC1	pepC2	pepX	pepE	pepE2	pepF	pepO	pepO2	pepO3
Lb. acidophilus NCFM	–	–	90	91	87	91	89	90	88	85	61	–
Lb. gasseri ATCC 33323	–	–	67	83	76	72	70	83	77	63	–	77
Lb. johnsonii NCC533	–	–	66	82	75	72	69	73	76	65	–	78
Lb. delbrueckii subsp. *bulgaricus* ATCC BAA-365	–	–	71	–	53	70	71	–	–	58	–	68
Lb. . casei subsp. casei ATCC334	–	–	62	59	–	–	–	–	54	–	–	–
Lc. lactis IL1403	–	–	–	–	–	–	–	–	–	–	–	–

Lb., *Lactobacillus*; *Lc.*, *Lactococcus.* All values are shown as% homology. Values below 50% were omitted (–).

through inhibition of a neurotransmitter in the central nervous system (Stanton, 1963). Lactic acid bacteria that produced GABA were isolated from commercial cheese starters and identified as *L. lactis* subsp. *lactis* (Nomura *et al.*, 1998). Glutamate decarboxylase catalyses the decarboxylation of glutamate into GABA. Oral administration of GABA to hypertensive patients decreased their blood pressures. A randomized, placebo-controlled study was recently performed on Japanese subjects (Inoue *et al.*, 2003). Thirty-nine mildly hypertensive patients (16 women and 23 men) aged 28–81 years were enrolled and took a fermented milk product containing 10–12 mg of GABA or placebo sample every day for 12 weeks. A significant decrease in blood pressure within 2–4 weeks was detected in the group taking GABA; decreases in blood pressure throughout the 12 weeks were maintained.

A recent study showed production of some ACE inhibitory peptides in milk fermented with *Lactobacillus delbrueckii* subsp. *bulgaricus* SS1, *Lactococcus lactis* subsp. *cremoris* FT4 (Gobbetti *et al.*, 2000), and others from cheese whey (Amhar *et al.*, 1996; Saito *et al.*, 1997). The clinical efficacy for these peptides are still unknown, but the fermented milk products may have potential in controlling blood pressure in subjects with high blood pressure.

Conclusion

In this chapter, antihypertensive effect and ACE inhibitory peptides in milk fermented with *L. helveticus* were reviewed. Among them, the well characterized ACE inhibitory peptides, VPP and IPP, released in *L. helveticus* fermented milk were discussed in detail. ACE inhibitory peptides isolated from other kinds of fermented milk were also introduced. A fermented milk product containing GABA produced by certain lactic acid bacteria showed significant clinical effects on Japanese subjects. Biologically functional fermented milk exerting a mild influence on hypertensive subjects without adverse effects have enormous potential in reducing blood pressure and risk of cardiovascular disease. In Japan, some food products containing antihypertensive peptides and GABA, with proven antihypertensive effects in clinical studies, are already in the Japanese market with approval for functional foods, Foods for Specified Health Use (FOSHU). For a better understanding of the *in vivo* antihypertensive effect, a pharmacokinetic analysis and the mode of action are needed. In addition, understanding the regulatory metabolism for processing antihypertensive peptides and other components in lactic acid bacteria are very important in controlling the manufacturing process.

References

Altermann, E., Russell, W.M., Azcarate-Peril, M.A., Barrangou, R., Buck, B.L., McAuliffe, O., Souther, N., Dobson, A., Duong, T., Callanan, M., *et al.* (2005). Complete genome sequence of the probiotic lactic acid bacterium *Lactobacillus acidophilus* NCFM. Proc. Natl. Acad. Sci. U.S.A. *102*, 3906–3912.

Amhar, A., Saito, T., Aimar, M.V., and Itoh, T. (1996). New derivation of the inhibitory activity against angiotensin converting enzyme (ACE) from sweet cheese whey. Tohoku J. Agric. Res. *47*, 1–8.

Bellamy, W., Takase, M., Yamauchi, K., Shimamura, S., and Tomita, M. (1992). Identification of the bactericidal domain of lactoferrin. Biochim. Biophys. Acta *1121*, 130–136.

Caldwell, P.R., Seegal, B.C., Hsu, K.C., Das, M., and Soffer, R.L. (1976). Angiotensin-converting enzyme: vascular endothelial localization. Science *191*, 1050–1051.

Christensen, J.E., Dudley, E.G. Pederson, J.A., and Steele, J.L. (1999). Peptidases and amino acid catabolism in lactic acid bacteria. Antonie van Leeuwenhoek *76*, 217–246.

DOE Joint Genome Institute. (2006). *Lactobacillus gasseri* ATCC 33323, whole genome sequenc. http://www.ncbi.nlm.nih.gov/nuscore/CP00043

FitzGerald, R.J. (1998). Potential uses of caseinophosphopeptides. Int. Dairy J. *8*, 451–457.

Fujita, H., and Yoshikawa, M. (1999). LKPNM: a prodrug-type ACE-inhibitory peptide derived from fish protein. Immunopharmacology *15*, 123–127.

Gobbetti, M., Ferranti, P., Smacchi, E., Gofredi, F., and Addeo, F. (2000). Production of angiotensin-I-converting-enzyme-inhibitory peptides in fermented milks started by *Lactobacillus delbrueckii* subsp. *bulgaricus* SS1 and *Lactococcus lactis* subsp. *cremoris* FT4. Appl. Environ. Microbiol. *66*, 3898.

Hata, Y., Yamamoto, M., Ohni, M., Nakajima, K., Nakamura, Y., and Takano, T. (1996). A placebo-controlled study of the effect of sour milk on blood pressure in hypertensive subjects. Am. J. Clin. Nutr. *64*, 767–771.

den Hengst, C.D., van Hijum, S.A.F.T., Geurts, J.M.W., Nauta, A., Kok, J., and Kuipers O.P. (2005). The *Lactococcus lactis codY* regulon identification of a conserved cis-regulatoru element. J. Biol. Chem. *280*, 34332–34342.

Inoue, K., Shirai, T., Ochiai, H., Kasao, M., Hayakawa, K., Kimura, M., and Sansawa, H. (2003). Blood-pressure-lowering effect of a novel fermented milk containing γ-aminobutyric acid (GABA) in mild hypertensives. Eur. J. Clin. Nutr. 57, 490–495.

Kajimoto, O., Aihara, K., Hirata, H., Takahashi, R., and Nakamura, Y. (2001a). Hypotensive effects of tablets containing 'lactotripeptides (VPP, IPP)'. J. Nutr. Food *4*, 51–61 (in Japanese).

Kajimoto, O., Aihara, K., Hirata, H., Takahashi, R., and Nakamura, Y. (2001b). Safety evaluation of excessive intake of the tablet containing 'lactotripeptides (VPP, IPP)' in healthy volunteers. J. Nutr. Food *4*, 37–46 (in Japanese).

Kawasaki, T., Seki, E., Osajima, K., Yoshida, M., Asada, K., Matsui, T., and Osajima, Y. (2000). Antihypertensive effect of valyl-tyrosine, a short chain peptide derived from sardine muscle hydrolysate, on mild hypertensive subjects. J. Hum. Hypertens. *14*, 519–523.

Kiwaki, M., Ikemura, H., Shimizu-Kadota, M., and Hirashima, A. (1989). Molecular characterization of a cell wall-associated proteinase gene from *Streptococcus lactis* NCDO763. Mol. Microbiol. *3*, 359–369.

Kohmura, M., Nio, N., Kubo, K., Minoshima, Y., Munekata, E. and Ariyoshi, Y. (1989). Inhibition of angiotensin-converting enzyme by synthetic peptides of human β-casein. Agric. Biol. Chem. *53*, 2107–2114.

Kok, J., and Venema, G. (1988). Genetics of proteinases of lactic acid bacteria. Biochimie *70*, 475–488.

Kunji, E.R., Mierau, I., Hagting, A., Poolman, B., and Konings, W.N. (1996). The proteolytic systems of lactic acid bacteria. Antonie van Leeuwenhoek *70*, 187–221.

Matsui, T., Li, C.H., and Osajima, Y. (1999). Preparation and characterization of novel bioactive peptides responsible for angiotensin I-converting enzyme inhibition from wheat germ. J. Pept. Sci. *5*, 289–297.

Maruyama, S., Mitachi, H., Tanaka, H., Tomizuka, N., and Suzuki, H. (1987a). Studies on the active site and antihypertensive activity of angiotensin I-converting enzyme Inhibitors derived from casein. Agric. Biol. Chem. *51*, 1581–1586.

Maruyama, S., Mitachi, H., Awaya, J., and Suzuki, H. (1987b). Angiotensin I-converting enzyme inhibitory activity of the c-terminal hexapeptide of α_{s1}-casein. Agric. Biol. Chem. *51*, 2557–2561.

Meisel, H., and Bockelmann, W. (1999). Bioactive peptides encrypted in milk proteins: proteolytic activation and thropho-functional properties. Antonie van Leeuwenhoek *76*, 207–215.

Migliore-Samour, D., Floćh, F., and Jollès, P. (1989). Biologically active casein peptides implicated in immunomodulation. J. Dairy Res. *56*, 357–362.

Miguel, M., Recio, I., Gomez-Ruiz, J.A., Ramos, M., and Lopez-Fandino, R. (2004). Angiotensin I-converting enzyme inhibitory activity of peptides derived from egg white proteins by enzymatic hydrolysis. J. Food Prot. *67*, 1914–1920.

Miyoshi, S., Ishikawa, H., Kaneko, T., Fukui, F., Tanaka, H., and Maruyama, S. (1991). Structures and activity of angiotensin-converting enzyme inhibitors in an alpha-zein hydrolysate. Agric. Biol. Chem. *55*, 1313–1328.

Nakamura, Y., Yamamoto, N., Sakai, K., Okubo, A., Yamazaki, S., and Takano, T. (1995a). Purification and characterization of angiotensin I-converting enzyme inhibitors from sour milk. J. Dairy Sci. *78*, 777–783.

Nakamura, Y., Yamamoto, N., Sakai, K., and Takano, T. (1995b). Antihypertensive effect of sour milk and peptides isolated from it that are inhibitors to angiotensin I-converting enzyme. J. Dairy Sci. *78*, 1253–1257.

Nakamura, Y., Kajimoto, O., Aihara, K., Mizutani, J., Ikeda, N., Nishimura, A., and Kajimoto, Y. (2004). Effects of the liquid yogurts containing 'lactotripeptide (VPP,IPP)' on high-normal blood pressure. J. Nutr. Food *7*, 123–137 (in Japanese).

Nomura, M., Kimoto, H., and Someya, Y. (1998). Production of γ-aminobutyric acid by cheese starters during cheese ripening. J. Dairy Sci. *81*, 1486–1491.

Ono, H., Yamamoto, N., Maeno, M., and Takano, T. (1997). Purification and characterization of a cell-wall associated proteinase from *Lactobacillus helveticus* CP53. Milchwissenschaft *52*, 373–377.

Pederson, J.A., Mileski, G.J., Weimer, B.C., and Steele, J.L. (1999). Genetic characterization of a cell envelope-associated proteinase from *Lactobacillus helveticus* CNRZ32. J. Bacteriol. *181*, 4592–4597.

Pridmore, R.D., Berger, B., Desiere, F., Vilanova, D., Barretto, C., Pittet, A.C., Zwahlen, M.C., Rouvet, M., Altermann, E., Barrangou, *et al.* (2004). The genome sequenc e of the probiotic intestinal bacterium *Lactobacillus johnsonii* NCC 533. Proc. Natl. Acad. Sci. USA *101*, 2512–2517.

Pritchard, G.G., and Coolbear, T. (1993). The physiology and biochemistry of the proteolytic system in lactic acid bacteria. FEMS Microbiol. Rev. *12*, 179–206.

Saito, T., Abubakar, A., Itoh, T., Arai, I., and Aimar, M.V. (1997). Development of a new type of fermented cheese whey beverage with inhibitory effects against angiotensin converting enzyme. Tohoku J. Agric. Res. *48*, 15–23.

Seppo, L., Kerojoki, O., Suomalainen, T., and Korpela, R. (2002). The effect of a *Lactobacillus helveticus* LBK-16 H fermented milk on hypertension – a pilot study on humans. Milchwissenschaft *57*, 124–127.

Seppo, L., Jauhiainen, T., Poussa, T., and Korpela, R. (2003). A fermented milk high in bioactive peptides has a blood pressure-lowering effect in hypertensive subjects. Am. J. Clin. Nutr. *77*, 326–330.

Smid, E.J., Poolman, B., and Konings, W.N. (1991). Casein utilization by lactococci. Appl. Environ. Microbiol. *57*, 2447–2452.

Stanton, H.C. (1963). Mode of action of gamma amino butyric acid on the cardiovascular system. Arch. Int. Pharmacodyn. Ther. *143*, 195–204.

Takano, T. (2002). Anti-hypertensive activity of fermented dairy products containing biogenic peptides. Antonie van Leeuwenhoek *82*, 333–340.

Tan, P.S., Poolman, B., and Konings, W.N. (1993). Proteolytic enzymes of *Lactococcus lactis*. J. Dairy Res. *60*, 269–286.

Teschemacher, H., Koch, G., and Brantl, V. (1997). Milk protein-derived opioid receptor ligands. Biopolymers *43*, 99–117.

Ueno, K., Mizuno, S., and Yamamoto, N. (2004). Purification and characterization of an endopeptidase that has an important role in the carboxyl terminal processing of antihypertensive peptides in *Lactobacillus helveticus* CM4. Lett. Appl. Microbiol. *39*, 313–318.

de Vos, W.M., Vos, P., de Haard, H., and Boerrigter, I. (1989). Cloning and expression of the *Lactococcus lactis* subsp. *cremoris* SK11 gene encoding an extracellular serine proteinase. Gene *85*, 169–176.

Yamamoto, N., Akino, A., and Takano, T. (1993). Purification and specificity of a cell-wall-associated proteinase from *Lactobacillus helveticus* CP790. J. Biochem. *114*, 740–745.

Yamamoto, N., Akino., A., and Takano, T. (1994a). Antihypertensive effect of the peptides derived from casein by an extracellular proteinase from *Lactobacillus helveticus* CP790. J. Dairy Sci. *77*, 917–922.

Yamamoto, N., Akino, A., and Takano, T. (1994b). Antihypertensive effect of different kinds of fermented milk in spontaneously hypertensive rats. Biosci. Biotech. Biochem. *58*, 776–778.

Yamamoto, N., Akino, A., Takano, T., and Shishido, K. (1995). Presence of active and inactive molecules of a cell wall-associated proteinase in *Lactobacillus helveticus* CP790. Appl. Environ. Microbiol. *61*, 698–701.

Yamamoto, N., Ishida, Y., Kawakami, N., and Yada, H. (1997a). *Lactobacillus helveticus* bacterium having high capability of producing tripeptide, fermented milk product, and process for preparing the same. EU Patent 1016709A1.

Yamamoto, N., and Takano, T. (1997b). Maturation protein need for activation of an extracellular proteinase in *Lactobacillus helveticus* CP790. J. Dairy Sci. *80*, 1949–1954.

Yamamoto, N., Ono, H., Maeno, M., and Takano, T. (1998). Classification of *Lactobacillus helveticus* strains by immunological differences in extracellular proteinases. Biosci. Biotech. Biochem. *62*, 1228–1230.

Yamamoto, N., Shinoda, T., and Takano, T. (1999). Molecular cloning and sequence analysis of a gene encoding an extracellular proteinase from *Lactobacillus helveticus* CP790. Biosci. Biotech. Biochem. *64*, 1217–1222.

Yamamoto, N., Ejiri, M., and Mizuno, S. (2003). Biogenic peptides and their potential use. Current Pharma. Des. *9*, 1345–1355.

Yamamoto, N., Shinoda, T., and Mizuno, S. (2004). Cloning and expression of an endopeptidase gene from *Lactobacillus helveticus* CM4 involved in processing antihypertensive peptides. Milchwissenschaft *59*, 593–597.

Lactobacillus gasseri OLL2716 (LG21): Anti-*Helicobacter pylori* Lactic Acid Bacterium

16

Katsunori Kimura

Abstract

Lactobacillus gasseri OLL2716 (LG21) was isolated from the faeces of a healthy human. LG21 was selected as a probiotic against *H. pylori* infection by screening more than 200 *Lactobacillus* strains by *in vitro* assays and animal studies. In clinical studies, the ingestion of yogurt containing LG21 decreased the number of *H. pylori* and IL-8 concentration in the gastric mucosa, and alleviated the mucosal inflammation in humans infected with *H. pylori*. LG21 was shown to be effective as a probiotic against *H. pylori* infection.

Introduction

The evidence of a relationship between *Helicobacter pylori* and gastric diseases has gradually become clear since Warren and Marshall succeeded in culturing this bacterium in 1982. *H. pylori* is associated with the pathogenesis of chronic gastritis, peptic ulcers and gastric cancer (Uemura *et al.*, 2001; Suerbaum *et al.*, 2002). Since the infection rate of *H. pylori* is very high, about 50% of Japanese, it is difficult to eradicate *H. pylori* in all people infected. Although eradication therapies such as proton pomp inhibitor-based triple regimens are available at present, the eradication therapies using antibiotics are often accompanied by side effects and not always successful because of antibiotic-resistant strains. In addition, the number of antibiotic-resistant strains has recently been increasing (Kato *et al.*, 2004; Masuda *et al.*, 2004).

Probiotics have been proposed as safe alternatives to antibiotic therapy for *H. pylori* infection (Kabir *et al.*, 1997; Aiba *et al.*, 1998). We screened more than 200 *Lactobacillus* strains to select a strain that exhibited excellent anti-*H. pylori* activities by *in vitro* assays and animal studies, and selected *Lactobacillus gasseri* OLL2716 (LG21) as the most suitable strain for use as a probiotic against *H. pylori* infection. In clinical studies, the ingestion of yogurt containing LG21 decreases the number of *H. pylori* and alleviates the mucosal inflammation in humans infected with *H. pylori*.

Selection of a probiotic for *H. pylori* infection

Helicobacter pylori is a spiral Gram-negative bacterium with some unipolar flagella. This bacterium displays strong urease activity, and thus can catalyse urea in the stomach to ammonia and carbon dioxide. Because of this ammonia production, *H. pylori* can survive under strong acidic conditions. It is well known that lactic acid bacteria inhibit the growth of *H. pylori in vitro* (Bhatia *et al.*, 1989; Midolo *et al.*, 1995). It was reported that orally administered lactobacilli eradicated *H. pylori* using a gnotobiotic murine model (Aiba *et al.*, 1998).

To select a probiotic for *H. pylori* infection in humans, anti-*H. pylori* abilities (resistance to gastric juice, proliferation under acidic conditions, adherence to cultured gastric epithelial cells, suppression of *H. pylori* during co-fermentation) of more than 200 *Lactobacillus* strains were examined *in vitro*. *Lactobacillus gasseri* OLL2716 (LG21), *Lactobacillus gasseri* No. 6 and *Lactobacillus salivarius* WB1004 were selected as lactic acid bacteria with excellent anti-*H. pylori* abilities

(Kimura, 2004). Following *in vitro* experiments, the effects of the administration of the selected three *Lactobacillus* strains (LG21, *L. gasseri* No. 6 and *L. salivarius* WB1004) on mice infected with *H. pylori* were examined. Orally administered LG21 and *L. salivarius* WB1004 eradicated *H. pylori* in mice and the anti-*H. pylori* IgG levels of the LG21-administed mice were the lowest (Kimura, 2004). Therefore, LG21 was selected as a promising probiotic against *H. pylori* infection. LG21 also inhibited the growth of clarithromycin-resistant *H. pylori in vitro* and eradicated clarithromycin-resistant *H. pylori* in mice (Ushiyama *et al.*, 2003).

Effects of LG21 on *H. pylori* infection in humans

Effects of LG21 on *H. pylori*-infected healthy humans

The effects of yogurt containing LG21 on *H. pylori* infection in healthy humans were examined (Sakamoto *et al.*, 2001). Thirty-one subjects infected with *H. pylori* ingested 90 g of normal yogurt containing no LG21 twice daily for 8 weeks. This 8-week period was followed by a 1-week period without yogurt ingestion. This in turn was followed by an 8-week period during which the subjects ingested 90 g of yogurt containing 10^9 cfu of LG21 twice daily. The urea breath test and assays of serum pepsinogen I/II ratios were performed to measure the population of *H. pylori* and to evaluate the degree of mucosal inflammation in the stomach, respectively, at three points: prior to ingestion of yogurt, after the first 8-week period, and after the second 8-week period. In addition, gastric biopsy specimens were taken from six subjects for quantitative culture both before and after ingestion of LG21 yogurt. It is known that the urea breath test provides semi-quantitative assessments of the density of *H. pylori* colonization of gastric mucosa (Debongnie *et al.* 1991; Perri *et al.* 1998). A high $\Delta^{13}C$ value indicates a large population of *H. pylori* in the stomach. The serum pepsinogen I/II ratio is reduced in *H. pylori*-positive patients with gastritis and gastric ulcer (Wagner *et al.* 1994; Furuta *et al.* 1997) and is correlated with the activity of gastritis (Samloff *et al.* 1982). A low pepsinogen I/II ratio indicates severe mucosal inflammation in the stomach. The $\Delta^{13}C$ value for the urea breath test did not significantly change after ingestion of normal yogurt, but did significantly decrease following ingestion of LG21 yogurt (Fig. 16.1). The serum pepsinogen I/II ratio did not significantly change after ingestion of normal yogurt, but did significantly increase following ingestion of LG21 yogurt (Fig. 16.2). The number of *H. pylori* in the gastric biopsy specimens from all six subjects decreased after ingestion of LG21 yogurt. Similarly, long-term (12 months) ingestion of LG21 yogurt significantly decreased the $\Delta^{13}C$ value in the urea breath test and significantly increased the serum pepsinogen I/II ratio (Kimura, 2004).

Accumulating evidence indicates that IL-8 plays a major role in the mucosal inflammation caused by *H. pylori* infection. Since LG21 suppressed *H. pylori*-induced IL-8 production *in vitro* (Ushiyama *et al.*, 2003; Tamura *et al.*, 2006), the effect of LG21 on IL-8 production in the gastric mucosa of *H. pylori* infected subjects was examined in a study with a randomized double-blind placebo-control design. Twenty-five subjects infected with *H. pylori* ingested 120 g of yogurt containing 10^9 cfu of LG21 or placebo yogurt twice daily for 8 weeks. Biopsy specimens were taken both before and after ingestion of each yogurt to measure IL-8 levels in the gastric mucosa of subjects. IL-8 concentrations in the gastric mucosa of the LG21 treated group significantly decreased after ingestion of LG21 yogurt (580 ± 527 pg/mg protein vs. 387 ± 192 pg/mg protein; $P = 0.02$). In contrast, IL-8 concentrations in the gastric mucosa of the placebo treated group did not significantly change after ingestion of the placebo yogurt (519 ± 444 pg/mg protein vs. 526 ± 433 pg/mg protein; $P = 0.54$) (Tamura *et al.*, 2006). These results suggest that ingestion of LG21 yogurt decreases the number of *H. pylori* and alleviates the mucosal inflammation in the stomach.

Effects of LG21 on *H. pylori*-infected patients with upper gastrointestinal symptoms

The effects of yogurt containing LG21 on *H. pylori* infected patients with upper gastrointestinal

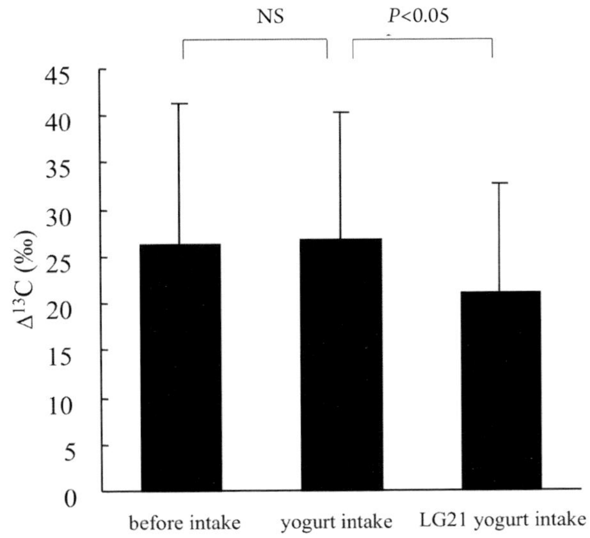

Figure 16.1 Effect of yogurt containing LG21 on $\Delta^{13}C$ value of subjects infected with *H. pylori.*

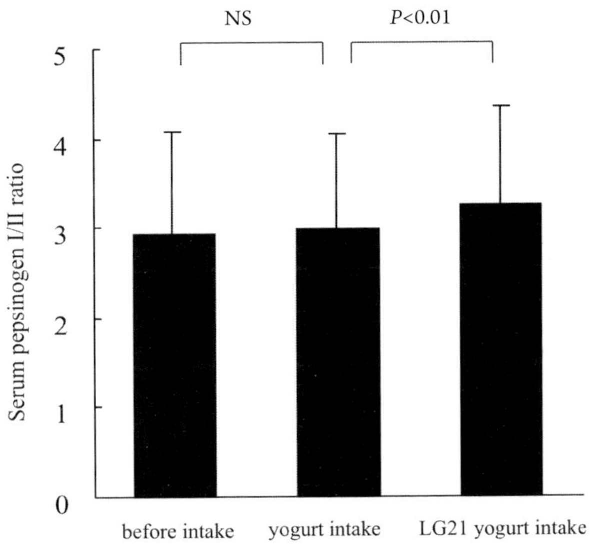

Figure 16.2 Effect of yogurt containing LG21 on serum pepsinogen I/II ratio of subjects infected with *H. pylori.*

symptoms were examined (Miyaji *et al.,* 2006). Thirty-nine *H. pylori* infected patients with upper gastrointestinal symptoms were divided into two groups. Yogurt containing LG21 or placebo yogurt was ingested for 12 weeks in either group. The serum pepsinogen I/II ratio significantly increased in patients who ingested LG21 yogurt and this increase lasted for 4 weeks even after cessation of yogurt ingestion. However, the positive effects of the LG21 yogurt were not observed in the placebo treated group. These results indicate that the ingestion of LG21 yogurt improves gastric mucosal inflammation even in *H. pylori* infected patients with upper gastrointestinal symptoms and LG21 yogurt is useful food for the therapy of gastric inflammation induced by *H. pylori* infection.

Mechanisms of therapeutic effects of LG21 on *H. pylori* infection

The exact mechanisms of therapeutic effects of LG21 on *H. pylori* infection are still unclear. LG21 displays high resistance to gastric juice, grows well under acidic conditions, and can be detected in gastric mucus layer of subjects who have ingested LG21 (Fujimura *et al.,* 2006). It is known that undissociated lactic acid exerts an inhibitory effect on bacteria by penetrating the membranes and lowering the intracellular pH. Lactic acid exerted a stronger inhibitory effect on the growth of *H. pylori* and the activity of urease of *H. pylori* than hydrochloric acid at the same pH (Aiba *et al.,* 1998). LG21 produced large amounts of lactic acid and inhibited the growth of *H. pylori* in *in vitro* experiments (Kimura, 2004). In addition, LG21 suppressed *H. pylori*-induced IL-8 production in both gastric cell line and gastric mucosa of *H. pylori* infected subjects (Ushiyama *et al.,* 2003; Tamura *et al.,* 2006). Considering these results, a decrease in *H. pylori* resulting from antimicrobial activity of lactic acid and suppression of *H. pylori*-induced IL-8 production by LG21 appear to be involved in the therapeutic effects of LG21 on *H. pylori* infection.

Conclusion

Patients who are *H. pylori*-positive and have peptic ulcers should receive eradication therapy. However, for many people infected with *H. pylori,* probiotic therapy may be a safe and simple alternative to antibiotic therapy. The density of *H. pylori* colonization in the stomach is of importance in the pathogenesis of infection associated with this bacterium. It was reported that there is a correlation between *H. pylori* density and gastric inflammation or duodenal ulceration (Atherton *et al.,* 1996; Furuta *et al.,* 1997). Since the ingestion of LG21 yogurt suppresses *H. pylori* in addition to alleviating inflammation in the stomach, LG21 may reduce the risk of the *H. pylori*-induced gastrointestinal diseases.

References

Atherton, J.C., Tham, K.T., Peek, R.M., Cover, T.L., and Blaser, M.J. (1996). Density of *Helicobacter pylori* infection *in vivo* as assessed by quantitative culture and histology. J. Infect. Dis. *174,* 552–556.

Aiba, Y., Suzuki, N., Kabir, A.M.A., Takagi, A., and Koga, Y. (1998). Lactic acid-mediated suppression of *Helicobacter pylori* by the oral administration of *Lactobacillus salivarius* as a probiotic in a gnotobiotic murine model. Am. J. Gastroenterol. *93,* 2097–2101.

Bhatia, S.J., Kochar, N., Abraham P., Nair, N.G., and Mehta, A.P. (1989). *Lactobacillus acidophilus* inhibits growth of *Campylobacter pylori in vitro*. J. Clin. Microbiol. *27,* 2328–2330.

Debongnie, J.C., Pauwels, S., Raat, A., de Meeus, Y., Haot, J., and Mainguet, P. (1991). Quantification of *Helicobacter pylori* infection in gastritis and ulcer disease using a simple and rapid carbon-14-urea breath test. J. Nucl. Med. *32,* 1192–1198.

Fujimura, S., Kato, S., Oda, M., Miyahara, M., Ito, Y., Kimura, K., Kawamura, T., Ohnuma, M., Tateno, H., and Watanabe, A. (2006). Detection of *Lactobacillus gasseri* OLL2716 strain administered with yogurt drink in gastric mucus layer in humans. Lett. Appl. Microbiol. *45,* 578–581.

Furuta, T., Kaneko, E., Baba, S., Arai, H., and Futami, H. (1997). Percentage changes in serum pepsinogens are useful as indices of eradication of *Helicobacter pylori.* Am. J. Gastroenterol. *92,* 84–88.

Kabir, A.M.A., Aiba, Y., Takagi, A., Kamiya, S., Miwa, T., and Koga, Y. (1997). Prevention of *Helicobacter pylori* infection by lactobacilli in a gnotobiotic murine model. Gut. *41,* 49–55.

Kato, S., Konno, M., Maisawa, S., Tajiri, H., Yoshimura, N., Shimizu, T., Toyoda, S., Nakayama, Y., and Iinuma, K. (2004). Results of triple eradication therapy in Japanese children: a retrospective multicenter study. J. Gastroenterol. *9,* 838–843.

Kimura, K. (2004). Health benefits of probiotics: Probiotics for *Helicobacter pylori* infection. Food Sci. Technol. Res. *10,* 1–5.

Masuda, H., Hiyama, T., Yoshihara, M., Tanaka, S., Haruma, K., and Chayama, K. (2004). Characteristics and trends of clarithromycin-resistant *Helicobacter pylori* isolates in Japan over a decade. Pathobiology *71,* 159–163.

Midolo, P.D., Lambert, J.R., Hull, R., Luo, F., and Grayson, M.L. (1995). *In vitro* inhibition of *Helicobacter pylori* NCTC 11637 by organic acids and lactic acid bacteria. J. Appl. Bacteriol. *79,* 475–479.

Miyaji, M., Yasue, N., Hayakawa, T., Higashi, K., Kimura, M., Tsukada, K., Takeuchi, T., Uchida, M., Kume, A., and Kohno, T. (2006). Yogurt containing *Lactobacillus gasseri* OLL2716 (LG21) ameliorates gastric mucosal inflammation in *Helicobacter pylori* infected patients with upper gastrointestinal symptoms. Jpn. Pharmacol. Ther. *34,* 91–96.

Perri, F., Clemente, R., Pastore, M., Quitadamo, M., Festa, V., Bisceglia, M., Li Bergoli, M., Lauriola, G., Leandro, G., Ghoos, Y., *et al.* (1998). The ^{13}C-urea breath test as a predictor of intragastric bacterial load and severity of *Helicobacter pylori* gastritis. Scand. J. Clin. Lab. Invest. *58,* 19–28.

Sakamoto, I., Igarashi, M., Kimura, K., Takagi, A., Miwa, T., and Koga, Y. (2001). Suppressive effect of *Lactobacillus gasseri* OLL2716 (LG21) on *Helicobacter pylori* infection in humans. J. Antimicrob. Chemother. *47*, 709–710.

Samloff, I.M., Varis, K., Ihamaki, T., Siurala, M., and Rotter, J.I. (1982). Relationships among serum pepsinogen I, serum pepsinogen II, and gastric mucosal histology. Gastroenterology *83*, 204–209.

Suerbaum, S. and Michetti, P. (2002). *Helicobacter pylori* infection. N. Engl. J. Med. *347*, 1175–1186.

Tamura, A., Kumai, H., Nakamichi, N., Sugiyama, T., Deguchi, R., Takagi, A., and Koga, Y. (2006). Suppression of *Helicobacter pylori*-induced interleukin-8 production *in vitro* and within the gastric mucosa by a live *Lactobacillus* strain. J. Gastroenterol. Hepatol. *21*, 1399–1406.

Uemura, N., Okamoto, S., Yamamoto, S., Matsumura, N., Yamaguchi, S., Yamakido, M., Taniyama, K., Sasaki, N., and Schlemper, R.J. (2001). *Helicobacter pylori* infection and the development of gastric cancer. N. Engl. J. Med. *345*, 784–789.

Ushiyama, A., Tanaka, K., Aiba, Y., Shiba, T., Takagi, A., Mine, T., and Koga, Y. (2003). *Lactobacillus gasseri* OLL2716 as a probiotic in clarithromycin-resistant *Helicobacter pylori* infection. J. Gastroenterol. Hepatol. *18*, 986–991.

Wagner, S., Haruma, K., Gladziwa, U., Soudah, B., Gebel, M., Bleck, J., Schmidt, H., and Manns, M. (1994). *Helicobacter pylori* infection and serum pepsinogen A, pepsinogen C, and gastrin in gastritis and peptic ulcer: significance of inflammation and effect of bacterial eradication. Am. J. Gastroenterol. *89*, 1211–1218.

Effects and Mechanisms of Probiotics on the Prevention and Treatment of Allergic Rhinitis

17

Toshitaka Odamaki,* Noriyuki Iwabuchi and Jin-zhong Xiao

Abstract

Studies on probiotic bacteria in the treatment or prevention of allergic rhinitis have shown encouraging results such as improved symptoms, reduced use of relief medication and modulated immunological parameters. Bacterial cells were shown to be sampled by intestinal immunocompetent cells and hence exerted diverse immunomodulatory effects on the hosts. Clinical studies using *Bifidobacterium longum* BB536 to prevent or treat allergic reactions to Japanese cedar pollen demonstrated possible involvement of gut microbiota in sensitization to allergens and development of symptoms, and the potential of probiotics in stabilizing the microbiota. Based on these studies, we suggest two possible mechanisms for the antiallergic activity of probiotics: (1) immunomodulatory effects via bacterial cell components mediated by intestinal antigen-presenting cells (biogenic effects); (2) immunomodulatory effects via generating or stabilizing a balanced gut microbiota (probiotic effects). These studies suggest that probiotics may serve as an alternative treatment for allergic rhinitis, although further studies are needed to determine this conclusively.

Introduction

The prevalence of atopic disorders such as eczema, allergic rhinitis, and asthma has rapidly increased worldwide over the past several decades, especially in the industrialized countries. Allergic rhinitis is a common disorder affecting a large percentage of the population, which can be classified as seasonal allergic rhinitis or perennial allergic rhinitis depending on whether symptoms manifest at defined yearly intervals or throughout the year, respectively. Estimates of the prevalence of allergic rhinitis range from 9% to as high as 42% of population in different countries (Settipane, 2001; Heinrich *et al.*, 2002). Recently, Baba and Nakae (2008) performed a nationwide epidemiological survey on the prevalence of allergic rhinitis in Japan, which showed that the prevalence has increased from 26.5% to 39.4% of Japanese population (with 26.5% for seasonal allergic rhinitis and 23.4% for perennial allergic rhinitis) over the past ten years, in comparison with a similar survey performed in 1998. Drugs such as antihistamines and local steroids are commonly used to relieve symptoms of allergic rhinitis but, as of yet, there are no treatments that lead to complete resolution of the disorder.

Association between the composition of intestinal microflora and the development of allergies has been suggested from numerous studies. Studies on the composition of intestinal microflora in both allergic and nonallergic 2-year-old children have indicated that the prevalence of bifidobacteria is lower in allergic infants, whereas counts of *Staphylococcus aureus* and enterobacteria are higher (Bjorksten 1999). In comparison with healthy infants, babies with allergies harbour less enterococci during the first month of life and less bifidobacteria during the first year of life (Bjorksten 2001). Meanwhile, numerous human and animal studies have shown that probiotic bacteria, such as lactobacilli and bifidobacteria, are potentially beneficial in the prevention and/or treatment of allergic diseases (Majama and

*corresponding author

Isolauri, 1997; Matsuzaki *et al.*, 1998; Kalliomaki *et al.*, 2001).

In this paper, we present clinical results of the antiallergic effects of probiotics to allergic rhinitis, focusing specifically on probiotic strain *Bifidobacterium longum* BB536 in the treatment and prevention of allergies to Japanese cedar (*Cryptomeria japonica*) pollen (JCP). We also attempt to provide insight into the mechanisms involved in the antiallergic effects of probiotics.

Clinical effects on seasonal allergic rhinitis

In Japan, the most common seasonal allergic rhinitis is Japanese cedar pollinosis (JCPsis), which is an immunoglobulin E (IgE)-mediated type I allergy caused by exposure to JCP. Typically, the JCP season begins in early February and ends the latter part of April. Characteristic symptoms of JCPsis are sneezing, rhinorrhea, nasal blockage, itching of the eyes, etc. Over recent decades, the prevalence of this allergy has been increasing and now affects >20% of the Japanese population (Baba and Nakae, 2008). Because JCPsis is an important public health issue in Japan, many studies from Japan have focused on evaluating the effects of functional foods and supplements, including probiotics, in the prevention and treatment of JCPsis.

Studies on lactobacilli

Ishida *et al.* (2005) conducted a placebo-controlled single-blind study over a 2-year period (2002–2003) by giving heat-killed *L. acidophilus* strain L-92 in milk during the JCP seasons. Compared with the milk-only placebo group, improvement of ocular symptoms and disruption of normal life activities were observed in the probiotic group, but no significant effects were found in blood parameters. In a randomized controlled trial, Fujiwara *et al.* (2005) reported that the intake of yogurt fermented with *L. paracasei* strain KW3110 was associated with repression of T-helper type (Th) 2 cell generation and eosinophil activation during the JCP season of 2003, but no clinical effect was observed. From a trial performed during the JCP season of 2005, Tamura *et al.* (2007) found that intake of fermented milk containing *L. casei* strain Shirota did not prevent allergic symptoms but delayed the occurrence of allergic symptom in JCPsis patients. No effect was observed on blood parameters. On the other hand, Ogawa *et al.* (2007) reported that administration of a symbiotic containing *L. casei* subsp. *casei* strain JCM 1134^{T} and dextran to JCPsis subjects in the season of 2005 tended to suppress changes in the nasal and ocular symptom scores and changes of some blood parameters. These studies assessed the effects of probiotic *Lactobacillus* strains in the treatment of JCPsis. However, the impact of these strains on the development of symptoms was inconsistent and thus a determination of their efficacies is difficult. These inconsistent results might be a result of different strains being tested, changes in symptom development during different seasons, and the limited numbers of subjects in some of the studies.

A study was conducted in Finland to test the effect of *L. rhamnosus* GG, a strain reported to be effective in the early prevention of atopic dermatitis (Kalliomaki *et al.*, 2002), in the treatment of birch pollen allergy during the pollen season. No observable change in symptoms of the patients was seen and there was no reduction in the use of medication during the birch pollen season. However, increased allergen specific IgA levels in saliva were observed (Helin *et al.*, 2002; Piirainen *et al.*, 2008).

Studies on bifidobacteria

We have conducted several randomized controlled trials to assess *Bifidobacterium longum* BB536, a probiotic strain shown to be effective in the prevention and treatment of a broad spectrum of animal and/or human disorders such as constipation, intestinal infections, colon cancer, and allergic diseases (Xiao, 2008). During the JCP season of 2004, JCPsis subjects were given yogurt supplemented with BB536 daily for a total of 14 weeks. Subjective symptoms were alleviated, the increase of blood eosinophil ratios was diminished, and the decrease of gamma-interferon (IFN-γ) levels was suppressed in the BB536 group compared with the placebo group (Xiao *et al.*, 2006a). As 2004 was a mild season for JCP dispersion and the effects of BB536 were tested using yogurt containing other lactic acid bacteria, the efficacy of BB536 alone in a normal or heavy pollen season needed to be investigated. We therefore performed another trial using lyophilized BB536 powder during the JCP season of 2005, a heavy JCP season, to confirm the efficacy of BB536 in the treatment of JCPsis (Xiao *et al.*, 2006b). Forty four JCPsis subjects received either BB536 (approximately 1×10^{11} cfu per day) or placebo daily beginning approximately 4 weeks before the onset of the pollen season through the end of the pollen season for a total of 13 weeks. BB536 intake was associated with a significant reduction in number of subjects prematurely terminated owing to severe symptoms and pollinosis medication (two subjects in the BB536 group vs. nine subjects in the placebo group, P=0.0057). Significant decrease of the subjective symptom scores of rhinorrhea and nasal blockage was observed in BB536 group compared with placebo group. Comparison medical symptom scores showed marked improvements of composite and each symptom scores following BB536 intake (Fig. 17.1A and B). A Th2-skewed immunoresponse, as indicated by the significant decrease of plasma IFN-γ and by the significant increase of thymus- and activation-regulated chemokine (TARC) and JCP-specific IgE levels, was observed in the JCPsis subjects during periods of pollen dispersion. BB536 tended to suppress elevations of JCP-specific IgE and significantly suppressed TARC levels (Fig. 17.1C,D).

In the studies during the pollen seasons, there were many variables such as differences in pollen amounts and symptom severities within or between seasons. In order to overcome the problems associated with these field-based

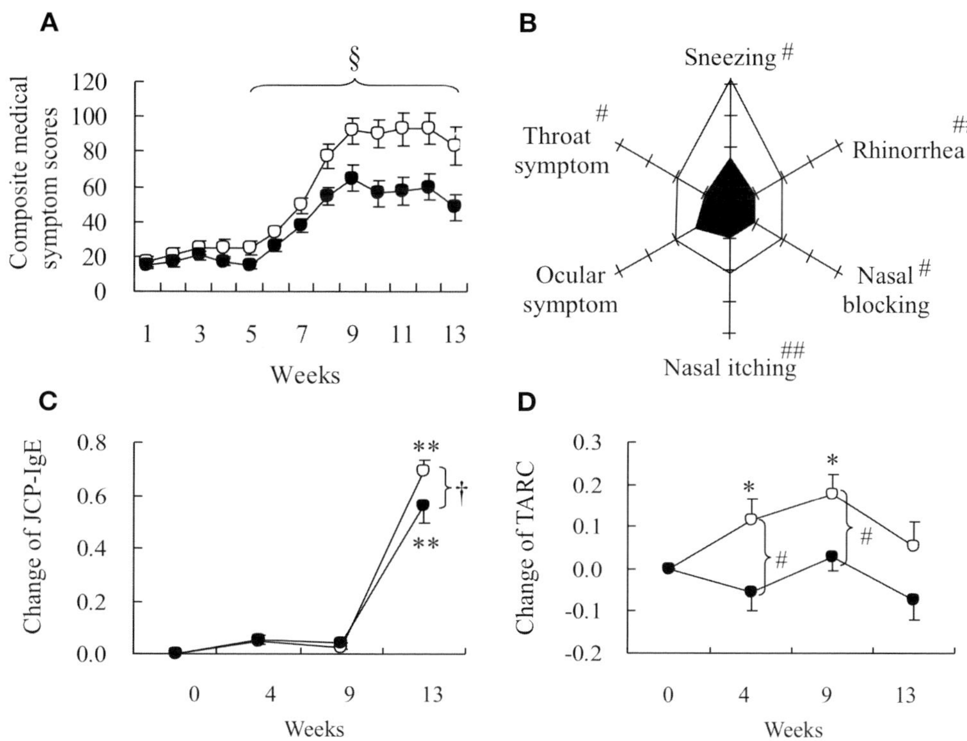

Figure 17.1 Symptom scores and blood parameters during the pollen season for BB536 (filled circles or fields) and placebo (open circles or fields) groups. (A) Means of composite medical symptom scores (sum of each of medical symptom score as indicated in B); (B) mean values of area-under-the-curves (AUCs) for each of the medical symptom scores; (C) changes of blood JCP (Japanese ceder pollen)-specific IgE titres; (D) changes of plasma TARC (thymus- and activation-regulated chemokine) levels. Error bars show SEM. §P<0.05, compared between groups on the scores of 5–13 weeks with the two way repeated ANOVA; †P< 0.1, *P<0.05, **P<0.01, compared with baseline by the Wilcoxon signed-rank test; # P<0.05, ## P<0.01, compared between groups with the Mann–Whitney U-test.

studies, we assessed the efficacy of BB536 intake on symptoms under controlled exposures to JCP in an environmental exposure unit outside of the normal JCP season in a double-blind, two-way crossover design (Xiao *et al.,* 2007b). After a 1-week run-in period, 24 JCPsis subjects ($n = 24$) were randomly assigned to receive BB536 powder (approximately 1×10^{11} cfu per day) or placebo daily for 4 weeks in a double-blind, two–way crossover design with a 2-week washout period. At the end of each intake period, subjects received controlled JCP exposure for 4 hours in the exposure unit. In comparison with the placebo, BB536 intake significantly reduced ocular symptom scores during JCP exposures (Fig. 17.2A). Evaluating delayed symptoms after exposures indicated that scores for disruption of normal activities were significantly lower in the BB536 group compared with the placebo group. Use of medication was markedly reduced by BB536 intake (Fig. 17.2B).

In summary, studies during the pollen season or outside of the normal pollen season showed significantly lower levels of allergic rhinitis symptoms, lower use of relief medication and markedly impacted immunological outcomes with the administration of BB536. The results of these studies indicate that BB536 is effective at preventing and treating the symptoms of JCP allergy, possibly through modulation of the Th2-skewed immune response.

Clinical effects on perennial allergic rhinitis

Several randomized controlled trials have been conducted to determine the effects of probiotic lactobacilli in the treatment of perennial allergic rhinitis. In the study by Wang *et al.* (2005), patients were given fermented milk supplemented with *L. paracasei* strain LP-33 daily for 30 days. Scores for overall quality of life were significantly improved in the LP-33 group compared with the placebo group. In a subsequent study by the same group, the overall quality of life of patients was significantly improved by living cells as well as by heat-killed cells of LP-33 (Peng and Tsu, 2005). Ishida *et al.* (2005) reported that orally administered *L. acidophilus* strain L-92 improved the nasal symptom-medication scores, but had no measurable effect on blood parameters. Giovannini *et al.* (2007) examined the effect of long-term consumption of fermented milk containing a specific *L. casei* in preschool children suffering from allergic asthma and/or rhinitis. The number and duration of rhinitis episodes were significantly reduced by the probiotic intake. These studies of the effect of probiotic lactobacilli on perennial allergic rhinitis resulted in lower symptom scores and/or reduced use of medication suggesting probiotic lactobacilli may serve as an alternative treatment for allergic rhinitis.

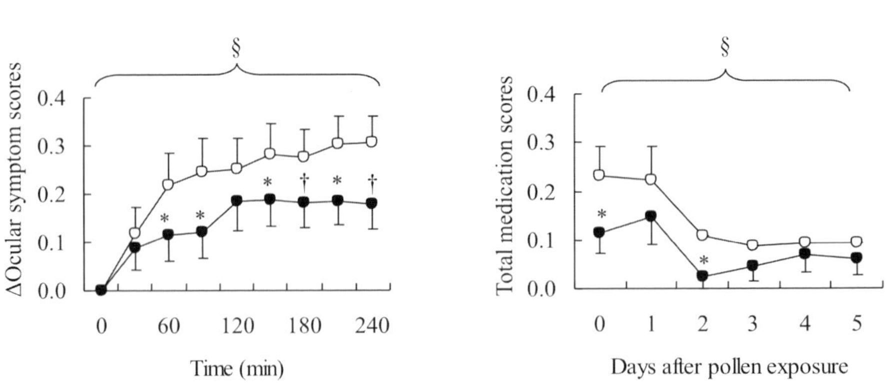

Figure 17.2 Symptom and medication scores during and after pollen exposure in the environmental exposure unit for BB536 (filled circles) and placebo (open circles) groups. (A) Means of changes of ocular symptom scores during pollen exposure; (B) means of total medication scores (sum of scores for oral medicines, nasal sprays and eye drops) during post-exposure. Error bars show SEM. † $P<0.1$, *$P<0.05$, compared between groups with Wilcoxon signed-rank test; § $P<0.05$, compared between groups on area under curves (AUCs) with Wilcoxon signed-rank test.

Studies on the antiallergic mechanism

Immunomodulatory effects of bacterial cellular constituents

Type I allergies such as allergic rhinitis and hay fever are caused by characteristic immune responses to allergens primarily mediated by Th2 cells (Lee *et al.*, 1988; Romagnani, 1994). Th2 cells synthesize high levels of interleukin (IL)-4, IL-5 and IL-13, which leads to the production of allergen-specific IgE and the release of histamine and leukotriene from mast cells (Romagnani, 1994). Conversely, Th1 cells suppress Th2 immune responses by secreting IFN-γ. IL-4 induces class switching to IgG1 and IgE, whereas IFN-γ is involved in IgG2a class switching (Kapsenberg *et al.*, 1992). Therefore, estimation of IgE production and Th1/Th2 cytokine balance is suggested to be important in the study of type I allergic disorders (Kidd, 2003). Indeed, some probiotic strains have been shown to promote production of Th1-skewing cytokines such as IL-12 and IFN-γ and to suppress production of IgE and Th2 cytokines (Shida *et al.*, 1998; Fujiwara *et al.*, 2004; Mohamadzadeh *et al.*, 2005; Pochard *et al.*, 2005). These studies also indicate that probiotics induce IL-12 production by antigen-presenting cells (APCs) such as macrophages and dendritic cells (DCs).

An alternative to the Th1/Th2 balance theory is that regulatory T-cells control Th1 and Th2 responses, largely via the production of IL-10 and TGF-β, and hence the development of both atopic and autoimmune diseases (Rautava *et al.*, 2005; Rook and Brunet, 2005). The regulatory function in the immune response plays a key role in preventing allergic disorders, and regulatory cytokines such as IL-10 and TGF-β are important candidates for being the agents involved in the immunomodulatory effect of probiotics. Several studies have shown that the Th2 suppressive effect of probiotics is mediated by immune-regulatory functions linked to the induction of regulatory cytokines such as IL-10 and transforming growth factor β (TGF-β) (von der Weid *et al.*, 2001; Niers *et al.*, 2005). It has been reported that macrophages stimulated with certain lactic acid bacteria and bifidobacteria produce IL-10 (He *et al.*, 2002; Niers *et al.*, 2005), and that TGF-β (Th3) is induced with *L. paracasei* in mixed lymphocyte reactions (von der Weid *et al.*, 2001). Further, it has been reported that neonatal application of probiotics inhibited allergic sensitization and airway disease in a murine model of asthma by induction of T regulatory cells associated with increased TGF-β production (Feleszko *et al.*, 2007).

We investigated *in vitro* cytokine and IgE production by murine splenic cells (Iwabuchi *et al.*, 2007). It was found that BB536 suppressed IgE and IL-4 production by ovalbumin-sensitized splenic cells without abundant induction of Th1-skewing cytokine production, such as IL-12 and IFN-γ in comparison with lactic acid bacteria (Fig. 17.3). These results suggested that BB536 suppressed Th2 immune responses partially independent of Th1-skewing cytokines mediated by APCs such as DCs. Similarly, Ohno *et al.* (2005) reported that *B. bifidum* suppressed antigen-specific IgE production *in vivo* via a different mechanism from those of Th1 inducible bacteria.

Recently, it has been discovered that chemokines play an important role in the inflammatory network responsible for allergy symptoms (Romagnani *et al.*, 2001). Th2-attracting chemokines, such as TARC (CCL17) and macrophage-derived chemokine (CCL22), selectively attract Th2 memory T cells. It has been reported that plasma TARC and macrophage-derived chemokine levels show significant correlation with the severity of atopic dermatitis (Fujisawa *et al.*, 2002) and anti-TARC antibodies diminished the degree of airway hyper-responsiveness in a murine model of allergic asthma (Kawasaki *et al.*, 2001). These observations suggest that Th2-attracting chemokines might be pivotal in the development of allergic disorders. In the trial on JCPsis subjects, BB536 significantly suppressed the increase of plasma TARC associated with pollen dispersion (Xiao *et al.*, 2006b, 2007a). In *in vitro* studies, we showed that heat-killed BB536 suppressed the production of Th2-attracting chemokines by APCs induced with T cell–APC interactions (Iwabuchi *et al.*, in press). Given the crucial role of Th2-attracting chemokines in the development of allergic disorders, it is likely that the alleviation of subjective symptoms of JCPsis by BB536 might be mediated by suppression of TARC production induced by

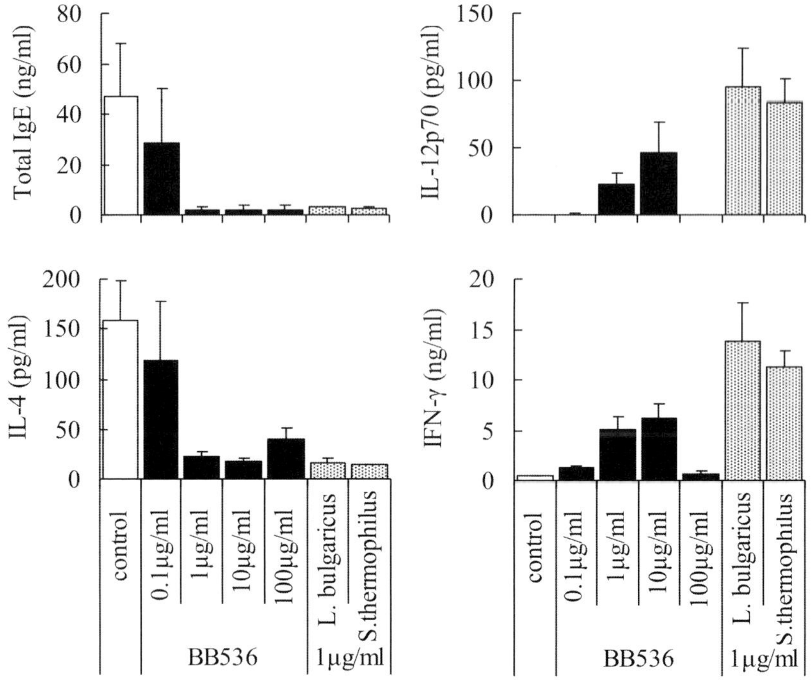

Figure 17.3 Effect of heat-killed microorganisms on ovalbumin-induced total IgE, IL-4, IL-12p70, and IFN-γ production by ovalbumin-sensitized BALB/c splenic cells. Splenic cells from ovalbumin-sensitized mice were cultured with 100 μg/ml ovalbumin in the absence (control) or presence of heat-killed bacterial cells (0.1–100 μg/ml). The levels of total IgE in supernatants on day 14 and cytokines on day 7 were measured using ELISA. Data are shown as mean ± SD of three independent experiments. Reprinted with permission from the Center for Academic Publications Japan.

interactions between cells infiltrated in inflammatory regions.

Some bacterial cell components such as peptidoglycan and lipoteichoic acid have been shown to have immunomodulatory effects (Matsuguchi *et al.*, 2003; Michelsen *et al.*, 2001). We have investigated the effects of a BB536 cell wall constituent on the production of IgE and cytokines by ovalbumin-sensitized splenic cells. Although the BB536 cell wall constituent failed to induce Th1-inducing cytokines, it did have a suppressive effect on IgE and IL-4 production (Iwabuchi *et al.*, 2007). This result suggests that constituents of the BB536 cell wall are involved in the suppressive effects of BB536 on the Th2 immune responses. On the other hand, we have identified an immunostimulatory sequence (5′-GCGTCGGTTTCGGTGCTCAC-3′) BL07S from the genome of BB536, which was shown to stimulate murine B-lymphocytes, induce Th1-skewing cytokines and inhibit IgE production *in vitro* (Takahashi *et al.*, 2006b). Oral administration of BL07S suppressed serum ovalbumin-specific IgE levels in a murine model, suggesting its potential for reducing allergic effects (Takahashi *et al.*, 2006a).

Since intestinal bacteria interact with intestinal immunocompetent cells in peyer's patches after transport through M cells, it has been proposed that probiotics exert an immunomodulatory effect by influencing the cytokine production of intestinal immunocompetent cells. DCs are considered the most powerful APCs and play a central role in the initiation of primary immune responses and in the enhancement of secondary immune responses. It has been reported that a potential dysregulation at the level of the DCs in allergic patients is responsible for the establishment of the allergic status (Hammad *et al.*, 2001). It has been proposed that DCs localized in gut-associated lymphoid tissue can sample bacteria from the intestinal lumen and modify the immune homeostasis (Rescigno *et al.*, 2001). DCs in the gut may be modulated by commensal intestinal bacteria and microorganisms administered orally, such as probiotics. It has been reported that some microbial cell components can regulate DC responses, and that DCs interact with microbes using pattern-recognition receptors such as toll-like receptors (Michelsen *et al.*, 2001). However, Christensen *et al.* (2002) showed that different species of lactobacilli exerted very different activation patterns on the DCs and certain strains of *L. reuteri* inhibited the activities of the other species. BB536 was found to induce the maturation of immature bone marrow-derived dendritic cells (BM-DCs) and to suppress antigen-induced IL-4 production in the BM-DCs-T cells co-culturing system (Iwabuchi *et al.*, 2007).

These results demonstrate that cell components of probiotics can be sampled by intestinal immunocompetent cells, such as DC, resulting in immunomodulatory effects on the hosts. In the case of BB536, both cell wall constituents and genomic immunostimulatory sequences are probably involved in the immunomodulatory effects.

Immunomodulatory effects via microbiota

One explanation for the prevalence of allergic disease is the 'hygiene hypothesis', which postulates that decreased opportunities for exposure to immunostimulating pathogens in early childhood cause increased prevalence of allergic diseases (Strachan, 1989). Noverr and Huffnagle (2005) proposed the 'microflora hypothesis', which suggests that perturbations in gastrointestinal microbiota due to antibiotic use and dietary differences in 'industrialized' countries have disrupted the normal microbiota-mediated mechanisms of immunological tolerance in the mucosa leading to increased incidence of allergic airway disease. Associations of intestinal microbiota with the development of a normal immune system and reductions in allergic risk have been proposed from many studies (Sudo *et al.*, 1997; Bjorksten *et al.*, 1999, 2001; Oyama *et al.*, 2001; Seppo *et al.*, 2005; Smehilová *et al.*, 2008). As recently reviewed by Penders *et al.* (2007), there are numerous studies that show differences in the gut microbiota between allergic and non-allergic infants such as lower counts of bifidobacteria and higher counts of clostridia, *Enterobacteriaceae* and staphylococci. Some reports also describe differences in the bifidobacterial species composition between allergic and non-allergic infants (Ouwehand *et al.*, 2001; Grönlund *et al.*, 2007; Stsepetova *et al.*, 2007; Suzuki *et al.*, 2007; Gore *et al.*, 2008).

Although these studies all provide valuable information indicating an association between the composition of gut microbiota and atopic sensitization or symptoms, no specific harmful or protective microbes have yet been identified. However, the use of probiotics in the treatment and prevention of atopic diseases may demonstrate a causative effect of bacteria and yield more insight into the underlying mechanisms. In the clinical trial during the 2004 JCP season in which yogurt with or without BB536 was administered to JCPsis subjects, an analysis of the faecal microbiota using terminal-restriction fragment length polymorphism (T-RFLP) and real-time PCR methods indicated that cell numbers of the *Bacteroides (Ba.) fragilis* group fluctuated significantly during the pollen season (Odamaki *et al.*, 2007a). It is unclear, however, whether this fluctuation in microbiota was associated with the symptoms of JCPsis, since all subjects ingested yogurt containing at least ordinary lactic acid bacteria and the change of microbiota was not compared with healthy non-JCPsis subjects. A similar analysis was performed in the trial of 2005 in which JCPsis subjects were given placebo (placebo group) or lyophilized BB536 powder (BB536 group). In this trial, however, healthy non-JCPsis subjects were given placebo powder (healthy group) for comparison (Odamaki *et al.*, 2007b). Principal component analysis (PCA) on T-RFLP fragments indicated no marked difference in the bacterial components of microbiota between JCPsis subjects and healthy subjects before the pollen season (Fig. 17.4, week 0). However, there seemed to be dispersion of the microbiota between placebo and healthy groups at the latter stage of the JCP season (Fig. 17.4, week 13). There was a distinct aggregation for placebo subjects (on the left side of the PCA plot) compared with healthy subjects (on the right side of the PCA plot), while the BB536 group was found to straddle over these groups. Further, correlation analysis indicated

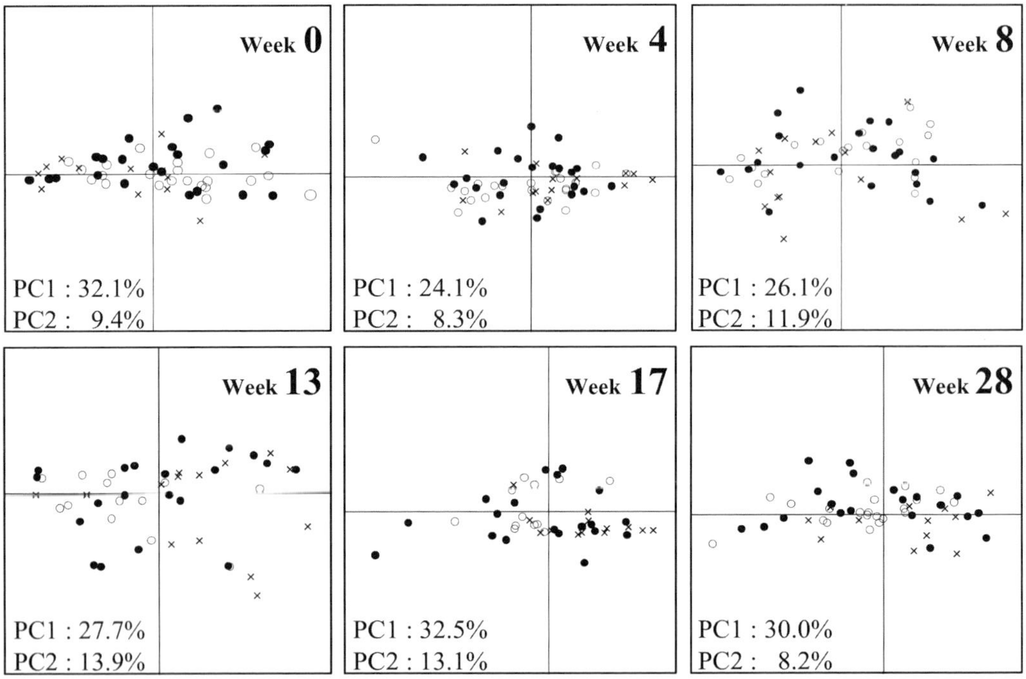

Figure 17.4 Principal component analysis (PCA) based on T-RFLP data sets from each subject before (week 0), during (weeks 4, 8 and 13), after intervention (week 17) and out of JCP season (week 28). PCA based on T-RFLP data sets generated by combined *Alu* I, *Hae* III, *Hha* I, *Msp* I and *Rsa* I + *Xsp* I digestion of faecal samples from JCPsis and healthy subjects. PC1, PC2: per cent variation covered by each principal component. Open circles, placebo; filled circles, BB536; crosses, healthy. Reprinted with permission from the Society for General Microbiology.

that subjects on the left side of the PCA plot experienced more severe symptoms. Database analysis revealed that the major terminal-restriction fragment contributing to the left-side plotting on PCA was from the *Ba. fragilis* group (Odamaki *et al.*, 2007b). Real-time PCR indicated that the 16S rRNA gene copy number of the *Ba. fragilis* group significantly increased from the baseline (week 0) in the both JCPsis groups during the pollen season (Fig. 17.5, week 13). This increase was not seen in the healthy group. Significant inter-group differences were also seen for both the placebo and BB536 groups compared with the healthy groups at week 13. Further, significantly higher 16s rRNA gene copy number of the *Ba. fragilis* group was found in the placebo group compared with the BB536 group at week 13, which indicated that probiotic administration suppressed changes of the microbiota (Fig. 17.5).

Some studies have shown an association of *Bacteroides* with an increased risk of atopic diseases. Kirjavainen *et al.* (2002) found a direct correlation between serum IgE titres and *Bacteroides* counts in gut microbiota of high-risk subjects with atopic disorders. Bifidobacterial supplementation prevented increases of *Bacteroides* and *Escherichia coli* during weaning. Fukuda *et al.* (2004) found that IgG titres against *Ba. vulgatus* were significantly higher among school children with any two of the allergic symptoms of asthma, rhinitis, eczema or food allergy than among non-allergic groups. The population of *Bacteroidaceae* was found to be significantly higher in the allergic group at the ages of 1 month and 2 months than in the non-allergic group, while no statistically significant difference was observed for the other bacterial populations (Songjinda *et al.*, 2007). Vael *et al.* (2008) suggested that *Ba. fragilis* colonization at age 3 weeks is an early indicator of possible asthma later in life. Recently, Suzuki *et al.* (2008) reported that allergic infants at 2 years of age had higher counts of *Bacteroides* and higher ratios of *Bacteroides* to

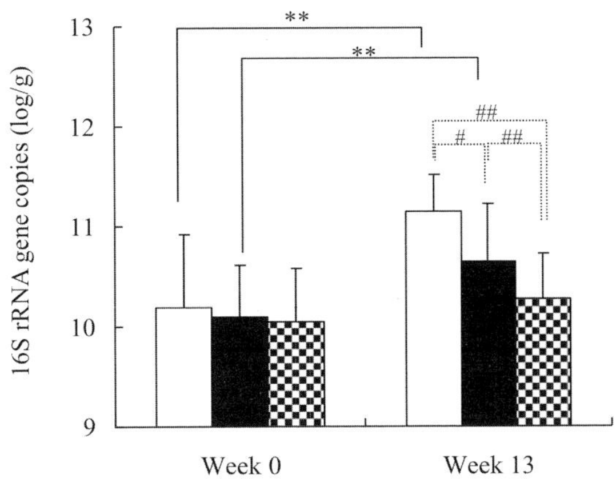

Figure 17.5 Real-time polymerase chain reaction (PCR) quantification of 16S rRNA gene copies of the *Bacteroides fragilis* group in the faecal microbiota before and after the pollen season. Gene copies of 16S rRNA (mean ± SD) were determined as log10 cells/g wet weight of faeces with plasmid DNA standard. $^{**}P < 0.01$, compared with baseline (week 0) by the Wilcoxon test; $\#P < 0.05$; $\#\#P < 0.01$, compared between groups with the Mann–Whitney *U*-test. Open bar, placebo group; filled bar, BB536 group; dotted bar, healthy group.

Bifidobacterium than non-allergic infants. In addition, we showed that strains of *Ba. fragilis* induced significantly more IL-6, an inflammatory cytokine, but significantly fewer Th1-type cytokines (IFN-γ and IL-12) compared with *Bifidobacterium* sp. in human peripheral blood mononuclear cells (Odamaki *et al.*, 2007a).

The results of the 2004 and 2005 studies suggest that the fluctuation of microbiota, e.g. increased prevalence of the *Ba. fragilis* group, during the pollen season may be associated with pollen sensitization or development of symptoms.

The genus *Bacteroides* is one of the predominant intestinal bacteria in humans. However, the taxonomy of the genus *Bacteroides* has undergone significant changes in the past few years (Sakamoto *et al.*, 2007) which has led to a redefinition of the genus *Bacteroides* and application of molecular biological techniques has led to the identification of several novel species (Bakir *et al.*, 2006a,b,c; Hayashi *et al.*, 2007; Kitahara *et al.*, 2005). The effects of intestinal bacteria on the development of allergies are suggested to be species-dependent (Penders *et al.*, 2008). We therefore designed 14 specific primer-pairs to detect each species of the *Ba. fragilis* group that have been isolated from and identified in human faeces. We then used these primers to investigate the distribution of each of these species in both JCPsis and healthy individuals by real-time PCR to evaluate possible associations of each species with development of JCPsis (Odamaki *et al.*, 2008).

We observed some differences in the distributions of each species of the *Ba. fragilis* group in the JCPsis groups compared with the healthy group before the pollen season. In particular, cell numbers of *Ba. fragilis* and *Ba. intestinalis* were significantly higher in the JCPsis group than in the healthy group (Table 17.1, week 0). Significant positive correlations were found for the cell numbers of these two species with both composite symptom scores (sum of each symptom score as indicated in Fig. 17.1A) and JCPsis-specific IgE levels (Fig. 17.1B) (Odamaki *et al.*, 2008). Compared with the pre-pollen season (week 0), some species of the *Ba. fragilis* group increased significantly during the pollen season. The numbers of species with increased cell numbers during the pollen season were 9, six and two in placebo, BB536 and healthy groups, respectively. Significantly increased cell numbers of *Ba. fragilis* and *Ba. intestinalis* were found only in the placebo group (Table 17.1, week 13). Because there are significant differences in cell numbers of *Ba. fragilis* and *Ba. intestinalis* between JCPsis individuals and healthy individuals before the pollen season, we suggest that the prevalence of *Ba. fragilis* and *Ba. intestinalis* might represent risk factors or

Table 17.1 Cell numbers of each species of the *Bacteroides fragilis* group in individuals with Japanese cedar pollinosis before and after the pollen season

		Mean (log/g) ± SD (prevalence)[1]			
Species	Time[2]	Placebo (*n*=13)		BB536 (*n* = 20)	Healthy (*n* – 14)
Ba. fragilis	Before (week 0)	8.80 ± 0.78 (12/13)	##	8.30 ± 0.51 (14/20)	8.19 ± 0.37 (5/14)
	After (week 13)	9.43 ± 0.78 (12/13)	*, ##, $$	9.12 ± 0.65 (10/20)	8.36 ± 0.79 (6/14)
Ba. intestinalis	Before (week 0)	8.13 ± 0.41 (3/13)		8.60 ± 0.83 (9/20)	8.54 ± 0.00 (1/14)
	After (week 13)	8.87 ± 1.45 (9/13)	*, ##	9.37 ± 0.93 (7/20)	7.70 ± 0.10 (2/14)

[1]Cell numbers (mean ± SD) were determined as $\log_{10}$ cells per gram wet weight of faeces, with prevalence given in parentheses as a number of subjects among the total subjects that each species was detected. Detection limits for each bacterial species by real-time PCR were 10^6/g wet weight of faeces.

[2]Before, pollen season (week 0); after, at the last stage of pollen season (week 13).

**P*<0.05.

***P*<0.01, significant intra-group difference from week 0 by the Wilcoxon signed-rank test.

##, *P*<0.01, compared with healthy group.

$$, *P*<0.01, compared between placebo and BB536 groups at each time point by Kruskal–Wallis H-test with Bonferroni correction. Both statistical analyses were conducted on cell numbers after logarithmic transformation, by substituting data with $\log_{10}$ 6 for individuals under detection limits.

indicators for JCPsis and these species might be involved in promoting sensitization to pollen and/or symptom development. In contrast, no significant changes were observed in cell numbers of *Ba. fragilis* or *Ba. intestinalis* in the BB536 group during the pollen season. These results suggest that BB536 may play a role in stabilizing the microbiota, which might in turn exert suppressive effects on sensitization to pollen and/or symptom development.

These results provide convincing evidence for the role of gut microbiota in the sensitization and symptom development of allergic disorders and suggest that a possible mechanism for the antiallergic effects of probiotics is a result of stabilizing the intestinal microbiota.

Conclusion

The use of probiotics in the prevention and treatment of allergic disorders appears promising, and there is accumulating experimental evidence from both animal and human studies to support their effectiveness. However, the efficacy of probiotics seems to be species- and, more specifically, to be strain-dependent. Based on the studies of BB536, we suggest the following mechanisms for the antiallergic activity of probiotics: (1) immunomodulatory effects via bacterial cell components mediated by intestinal antigen-presenting cells (biogenic effects); or (2) immunomodulatory effects owing to generating or stabilizing a balanced gut microbiota (probiotic effects). Probiotics may become alternative therapeutic agents for the prevention and treatment of allergic disorders because of their wide availability, minimal adverse effects, and relatively low cost although further studies are needed to confirm their efficacy as well as their mode of action.

References

Baba, K., and Nakae, K. (2008). National-wide epidemiological survey on the prevalence of allergic rhinitis in 2008 (in comparison with 1998): a survey on otolaryngologists and their families. Prog. Med. 28, 2001–2012.

Bakir, M.A., Kitahara, M., Sakamoto, M., Matsumoto, M., and Benno, Y. (2006a). Bacteroides intestinalis sp. nov., isolated from human faeces. Int. J. Syst. Evol. Microbiol. 56, 151–154, 931–935, 1639–1643.

Bjorksten, B., Naaber, P., Sepp, E., and Mikelsaar, M. (1999). The intestinal microflora in allergic Estonian and Swedish 2-year-old children. Clin. Exp. Allergy *108,* 342–346.

Bjorksten, B., Sepp, E., Julge, K., Voor, T., and Mikelsaar, M. (2001). Allergy development and the intestinal microflora during the first year of life. J. Allergy Clin. Immunol. *108,* 516–520.

Christensen, H.R., Frøkiær, H., and Pestka, J.J. (2002). Lactobacilli differentially modulate expression of cytokines and maturation surface markers in murine dendritic cells. J. Immunol. *168,* 171–178.

Feleszko, W., Jaworska, J., Rha, R.D., Steinhausen, S., Avagyan, A., Jaudszus, A., Ahrens, B., Groneberg, D.A., Wahn, U., and Hamelmann, E. (2007). Probiotic-induced suppression of allergic sensitization and airway inflammation is associated with an increase of T regulatory-dependent mechanisms in a murine model of asthma. Clin. Exp. Allergy *37,* 498–505.

Fujisawa, T., Fujisawa, R., Kato, Y., Nakayama, T., Morita, A., Katsumata, H., Nishimori, H., Iguchi, K., Kamiya, H., and Gray, P.W. (2002). Presence of high contents of thymus and activation-regulated chemokine in platelets and elevated plasma levels of thymus and activation-regulated chemokine and macrophage-derived chemokine in patients with atopic dermatitis. J. Allergy Clin. Immunol. *110,* 139–146.

Fujiwara, D., Inoue, S., Wakabayashi, H., and Fujii, T. (2004). The anti-allergic effects of lactic acid bacteria are strain dependent and mediated by effects on both Th1/Th2 cytokine expression and balance. Int. Arch. Allergy Immunol. *135,* 205–215.

Fujiwara, D., Wakabayashi, H., Watanabe, H., Nishida, S., and Iino, H. (2005). A double-blind trial of *Lactobacillus paracasei* strain KW3110 administration for immunomodulation in patients with pollen allergy. Allergol. Int. *54,* 143–149.

Fukuda, S., Ishikawa, H., Koga, Y., Aiba, Y., Nakashima, K., Cheng, L. & Shirakawa, T. (2004). Allergic symptoms and microflora in schoolchildren. J. Adolesc. Health *35,* 156–158.

Giovannini, M., Agostoni, C., Riva, E., Salvini, F., Ruscitto, A., Zuccotti, G.V., Radaelli, G., and Felicita Study Group. (2007). A randomized prospective double blind controlled trial on effects of long-term consumption of fermented milk containing *Lactobacillus casei* in pre-school children with allergic asthma and/or rhinitis. Pediatr. Res. *62,* 215–20.

Gore, C., Munro, K., Lay, C., Bibiloni, R., Morris, J., Woodcock, A., Custovic, A., and Tannock, G.W. (2008). *Bifidobacterium pseudocatenulatum* is associated with atopic eczema: a nested case-control study investigating the fecal microbiota of infants. J. Allergy Clin. Immunol. *121,* 135–140.

Grönlund, M.M., Gueimonde, M., Laitinen, K., Kociubinski, G., Grönroos, T., Salminen, S., and Isolauri, E. (2007). Maternal breast-milk and intestinal bifidobacteria guide the compositional development of the *Bifidobacterium microbiota* in infants at risk of allergic disease. Clin. Exp. Allergy *37,* 1764–1772.

Hammad, H., Charbonnier, A.S., Duez, C., Jacquet, A., Stewart, G.A., Tonnel, A.B., and Pestel, J. (2001). Th2 polarization by Der p 1-pulsed monocyte-derived dendritic cells is due to the allergic status of the donors. Blood *98,* 1135–1141.

Hayashi, H., Shibata, K., Bakir, M.A., Sakamoto, M., Tomita, S., and Benno, Y. (2007). *Bacteroides coprophilus* sp. *nov.*, isolated from human faeces. Int. J. Syst. Evol. Microbiol. *57,* 1323–1326.

He, F., Morita, H., Ouwehand, A.C., Hosoda, M., Hiramatsu, M., Kurisaki, J., Isorauri, E., Benno, Y., and Salminen, S. (2002). Stimulation of selection of pro-inflammatory cytokines by *Bifidobacterium* strains. Microbiol. Immunol. *46,* 781–785.

Heinrich, J., Richter, K., Frye, C., Meyer, I., Wölke, G., Wjst, M., Nowak, D., Magnussen, H., and Wichmann, H.E. (2002). European community respiratory health survey in adults (ECRHS). Pneumologie *56,* 297–303.

Helin, T., Haahtela, S., and Haahtela, T. (2002). No effect of oral treatment with an intestinal bacterial strain, *Lactobacillus rhamnosus* (ATCC 53103) on birch-pollen allergy: a placebo-controlled double-blind study. Allergy *57,* 243–246.

Ishida, Y., Nakamura, F., Kanzato, H., Sawada, D., Hirata, H., Nishimura, A., Kajimoto, O., and Fujiwara, S. (2005). Clinical effects of *Lactobacillus acidophilus* strain L-92 on perennial allergic rhinitis: a double-blind, placebo-controlled study. J. Dairy Sci. *88,* 527–533.

Ishida, Y., Nakamura, F., Kanzato, H., Sawada, D., Yamamoto, N., Kagata, H., Oh-Ida, M., Takeuchi, H., and Fujiwara, S. (2005). Effect of milk fermented with *Lactobacillus acidophilus* strain L-92 on symptoms of Japanese cedar pollen allergy: a randomized placebo-controlled trial. Biosci. Biotechnol. Biochem. *69,* 1652–1660.

Iwabuchi, N., Takahashi, N., Xiao, J.Z., Miyaji, K., and Iwatsuki, K. (2007). *In vitro* Th1 cytokine-independent Th2 suppressive effects of bifidobacteria. Microbiol. Immunol. *51,* 649–660.

Iwabuchi, N., Takahashi, N., Xiao, J.Z., Yonezawa, S., Yaeshima, T., Iwatsuki, K., and Hachimura, S. (2009). Suppressive effects of *Bifidobacterium longum* on production of Th2-attracting chemokines induced with T cell-antigen-presenting cell interactions. FEMS Immunol. Med. Microbiol. *55,* 324–334.

Kalliomaki, M., Salminen, S., Arvilommi, H., Kero, P., Koskinen, P., and Isolauri, E. (2001). Probiotics in primary prevention of atopic disease: a randomised placebo-controlled trial. Lancet *357,* 1076–1079.

Kapsenberg, M.L., Jansen, H.M., Bos, J.D., and Wierenga, E.A. (1992). Role of type 1 and type 2 T helper cells in allergic diseases. Curr. Opin. Immunol. *4* 788–793.

Kawasaki, S., Takizawa, H., Yoneyama, H., Nakayama, T., Fujisawa, R., Izumizaki, M., Imai, T., Yoshie, O., Homma, I., and Yamamoto, K. (2001). Intervention of thymus and activation-regulated chemokine attenuates the development of allergic airway inflammation and hyperresponsiveness in mice. J. Immunol. *166,* 2055–2062.

Kidd, P. (2003). Th1/Th2 balance: the hypothesis, its limitations, and implications for health and disease. Altern. Med. Rev. *83,* 223–246.

Kirjavainen, P.V., Arvola, T., Salminen, S.J., and Isolauri, E. (2002). Aberrant composition of gut microbiota of allergic infants: a target of bifidobacterial therapy at weaning? Gut *51,* 51–55.

Kitahara, M., Sakamoto, M., Ike, M., Sakata, S., and Benno, Y. (2005). *Bacteroides plebeius* sp. *nov*. and *Bacteroides coprocola* sp. *nov*., isolated from human faeces. Int. J. Syst. Evol. Microbiol. *55,* 2143–2147.

Lee, B.W., Geha, R.S., and Leung, D.Y. (1988). IgE response and its regulation in allergic diseases. Pediatr. Clin. North Am. *35,* 953–967.

Majamaa, H., and Isolauri, E. (1997). Probiotics: a novel approach in the management of food allergy. J. Allergy Clin. Immunol. *99,* 179–185.

Matsuguchi, T., Takagi, A., Matsuzaki, T., Nagaoka, M., Ishikawa, K., Yokokura, T., and Yoshikai, Y. (2003). Lipoteichoic acids from *Lactobacillus* strains elicit strong tumor necrosis factor alpha-inducing activities in macrophages through Toll-like receptor 2. Clin. Diagn. Lab. Immunol. *10,* 259–266.

Matsuzaki, T., Yamazaki, R., Hashimoto, S., and Yokokura, T. (1998). The effect of oral feeding of *Lactobacillus casei* strain Shirota on immunoglobulin E production in mice. J. Dairy Sci. *81,* 48–53.

Michelsen, K.S., Aicher, A., Mohaupt, M., Hartung, T., Dimmeler, S., Kirschning, C.J., and Schumann, R.R. (2001). The role of toll-like receptors (TLRs) in bacteria-induced maturation of murine dendritic cells (DCs). J. Biol. Chem. *276,* 25680–25686.

Mohamadzadeh, M., Olson, S., Kalina, W.V., Ruthel, G., Demmin, G.L., Warfield, K.L., Bavari, S., and Klaenhammer, T.R. (2005). Lactobacilli activate human dendritic cells that skew T cells toward T helper 1 polarization. Proc. Natl. Acad. Sci. USA. *102,* 2880–2885.

Niers, L.E., Timmerman, H.M., Rijkers, G.T., van Bleek, G., van Uden, N.O., Knol, E.F., Kapsenberg, M.L., Kimpen, J.L., and Hoekstra, M.O. (2005). Identification of strong interleukin-10 inducing lactic acid bacteria which down-regulate T helper type 2 cytokines. Clin. Exp. Allergy *35,* 1481–1489.

Noverr, M.C., and Huffnagle, G.B. (2005). The 'microflora hypothesis' of allergic diseases. Clin. Exp. Allergy *35,* 1511–1520.

Odamaki, T., Xiao, J.Z., Iwabuchi, N., Sakamoto, M., Takahashi, N., Kondo, S., Iwatsuki, K., Kokubo, S., Togashi, H., Enomoto, T., *et al.* (2007). Fluctuation of fecal microbiota in individuals with Japanese cedar pollinosis during pollen season. J. Invest. Allergol. Clin. Immunol. *17,* 92–100.

Odamaki, T., Xiao, J.Z., Iwabuchi, N., Sakamoto, M., Takahashi, N., Kondo, S., Miyaji, K., Iwatsuki, K., Togashi, H., Enomoto, T., and Benno, Y. (2007). Influence of *Bifidobacterium longum* BB536 intake on faecal microbiota in individuals with Japanese cedar pollinosis during the pollen season. J. Med. Microbiol. *56,* 1301–1308.

Odamaki, T., Xiao, J.Z., Sakamoto, M., Kondo, S., Yaeshima, T., Iwatsuki, K., Togashi, H., Enomoto, T., and Benno, Y. (2008). Distribution of different species of the *Bacteroides fragilis* group in individuals with Japanese cedar pollinosis. Appl. Environ. Microbiol. *74,* 6814–7.

Ogawa, T., Hashikawa, S., Asai, Y., Sakamoto, H., Yasuda, K., and Makimura, Y. (2006). A new synbiotic, *Lactobacillus casei* subsp. *casei* together with dextran, reduces murine and human allergic reaction. FEMS Immunol. Med. Microbiol. *46,* 400–409.

Ohno, H., Tsunemine, S., Isa, Y., Shimakawa, M., and Yamamura, H. (2005). Oral administration of *Bifidobacterium bifidum* G9-1 suppresses total and antigen specific immunoglobulin E production in mice. Biol. Pharm. Bull. *28,* 1462–1466.

Ouwehand, A.C., Isolauri, E., He, F., Hashimoto, H., Benno, Y., and Salminen, S. (2001). Differences in *Bifidobacterium flora* composition in allergic and healthy infants. J. Allergy Clin. Immunol. *108,* 144–145.

Oyama, N., Sudo, N., Sogawa, H., and Kubo, C. (2001). Antibiotic use during infancy promote a shift in the T(H)1/T(H)2 balance toward T(H)2-dominant immunity in mice. J. Allergy Clin. Immunol. *107,* 153–159.

Penders, J., Stobberingh, E.E., van den Brandt, P.A., and Thijs, C. (2007). The role of the intestinal microbiota in the development of atopic disorders. Allergy *62,* 1223–1236.

Peng, G.C., and Hsu, C.H. (2005). The efficacy and safety of heat-killed *Lactobacillus paracasei* for treatment of perennial allergic rhinitis induced by house-dust mite. Pediatr. Allergy Immunol. *16,* 433–438.

Piirainen, L., Haahtela, S., Helin, T., Korpela, R., Haahtela, T., and Vaarala, O. (2008). Effect of *Lactobacillus rhamnosus* GG on rBet v1 and rMal d1 specific IgA in the saliva of patients with birch pollen allergy. Ann. Allergy Asthma Immunol. *100,* 338–342.

Pochard, P., Hammad, H., Ratajczak, C., Charbonnier-Hatzfeld, A.S., Just, N., Tonnel, A.B., and Pestel, J. (2005). Direct regulatory immune activity of lactic acid bacteria on Der p 1-pulsed dendritic cells from allergic patients. J. Allergy Clin. Immunol. *116,* 198–204.

Rautava, S., Kalliomäki, M., and Isolauri, E. (2005). New therapeutic strategy for combating the increasing burden of allergic disease: Probiotics-A Nutrition, Allergy, Mucosal Immunology and Intestinal Microbiota (NAMI) Research Group report. J. Allergy Clin. Immunol. *116,* 31–37.

Rescigno, M., Rotta, G., Valzasina, B., and Ricciardi-Castagnoli, P. (2001). Dendritic cells shuttle microbes across gut epithelial monolayers. Immunobiology *204,* 572–581.

Romagnani, S. (1994). Regulation of the development of type 2 T-helper cells in allergy. Curr. Opin. Immunol. *6* 838–846.

Romagnani, S. (2001). Cytokines and chemoattractants in allergic inflammation. Mol. Immunol. *38,* 881–885.

Rook, G.A., and Brunet, L.R. (2005). Old friends for breakfast. Clin. Exp. Allergy *35,* 841–842.

Sakamoto, M. (2007). Recent classification of the genus *Bacteroides* and related taxa. Microbiol. Cult. Coll. *23,* 1–16 (in Japanese).

Sepp, E., Julge, K., Mikelsaar, M., and Björkstén, B. (2005). Intestinal microbiota and immunoglobulin E responses in 5-year-old Estonian children. Clin. Exp. Allergy *35,* 1141–1146.

Settipane, R.A. (2001). Demographics and epidemiology of allergic and nonallergic rhinitis. Allergy Asthma Proc. *22,* 185–189.

Shida, K., Makino, K., Morishita, A., Takamizawa, K., Hachimura, S., Ametani, A., Sato, T., Kumagai, Y., Habu, S., and Kaminogawa, S. (1998). *Lactobacillus casei* inhibits antigen-induced IgE secretion through regulation of cytokine production in murine splenocyte cultures. Int. Arch. Allergy Immunol. *115,* 278–287.

Smehilová, M., Vlková, E., Nevoral, J., Flajsmanová, K., Killer, J., and Rada, V. (2008). Comparison of intestinal microflora in healthy infants and infants with allergic colitis. Folia Microbiol. (Praha). *53,* 255–258.

Songjinda, P., Nakayama, J., Tateyama, A., Tanaka, S., Tsubouchi, M., Kiyohara, C., Shirakawa, T., and Sonomoto, K. (2007). Differences in developing intestinal microbiota between allergic and non-allergic infants: a pilot study in Japan. Biosci. Biotechnol. Biochem. *71,* 2338–2342.

Stsepetova, J., Sepp, E., Julge, K., Vaughan, E., Mikelsaar, and M., de Vos, W.M. (2007). Molecularly assessed shifts of *Bifidobacterium* subsp. and less diverse microbial communities are characteristic of 5-year-old allergic children. FEMS Immunol. Med. Microbiol. *51,* 260–269.

Strachan, D.P. (1989). Hay fever, hygiene, and household size. BMJ *299,* 1259–1260.

Sudo, N., Sawamura, S., Tanaka, K., Aiba, Y., Kubo, C., and Koga, Y. (1997). The requirement of intestinal bacterial flora for the development of IgE production system fully susceptible to oral tolerance induction. J. Immunol. *159,* 1739–1745.

Suzuki, S., Shimojo, N., Tajiri, Y., Kumemura, M., and Kohno, Y. (2007). Differences in the composition of intestinal *Bifidobacterium* species and the development of allergic diseases in infants in rural Japan. Clin. Exp. Allergy *37,* 506–511.

Suzuki, S., Shimojo, N., Tajiri, Y., Kumemura, M., and Kohno, Y. (2008). A quantitative and relative increase in intestinal bacteroides in allergic infants in rural Japan. Asian Pac. J. Allergy Immunol. *26,* 113–119.

Takahashi, N., Kitazawa, H., Iwabuchi, N., Xiao, J.Z., Miyaji, K., Iwatsuki, K., and Saito, T. (2006a). Oral administration of an immunostimulatory DNA sequence from *Bifidobacterium longum* improves Th1/Th2 balance in a murine model. Biosci. Biotechnol. Biochem. *70,* 2013–2017.

Takahashi, N., Kitazawa, H., Shimosato, T., Iwabuchi, N., Xiao, J.Z., Iwatsuki, K., Kokubo, S., and Saito, T. (2006b). An immunostimulatory DNA sequence from a probiotic strain of *Bifidobacterium longum* inhibits IgE production *in vitro*. FEMS Immunol. Med. Microbiol. *46,* 461–469.

Tamura, M., Shikina, T., Morihana, T., Hayama, M., Kajimoto, O., Sakamoto, A., Kajimoto, Y., Watanabe, O., Nonaka, C., Shida, K., *et al.* (2007). Effects of probiotics on allergic rhinitis induced by Japanese cedar pollen: randomized double-blind, placebo-controlled clinical trial. Int. Arch. Allergy Immunol. *143,* 75–82.

Vael, C., Nelen, V., Verhulst, S.L., Goossens, H., and Desager, K.N. (2008). Early intestinal *Bacteroides fragilis* colonisation and development of asthma. BMC Pulm. Med. *26,* 8–19.

Wang, M.F., Lin, H.C., Wang, Y.Y., and Hsu, C.H. (2004). Treatment of perennial allergic rhinitis with lactic acid bacteria. Pediatr. Allergy Immunol. 5 152–158.

Xiao, J.Z. (2008). *Bifidobacterium longum* BB536. In Handbook of Probiotics and Prebiotics, Lee, Y.K., and Salminen, S., eds. (John Wiley & Sons, Inc., Hoboken, New Jersey), pp. 488–491.

Xiao, J.Z., Kondo, S., Yanagisawa, N., Takahashi, N., Odamaki, T., Iwabuchi, N., Iwatsuki, K., Kokubo, S., Togashi, H., Enomoto, K., *et al.* (2006a). Effect of probiotic *Bifidobacterium longum* BB536 in relieving clinical symptoms and modulating plasma cytokine levels of Japanese cedar pollinosis during the pollen season, a randomized, double-blind, placebo-controlled trial. J. Invest. Allergol. Clin. Immunol. *16,* 86–93.

Xiao, J.Z., Kondo, S., Yanagisawa, N., Takahashi, N., Odamaki, T., Iwabuchi, N., Miyaji, K., Iwatsuki, K., Togashi, H., Enomoto, K., *et al.* (2006b). Probiotics in the treatment of Japanese cedar pollinosis: a double-blind placebo-controlled trial. Clin. Exp. Allergy *36,* 1425–1435.

Xiao, J.Z., Kondo, S., Takahashi, N., Odamaki, T., Iwabuchi, N., Miyaji, K., Iwatsuki, K., and Enomoto, T. (2007a). Changes in plasma TARC levels during Japanese cedar pollen season and relationships with symptom development. Int. Arch. Allergy Immunol. *144,* 123–127.

Xiao, J.Z., Kondo, S., Yanagisawa, N., Miyaji, K., Enomoto, K., Sakota, T., Iwatsuki, K., and Enomoto, T. (2007b). Clinical efficacy of probiotic *Bifidobacterium longum* for the treatment of symptoms of Japanese cedar pollen allergy in subjects evaluated in an environmental exposure unit. Allergol. Int. *56,* 67–75.

von der Weid, T., Bulliard, C., and Schiffrin, E.J. (2001). Induction by a lactic acid bacterium of a population of CD4(+) T cells with low proliferative capacity that produce transforming growth factor beta and interleukin-10. Clin. Diagn. Lab. Immunol. *8,* 695–701.

Probiotics Health Claims in Japan and Europe

18

Yoichi Fukushima* and Eva Hurt

Abstract

Scientific evidence of health benefit of food is accumulating and health claims on food are highlighted. Food for Specified Health Use (FOSHU) was established in Japan in 1991 as the first systematic regulatory system for food health claims, and 799 FOSHU products are approved as of November 2008. Probiotics are food components scientifically well-documented, which may deliver health benefits including gut microbiota balancing, normalizing regularity, gut comfort, immune boosting, anti-infection, anti-diarrhoea, anti-allergy, and skin benefits. In FOSHU, 73 products with 13 probiotic strains and/or combination of strains were approved under the 'modify gastrointestinal (GI) conditions' claim category, whose criteria for scientific substantiation is clearly defined in Japan. In the European Union, a new Nutrition and Health Claims Regulation harmonizes since 2007 the requirements on claims on food products, and the regulation will be fully applicable in 2010. Currently in 2009, the European Food Safety Authority (EFSA) is evaluating several thousand diet–health relationships, which will lead to a Community list of permitted claims. The 27 EU member governments have submitted the claims, including, in total, 268 proposed diet–health relationships involving probiotics. This chapter will review the health claim systems in Japan and in the European Union based on the current status in April 2009.

Introduction

The term probiotics, defined as live microorganisms, which when administered in adequate amounts confer a health benefit on the host [Joint Food and Agriculture Organization/World Health Organization (FAO/WHO), 2002], was introduced in 1991 (Fuller, 1991). The FAO/WHO Expert Consultation reported that there is emerging evidence to indicate that probiotics can be taken by otherwise healthy people as a means to prevent certain diseases and modulate host immunity. They also recommended that probiotic products shown to confer defined health benefits on the host should be permitted to describe these specific health benefits (Joint FAO/WHO, 2001).

The concept of 'probiotics' is old, in fact, much older than the development of probiotics terminology. For example, the first probiotic product in the world, the lactic acid bacteria drink YAKULT containing viable *Lactobacillus casei* Shirota, was, introduced in Japan in 1935. Cow's milk with viable *Bifidobacterium longum* BB536 was first introduced in Japan in 1977. Infant formula powder containing *Bifidobacterium animalis* spp. *lactis* Bb-12 was launched in 1993, also in Japan, and is now widely used around the world. In Japan, Food for Specified Health Use (FOSHU) health claims on modifying gastrointestinal (GI) conditions were first approved on Onakahe GG fermented milk with *Lactobacillus rhamnosus* GG probiotics in 1996. In the late 1990s, probiotics launch was boomed in the European market through introduction of brands such as YAKULT, LC1 containing *Lactobacillus johnsonii* La1 (NCC533), ACTIMEL containing *Lactobacillus casei* DN-114 001, and BIO (ACTIVIA) containing *Bifidobacterium animalis* spp. *lactis* DN-173

*corresponding author

010. The boom on probiotics came in Japan in 2004 through the launch of products such as LG21 yogurt with *Lactobacillus gasseri* OLL2716 and LC1. Some common probiotics and their health benefits shown in humans are summarized in Table 18.1.

Recently, attempts to reveal the physiological function of food and to substantiate health claims have accelerated in food industries. Research groups in the academy and industry have played key roles in establishing the concept of 'functional food' in Japan. Since FOSHU health claim regulation system started in 1991, functional food market has been very active in Japan, and 799 FOSHU products are approved as of Nov. 2008. Food with Nutrient Function Claim (FNFC) was also launched to assess claims on vitamin and minerals. This emergence of regulatory bodies is attributable not only to the claim system itself but also to the Japanese consumers who appreciate new healthy foods with proven health benefits and to food industries highly active in launching and communicating on new products. Probiotics are one of the key functional components for FOSHU and 73 FOSHU products with probiotic strains are currently approved.

The importance of intestinal microflora and bifidobacteria and their health benefits were hypothesized in 1969 by Mitsuoka (1969), who described the function of intestinal flora and its influence on the host including exclusion of pathogens, stimulation of the immune response, anti-diarrhoea effects, anti-constipation effects, and assistance to absorption/digestion of nutrients. These postulated scientific concepts strongly contributed to activating the research on probiotics/prebiotics and microflora and prompted the establishment of health claim criteria to provide a road map for the process.

In the European Union (EU), a new Nutrition and Health Claims Regulation harmonizes since 2007 the requirements on claims on food products across the 27 EU countries. Health claims can be made on a food category, a food or a food constituent, e.g. a nutrient, another substance or a combination of nutrients and substances. All claims have to be authorized for use after a scientific assessment of the highest possible standard and may not be false or misleading, or refer to prevention, treatment or cure of a disease. During a transitional period until 2010, health claims based on generally accepted scientific evidence are under central scientific evaluation, but can be made under manufacturers' own responsibility. In 2009, the European Food Safety Authority (EFSA) is evaluating several thousand diet–health relationships, which will lead to Community list of permitted claims. The 27 EU member governments have submitted the claims based on a list proposed by EU industry associations and food and food supplement companies, including, in total, 268 proposed diet–health relationships involving probiotics. In April 2009, no health claims have been approved yet, and it is expected that the list of permitted health claims will be much shorter than the submitted claims at more than 4000. For other types of health claims, it is noted that EFSA is publishing more negative than positive scientific opinions.

Emerging science shows increasingly the health benefits of food. Globally, approximately 500 scientific articles per year alone show the health benefits of probiotics, and food products claiming health benefit are increasing on the market. Probiotics with health benefits shown by human trial(s) are summarized in Table 18.1. Digestive health is one of the major benefit areas of probiotics, and many probiotics have been proven on the benefit for microbiota balancing and improved gut transit, which is sufficient evidence to make a health claim relating to digestive health in Japan under FOSHU regulation. In addition, scientific evidence in human is accumulating on many other benefits of probiotics such as immune boost (e.g. increased phagocytotic cell and natural killer (NK) cell activity, and intestinal IgA secretion), anti-infection (e.g. protection against common cold), diarrhoea prevention and therapy, anti-gastritis through anti-*Helicobacter pylori* activity, anti-allergy effect for prevention and conditioning of atopic dermatitis in infants and reduction of symptoms of pollen allergy, vaginal and urogenal health, oral health, and benefits for skin health. These benefits may well spur many new health claims in the future. In order to distinguish well substantiated health claims from vague promises, it is necessary to introduce robust regulatory systems, however. This chapter will review the health claim systems in Japan and Europe with regard to the current status in April 2009.

Table 18.1 Probiotics and proven health benefit in humans

Probiotic strain(s)	Strain supplier	in FOSHU	CIAA list 2007	Health benefit category: Digestive health	Protection	Others	Reference
Single strain							
Bacillus subtilis K-2	Asahimatsu Foods (Japan)	✓		✓			Mitsui *et al.* (2006)
Bifidobacterium animalis Lafti B94 (CBS118.529)	DSM (Netherland)		✓	✓			Su *et al.* (2005)
Bifidobacterium animalis spp. *lactis* Bb-12® (BL818) (NCC2818)	Chr.Hansen (Denmark)	✓	✓	✓	✓[a,d,i]		Saavedra *et al.* (1994, 2004), Fukushima *et al.* (1997, 1998), Nishida *et al.* (2004), Katayama *et al.* (2004), Wang *et al.* (2004) Weizman *et al.* (2005), Mao *et al.* (2008), Isolauri *et al.* (2008)
Bifidobacterium animalis spp. lactis CNCM I-2494 / DN-173 010	Danone (France)		✓	✓			Bouvier *et al.* (2001), Marteau *et al.* (2002)
Bifidobacterium breve Yakult (BbY)	Yakult (Japan)	✓	✓	✓			Tanaka *et al.* (1980, 1994), Ishikawa *et al.* (2003), Nomoto (2005)
Bifidobacterium infantis 35624 (Bifantis®)	Alimentary Health (Ireland)		✓			✓[p]	Whorwell *et al.* (2006)
Bifidobacterium lactis HN019 AGAL NM97/09513 (DR10)	Danisco (Denmark)		✓	✓	✓[i]		Ahmed *et al.* (2007), Gill *et al.* (2001a,b)
Bifidobacterium longum BB536 (BL999) (NCC3001)	Morinaga (Japan)		✓	✓	✓[a]		Ogata *et al.* (1999), Xiao *et al.* (2006), Odamaki *et al.* (2007)
Enterococcus faecium SF68 (NCIMB 10415) (NCC2768)	Cerbios Pharma (Switzerland)		✓		✓[d]		Cremonini *et al.* (2002)
Escherichia coli Nissle 1917	Ardeypharm (Germany)		✓			✓[b]	Schultz *et al.* (2008)
Lactobacillus acidophilus L92	Calpis (Japan)		✓	✓	✓[a*]		Ishida *et al.* (2005a,b)
Lactobacillus acidophilus LA14	Danisco (Denmark)		✓			✓[v]	Hilton *et al.* (1992)
Lactobacillus acidophilus NCFM ATCC SD5221	Danisco (Denmark)		✓	✓			Sandres *et al.* (2001)
Lactobacillus brevis KB290	Kagome Rabio (Japan)				✓[i]		Kishi *et al.* (1996)
Lactobacillus casei CNCM I-1518 / DN-114 001	Danone (France)		✓		✓[i,d]		Pedone *et al.* (2000), Agarwal *et al.* (2001) Marcos *et al.* (2006), Turchet *et al.* (2003), Cobo Sanz *et al.* (2006)
Lactobacillus casei NY13101	Nisshin York (Japan)	✓		✓			Azuma *et al.* (2000)

Table 18.1 continued

Probiotic strain(s)	Strain supplier	in FOSHU	CIAA list 2007	Health benefit category: Digestive health	Protection	Others	Reference
Lactobacillus casei Shirota (LcS)	Yakult (Japan)	✓	✓	✓	✓[c,i]		Tanaka *et al.* (1991), Aso *et al.* (1995), Nagao *et al.* (2000), Ohashi *et al.* (2002), Koebrick *et al.* (2003), Matsumoto *et al.* (2006), Takeda *et al.* (2007), Tamura *et al.* (2007), Ivory *et al.* (2008),
Lactobacillus casei SBR1202	Kagome Rabio (Japan)	✓		✓			Mori *et al.* (2002)
Lactobacillus fermentum CECT5716	Puleva Biotech (Spain)		✓		✓[i]		Olivares *et al.* (2007)
Lactobacillus fermentum VRI-033 PCC (NM 02/31047)	Probiomics (Australia)				✓[a]		Weston *et al.* (2005), Prescott *et al.* (2005)
Lactobacillus gasseri OLL2716 (LG21)	Meiji Dairies (Japan)				✓[h]		Sakamoto *et al.* (2001)
Lactobacillus johnsonii NCC 533 (La1) (CNCM I-2115)	Nestlé (Switzerland)	✓	✓	✓	✓[h,i]	✓[s]	Link-Amster *et al* (1994), Schiffrin *et al.* (1995), Michetti *et al.* (1999), Felley *et al.* (2001), Fukushima (2007), Fukushima *et al.* (2007), Yamano *et al.* (2006), Gotteland *et al.* (2008), Peguet-Navarro *et al.* (2008)
Lactobacillus lactis spp *lactis* FC	Fujicco (Japan)						Taod *et al.* (2005)
Lactobacillus paracasei 33	GenMont Biotech (Taiwan)		✓		✓[a]		Wang MF *et al.* (2004)
Lactobacillus paracasei KW3110*	Kirin (Japan)			✓	✓[a]		Fujiwara *et al.* (2005), Nishida *et al.* (2008)
Lactobacillus paracasei NCC 2461 (ST11) (CNCM I-2116)	Nestlé (Switzerlnad)		✓		✓[d,i]		Bunout *et al.* (2004), Sarker *et al.* (2005)
Lactobacillus paracasei spp. *paracasei* CRL-431	Chr.Hansen (Denmark)		✓		✓[i]		de Vrese *et al.* (2005a)
Lactobacillus plantarum 299v	Probi AB (Sweden)		✓		✓[d,p]		Niedzielin *et al.* (2001), Klarin *et al.* (2008), Wullt *et al.* (2007)
Lactobacillus pluntarum ONC141 (LP1)	Otsuka (Japan)			✓			Kumemura *et al.* (2001)
Lactobacillus reuteri ATCC 55730 (SD2112)	BioGaia (Sweden)		✓		✓[a,d,i]	✓[o,p]	Shornikova *et al.* (1997), Valeur *et al.* (2004), Tubelius *et al.* (2005), Imase *et al.* (2007), Nikawa *et al.* (2004), Abrahamsson *et al.* (2007), Savino *et al.* (2007)

Lactobacillus rhamnosus ATCC53103 (LGG®) (LPR) (NCC4007)	Valio (Finland)	✓	✓	✓	✓[a,d,i]	✓[o]	Hosoda *et al.* (1994), Majamaa *et al.* (1995), Hilton *et al.* (1997), Kalliomäki *et al.* (2001, 2003, 2007), Hatakka *et al.* (2001), Nase *et al.* (2001), Szajewska *et al.* (2001), de Vrese *et al.* (2004), Canani *et al.* (2007), Kawase *et al.* (2008), Kukkonen *et al.* (2006), Kajander *et al.* (2008)
Lactobacillus rhamnosus HN001 AGAL NM97/09514 (DR20)	Danisco (Denmark)		✓	✓	✓[a,i]		Sheih *et al.* (2001), Sistek *et al.* (2006), Tannock *et al.* (2000), Wickens *et al.* (2008)
Lactobacillus salivarius T12711	Frente International (Japan)					✓[o]	Matsuoka *et al.* (2005)
Saccharomyces boulardii PXN68TM	Biocodex (France)		✓		✓[d]		McFarland *et al.* (2007)
Streptococcus salivarius K12 (Blis) (ATCC BAA-1024)	BLIS Technologies (New Zealand)		✓		✓[i]	✓[o]	Burton *et al.* (2006), Power *et al.* (2008)
Combination of strains							
Bifidobacterium animalis spp. *lactis* Bb-12® *Lactobacillus* LA-5®	Chr.Hansen (Denmark)		✓	✓	✓[b,d]		Black *et al.* (1991), Wildt *et al.* (2006), Sheu *et al.* (2006),
Lactobacillus acidofilus CK92 (L92) *L. helbeticaus*CK60	Calpis (Japan)	✓		✓			Ikeda *et al.* (2000a,b)
Lactobacillus acidophilus LB (Lactéol) *	Axcan Pharma (France)		✓		✓[d]		Salazar-Lindo *et al.* (2007), Xiao *et al.* (2003)
Lactobacillus delbrueckii subsp. *bulgaricus* 2038 *S. salivarius* subsp. *thermophilus* 1131	Meiji Dairies (Japan)	✓		✓		✓[s]	Hara *et al.* (1993), Kinumaki *et al.* (2001)
Lactobacillus gasseri SBT2055 *Bifidobacterium longum* SBT2928	Snow Brand (Japan)	✓		✓			Izawa *et al.* (2008)
Lactobacillus gasseri PA16/8 *Bifidobacteirum bifidum* MF 20/5 *Bifidobacterium longum* SP07/3	Merck (USA)		✓		✓[i]		de Vrese *et al.* (2005b, 2006), Winkler *et al.* (2005)
Lactobacillus gasseri CECT5714 *Lactobacillus coryniformis* CECT5711	Puleva Biotech (Spain)		✓		✓[i]		Olivares *et al.* (2006a,b), Jiminez *et al.* (2008)

Table 18.1 continued

Probiotic strain(s)	Strain supplier	in FOSHU	CIAA list 2007	Health benefit category: Digestive health	Protection	Others	Reference
Lactobacillus acidophilus I-1722 (Rosell-52) *Bifidobacterium longum* CNCM I-3470 (Rosell-175)	Lallemand (Canada)		✓	✓			Diop *et al.* (2008)
Lactobacillus rhamnosus 19070-2 *Lactobacillus reuteri* DSM12246	Chr.Hansen (Denmark)		✓		✓[a,d]		Rosenfeldt *et al.* (2002a,b, 2003, 2004)
Lactobacillus rhamnosus GR 1 (ATCC 55826) *Lactobacillus reuteri* RC-14 (ATCC 55845)	Urex Biotech (Canada)		✓			✓[v]	Reid (2008)
Propionibacterium freudenreichii SI 41 *Propionibacterium freudenreichii* SI 26 *Propio-Fidus*®	Standa (France)		✓	✓			Bougle *et al.* (1999)
VSL#3 (8 strains mixture including lactobacilli and bifidobacteria)	Sigma-Tau (USA)		✓			✓[b,p]	Chapman *et al.* (2006)

*Heat-killed bacteria

[a]anti-allergy

[b]anti-inflammatory bowel diseases (IBD)

[c]anti-cancer

[d]anti-diarrhea

[h]anti-Helicobactor pylori

[i]immune boosting/anti-infection

[o]oral health

[p]anti-irritable bowel syndrome (IBS)

[s]skin health

[v]verginal flora/urogential protection

Functional food and health claims in Japan

History of functional food and FOSHU

Food has three functions: (1) the nutrient function needed to maintain life or growth of the body, (2) the taste function provided by some components of food interacting with the sensory system, and (3) the body controlling function related to the body's defences or modification of the body conditions contributing to health maintenance and prevention of diseases (Arai *et al.*, 2001). The concept 'functional food' targets the third function of food, including the body's defences (anti-allergy and immune reinforcement), prevention and restoration (hypertension, diabetes and metabolic deficiencies in nerve, gut and internal secretions), and protection from ageing (control of fat peroxidization). It was proposed in 1984 based on research projects supported by the Japanese government. The academic societies that participated in this series of government funded projects helped define the functional food concept and generated the basic understanding of food functionality, which then gave rise to the evaluation system for health claims on functional foods.

The Ministry of Health and Welfare, the former Ministry of Health, Labour, and Welfare (MHLW), issued the Investigative Commission Report on Functional Food in 1990. In 1991, the Food for Specified Health Uses (FOSHU) system was inaugurated as the world's first approval system on health claim labelling for food products. The first FOSHU product, hypoallergenic rice, was approved in 1993. In 1997, tablet shaped products also became eligible for FOSHU approval. In 2001, the 'Food with Health Claims (FHC)' system was introduced. The FOSHU system was retained but re-positioned under a category of FHC, and a new category 'Food with Nutrient Function Claims (FNFC)' was created beside FOSHU (Fig. 18.1). FHC refers to foods that comply with the specifications and standards established by the MHLW and are labelled with certain nutritional or health functions. These foods are categorized into two groups, according to differences in purpose and function: FNFC, foods that are labelled with the functions of nutritional ingredients (vitamins and minerals), and FOSHU, foods officially approved to claim physiological effects on the human body. In 2005, it became possible to claim reduction of disease risk under FOSHU regulation in cases where reduction of disease risk is clinically and nutritionally established for an ingredient. Hence,

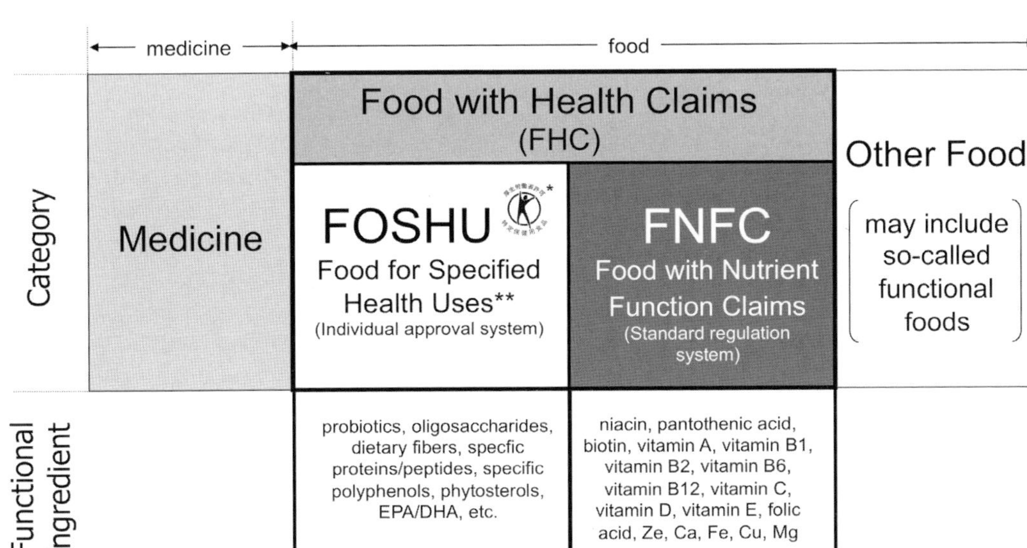

Figure 18.1 Food with Health Claims including FOSHU in Japan. *Seal for FOSHU approval by the Ministry of Health, Labor and Welfare (currently, Consumer Affairs Agency). **Regular FOSHU, Standard FOSHU, Qualified FOSHU and Reduction of Disease Risk FOSHU are included.

two risk reduction claims for osteoporosis and neural tube defect were allowed on calcium and folic acid, respectively. In addition to 'regular' FOSHU, Qualified FOSHU and Standardized FOSHU, were introduced to facilitate applications for FOSHU approval (http://www.mhlw.go.jp/english/topics/foodsafety/fhc/index.html) (Fig. 18.1). The system was relegated from MHLW to the Consumer Affairs Agency in 2009.

Food with Nutrient Function Claim (FNFC)

FNFC refers to all food that is labelled with nutrient function claims specified by the MHLW. So far, the standards and specifications for indication of nutritional function have been established for 12 vitamins and five minerals (niacin, pantothenic acid, biotin, vitamin A, vitamin B1, vitamin B2, vitamin B6, vitamin B12, vitamin C, vitamin D, vitamin E, folic acid, zinc, calcium, iron, copper, and magnesium) (Fig. 18.1). These foods may be freely manufactured and distributed without permission from or notification to the national government, provided that they meet the established standards and specifications. The recommended daily intake of the nutritional ingredient must be within a specified range, and in addition to the nutrient functional claims, warning indications must be displayed on the packaging.

For instance, food product with zinc at 2.1–25 mg/day may be described as FNFC with the fixed nutrient function claim:

> a nutrient necessary for the maintenance of normal taste and healthy skin and mucous membranes. It is involved in the metabolism of proteins and nucleic acids and helps maintain health

together with the warning indication

> increased intake of this product will not result in curing of diseases or promotion of health. Too high intake of zinc may inhibit absorption of copper. Please comply with the advisable daily intake. Infants and young children should avoid the use of this product.

Niacin, pantothenic acid, biotin, vitamin A, vitamin B1, vitamin B2, vitamin B6, vitamin B12, and vitamin C are also allowed the claim 'helps maintain healthy skin and mucosa.' Other claims that are allowed include the anti-oxidant effect of vitamin C and E, the energy generation effect of vitamin B1 and B6, the red blood cell formation effect of iron, copper, vitamin B12 and folic acid, and the bone health effect of calcium, magnesium and vitamin D. For FNFC, the wording of the claim and warning are fixed.

Health claim under FOSHU

FOSHU refers to foods containing ingredients with a function in health officially approved to claim to have beneficial physiological effects on the human body. FOSHU is intended to be consumed for the maintenance/promotion of health or for special health uses by people who wish to control various health conditions, including blood pressure or blood cholesterol. In order to commercialize food as FOSHU, assessment of its safety and effectiveness is required, and the claim must be approved by the government. FOSHU health claims are approved by the government, product by product, based on applications submitted by food manufacturers. FOSHU approved product can carry a seal of FOSHU approval on its packaging.

There have been nine health claim (Specified Health Uses) categories: foods to modify gastrointestinal conditions (334 products), foods related to blood pressure (100 products), foods related to triacylglycerol (56 products), foods related to blood cholesterol levels (125 products), foods related to blood sugar levels (113 products), foods related to osteogenesis/foods related to mineral absorption (39 products), foods related to dental hygiene (37 products), and foods related to reduction of disease risk for osteoporosis (six products). In total, 799 products, are approved under FOSHU as of Nov 2008. The following 91 ingredients exhibiting health functions are used in the approved FOSHU products: 10 single and three combinations of probiotic strains, 9 oligosaccharides, 12 dietary fibres, 9 anti-hypertensive peptides, 7 specific proteins, four specific polyphenols, phytosterols, EPA/DHA, medium-chain fatty acids (MCT), xylitol and sugar alcohols, casein phosphopeptide amorphous calcium phosphate nanocomplexes (CPP-ACP). The ingredients are found in many food product categories including beverages, soup, snacks, noodles, candy, fermented milk, table sugar, cocking oil, and tablets. In FOSHU, the target age group need not be identified, but most products are targeting adult health.

The Pharmaceutical Affairs Law decrees that a product aimed at influencing a body function or structure should be categorized as a medicine, and medical terms such as cure and prevention can not be communicated in food categories. In the FOSHU system, the applicant may submit the wording of the health claim as long as they can support it with scientific evidence, however, the wordings are strictly regulated by the government and it is not easy to obtain FOSHU claim approval out of nine existing claim categories with a limited range of wording based on the past approval. For example, the following types of claims are currently approved under FOSHU:

> This product uses xxx to increase the number of bifidobacteria and thus helps maintain good GI environment.

> The product is suitable for those who tend to have high blood pressure.

> This product contains xxx, which helps lower the serum cholesterol level. It is thus suitable for those concerned about cholesterol.

> xxx helps the absorption of calcium to build strong bones.

> This product uses sweeteners (xxx), which do not cause tooth decay. It also contains xxx that increases recalcification, and therefore helps maintain strong and healthy teeth.

Health claim categories such as immune boosting or fatigue reduction have not been approved under FOSHU, because of the hurdle of drug-food differentiation and insufficient consensus with regard to the scientific evaluation criteria.

Standard FOSHU is granted when a food meets the standards and specifications sufficient for FOSHU approval, including accumulation of scientific evidence. For example, three dietary fibres including indigestible dextrin (3–8 g/day), polydextrose (7–8 g/day), hydrolysed guar gum (5–12 g/day) are allowed to use the fixed health claim 'contains fibre xx and maintains GI condition', whilst six indigestible oligosaccharides, including soy oligosaccharides (2–6 g/day), fructooligosaccharides (3–8 g/day), lactosucrose (2–8 g/day), galactooligosaccharides (2–5 g/day), xylooligosaccharides (1–3 g/day), and isomaltooligosaccharides (10 g/day) are allowed to use the fixed health claim 'contains xx OS, which increase intestinal bifidobacteria and maintain good intestinal environment and GI condition'. Standard FOSHU claim on products with indigestible dextrin for 'blood glucose level' category is likely to be approved shortly.

The reduction of disease risk claim is also permitted when the reduction of disease risk is clinically and nutritionally established for the ingredient. Two ingredients with fixed health claims, calcium and folic acid, have been allowed to carry a 'Reduction of Disease Risk Claim', on osteoporosis and neural tube defect, respectively. Six calcium products are allowed to carry the fixed FOSHU claim:

> Intake of proper amount of calcium contained in healthy meals with appropriate exercise may support healthy bones in young women and reduce the risk of osteoporosis during ageing

Folic acid containing products, in turn, are permitted the fixed claim:

> Intake of proper amount of folic acid contained in healthy meals may support women to bear healthy babies by reducing the risk of neural tube defect, such as spondyloschisis, during fetal development

However, so far, no FOSHU product containing folic acid has applied for this claim.

FOSHU application and approval process

FOSHU approval requires that the effectiveness of the ingredient on human health is clearly proven in the absence of any safety concerns (animal toxicity tests, confirmation of effects in case of excess intake, etc.). In addition, the product must use nutritionally appropriate ingredients (e.g. no excessive use of salt, etc.), guarantee compatibility with product specifications by the time of consumption, and use established quality control methods, such as specification of the product and ingredients, the manufacturing processes, and the methods of analysis. At least one human trial in healthy subjects is required to demonstrate the efficacy of the final product. For safety information, results of one human trial using at minimum three times higher than the recommended dose of final product for several weeks should be provided. The results of the studies should be published in a peer-reviewed scientific journal to ensure their validity from an objective and scientific point of view. The subjects of the human studies should preferably be Japanese, but results obtained in non-Japanese subjects can also be utilized when obtained in subjects whose dietary habits, lifestyle, and physiques match those of the Japanese (e.g. the food related dental hygiene category).

If a food does not meet the strict criteria required for 'Regular' FOSHU, it may be approved under the new system for 'Qualified FOSHU', with limited conditions. This category is reserved for claims on foods with a health function not fully substantiated by scientific evidence that meets the standards of Regular FOSHU. Unlike Regular FOSHU, both the Reduction of Disease Risk FOSHU and Standard FOSHU do not require human trials on the final product for efficacy confirmation, because their scientific evidence for the incorporated functional ingredients are recognized as sufficient for claim, however it is still required to conduct a human trial for safety confirmation on the final product.

The application process for FOSHU is summarized in Fig. 18.2. The applicants (food companies) submit the FOSHU application documents to the government and the local health centre. Normally, the application is not made public before approval. After a hearing session with the applicants at the government, the application is evaluated in the absence of the applicants by the two evaluation committees composed of specialists from medical, nutritional, food hygiene, and pharmaceutical fields. Before completing the approval process, the National Institute of Health and Nutrition examines the amount of the functional component in the final product samples by using the method of analysis indicated by the applicant in the FOSHU application documents. It takes at least 6 months for the MHLW to approve the product as FOSHU through this evaluation process. Since 2003, all FOSHU applications with novel functional components must be evaluated by the Food Safety Committee, which increases the duration of the FOSHU application process by six months.

The active component should be described in terms of its physical, chemical, and biological characteristics, and the qualitative and quantitative test methods for evaluation of the efficacy and

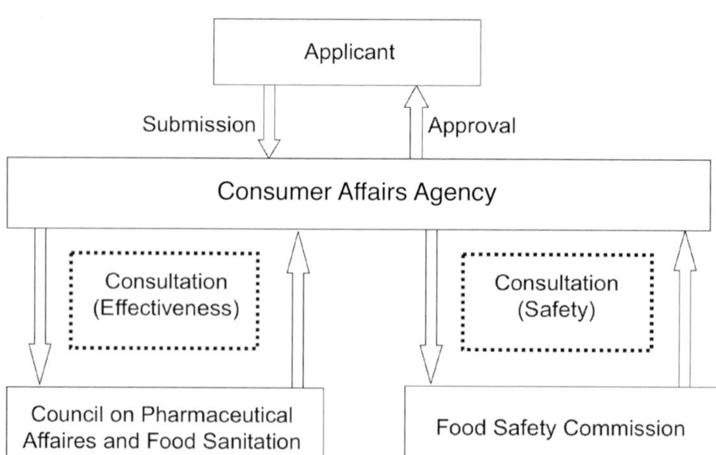

Figure 18.2 Application process for FOSHU.

quality assurance should be specified. For probiotics, whose related functional component is the probiotic bacterial strain, a specific method of quantification should be developed before application. These methods include culturing in specific selective media or PCR using specific DNA primers. The manufacturing process related to quality assurance should also be disclosed. Most lactic acid bacterial species used as probiotics are safe and recognized as GRAS (generally recognized as safe with long history of consumption) (Salminen *et al.*, 1998; Mogensen *et al.*, 2002). However, new products with probiotics are now requested by the government to pass through the safety committee before submission of the application. The safety evidence submitted by the FOSHU applicants for probiotic products generally includes the results of a human study with 3-fold excessive intake for 2–4 weeks using the final product and an acute toxicity study in rodents.

FOSHU claim and criteria on probiotics

A health claim category 'modify gastrointestinal (GI) conditions' is a major FOSHU claim category, under that many FOSHU claims on products with dietary fibres, prebiotics, as well as probiotics have been approved.. In 1998, the government issued guidelines on how to compile and present scientific evidence when applying for FOSHU for some ingredients exhibiting health functions including indigestible oligosaccharides (OS). Although, there are no written guidelines for FOSHU application on probiotics, the guidelines for OS are instructive on how to establish the evidence or 'modify GI conditions' for probiotics. The health benefit of OS is basically exerted by the following mechanism; orally fed OS reaches the colon, where beneficial microorganisms such as bifidobacteria multiply by feeding on the OS, whilst harmful bacteria are suppressed. Hence, the balance of intestinal flora is improved, with beneficial consequences on faecal conditions and defecation. Overall, these effects can be referred to as 'modifying GI conditions'. Based on these guidelines, the working FOSHU criteria for human trails on a final probiotic product are now as follows:

- improvement of intestinal flora showing an increase in beneficial bacteria such as bifidobacteria (number and/or %) and decrease in harmful bacteria such as *Clostridium perfringens* (number and/or %);
- increased defecation frequency per week in subjects with mild constipation; i.e. frequency less than 7 days/week before intervention;
- detection of the probiotic strain in faeces; and
- no side-effects on general health status in excess dose study on the final product.

Changes in bacterial metabolite production, moisture content, pH, and putrefactive enzymes in faeces are also supportive evidence for microflora balancing. The criteria for GI conditions are well accepted by the Japanese academy and authorities based on accumulated research works on intestinal microbiota, showing that an increase in bifidobacteria in the gut is related to some events such as faecal moisturising, faecal mass increase, acidification of the gut environment, which result in an increase in defecation frequency and a decrease in harmful bacteria in the faeces. Overall, these benefits are permitted to be described as GI environment improvement (Mitsuoka 1969, 2006). It is currently recommended that a human trial on probiotics should be a randomized double blind placebo controlled study, preferably performed as a cross-over design with 3 weeks each of intake and wash out periods.

The only FOSHU health claim currently allowed on probiotic products relates to modifying GI conditions (Table 18.2). A typical example of such claim is 'This product is produced by using fermented yogurt with xxx strain, which reaches one's intestines in an active state (or still alive when it arrives in our intestines) so as to help increase intestinal bifidobacteria and lactobacilli (or good bacillus) and decrease bad ones, and this promotes the maintenance of a good intestinal environment and regulates the gastrointestinal conditions'. Other possible wordings include 'Keeps the intestines healthy', 'Regulates the balance of intestinal microflora', 'Useful to improve defecation', and 'Suitable for those concerned about gut health'. The effective dose should be described in the FOSHU

Table 18.2 FOSHU product with probiotics in Japan

Probiotic strains	Number of FOSHU approved product	Dose (min. × 10^9 cfu/day)	Product name	Product type	Manufacturer (strain supplier)	FOSHU health claim (example)
Bacillus subtilis K-2	1	3.8 (spore)	ONAKA NATTO	Fermented soy bean	Asahimatsu Foods	This product contains *Bacillus subtilis* K-2, which helps increasing the number of bifidobacteria. It helps maintenance of an excellent GI environment, and is suitable for people who are concerned about their GI condition
Bifidobacterium breve Yakult	10	10	MILMIL, BIFIEL	Fermented milk	Yakult Honsha	*B. breve* Yakult reaches in the intestines in an active state. It increases useful bacteria and helps regulation of the balance of intestinal microflora that lead to maintain a good GI condition
Bifidobacterium lactis BB-12	1	1.0	ONAKA NO GENKI WO OHEN BB-12 YOGULT	Fermented milk	Azumino Shokuhin Kobo	*Bifidobacterium lactis* BB-12 that reaches into intestines as a viable form improves the gastrointestinal (GI) condition and adjust a bowel movement
	1	1.0	KOIWAI SEINYU 100% YOGURT	Fermented milk	Koiwai Dairy Products	Bifidobacteria contained in this product reaches in the intestines in an active state. It improves GI condition and keeps the intestines healthy
	3	1.0	YOTSUBA HOKKAIDO TOKACHI PLAIN YOGURT SEINYU 100	Fermented milk	Yotsuba	*Bifidobacterium lactis* Bb-12 contained in this products help to improve intestinal environment
	1	1.0	IKARUGA YOGURT BB-12	Fermented milk	Ikaruga Milk	*Bifidobacterium lactis* Bb-12 in this product reaches the intestine alive, and is useful in improving intestinal environment.
Bifidobacterium lactis FK120 *	2	1.0	NOMU DENMARK YOGURT	Fermented milk	Fukushima Milk	Bifidobacteria contained in this product reaches in the intestines in an active state. It helps growth of good bacillus and reduction of bad ones. This improves a GI condition and keeps the intestines healthy.

Bifidobacterium lactis LKM512 *	2	1.0	ONAKA NI OISHII YOGURT	Fermented milk	Kyodo Milk Industry	This product helps regulation and maintenance of a comfortable GI condition.
Bifidobacterium longum BB536	6	2.0	BIFIDUS PLAIN YOGURT	Fermented milk	Morinaga Milk Industries	This product contains living bifidobacteria and helps growth of intestinal bifidobacteria. It helps maintenance of a good intestinal environment and regulation of the GI condition.
Lactobacillus acidofilus CK92 and *L. helveticus* CK60	7	1.0/1.0	CALPIS KIDS	Fermented milk	Danone Japan	This product improves the GI environment and keeps the intestines healthy.
Lactobacillus casei NY1301	2	15	PILKUL	Fermented milk	Nisshin York	This helps improvement of the GI environment and keeps the intestines healthy.
Lactobacillus casei Shirota YIT9029	24	15	YAKULT, JOE, SOFUHL	Fermented milk	Yakult Honsha	*L. casei* Shirota reaches the intestines in an active state. It increases the number of useful bacteria in the intestines and helps regulation of the balance of intestinal microflora that maintain good GI condition
Lactobacillus delbrueckii subsp. *bulgaricus* 2038 and *S. salivarius* subsp. *thermophilus* 1131	6	1.0/10	MEIJI BLUGARIA YOGURT LB81	Fermented milk	Meiji Dairies	This yogurt, with its *Lactobacillus* LB81, helps regulation of the balance of intestinal microflora and maintenance of a good GI condition
Lactobacillus gasseri SBT2055 and *Bifidobacterium longum* SBT2928	2	0.5/1.0	NATURE PRO GB	Fermented milk	Nippon Milk Community	*Lactobacillus gasseri* SP strain and *Bifidobacterium longum* SP strain contained in this products help to improve intestinal environment
Lactobacillus johnsonii La1	2	1.0	NESTLÉ LC1 YOGURT	Fermented milk	Nestlé Japan	*Lactobacillus johnsonii* La1 reaches in the intestines in an active state and adheres onto intestinal wall. This decreases harmful bacteria such as Welch bacillus *Clostridium perfringens* competitively in the intestines and increases beneficial bacteria such as bifidobacteria. This product is helpful to maintain of a good GI condition
Lactobacillus rhamnosus GG	2	14	YOGURT ONAKA-HE GG!	Fermented milk	Takanashi Milk Products	This product is a yogurt drink fermented with *Lactobacillus* GG, which reaches one's intestines in an active state so as to help growth of intestinal bifidobacteria and regulation of the GI condition

*Same as *B. lactis* BB-12

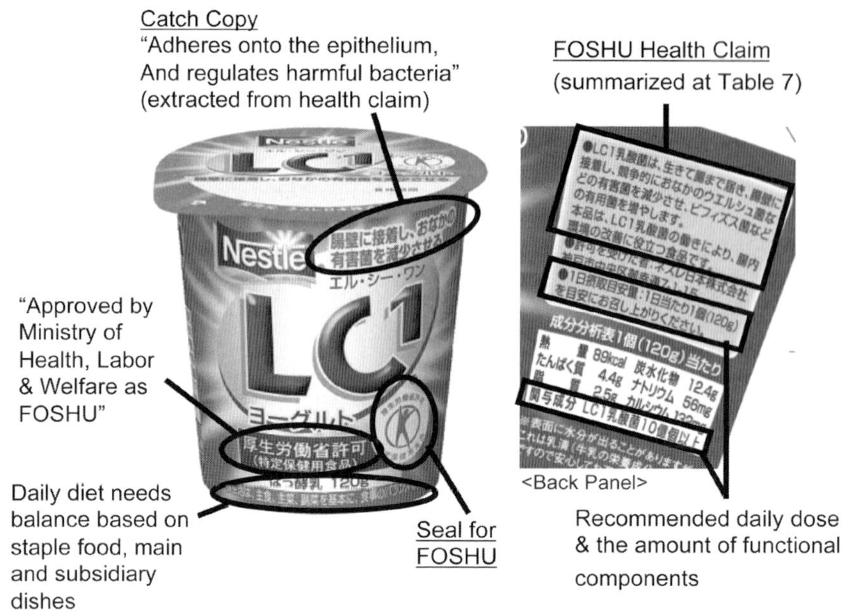

Figure 18.3 Package example for a FOSHU product in 2006.

application document, and it is mandatory to indicate it on the approved product package. Fig. 18.3 shows an example of packaging of a probiotic product with FOSHU approval. The package design must be submitted to the government as a part of the FOSHU application and be evaluated by the government before approval. Displayed on the pack of a FOSHU product should also be 'Food with Health Claims (Food for Specified Health Uses),' the nutritional composition stating the effective level of the related functional component, the approved health claim, the daily standard dosage, the intake manner, and the caution of excessive intake. In addition, the package may contain the statement 'Approved by the government' and display the seal for FOSHU. It may also describe a simplified health claim 'catch copy (phrase)' where words are extracted from the approved health claim. The catch copy (phrase) is also evaluated by the government. Recently, it has become mandatory to display the message 'Daily diet needs balance based on staple food, main and subsidiary dishes' on the pack.

Functional food and health claims in Europe

The term 'Functional Food' is not defined in Europe. A working definition has been used in the EU Project Functional Foods in Europe (Diplock *et al.*, 1999) but Functional Food is generally understood as 'food with a health claim'. Although the food and diet culture in Europe may be regarded as more conservative than in Asia, the relationship between nutrition and health has gained increasing public acceptance over the last decade. Consumers have become more and more health conscious and industry wishes to take advantage of the evolution in science and invest in innovation in this area.

Consequently, it is the use of health claims, not the use of functional foods, which is regulated in the 27 countries that form the EU. The European Commission, the central institution of the EU political system, had preliminary discussions concerning a future claims regulatory system since the early 1990s. It was, however, not until 2003 that an official draft was presented to the European Parliament and the member states. Divergent views and lengthy debates, in particular on the level of scientific substantiation, data protection and not least, the introduction of nutrient profiles as a condition for making nutrition and health claims, delayed the final adoption of this Regulation until late 2006.

Before the adoption of the EU Regulation in 2006, there was a legal void with regard to health claims. The EU framework labelling legislation prohibited 'attributing to any foodstuff the property of preventing, treating or curing a human disease or referring to such properties' (EC, 2000). This was interpreted differently in each of the 15 EU member states in 2000. Therefore, it may have occurred that a claim, which was permitted in one country, may have been prohibited in another and vice versa (Hurt, 2002).

The most difficult type of claim to regulate was the Disease Risk Reduction Claim. It was often regarded as medicinal claim and therefore prohibited. It was included in the final Regulation subject to individual application. There was a greater acceptance of so called functional claims, although the term was not used as such. Nutrition Claim mostly follow Codex Alimentarius rules.

In the 1990s, in the absence of a specific European legislation on health claims, the food industry became itself a stakeholder in self–regulatory systems in a number of EU countries, including the UK, Sweden, France, Belgium, Netherlands, Spain, and Finland. It was observed that there was an 'underlying consistency of approach, but analysis shows that differences were sometimes minor but sometimes major' (Richardson *et al.*, 2003). These innovative initiatives were introduced to facilitate the use of health claims. They included the adoption of guidelines and Codes of Practice in various member states of the EU and of one industry-led European Code of Practice by CIAA (European Food and Drink Association). In most of these countries, a partnership of industry experts, enforcement authorities and consumer representatives developed the rules for the scientific substantiation, communication and presentation of health claims to the scientific community and government authorities. The most prominent examples of these European self-regulatory systems are the Swedish Self Regulation Code and the UK Joint Health Claims Initiative.

Self-regulatory initiatives in Europe

The Swedish food sector's code of practice

In Sweden, the use of health claims within the scope of the Food Sector's Code of Practice – Health Claims in the Labelling and Marketing of Food Products has been possible since 1990. Since its introduction, the Code, which was established in close contact with the competent authorities, has been revised three times, and the last version of the Code was launched in September 2004. Principals of the Code were the Swedish Food Federation and the Swedish Food Retailers Federation. The Swedish Nutrition Foundation was the co-ordinating and advisory body within the Code. The Code consisted of five parts: Rules, expert advice, follow-up, evaluation and information.

Under the Swedish Health Claims Code, two probiotic claims were considered, a fruit drink with *Lactobacillus plantarum* 299v in 2003, and a milk-based drink containing *L. rhamnosus* GG, *L. rhamnosus* Lc-705, *P. freudenreichii* spp., *Shermani* JS, *Bifidobacterium* strain in 2007.

The Swedish Code has gradually ceased to exist in its present form. However, during the implementation of the EU Nutrition and Health Claims Regulation, parts of the Code were used as guidelines on how to apply health claims in a responsible manner on a national level.

Joint Health Claims Initiative (UK)

The Joint Health Claims Initiative (JHCI) was established in 1997 as a joint venture between consumer organizations, enforcement authorities and industry bodies to adopt a Code of Practice for the use of health claims. The initiative aimed to create a level playing field for industry and control bodies and to increase consumer protection by evaluating the science behind generic and innovative health claims.

The JHCI Code of Practice defined the source and the nature of scientific evidence requesting a systematic review of all available scientific evidence relating to the validity of the claim (Art 8.4.1). Once evaluated by the JHCI Expert Committee, the proposed claim was submitted for approval to the JHCI Council. The JHCI

Directors and Council agreed in 2007 that the new EU 'Health Claims' Regulation for foods was about to supplant this area of activity (JHCI Press Release 22.3.2007). JHCI has not approved any probiotic claims.

EU regulation on Nutrition and Health Claims (NHC) in foodstuffs

The EU Regulation on nutrition and health made on foods (Regulation (EC) 1924/2006) lays down harmonized rules for the use of nutrition claims (such as 'low fat', 'high fibre') and health claims (e.g. 'helps lower cholesterol') on foodstuffs based on nutrient profiles, which are to be defined at a later stage. The Health Claims Regulation will ensure that any claim made on a food label in the EU is clear, accurate and substantiated. In doing so, it will enable consumers to make informed and meaningful choices when it comes to food and drinks. This should also contribute to a higher level of health protection, as it ties in with the European Commission's campaign for healthier lifestyle choices by allowing citizens to know exactly what they are consuming.

Health claims under the EU regulation

Permitted claims

The European Regulation defines four categories of health claims and their authorization (Fig. 18.4). For health claims which fall under Article 14 of the Regulation, i.e.:

- health claims referring to reduction of disease risk (meaning any health claim that states, suggests or implies that the consumption of a food category, a food or one of its constituents significantly reduces a risk factor in the development of human disease); and
- health claims referring to children's development and health.

an individual authorization is needed. Applications for authorization are subject to scientific evaluation by EFSA, prior to their inclusion in a Community list of permitted claims.

Health claims that fall under Article 13.1 of the Regulation (other than those referring to the reduction of a disease risk and to children's development and health), are generally accepted and well understood, i.e health claims describing or referring to:

- the role of a nutrient or other substance in growth, development and the functions of the body, or
- psychological and behavioural functions, or
- slimming or weight control

These have to be included in the European Community list of permitted claims (to be adopted). EFSA is currently (in 2009) evaluating several thousand diet–health relationships, which were submitted by the 27 member states, often prepared by EU industry associations or individual companies. This includes diet–health relationships of probiotics.

For health claims (other than those referring to the reduction of a disease risk and to children's development and health) which have not been included in the Community list of permitted

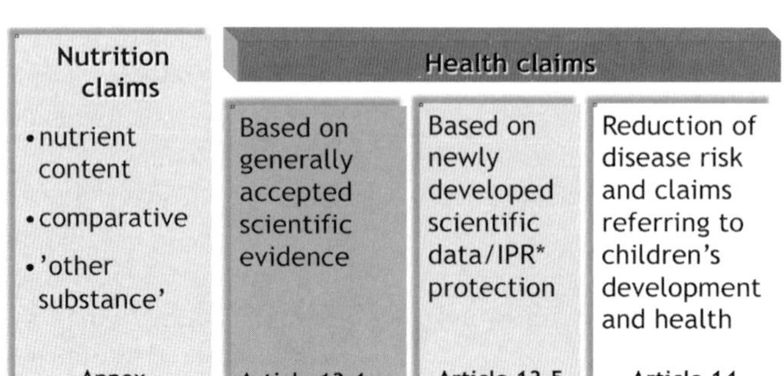

Figure 18.4 Classification of EU claims on foods.

claims (Article 13.5) based on newly developed scientific evidence, and/or which included a request for the intellectual property right (IPR) protection of proprietary data, an individual application has to be filed.

When an individual application for authorization is made, the applicant sends a full dossier to the national authority of an EU Member State, which informs EFSA. EFSA makes the summary available to the public, usually on its website (www.efsa.europa.eu). Within 5 months – unless supplementary information is sought from the applicant, in which case the time limit is extended by 2 months – EFSA adopts a favourable or unfavourable opinion, which is again published and forwarded the European Commission, the 27 member states and the applicant. Within two months, the European Commission has to submit a draft authorization for adoption. The final decision is published in the Official Journal of the European Union.

Conditions for making claims

The Regulation contains other challenging provisions and interpretations: the introduction of Nutrient Profiles, in 2009, as a condition for making a claim has led to much debate and polemic about the role of food products in a balanced diet. The decision to classify simple 'contains' claims, such as 'contains probiotic' as a health, not a nutrition or ingredient claim, will severely limit manufacturers' ability to make claims on the benefits of probiotics because of the additional conditions, which have to be fulfilled. (JHCI, 2001)

When making health claims the following labelling obligations apply:

- a statement on the label indicating the importance of a varied and balanced diet and a healthy lifestyle;
- quantity of the food and pattern of consumption required to obtain the claimed beneficial effect;
- where appropriate, a statement addressed to those persons who should avoid consuming the food; and
- where appropriate, a warning statement.

Scientific substantiation in Europe

The EU Nutrition and Health Claim (NHC) Regulation is silent on specific rules for the substantiation of well-established and well-understood health claims (known as 'Art 13.1 claims'). On the other hand, the technical rules for the preparation of an application for other health claims (known as 'Article 13.5' or 'Article 14 claims') are well documented (EC, 2008; EFSA 2007a; EFSA 2007b).

In the EU system, scientific substantiation must include all scientific data, published and unpublished, in favour and not in favour together with a comprehensive rwweview of the data from human studies to demonstrate that the claim is substantiated. The totality of the scientific data must be taken into account and by weighing the evidence shall demonstrate:

- how the claimed benefit is beneficial for human health;
- cause and effect relationship;
- quantity of food and pattern of consumption required; and
- specific study group in which evidence was obtained.

Foods or food constituents have to be fully characterized to enable EFSA to evaluate the relevance of the submitted studies.

The organization of pertinent data must follow the order: human data classified according to hierarchy of study design (human intervention studies, human observational studies, other human studies), non-human data (animal, ex-vivo or *in vitro*). Data from studies in humans addressing the relationship between consumption of the food and the claimed effect is required for substantiation of a health claim. This must include a comprehensive review of the data. The framework of the health claims Regulation does not foresee a safety evaluation of the food nor a decision whether the food is classified legally as a foodstuff.

The European Community list of generally accepted health claims (Article 13.1 Health Claims)

The EFSA, the central body providing scientific advice to European decision makers is evaluating, in 2009, several thousand diet–health relationships that the 27 member states of the EU have submitted to the European Commission. In January 2009, EFSA published its own list of health claims received from the European Commission for assessment. In total, EFSA has received 4185 main health claim entries, taking into account the conditions of use and references available for around 10,000 similar health claims. Each entry comprises a food component, a health relationship, and an example of wording.

It had been widely understood among industry that EFSA should approach scientific substantiation on Article 13.1 claims in a different way to its approach to Article 14 and 13.5 but it would now appear that the same EFSA process of evaluation, in particular with regard to the requirement of human intervention studies will be applied to Article 13.1 generic claims. It is very likely that the list of permitted health claims will not be available in January 2010, foreseen in legislation, because of the sheer quantity of claims submitted to the authorities and the complexity of the task.

Probiotic claims in Europe

At present, the 27 EU member governments have submitted the claims to the central scientific body EFSA based on a list prepared by EU industry associations and food and food supplement companies, including, in total, 268 proposed diet–health relationships involving probiotics after pre-screening for potential inclusion in the list of permitted claims. However, only 102 will directly qualify for evaluation by EFSA and scientific opinions are expected in July and November 2009, respectively. The most common benefit areas for probiotics are digestive health, intestinal flora, gut health, digestive system and immune system, but final evaluation and adoption are still outstanding.

To date, there has only been a small number of individual applications involving probiotic strains in food and as dietary supplement and none has so far received a positive evaluation from EFSA. It was EFSA's opinion in all cases that cause and effect relationships between consumption and claimed effect could not be established.

Discussion

Regulatory authorities around the world have started to introduce rules for health claim approval to allow the communication of health benefits of foods, if they are fully scientifically substantiated and the communication is proportionate to the scientific evidence behind a claim and well understood by consumers. Such a robust regulatory system allows consumers to maximize their health benefit through consumption of food that is truly functional. It allows enforcers and consumers to distinguish between well-established benefits and vague promises. Finally it allows the food industry to leverage their investment in research and development by effectively communicating with the consumer at all levels from packaging to communication. The Japanese and European regulatory systems are set up to fulfil all of the above objectives using each a legislative approach, which is rooted in their respective political, scientific, and even food cultures.

In Japan, the FOSHU health claim system started in 1991, whereas the harmonized EU rules on scientific substantiation and communication only entered into force in 2007 after two decades of intensive debates. The EU will establish and maintain a Register of Nutrition and Health Claims with authorized and rejected health claims. Differences in the health claim system between Japan and the EU are summarized in Table 18.3. Japan started the health claim system named FOSHU and the MHLW, currently regulated to the Consumer Affairs Agency, approved health claim on food products using functional food components. The system of function for micronutrients with fixed claim (FNHC) were established later. The EU system distinguishes not only between health claims that are well established and those based on emerging science or related to the reduction of disease risk, but it also requires an individual application for authorization for claims referring to children's health and development. This separation is not seen in the Japanese system. As scientific substantiation, human trials are required in both systems, however, the underlying principles are slightly different. FOSHU is the approval system for health claim on product, and the applicants should submit human trial results on the final product as scientific evidences. The human

Table 18.3 Summary of health claim systems in Japan and EU

	Japan		European Union (27 Member States)
Health claim system	Foods for Specified Health Uses (FOSHU)	Food with Nutrient Function Claim (FNFC)	Regulation on nutrition and health claims (NHC) made on foods
Newly developed scientific data	Regular FOSHU/ Qualified FOSHU	–	Art. 13.5 (food, nutrients or other substances with/without IP right protection)*
Generally accepted scientific evidence	Standard FOSHU (three oligosaccharides and six dietary fibres)	FNFC (12 vitamins and 5 minerals)	Art. 13.1 (food, nutrients or other substances)*
Reduction of disease risk claim	Reduction of Disease Risk FOSHU (Ca and folic acid with fixed claim)	–	Art. 14 (food, nutirents or other substances with/without IP right protection) *
Children use	Not specified	–	Art. 14 (functional component and micronutrients)*
Initial launch	1991	2001	2006 (fully applicable after 2010)
Definition	Foods officially approved to claim their physiological effects on the human body	Foods labelled with the nutrient function claims	Health claim means any claim that states, suggests or implies that a relationship exists between a food category, a food or one of its constituents and health
Health claim (wording)	Freely apply for Regular FOSHU (fixed claim on Standard and Reduction of Disease Risk FOSHU)	Fixed claim	Decision on case by case basis
	Category currently approved: GI condition, blood pressure, cholesterol, triacylglyceride, body fat, blood sugar, bone health, mineral absorption, oral health	Category currently approved: skin and mucosa health, vision, energy generation, red blood cell formation, antioxidant, bone/ teeth development etc.	Categories currently applied for: immune, cardiac, cognitive, and bowel functions, metabolisms, bone, antioxidant, muscle, blood pressure, weight management, teeth, cholesterol, digestive health, etc.
	Category currently rejected: immunity defence, fatigue		
Claim in food product/ ingredient	On food product	On nutrients	On ingredient (Art. 13.1); on food product or ingredient (Art. 13.5 and 14)
Functional component	Probiotics, oligosaccharides, dietary fibres, protein/peptides, phytochemicals, fatty acid/fat, sugar alcohols, minerals, etc.	Niacin, pantothenic acid, biotin, vitamin A, vitamin B1, vitamin B2, vitamin B6, vitamin B12, vitamin C, vitamin D, vitamin E, folic acid, Ze, Ca, Fe, Cu and Mg	The scientific evaluation body EFSA has received 4185 main health claim entries, taking into account the conditions of use and references available for around 10,000 similar health claims. Each entry comprises a food component, a health relationship and an example of wording

Table 18.3 continued

	Japan		European Union (27 Member States)
Claim Application	required for each food product	not required	Required for health claims based on newly developed science, for disease risk reduction claims and for claims referring to children's developments and health
Target population	People who need specific health benefit, age is not specified	General population	General population, children
Human trial	Required on each final product for efficacy and safety confirmation. For Reduction of Disease Risk and Standard FOSHUs, just safety confirmation on product is required.	not required	Required for scientific substantiation in view of inclusion in EU list of permitted claims.
Number of human trial required	At least one	–	Not specified
Subjects	Local population except for oral health	–	Not specified. Must be relevant for local population
Study protocol	Free but mostly fixed precedentally. Randomized controlled study is required for Regular FOSHU.	–	Free
Data publication	Required to publish in a peer reviewed journal	–	Preferable
Claim on probiotics	Allowed in 'modify GI conditions' category	–	Proposed benefit areas include digestive health, intestinal flora, gut health, digestive system and immune system
Claim approved on probiotics	73 products using 10 single strains and three strain combinations	–	None in April 2009
Remarks	–	–	Regulation will be fully applicable in 2010

Information as of April 2009.

trial is practically conducted in Japanese population, and the results should be published in a peer reviewed journal. According to the so far very limited experience, a positive scientific evaluation by EFSA requires most likely at least one human trial, whereby the study group must represent the target group or population, in other words being conducted in European subjects.

In the EU, an authorization for a health claim will not necessary be valid for a final food product or brand. Health claims can be made for a food category, a food or a food constituent. A claim once approved by the EU authorities will appear in the central Register of Nutrition and Health claims unless there is proprietary data protection in place. The authorized health claims can be used by anyone if the conditions of use are fulfilled. The law does not confer exclusive use of a health claim to a named applicant, but of the scientific data behind a health claim, subject to conditions, for a period of up to 5 years. Such issue on intellectual property never happens in Japan because the claim is approved always on the final product clearly belonged to the applicant.

Health claim and its criteria for FOSHU approval have been established through the collaboration among the industries, government and academy in Japan. In case of probiotics, just 'modify GI conditions' category is currently allowed for FOSHU claims, where both for improvement of faecal microbiota and mild defecation should be proven in healthy subjects. Faecal bifidobacteria count and defecation frequency are major biomarkers, and the criteria has been well accepted by Japanese scientists both in academy and industries owing to the accumulated research works on intestinal microbiota over 30 years initiated by Professor Mitsuoka. Japan Bifidus Foundation played a certain role in supporting the research on bifidobacteria and microbiota, and successfully promoted improved understanding of their scientific concept. Unlike medicine, food is intrinsically difficult to prove the health benefits, especially when the target population is healthy to begin with. Even in healthy people, however, a range of 'healthy-ness' could exist, and the FOSHU system has been trying to establish how to identify the healthy people with needs for specific health use (SHU) and provide biomarkers under a consensus of the academy, government and industries. Consumer behaviour also influences the stringency of the regulatory system. For example, Japanese consumers believe that food is naturally something beneficial for health, and people are confident to buy new functional foods, even if their benefits could be limited in a medicinal sense.

In the discussion around the new EU regulation, it is sometimes forgotten that the scientific evaluation is only one of the steps, albeit a significant and widely publicized one, in the EU authorization procedure. The final verdict, for example on the wording of a health claim, may need to take into account aspects other than agreement with the scientific evidence, such as for example consumer understanding. To date, no health claim on probiotics has been approved under the EU system. On the one hand, the Community list of permitted claims is not available yet, on the other hand, a few individual applications have received negative scientific opinions.

There is an argument that the new EU health claim legislation requiring high standard of scientific evidence would diminish the number of existing health claims on probiotic products currently in use and therefore, decrease the opportunity for consumers to choose such foods with health benefit, which are allowed in the FOSHU system in Japan. It is possible that a claim on the balancing of the intestinal flora, which is permitted not only in Japan but also in some countries such as China and Brazil will not be able to satisfy the European scientific and regulatory authorities. Such EU standard may influence other parts of the world, especially in the countries where health claim systems have not been established yet, as in for example Australia, that the authorities may copy the system generated in Brussels.

It is difficult to predict how many health claims will appear on the European Community list of permitted claims in 2010, and how many individual applications will be successful over time. At the moment the desired benefit categories are more widespread in the EU than in Japan. In the case of probiotics, the potential health claims are not limited to healthy gut but also include natural defences (e.g. immune boost, anti-allergy). Some FOSHU applications on immunity or natural defence were rejected by the government, according to the Pharmaceutical Affairs Law in Japan, where just medicine can influence the body structure and conditions, and such functional components should be registered as medicine. Based on the current discussion, immunity may be under the medicinal territory, even if the evidence of probiotics in food was submitted. The Japanese Association for Food Immunology was started in 2004, and is working on health claim criteria with academy and industries to extend future opportunities to reach a consensus on the new claim categories.

It is a great moment when the establishment of the health claim system on food becomes a world wide trend so as to grant legal certainty to operators and consumers. Thanks to the scientific community, this trend has been firmly established. Both Europe and Japan should balance the requirements for the highest level of scientific substantiation with the benefits of consumer-friendly communication. This would maximize benefits for consumers and grant at the same time a level playing field to the food industry and R&D communities. Without such a framework the consumer may miss the opportunity to find foodstuffs with proven health benefits without being misled. Further accumulation of scientific evidence should be a key for healthy development of the health claim system and ultimately make consumers healthier.

Acknowledgements

We would like to express our gratitude to Dr K. Makinen (Nestlé Research Centre) for reviewing this article.

References

Abrahamsson, T.R., Jakobsson, T., Böttcher, M.F., Fredrikson, M., Jenmalm, M.C., Björkstén, B., and Oldaeus, G. (2007). Probiotics in prevention of IgE-associated eczema: a double-blind, randomized, placebo-controlled trial. J. Allergy Clin. Immunol. *119,* 1174–1180.

Agarwal, K.N., Bhasin, S.K., Faridi, M.M.A., Mathur, M., and Gupta, S. (2001). *Lactobacillus casei* in the control of acute diarrhea – a pilot study. Indian Pediatr. *38,* 905–910.

Ahmed, M., Prasad, J., Gill, H., Stevenson, L., and Gopal, P. (2007). Impact of consumption of different levels of *Bifidobacterium lactis* HN019 on the intestinal microflora of elderly human subjects. J. Nutr. Health Aging *11,* 26–31.

Arai, S., Osawa, T., Ohigashi, H., Yoshikawa, M., Kaminogawa, S., Watanabe, M., Ogawa, T., Okubo, K., Watanabe, S., Nishino, H., *et al.* (2001). A mainstay of functional food science in Japan – history, present status, and future outlook. Biosci. Biotech. Biochem. *65,* 1–13.

Aso, Y., Akaza, H., Kotake, T., Tsukamoto, T., Imai, K., and Naito, S. (1995). Preventive effect of a *Lactobacillus casei* preparation on the recurrence of superficial bladder cancer in double-blind trial. Eur. Urol. *27,* 104–109.

Azuma, Y., Ito, K., Ohki, A., Ueda, K., Inoue, K., Sato, M., and Benno, Y. (2000). Effects of the fermented milk drink with *Lactobacillus gasseri* NY0509 and *Lactobacillus casei* NY1301 on defecation and fecal microflora in healthy volunteers. J. Nutr. Food 3, 63–72.

Black, F., Einarsson, K., Lidbeck, A., Orrhage, K., and Nord, C.E. (1991). Effect of lactic acid producing bacteria on the human intestinal microflora during ampicillin treatment. Scand. J. Infect. Dis. *23,* 247–254.

Bougle, D., Roland, N., Lebeurrier, F., and Arhan, P. (1999). Effect of propionibacteria supplementation on fecal bifidobacteria and segmental colonic transit time in healthy human subjects. Scand. J. Gastroenterol. *34,* 144–148.

Bouvier, M., Meance, S., Boulev, C., Berta, J.L., and Grimaud, J.C. (2001). Effects of consumption of a milk fermented by the probiotic *Bifidobaterium animalis* DN-173 010 on colonic transit time in healthy humans. Biosci. Microflora *20,* 43–48.

Bunout, D., Barrera, G., Hirsch, S., Gattas, V., de la Maza, M.P., Haschke, F., Steenhout, P., Klassen, P., Hager, C., Avendaño, M., *et al.* (2004). Effects of a nutritional supplement on the immune response and cytokine production in free-living elderly. J. Parenteral Enteral Nutr. *28,* 348–354.

Burton, J.P., Chilcott, C.N., Moore, C.J., Speiser, G., and Tagg, J.R. (2006). A preliminary study of the effect of probiotic *Streptococcus salivarius* K12 on oral malodour parameters. J. Appl. Microbiol. *100,* 754–764.

Canani, R.B., Cirillo, P., Terrin, G., Cesarano, L., Spagnuolo, M.I., De Vincenzo, A., Albano, F., Passariello, A., De Marco, G., Manguso, F., *et al.* (2007). Probiotics for treatment of acute diarrhoea in children: randomised clinical trial of five different preparations. BMJ *335,* 340.

Chapman, T.M., Plosker, G.L., and Figgitt, D.P. (2006). VSL#3 probiotic mixture: a review of its use in chronic inflammatory bowel diseases. Drugs *66,* 1371–1387.

Chouraqui, J.P., Van Egroo, L.D., and Fichot, M.C. (2004). Acidified milk formula supplemented with *Bifidobacterium lactis*: impact on infant diarrhea in residential care settings. J. Pediatr. Gastroenterol. Nutr. *38,* 288–292.

Cobo Sanz, J.M., Mateos, J.A., and Muñoz Conejo, A. (2006). Effect of *Lactobacillus casei* on the incidence of infectious conditions in children. Nutr. Hosp. *21,* 547–551 (in Spanish).

Codex (2002). Committee on Food Labeling. (Proposed) draft guidelines for use of nutirition and health claims. CCFL ALINORM 03/22 appendix VII.

Cremonini, F., Di Caro, S., Santarelli, L., Gabrielli, M., Candelli, M., Nista, E.C., Lupascu, A., Gasbarrini, G., and Gasbarrini, A. (2002). Probiotics in antibiotic-associated diarrhoea. Dig. Liver Dis. *34* (Suppl .2), S78–80.

Diop, L., Guillou, S., and Durand, H. (2008). Probiotic food supplement reduces stress-induced gastrointestinal symptoms in volunteers: a double-blind, placebo-controlled, randomized trial. Nutr. Res. *28,* 1–5.

Diplock, A.T., Aggett, P.J., Ashwell, M., Bornet, F., Fern, E.B., and Roberfroid, M.B. (1999). Scientific concepts of functional foods in Europe. Consensus document. Br. J. Nutr. *81* (Suppl. 1), S1–27.

EC (2000). Directive 2000/13/EC of the European Parliament and the Council of 20 March 2000 on the approximation of thet laws of the Member States relating to the labelling, presentation and advertising of foodstuffs.

EC (2008). Commission Regulation (EC) No 353/2008 of 18 April 2008 establishing implementing rules for applications for autorisation of health claims as provided for in Article 15 of Regulation (EC) No 1924/2006 of the European Parliament and of the Council.

EFSA (2007a). Scientific and Technical Guidance for the Preparation and Presentation of the Application for

Authorisation of a Health Claim, (Parma, Italy) 23 July 2007.
EFSA (2007b). Pre-Submission Guidance for Applicants intending to submit applications for authorisation of health claims made on foods, (Parma, Italy) 14 March 2007, last updated 21 December 2007.
Felley, C.P., Corthesy-Theulaz, I., Rivero, J.L., Sipponen, P., Kaufmann, M., Bauerfeind, P., Wiesel, P.H., Brassart, D., Pfeifer, A., Blum, A.L., *et al.* (2001). Favourable effect of an acidified milk (LC-1) on *Helicobacter pylori* gastritis in man. Eur. J. Gastroenterol. Hepatol. *13,* 25–29.
Fujiwara, D., Wakabayashi, H., Watanabe, H., Nishida, S., and Iino, H. (2005). A double-blind trial fo *Lactobacillus paracasei* strain KW3110 administration for immunomodulation in patients with pollen allergy. Allergol. Int. *54,* 143–149.
Fukushima, Y. (2007). Probiotics and natural defense function of the host. Biosci. Microflora *26,* 1–10.
Fukushima, Y., Li, S., Hara, H., Terada, A., and Mitsuoka, T. (1997). Effect of follow-up formula containing bifidobacteria (NAN BF) on fecal flora and fecal metabolites in healthy children. Biosci. Microflora *16,* 65–72.
Fukushima, Y., Kawata, Y., Hara, H., Terada, A., and Mitsuoka, T. (1998). Probiotic formula enhances intestinal immunoglobulin A production in healthy children. Int. J. Food Microbiol. *42,* 39–44.
Fukushima, Y., Miyaguchi, S., Yamano, T., Kaburagi, T., Iino, H., Ushida, K., and Sato, K. (2007). Improvement of nutritional status and incidence of infection in hospitalised, enterally fed elderly by feeding of fermented milk containing probiotic *Lactobacillus johnsonii* La1 (NCC533). Br. J. Nutr. *98,* 969–977.
Fuller, R. (1991). Probiotics in human medicine. Gut *32,* 439–442.
Gill, H.S., Rutherfurd, K.J., and Cross, M.L. (2001). Dietary probiotic supplementation enhances natural killer cell activity in the elderly: an investigation of age-related immunological changes. J. Clin. Immunol. *21,* 264–271.
Gill, H.S., Rutherfurd, K.J., Cross, M.L., and Gopal, P.K. (2001). Enhancement of immunity in the elderly by dietary supplementation with the probiotic *Bifidobacterium lactis* HN019. Am. J. Clin. Nutr. *74,* 833–839.
Gotteland, M., Andrews, M., Toledo, M., Muñoz, L., Caceres, P., Anziani, A., Wittig, E., Speisky, H., and Salazar, G. (2008). Modulation of *Helicobacter pylori* colonization with cranberry juice and *Lactobacillus johnsonii* La1 in children. Nutrition *24,* 421–426.
Hara, H., Terada, A., Takahashi, M., Kaneko, T., and Mitsuoka, T. (1993). Effects of yoghurt-administration on fecal flora and putrefactive metabolites of normal adults. Bifidus *6* 169–175 (in Japanese with English abstract).
Hatakka, K., Savilahti, E., Ponka, A., Meurman, J.H., Poussa, T., Nase, L., Saxelin, M., and Korpela, R. (2001). Effect of long term consumption of probiotic milk on infections in children attending day care centres: double blind, randomised trial. BMJ *322,* 1327–1329.
Hilton, E., Isenberg, H.D., Alperstein, P., France, K., and Borenstein, M.T. (1992). Ingestion of yogurt containing *Lactobacillus acidophilus* as prophylaxis for candidal vaginitis. Ann. Intern. Med. *116,* 353–357.
Hilton, E., Kolakowski, P., Singer, C., and Smith, M. (1997). Efficacy of *Lactobacillus* GG as a diarrheal preventive in travelers. J. Travel Med. *4* 41–43.
Hosoda, M., He, F., Hiramatsu, M., Hasimoto, H., and Benno, Y. (1994). Effects of *Lactobacillus* GG strains intake on fecal microflora and defication in healthy volunteers. Bifidus *8,* 21–28.
Hurt, E. (2002). International guidelines and experiences on health claims in Europe. Asia-pacific J. Clin. Nutr. *11* (Suppl.), S90–93.
Ikeda, N., Kaneko, K., Ejiri, M., Watabe, J., Suzuki, H., Mizutani, j., Ohshima, K., Kajimoto, O., and Takahashi, T. (2000). Effect of *Lactobacillus* beverage adinistration on defecation of healthy volunteers with constipation. J. Nutr. Food *3,* 59–68.
Ikeda, N., Suzuki, H., Kaneko, K., Watabe, J., and Mizutani, J. (2000). Effect of *Lactobacillus* beverage intake on fecal microflora of infants. J. Nutr. Food *3,* 69–74.
Imase, K., Tanaka, A., Tokunaga, K., Sugano, H., Ishida, H., and Takahashi, S. (2007). *Lactobacillus reuteri* tablets suppress *Helicobacter pylori* infection – a double-blind randomised placebo-controlled cross-over clinical study. J. Jpn. Associ. Infect. Dis. *81,* 387–393.
Ishida, Y., Nakamura, F., Kanzato, H., Sawada, D., Hirata, H., Nishimura, A., Kajimoto, O., and Fujiwara, S. (2005). Clinical effects of *Lactobacillus acidophilus* strain L-92 on perennial allergic rhinitis: a double-blind, placebo-controlled study. J. Dairy Sci. *88,* 527–533.
Ishida, Y., Nakamura, F., Kanzato, H., Sawada, D., Yamamoto, N., Kagata, H., Oh-Ida, M., Takeuchi, H., and Fujiwara, S. (2005). Effect of milk fermented with *Lactobacillus acidophilus* strain L-92 on symptoms of Japanese cedar pollen allergy: a randomized placebo-controlled trial. Biosci. Biotechnol. Biochem. *69,* 1652–1660.
Ishikawa, H., Akedo, I., Umesaki, Y., Tanaka, R., Imaoka, A., and Otani, T. (2003). Randomized controlled trial of the effect of bifidobacteria-fermented milk on ulcerative colitis. J. Am. Coll. Nutr. *22,* 56–63.
Isolauri, E., Arvola, T., Sutas, Y., Moilanen, E., and Salminen, S. (2000). Probiotics in the management of atopic eczema. Clin. Exp. Allergy *30,* 1604–1610.
Ivory, K., Chambers, S.J., Pin, C., Prieto, E., Arques, J.L., and Nicoletti, C. (2008). Oral delivery of *Lactobacillus casei* Shirota modifies allergen-induced immune responses in allergic rhinitis. Clin. Exp. Allergy *38,* 1282–1289.
Izawa, K., Noma, T., Yamamoto, M., Kimura, K., Ito, H., Taketomo, N., Numano, K., and Kawashima, M. (2008). Verifying the ability of yogurut prepared with LB81 lactic acid bacteria to improve skin function. J. Intest. Microbiol. *22,* 1–5.
Jimenez, E., Fernandez, L., Maldonado, A., Martin, R., Olivares, M., Xaus, J., and Rodriguez, J.M. (2008).

Oral administration of *Lactobacillus* strains isolated from breast milk as an alternative for the treatment of infectious mastitis during lactation. Appl. Environ. Microbiol. *74,* 4650–4655.

Joint FAO/WHO expert consultation (2001). Health and Nutritional Properties of Probiotics in Food including Powder Milk with Live Lactic Acid Bacteria. Córdoba, Argentina, 1–4 October 2001.

Joint FAO/WHO Working Group. (2002). Guidelines for the Evaluation of Probiotics in Food. London, ON, 30 April to 1 May 2002.

Joint Health Claims Initiative. (2000). JHCI Code of Practice on Health Claims on Foods. JHCI, Brackwell, UK.

Kajander, K., Myllyluoma, E., Rajilić-Stojanović, M., Kyrönpalo, S., Rasmussen, M., Järvenpää, S., Zoetendal, E.G. de Vos, W.M., Vapaatalo, H., and Korpela, R. (2008). Clinical trial: multispecies probiotic supplementation alleviates the symptoms of irritable bowel syndrome and stabilizes intestinal microbiota. Aliment. Pharmacol. Ther. *27,* 48–57.

Kalliomäki, M., Salminen, S., Arvilommi, H., Kero, P., Koskinen, P., and Isolauri, E. (2001). Probiotics in primary prevention of atopic disease: a randomised placebo-controlled trial. Lancet *357,* 1076–1079.

Kalliomäki, M., Salminen, S., Poussa, T., Arvilommi, H., and Isolauri, E. (2003). Probiotics and prevention of atopic disease: 4-year follow-up of a randomised placebo-controlled trial. Lancet *361,* 1869–1871.

Kalliomäki, M., Salminen, S., Poussa, T., and Isolauri, E. (2007). Probiotics during the first 7 years of life: a cumulative risk reduction of eczema in a randomized, placebo-controlled trial. J. Allergy Clin. Immunol. *119,* 1019–1021.

Katayama, K., Hamasaki, Y., Fuchu, H., Abiru, Y., Sugiyama, M., Yamaguchi, M., and Benno, Y. (2004). Effects of yogurt containing *Bifidobacterium lactis* BB-12 on defecation and fecal microflora in healthy volunteers, and its safety. J. Nutr. Food 7 (2), 73–81.

Kawase, M., He, F., Kubota, A., Hiramatsu, M., Saito, H., Ishii, T., Yasueda, H., and Akiyama, K. (2009). Effect of fermented milk prepared with two probiotic strains on Japanese cedar pollinosis in a double-blind placebo-controlled clinical study. Int. J. Food Microbiol. *128,* 429–434.

Kinumaki, A., Obara, Y., Kasuga, O., Onosawa, M., Misawa, H., Takagi, K., Chiba, E., Tani, K., Suzuki, Y., and Benno, Y. (2001). Effects of a new fermented milk prepared using *Lactobacillus gasseri* SBT2055 and *Bifidobacterium longum* SBT2928, both of human origin, on the defication frequency and fecal microflora of healthy human adults. Jpn. J. Lactic Acid Bacteria *12,* 92–101(in Japanese with English abstract).

Kishi, A., Uno, K., Matsubara, Y., Okuda, C., and Kishida, T. (1996). Effect of the oral administration of *Lactobacillus brevis* subsp. *coagulans* on interferon-alpha producing capacity in humans. J. Am. Coll. Nutr. *15,* 408–412.

Klarin, B., Wullt, M., Palmquist, I., Molin, G., Larsson, A., and Jeppsson, B. (2008). *Lactobacillus plantarum* 299v reduces colonisation of *Clostridium difficile* in critically ill patients treated with antibiotics. Acta Anaesthesiol. Scand. *52,* 1096–1102.

Koebnick, C., Wagner, I., Leitzmann, P., Stern, U., and Zunft, H.J. (2003). Probiotic beverage containing *Lactobacillus casei* Shirota improves gastrointestinal symptoms in patients with chronic constipation. Can. J. Gastroenterol. *17,* 655–659.

Kukkonen, K., Nieminen, T., Poussa, T., Savilahti, E., and Kuitunen, M. (2006). Effect of probiotics on vaccine antibody responses in infancy – a randomized placebo-controlled double-blind trial. Pediatr. Allergy Immunol. *17,* 416–421.

Kumemura, M., Toba, M., Sogawa, Y., Shimizu, S., and Kawaguchi, S. (2001). Physiological function of *Lactobacillus plantarum* fermented milk I. Survival of *Lactobacillus plantarum* species in the gastrointestinal tract and the effect on fecal microflora in humans. J. Intest. Microbiol. *15,* 15–20.

Link-Amster, H., Rochat, F., Saudan, K.Y., Mignot, O., and Aeschlimann, J.M. (1994). Modulation of a specific humoral immune response and changes in intestinal flora mediated through fermented milk intake. FEMS Immunol. Med. Microbiol. *10,* 55–63.

McFarland, L.V. (2007). Meta-analysis of probiotics for the prevention of traveler's diarrhea. Travel Med. Infect. Dis. *5* 97–105.

Majamaa, H., Isolauri, E., Saxelin, M., and Vesikari, T. (1995). Lactic acid bacteria in the tretment of acute rotavirus gastroenteritis. J. Pediatr. Gastroenterol. Nutr. *20,* 333–339.

Mao, M., Yu, T., Xiong, Y., Wang, Z., Liu, H., Gotteland, M., and Brunser, O. (2008). Effect of a lactose-free milk formula supplemented with bifidobacteria and streptococci on the recovery from acute diarrhoea. Asia Pac. J. Clin. Nutr. *17,* 30–34.

Marcos, A., Wärnberg, J., Nova, E., Gómez, S., Alvarez, A., Alvarez, R., Mateos, J.A., and Cobo, J.M. (2004). The effect of milk fermented by yogurt cultures plus *Lactobacillus casei* DN-114001 on the immune response of subjects under academic examination stress. Eur. J. Nutr. *43,* 381–389.

Marteau, P., Cuillerier, E., Meance, S., Gerhardt, M.F., Myara, A., Bouvier, M., Bouley, C., Tondu, F., Bommelaer, G., and Grimaud, J.C. (2002). *Bifidobacterium animalis* strain DN-173 010 shortens the colonic transit time in healthy women: a double-blind, randomized, controlled study. Aliment. Pharmacol. Ther. *16,* 587–593.

Matsumoto, K., Takada, T., Shimizu, K., Kado, Y., Kawakami, K., Makino, I., Yamaoka, Y., Hirano, K., Nishimura, A., Kajimoto, O., *et al.* (2006). The effects of a probiotic milk product containing *Lactobacillus casei* strain Shirota on the defecation frequency and the intestinal microflora of sub-optimal health state volunteers: a randomized placebo-controlled cross-over study. Biosci. Microflora *25,* 39–48.

Matsuoka, T., Sugano, N., Ito, K., and Koga, Y. (2005). Effect of Oral *Lactobacillus salivarius* TI 2711 (LS1) administration. Soc. Periodontol. *47,* 194–202.

Metchnikoff, E. (1907). The prolongation of life (Heinemann, London).

Michetti, P., Dorta, G., Wiesel, P.H., Brassart, D., Verdu, E., Herranz, M., Felley, C., Porta, N., Rouvet, M., Blum, A.L., *et al.* (1999). Effect of whey-based culture supernatant of *Lactobacillus acidophilus* (*johnsonii*) La1 on *Helicobacter pylori* infection in humans. Digestion *60,* 203–209.

Mitsui, N., Kajimoto, O., Tsukahara, M., Murasawa, H., Tamura, M., Nishimura, A., Kajimoto, Y., and Benno, Y. (2006). Effect of natto including bacillus subtilis k-2 (spore) on defecation and fecal microbiota, and safety of excessive ingestion in healthy volunteers. Jpn. Pharm. Therapeutics *34,* 135–148.

Mitsuoka, T. (1969). Intestinal flora and the host. Pharmacia *5,* 608–609.

Mitsuoka, T., ed. (2006). Probiotics, prebiotics, biogenics – Research related to intestinal microflora and its significance. (Japan Bifidus Foundation, Tokyo).

Mogensen, G., Salminen, S., O'Brien, J., Ouwehand, A., Holzapfel, W., Shortt, C., Fonden, R., Miller, G.D., Donohue, D., Playne, M., *et al.* (2002). Food microorganisms – health benefits, safety evaluation and strains with documented history of use in foods. IDF Bullitin *377,* 4–19.

Mori, K., Nakagawa, K., Yajima, N., Chikamatsu, H., and Murayama, T. (2002). Effect of the fermeted milk drink with *Lactobacillus casei* strain SBR 1202 and *Lactobvacillus helveticus* strain SBR1101 on the defecation, fecal properties, fecal microflora and safety in healthy volunteers. J. Nutr. Food *5,* 11–27.

Nagao, F., Nakayama, M., Muto, T., and Okumura, K. (2000). Effects of a fermented milk drink containing *Lactobacillus casei* strain Shirota on the immune system in healthy human subjects. Biosci. Biotech. Biochem. *64,* 2706–2708.

Nase, L., Hatakka, K., Savilahti, E., Saxelin, M., Ponka, A., Poussa, T., Korpela, R., and Meurman, J.H. (2001). Effect of long-term consumption of a probiotic bacterium, *Lactobacillus rhamnosus* GG, in milk on dental caries and caries risk in children. Caries Res. *35,* 412–420.

Niedzielin, K., Kordecki, H., and Birkenfeld, B. (2001). A controlled, double-blind, randomized study on the efficacy of *Lactobacillus plantarum* 299V in patients with irritable bowel syndrome. Eur. J. Gastroenterol. Hepatol. *13,* 1143–1147.

Nikawa, H., Makihira, S., Fukushima, H., Nishimura, H., Ozaki, Y., Ishida, K., Darmawan, S., Hamada, T., Hara, K., Matsumoto, A., *et al.* (2004). *Lactobacillus reuteri* in bovine milk fermented decreases the oral carriage of mutans streptococci. Int. J. Food Microbiol. *95,* 219–223.

Nishida, S., Gotou, M., Akutsu, S., Ono, M., Hitomi, Y., Nakamura, T., and Iino, H. (2004). Effect of yogurt containing *Bifidobacterium lactis* BB-12 on improvement of defecation and fecal microflora of healthy female adults. Milk Science *53,* 71–80.

Nishida, S., Michinaka, A., Nakashima, K., Iino, H., and Fujii, T. (2008). Evaluation of the probiotic potential of *Lactobacillus paracasei* KW3110 based on *in vitro* tests and oral administration tests in healthy adults. J. Gen. Appl. Microbiol. *54,* 267–276.

Nomoto, K. (2005). Prevention of infections by probiotics. J. Biosci. Bioeng. *100,* 583–592.

Odamaki, T., Xiao, J.Z., Iwabuchi, N., Sakamoto, M., Takahashi, N., Kondo, S., Miyaji, K., Iwatsuki, K., Togashi, H., Enomoto, T., *et al.* (2007). Influence of *Bifidobacterium longum* BB536 intake on faecal microbiota in individuals with Japanese cedar pollinosis during the pollen season. J. Med. Microbiol. *56,* 1301–1308.

Ogata, T. (1999). Effect of *Bifidobacterium longum* BB536 yogurt administration on the intestinal environment of healthy adults. Microbial. Ecol. Health Dis. *11,* 41–44.

Ohashi, Y., Nakai, S., Tsukamoto, T., Masumori, N., Akaza, H., Miyanaga, N., Kitamura, T., Kawabe, K., Kotake, T., Kuroda, M., *et al.* (2002). Habitual intake of lactic acid bacteria and risk reduction of bladder cancer. Urol. Int. *68,* 273–280.

Olivares, M., Diaz-Ropero, M.A., Gomez, N., Lara-Villoslada, F., Sierra, S., Maldonado, J.A., Martin, R., Lopez-Huertas, E., Rodriguez, J.M., and Xaus, J. (2006). Oral administration of two probiotic strains, *Lactobacillus gasseri* CECT5714 and *Lactobacillus coryniformis* CECT5711, enhances the intestinal function of healthy adults. Int. J. Food Microbiol. *107,* 104–111.

Olivares, M., Diaz-Ropero, M.P., Gomez, N., Lara-Villoslada, F., Sierra, S., Maldonado, J.A., Martin, R., Rodriguez, J.M., and Xaus, J. (2006). The consumption of two new probiotic strains, *Lactobacillus gasseri* CECT 5714 and *Lactobacillus coryniformis* CECT 5711, boosts the immune system of healthy humans. Int. Microbiol. 9 47–52.

Olivares, M., Díaz-Ropero, M.P., Sierra, S., Lara-Villoslada, F., Fonollá, J., Navas, M., Rodríguez, J.M., and Xaus, J. (2007). Oral intake of *Lactobacillus fermentum* CECT5716 enhances the effects of influenza vaccination. Nutrition *23,* 254–260.

Pedone, C.A., Arnaud, C.C., Postaire, E.R., Bouley, C.F., and Reinert, P. (2000). Multicentric study of the effect of milk fermented by *Lactobacillus casei* on the incidence of diarrhoea. Int. J. Clin. Pract. *54,* 568–571.

Peguet-Navarro, J., Dezutter-Dambuyant, C., Buetler, T., Leclaire, J., Smola, H., Blum, S., Bastien, P., Breton, L., and Gueniche, A. (2008). Supplementation with oral probiotic bacteria protects human cutaneous immune homeostasis after UV exposure-double blind, randomized, placebo controlled clinical trial. Eur. J. Dermatol. *18,* 504–511.

Power, D.A., Burton, J.P., Chilcott, C.N., Dawes, P.J., and Tagg, J.R. (2008). Preliminary investigations of the colonisation of upper respiratory tract tissues of infants using a paediatric formulation of the oral probiotic *Streptococcus salivarius* K12. Eur. J. Clin. Microbiol. Infect. Dis. *27,* 1261–1263.

Prescott, S.L., Dunstan, J.A., Hale, J., Breckler, L., Lehmann, H., Weston, S., and Richmond, P. (2005). Clinical effects of probiotics are associated with increased interferon-gamma responses in very young children with atopic dermatitis. Clin. Exp. Allergy *35,* 1557–1564.

Reid, G. (2008). Probiotic Lactobacilli for urogenital health in women. J. Clin. Gastroenterol. *42* (Suppl. 3), S234–236.

Richardson, D.P., Affertsholt, T., Asp, N.-G., Bruce, A., Grossklawer, R., Howlett, J., Pannemans, D., Ross, R., Verhagen, H., Viechtbauer, V. (2003). PASSCLAIM – Synthesis and review of existing processes. Eur. J. Nutr. *42* (Suppl .1), 96–111.

Rosenfeldt, V., Michaelsen, K.F., Jakobsen, M., Larsen, C.N., Møller, P.L., Pedersen, P., Tvede, M., Weyrehter, H., Valerius, N.H., and Paerregaard, A. (2002). Effect of probiotic *Lactobacillus* strains in young children hospitalized with acute diarrhea. Pediatr. Infect. Dis. J. *21,* 411–416.

Rosenfeldt, V., Michaelsen, K.F., Jakobsen, M., Larsen, C.N., Møller, P.L., Tvede, M., Weyrehter, H., Valerius, N.H., and Paerregaard, A. (2002). Effect of probiotic *Lactobacillus* strains on acute diarrhea in a cohort of nonhospitalized children attending day-care centers. Pediatr. Infect. Dis. J. *21,* 417–419.

Rosenfeldt, V., Benfeldt, E., Nielsen, S.D., Michaelsen, K.F., Jeppesen, D.L., Valerius, N.H., and Paerregaard, A. (2003). Effect of probiotic *Lactobacillus* strains in children with atopic dermatitis. J. Allergy Clin. Immunol. 111, 389–395.

Rosenfeldt, V., Benfeldt, E., Valerius, N.H., Paerregaard, A., and Michaelsen, K.F. (2004). Effect of probiotics on gastrointestinal symptoms and small intestinal permeability in children with atopic dermatitis. J. Pediatr. *145,* 612–616.

Saavedra, J.M., Bauman, N.A., Oung, I., Perman, J.A., and Yolken, R.H. (1994). Fedding of *Bificobacterium bificum* and *Streptococcus thermophilus* to infants in hospital for prevention of diarrhoea and shedding of rotavirus. Lancet *344,* 1046–1049.

Saavedra, J.M., Abi-Hanna, A., Moore, N., and Yolken, R.H. (2004). Long-term consumption of infant formulas containing live probiotic bacteria: tolerance and safety. Am. J. Clin. Nutr. *79,* 261–267.

Sakamoto, I., Igarashi, M., Kimura, K., Takagi, A., Miwa, T., and Koga, Y. (2001). Suppressive effect of *Lactobacillus gasseri* OLL2716 (LG21) on *Helicobacter pylori* infection in humans. J Antimicrob Chemother *27,* 709–710.

Salazar-Lindo, E., Figueroa-Quintanilla, D., Caciano, M.I., Reto-Valiente, V., Chauviere, G., and Colin, P. (2007). Lacteol Study Group. Effectiveness and safety of *Lactobacillus* LB in the treatment of mild acute diarrhea in children. J. Pediatr. Gastroenterol. Nutr. *44,* 571–576.

Salminen, S., von Wright, A., Morelli, L., Marteau, P., Brassart, D., de Vos, W.M., Fonden, R., Saxelin, M., Collins, K., Mogensen, G., *et al.* (1998). Demonstration of safety of probiotics – a review. Int. J. Food Microbiol. *44,* 93–106.

Sandres, M.E., and Klaenhammer, T.R. (2001). The scientific basis of *Lactobacillus acidophilus* NCFM functionality as a probiotic. J. Dairy Sci. *84,* 319–331.

Sarker, S.A., Sultana, S., Fuchs, G.J., Alam, N.H., Azim, T., Brüssow, H., and Hammarström, L. (2005). *Lactobacillus paracasei* strain ST11 has no effect on rotavirus but ameliorates the outcome of nonrotavirus diarrhea in children from Bangladesh. Pediatrics *116,* 221–228.

Savino, F., Pelle, E., Palumeri, E., Oggero, R., and Miniero, R. (2007). *Lactobacillus reuteri* (American Type Culture Collection Strain 55730) versus simethicone in the treatment of infantile colic: a prospective randomized study. Pediatrics *119,* 124–130.

Schiffrin, E.J., Rochat, F., Link-Amster, H., Aeschlimann, J.M., and Donnet-Hughes, A. (1995). Immunomodulation of human blood cells following the ingestion of lactic acid bacteria. J. Dairy Sci. *78,* 491–497.

Schultz, M. (2008). Clinical use of *E. coli* Nissle 1917 in inflammatory bowel disease. Inflamm. Bowel Dis. *14,* 1012–1018.

Sheih, Y.H., Chiang, B.L., Wang, L.H., Liao, C.K., and Gill, H.S. (2001). Systemic immunity-enhancing effects in healthy subjects following dietary consumption of the lactic acid bacterium *Lactobacillus rhamnosus* HN001. J. Am. Coll. Nutr. *20* (Suppl. 2),149–156.

Sheu, B.S., Cheng, H.C., Kao, A.W., Wang, S.T., Yang, Y.J., Yang, H.B., and Wu, J.J. (2006). Pretreatment with *Lactobacillus*- and *Bifidobacterium*-containing yogurt can improve the efficacy of quadruple therapy in eradicating residual *Helicobacter pylori* infection after failed triple therapy. Am. J. Clin. Nutr. *83,* 864–869.

Shornikova, A.V., Casas, I.A., Mykkänen, H., Salo, E., and Vesikari, T. (1997). Bacteriotherapy with *Lactobacillus reuteri* in rotavirus gastroenteritis. Pediatr. Infect. Dis. J. *16,* 1103–1107.

Sistek, D., Kelly, R., Wickens, K., Stanley, T., Fitzharris, P., and Crane, J. (2006). Is the effect of probiotics on atopic dermatitis confined to food sensitized children? Clin. Exp. Allergy *36,* 629–633.

Su, P., Henriksson, A., Tandianus, J.E., Park, J.H., Foong, F., and Dunn, N.W. (2005). Detection and quantification of *Bifidobacterium lactis* LAFTI B94 in human faecal samples from a consumption trial. FEMS Microbiol. Lett. *244,* 99–103.

Szajewska, H., and Mrukowicz, J.Z. (2001). Probiotics in the treatment and prevention of acute infectious diarrhea in infants and children: a systematic review of published randomized, double-blind, placebo-controlled trials. J. Pediatr. Gastroenterol. Nutr. *33,* S17–25.

Takeda, K. and Okumura, K. (2007). Effects of a fermented milk drink containing *Lactobacillus casei* strain Shirota on the human NK-cell activity. J. Nutr. *137* (3 Suppl. 2), 791S–793S.

Tamura, M., Shikina, T., Morihana, T., Hayama, M., Kajimoto, O., Sakamoto, A., Kajimoto, Y., Watanabe, O., Nonaka, C., Shida, K., *et al.* (2007). Effects of probiotics on allergic rhinitis induced by Japanese cedar pollen: randomized double-blind, placebo-controlled clinical trial. Int. Arch. Allergy Immunol. *143,* 75–82.

Tanaka, R. (1994). Clinical applications of *Bifidobacterium* in humans. In Research of Bifidobacteria (in Japanese), Mitsuoka, T., ed. (Japan Bifidus Foundation, Tokyo), pp. 221–228.

Tanaka, R., and Ohwaki, M. (1991). A controlled study of the effects of the ingestion of *Lactobacillus casei* fermented milk on the intestinal microflora, its microbial metabolism and the immune system of healthy humans. In Mitsuoka, T., ed., Intestinal Flora and Diet (in Japanese with English summary), Proceedings of 12th ICPR Symposium on Intestinal Flora, (Japan Scientific Societies Press, Tokyo), pp. 85–142.

Tanaka, R., Kan, T., Tejima, H., Kuroshima, T., Kodaira, S., Suzuki, S., Terashima, T., and Mutai, M. (1980). *Bifidobacterium* interplantation –intake for *B. bifidum* 4006 or *B. breve* 4007. Jpn. J. Clin. Pediatr. *33,* 2483–2492 (in Japanese).

Tannock, G.W., Munro, K., Harmsen, H.J., Welling, G.W., Smart, J., and Gopal, P.K. (2000). Analysis of the fecal microflora of human subjects consuming a probiotic product containing *Lactobacillus rhamnosus* DR20. Appl. Environ. Microbiol. *66,* 2578–2588.

Toda, T., Kosaka, H., Terai, M., Mori, H., Benno, Y., and Yamori, Y. (2005). Effect of fermented milk with *Lactococcus lactis* subsp. *cremoris* FC on defecation frequency and fecal microflora in healthy elderly volunteers. J. Jpn. Soc. Food Sci. Technol. *52,* 243–250.

Tubelius, P., Stan, V., and Zachrisson, A. (2005). Increasing work-place healthiness with the probiotic *Lactobacillus reuteri*: a randomised, double-blind placebo-controlled study. Environ. Health *4,* 25.

Turchet, P., Laurenzano, M., Auboiron, S., and Antoine, J.M. (2003). Effect of fermented milk containing the probiotic *Lactobacillus casei* DN-114001 on winter infections in free-living elderly subjects: a randomised, controlled pilot study. J. Nutr. Health Aging *7,* 75–77.

Valeur, N., Engel, P., Carbajal, N., Connolly, E., and Ladefoged, K. (2004). Colonization and immunomodulation by *Lactobacillus reuteri* ATCC 55730 in the human gastrointestinal tract. Appl. Environ. Microbiol. *70,* 1176–1181.

de Vrese, M., Rautenberg, P., Laue, C., Koopmans, M., Herremans, T., and Schrezenmeir, J. (2005a). Probiotic bacteria stimulate virus-specific neutralizing antibodies following a booster polio vaccination. Eur. J. Nutr. *44,* 406–413.

de Vrese, M., Winkler, P., Rautenberg, P., Harder, T., Noah, C., Laue, C., Ott, S., Hampe, J., Schreiber, S., Heller, K., *et al.* (2005b). Effect of *Lactobacillus gasseri* PA 16/8, *Bifidobacterium longum* SP 07/3, *B. bifidum* MF 20/5 aon common cold episodes: a double blind, randomized, controlled trial. Clin. Nutr. *24,* 481–491.

de Vrese, M., Winkler, P., Rautenberg, P., Harder, T., Noah, C., Laue, C., Ott, S., Hampe, J., Schreiber, S., Heller, K., *et al.* (2006). Probiotic bacteria reduced duration and severity but not the incidence of common cold episodes in a double blind, randomized, controlled trial. Vaccine *24,* 6670–6674.

Wang, K.Y., Li, S.N., Liu, C.S., Perng, D.S., Su, Y.C., Wu, D.C., Jan, C.M., Lai, C.H., Wang, T.N., and Wang, W.M. (2004). Effects of ingesting *Lactobacillus*- and *Bifidobacterium*-containing yogurt in subjects with colonized *Helicobacter pylori.* Am. J. Clin. Nutr. *80,* 737–744.

Wang, M.F., Lin, H.C., Wang, Y.Y., and Hsu, C.H. (2004). Treatment of perennial allergic rhinitis with lactic acid bacteria. Pediatr. Allergy Immunol. *15,* 152–158.

Weizman, Z., Asli, G., and Alsheikh, A. (2005). Effect of a probiotic infant formula on infections in child care centers: comparison of two probiotic agents. Pediatr. *115,* 5–9.

Weston, S., Halbert, A., Richmond, P., and Prescott, S.L. (2005). Effects of probiotics on atopic dermatitis: a randomised controlled trial. Arch. Dis. Child *90,* 892–897.

Whorwell, P.J., Altringer, L., Morel, J., Bond, Y., Charbonneau, D., O'Mahony, L., Kiely, B., Shanahan, F., and Quigley, E.M. (2006). Efficacy of an encapsulated probiotic *Bifidobacterium infantis* 35624 in women with irritable bowel syndrome. Am. J. Gastroenterol. *101,* 1581–1590.

Wickens, K., Black, P.N., Stanley, T.V., Mitchell, E., Fitzharris, P., Tannock, G.W., Purdie, G., and Crane, J. (2008). Probiotic Study Group. A differential effect of 2 probiotics in the prevention of eczema and atopy: a double-blind, randomized, placebo-controlled trial. J. Allergy Clin. Immunol. *122,* 788–794.

Wildt, S., Munck, L.K., Vinter-Jensen, L., Hanse, B.F., Nordgaard-Lassen, I., Christensen, S., Avnstroem, S., Rasmussen, S.N., and Rumessen, J.J. (2006). Probiotic treatment of collagenous colitis: A randomized, double-blind, placebo-controlled trial with *Lactobacillus acidophilus* and *Bifidobacterium animalis* subsp. *lactis.* Inflamm. Bowel Dis. *12,* 395–401.

Winkler, P., de Vrese, M., Laue, Ch., and Schrezenmeir, J. (2005). Effect of a dietary supplement containing probiotic bacteria plus vitamins and minerals on common cold infections and cellular immune parameters. Int. J. Clin. Pharmacol. Ther. *43,* 318–326.

Wullt, M., Johansson Hagslatt, M.L., Odenholt, I., and Berggren, A. (2007). *Lactobacillus plantarum* 299v enhances the concentrations of fecal short-chain fatty acids in patients with recurrent clostridium difficile-associated diarrhea. Dig. Dis. Sci. *52,* 2082–2086.

Xiao, S.D., Zhang, D.Z., Lu, H., Jiang, S.H., Liu, H.Y., Wang, G.S., Xu, G.M., Zhang, Z.B., Lin, G.J., and Wang, G.L. (2003). Multicenter, randomized, controlled trial of heat-killed *Lactobacillus acidophilus* LB in patients with chronic diarrhea. Adv. Ther. *20,* 253–260.

Xiao, J.Z., Kondo, S., Yanagisawa, N., Takahashi, N., Odamaki, T., Iwabuchi, N., Miyaji, K., Iwatsuki, K., Togashi, H., Enomoto, K., *et al.* (2006). Probiotics in the treatment of Japanese cedar pollinosis: a double-blind placebo-controlled trial. Clin. Exp. Allergy *36,* 1425–1435.

Yamano, T., Iino, H., Takada, M., Blum, S., Rochat, F., and Fukushima, Y. (2006). Improvement of the human intestinal flora by ingestion of the probiotic strain *Lactobacillus johnsonii* La1. Br. J. Nutr. *95,* 303–312.

Other Books of Interest

Index

A

B

T

U

V

W

Y